Ethnobotanical Insights Into Medicinal Plants

Sara Musaddiq
The Women University Multan, Pakistan

Imama Fayyaz
The Women University Multan, Pakistan

Kiran Mustafa
The Women University Multan, Pakistan & Government Graduate College (W), Khanewal, Pakistan

A volume in the Advances in Bioinformatics and Biomedical Engineering (ABBE) Book Series

IGI Global
PUBLISHER of TIMELY KNOWLEDGE

Published in the United States of America by
IGI Global
Medical Information Science Reference (an imprint of IGI Global)
701 E. Chocolate Avenue
Hershey PA, USA 17033
Tel: 717-533-8845
Fax: 717-533-8661
E-mail: cust@igi-global.com
Web site: http://www.igi-global.com

Library of Congress Cataloging-in-Publication Data

CIP DATA PROCESSING

Ethnobotanical Insights into Medicinal Plants
Sara Musaddiq, Imama Fayyaz, Kiran Mustafa
2024 Medical Information Science Reference

ISBN: 9798369361054(hc) | ISBN: 9798369361061(sc) | eISBN: 9798369361078

This book is published in the IGI Global book series Advances in Bioinformatics and Biomedical Engineering (ABBE) (ISSN: 2327-7033; eISSN: 2327-7041)

British Cataloguing in Publication Data
A Cataloguing in Publication record for this book is available from the British Library.

All work contributed to this book is new, previously-unpublished material.
The views expressed in this book are those of the authors, but not necessarily of the publisher.

For electronic access to this publication, please contact: eresources@igi-global.com.

Advances in Bioinformatics and Biomedical Engineering (ABBE) Book Series

ISSN:2327-7033
EISSN:2327-7041

MISSION

The fields of biology and medicine are constantly changing as research evolves and novel engineering applications and methods of data analysis are developed. Continued research in the areas of bioinformatics and biomedical engineering is essential to continuing to advance the available knowledge and tools available to medical and healthcare professionals.

The **Advances in Bioinformatics and Biomedical Engineering (ABBE) Book Series** publishes research on all areas of bioinformatics and bioengineering including the development and testing of new computational methods, the management and analysis of biological data, and the implementation of novel engineering applications in all areas of medicine and biology. Through showcasing the latest in bioinformatics and biomedical engineering research, ABBE aims to be an essential resource for healthcare and medical professionals.

COVERAGE

- Genetic Engineering
- Finite Elements
- Orthopedic Bioengineering
- Chemical Structures
- Biostatistics
- Biomechanical Engineering
- Molecular Engineering
- Databases
- Protein Structure
- Computational Biology

IGI Global is currently accepting manuscripts for publication within this series. To submit a proposal for a volume in this series, please contact our Acquisition Editors at Acquisitions@igi-global.com or visit: http://www.igi-global.com/publish/.

The Advances in Bioinformatics and Biomedical Engineering (ABBE) Book Series (ISSN 2327-7033) is published by IGI Global, 701 E. Chocolate Avenue, Hershey, PA 17033-1240, USA, www.igi-global.com. This series is composed of titles available for purchase individually; each title is edited to be contextually exclusive from any other title within the series. For pricing and ordering information please visit http://www.igi-global.com/book-series/advances-bioinformatics-biomedical-engineering/73671. Postmaster: Send all address changes to above address.

Titles in this Series

For a list of additional titles in this series, please visit:
http://www.igi-global.com/book-series/advances-bioinformatics-biomedical-engineering/73671

Applications of Parallel Data Processing for Biomedical Imaging
Rijwan Khan (Galgotias University, India) Indrajeet Kumar (Graphic Era Hill University, Dehradun, India) and Pushkar Praveen (Govind Ballabh Pant Institute of Engineering and Technology, India)
Medical Information Science Reference • copyright 2024 • 346pp • H/C (ISBN: 9798369324264) • US $415.00 (our price)

Applying Machine Learning Techniques to Bioinformatics Few-Shot and Zero-Shot Methods
Umesh Kumar Lilhore (School of Computing, Galgotias University, Greater Noida, India) Abhishek Kumar (Chandigarh University, India) Sarita Simaiya (Galgotia University, India) Narayan Vyas (Chandigarh University, India) and Vishal Dutt (Chandigarh University, India)
Medical Information Science Reference • copyright 2024 • 398pp • H/C (ISBN: 9798369318225) • US $340.00 (our price)

Biomedical Research Developments for Improved Healthcare
Pranav Kumar Prabhakar (Lovely Professional University, India)
Medical Information Science Reference • copyright 2024 • 393pp • H/C (ISBN: 9798369319222) • US $360.00 (our price)

Quantum Innovations at the Nexus of Biomedical Intelligence
Vishal Dutt (AVN Innovations Pvt. Ltd., India) Abhishek Kumar (Department of CSE, UIE, Chandigarh University, Punjab, India) Sachin Ahuja (Chandigarh University, India) Anupam Baliyan (Geeta University, India) and Narayan Vyas (AVN Innovations Pvt. Ltd., India)
Medical Information Science Reference • copyright 2024 • 287pp • H/C (ISBN: 9798369314791) • US $330.00 (our price)

Physiology and Function of Glial Cells in Health and Disease
Bilal El-Mansoury (Faculty of Sciences, Chouaib Doukkali University, Morocco) Omar El Hiba (Chouaib Doukkali University, Morocco) and Arumugam Radhakrishnan Jayakumar (University of Miami, USA)

For an entire list of titles in this series, please visit:
http://www.igi-global.com/book-series/advances-bioinformatics-biomedical-engineering/73671

IGI Global
PUBLISHER of TIMELY KNOWLEDGE

701 East Chocolate Avenue, Hershey, PA 17033, USA
Tel: 717-533-8845 x100 • Fax: 717-533-8661
E-Mail: cust@igi-global.com • www.igi-global.com

Table of Contents

Detailed Table of Contents

The human body's overproduction of free radicals can throw things out of balance and destroy the macromolecules like proteins, DNA, and lipids through oxidative damage. This damage is responsible for the pathogenesis of several human diseases, including cardiovascular diseases, cancers, and aging. Antioxidant phytochemicals can therefore be crucial in the management and prevention of chronic illnesses. Studies on humans and in vitro have shown that phytochemicals possess antioxidant properties. Many phytochemicals, especially polyphenols such anthocyanins, flavonoids, phenolic acids, and tannins, are known to have antioxidant and free radical scavenging properties. This chapter summarizes the phytochemical and antioxidant activity of the red coloured fruits such as grapes (Vitis vinifera), red guava (Psidium guajava), and watermelon (Citrullus lanatus), and this review highlights the recent progress on antioxidant phytochemicals, potential mechanisms in the prevention and treatment of cancer.

Plants are one of the essential assets of medicine. Papaver somniferum (scientific name) is native to Southeastern Europe and Asia. The poppy plant is a common herb that is known for its nutritive purposes. Papaver somniferum, which is generally called as the opium poppy, is a source of drugs that have historically been used to help those suffering from pain. It contains several phytochemicals, mainly isolated from seeds. It consists of a lot of alkaloids such as morphine, codeine, narcotine, etc. It can be used in a variety of food products and to make soaps, paints, and other products of industry. This chapter covers cultivation aspects, production, pharmaceutical, food, cosmetic, and ethnobotanical uses. It also consists of biogenic synthesis from poppy seeds.

G. Shobana, Srimad Andavan Arts and Science College, Trichy, India
N. Agnel Arul John, Srimad Andavan Arts and Science College, Trichy, India
Hari Haran Govindasamy, Srimad Andavan Arts and Science College, Trichy, India

The study aimed at evaluating the phytochemical profile and invitro antioxidant potentials of a three medicinal herbs including Calendula officinalis L, Lantana camara L., Desmodium gangeticum L. The aqueous extract of the selected plants were evaluated for phytochemical screening and antioxidant activities using DPPH, reducing power, hrdroxy radical, and nitric oxide radical scavenging assays. The present study exhibited higher yields of phytochemical constituents and noticeable antioxidative properties.

Javaria Kanwal, The Women University Multan, Pakistan
Mehwish Nawaz, The Women University Multan, Pakistan
Wajiha Noor, The Women University Multan, Pakistan
Mahrukh Khan, The Women University Multan, Pakistan
Hafsa Javaid, The Women University Multan, Pakistan
Abida Ashraf, Bahauddin Zakariya University, Pakistan
Nusrat Shafiq, Government College Women University, Faisalabad, Pakistan
Adnan Ali Qureshi, University of Engineering and Technology, Lahore, Pakistan

summary, numerous spice seeds are traditionally considered healing agents, and modern research has confirmed their benefits. The convergence of ethnobotanical knowledge and current scientific advancements has established the foundation for comprehensive integration into modern medicinal practices.

Dill, also known as Anethum graveolens, is a botanical species with an annual life cycle and a branching structure. It is a significant part of Ayurvedic medicine and is used as a spice, flavor enhancer, and essential oil supplier. Dill seeds have carminative, stomachic, and diuretic properties. The volatile components of dill seeds and plants are diverse, with carvone being the main odorant. Other substances found in seeds include steroids, phenolic acids, flavonoids, and coumarins. Dill is used in traditional herbal therapy to treat various ailments, including respiratory issues, stomach discomfort, digestive disorders, and low blood sugar and cholesterol. It also has pharmacological benefits, including hypolipidemic, anticancer, antidiabetic, antispasmodic, antibacterial, anti-inflammatory, antioxidant, and anti-infertility benefits. Numerous studies have confirmed these effects.

Corianderum sativum L. commonly known as coriander is an annual herb. It belongs to the family Apiaceae (Umbelliferae). It is a well-known herb all around the world because of its food, culinary, and medicinal uses. It is a source of many phytochemicals such as polyphenols, essential oil, reducing sugars, carotinoids, terpenoids, glycosides, sterols, isocoumarins, flavonoids, tannins, alkaloids, fatty acids, and coumarins. Coriander sativum also contains high nutritional values of carbohydrates, fibers, oils, proteins, along with wide range of vitamins, minerals, and trace elements. Due to the easy collection of the plant and its widespread uses and also its remarkable biological activities, this plant has become the food and medicine in many parts of the world. The primary objective of this study is to present comprehensive analyzed information on the botanical, chemical, pharmacological aspects of C. sativum along with its applications in the biogenic synthesis.

Aframomum melegueta (grains of paradise) is a herbaceous plant of West Africa, locally used as a spice and to treat common ailments like diarrhea, painful arthritis, snake bite, and scorpion sting. Various parts of the plant like stem, leaves, and roots have various phytochemicals like flavonoids, terpenoids, saponins, phenolic compounds, alkaloids cardiac glycosides, and saponins that give alligator pepper a great medicinal value. These phytochemicals exert anti-inflammatory, anti-cancer, anti-diabetic, antimicrobial, antioxidants, and hepatoprotective effects. However, more comprehensive studies need to be performed to reveal the medicinal purposes of alligator pepper. This chapter encapsulates findings from previous studies on the phytochemicals, ethnobotanical, pharmacological activities, and biogenic synthesis of chemicals of Aframomum melegueta.

Macadamia nuts, native to Australia, are a nutritional powerhouse with numerous health benefits. They are rich in heart-healthy monounsaturated fats, antioxidants, essential vitamins and minerals, and antioxidants. They help regulate blood sugar levels, provide a sensation of fullness, and have anti-inflammatory properties. Macadamia nuts also hydrate skin and may prevent signs of ageing. They are used in skin care products due to their hydrating and restorative properties. However, some people have severe allergic reactions, potentially leading to respiratory issues. Overall, Australian macadamia nuts offer a fun and health-conscious option for those seeking a holistic approach to well-being.

Preface

In the vibrant tapestry of nature, plants have long been revered for their medicinal properties, housing a treasure trove of natural compounds with diverse biological effects. From ancient remedies to modern pharmacology, the world of medicinal plants continues to astonish with its potential. It is with great pleasure that we introduce *Ethnobotanical Insights Into Medicinal Plants*, a comprehensive exploration into the multifaceted realm of plant-based medicine.

This book is a collaborative effort by esteemed editors Sara Musaddiq, Imama Fayyaz, and Kiran Mustafa, along with contributions from researchers and experts in the field. Our journey into the world of medicinal plants takes us through phytochemical analysis, ethnobotanical traditions, pharmacological studies, botanical descriptions, industrial and culinary applications, biogenic synthesis, and innovative uses of kitchen spice seeds.

The allure of kitchen spice seeds, beyond their culinary charm, lies in their rich history, cultural significance, and potential therapeutic benefits. We invite you to embark on a scholarly odyssey as we unravel the secrets hidden within these tiny yet powerful seeds. Whether you are a seasoned botanist, a curious ethnobotanist, a dedicated pharmacologist, or simply an enthusiast of natural remedies, this book offers insights and discoveries that will captivate your intellect and inspire further exploration.

Our call for chapters invites researchers, scholars, and practitioners to contribute their expertise and share their discoveries. We encourage diverse perspectives and approaches, ranging from traditional knowledge systems to cutting-edge scientific methodologies. Topics of interest include phytochemical analysis, ethnobotanical insights, pharmacological studies, botanical descriptions, industrial and culinary applications, biogenic synthesis, and innovative uses of kitchen spice seeds.

We believe that by delving into the rich tapestry of ethnobotanical knowledge, we not only enrich our understanding of nature's bounty but also pave the way for innovative applications and sustainable practices. Join us in this exciting journey of discovery and appreciation for the medicinal marvels hidden within our botanical heritage.

Chapter Overview

Chapter 1 delves into the critical role of antioxidant phytochemicals found in red-colored fruits such as grapes, red guava, and watermelon. These fruits, rich in polyphenols like anthocyanins, flavonoids, phenolic acids, and tannins, exhibit potent antioxidant and free radical scavenging properties. The chapter reviews recent advancements in understanding how these phytochemicals contribute to preventing and treating diseases like cancer, emphasizing their potential in antioxidant therapy.

Chapter 2 explores the multifaceted uses of the opium poppy plant, Papaver somniferum. Beyond its historical significance in pain relief, this chapter delves into its cultivation, pharmaceutical applications, nutritive properties, and industrial uses. It provides insights into the diverse alkaloids present in poppy seeds, including morphine and codeine, and discusses biogenic synthesis possibilities from these seeds, offering a comprehensive view of the plant's contributions across various sectors.

Chapter 3 focuses on the phytochemical and antioxidant potential of medicinal herbs like Calendula officinalis, Lantana camara, and Desmodium gangeticum. Through in vitro studies, this chapter highlights the abundance of phytochemical constituents in these herbs and their notable antioxidative properties. Readers gain a deeper understanding of these herbs' therapeutic potential and their relevance in antioxidant-based treatments.

Chapter 4 sheds light on celery's bioactive compounds and their pharmacological significance. From anticancer and antihypertensive properties to antimicrobial and antioxidant effects, celery emerges as a versatile plant with diverse biological activities. The chapter also explores celery's role in eco-friendly nanomaterial synthesis, showcasing its potential applications beyond traditional medicine.

Chapter 5 delves into the ethnopharmacological landscape of diabetes treatment in Morocco. Through an exhaustive catalog of plant species and their utilization in diabetes management, this chapter offers valuable insights into traditional herbal remedies. However, it also emphasizes the need for further research to identify active compounds and validate their efficacy in modern healthcare practices.

Chapter 6 bridges ancient traditions with modern scientific insights, focusing on spice seeds' ethnobotanical and pharmacological properties. By integrating traditional knowledge with recent research findings, this chapter highlights the therapeutic potential of various spice seeds and their applications in contemporary medicine, offering a holistic view of their healing properties.

Chapter 7 delves into the botanical, chemical, and pharmacological aspects of dill, presenting a comprehensive analysis of its therapeutic properties. From carminative and stomachic effects to hypolipidemic and anti-inflammatory benefits, dill emerges as a potent herbal remedy with a rich history in traditional medicine and modern pharmacology.

Chapter 8 provides an in-depth exploration of Corianderum sativum, commonly known as coriander, covering its botanical, chemical, and pharmacological characteristics. With a wide array of phytochemicals and nutritional values, coriander showcases its significance as both a culinary spice and a medicinal herb, offering diverse health benefits and applications in biogenic synthesis.

Chapter 9 focuses on Aframomum melegueta, or grains of paradise, highlighting its phytochemical composition, pharmacological activities, and potential biogenic synthesis applications. This chapter provides a comprehensive overview of the medicinal properties of this West African herb, emphasizing the need for further research to unlock its full therapeutic potential.

Chapter 10 concludes the book with a detailed exploration of Australian macadamia nuts, showcasing their nutritional benefits, skincare applications, and potential allergic reactions. Through a holistic examination of macadamia nuts' health benefits and uses, this chapter offers a fitting conclusion to the diverse array of plants discussed in this book.

Each chapter contributes unique insights into the ethnobotanical and medicinal properties of various plants, offering readers a comprehensive understanding of their therapeutic potential and applications in modern healthcare and industry.

Conclusion

In the grand tapestry of nature, plants have always held a revered place, not just for their beauty but for their remarkable medicinal properties. This book, *Ethnobotanical Insights Into Medicinal Plants*, is our tribute to the profound wisdom and healing potential of botanical treasures. We, as editors Sara Musaddiq, Imama Fayyaz, and Kiran Mustafa, alongside our esteemed contributors, have embarked on a journey through the intricate world of medicinal plants, from ancient remedies to cutting-edge pharmacology.

Our preface sets the stage for an exploration that spans phytochemical analysis, ethnobotanical traditions, pharmacological studies, botanical descriptions, industrial and culinary applications, biogenic synthesis, and the fascinating realm of kitchen spice seeds. These chapters represent a collaborative effort to unravel the secrets hidden within nature's bounty, inviting readers on a scholarly odyssey that transcends disciplines and delves deep into the therapeutic wonders of plants.

Chapter by chapter, we delve into the intricate details of plants' medicinal properties. Chapter 1 uncovers the antioxidant potential of red-colored fruits, while Chapter 2 illuminates the versatile uses of the opium poppy plant. Chapter 3 explores the antioxidative properties of medicinal herbs, and Chapter 4 showcases celery's diverse biological activities. Chapters 5 and 6 offer insights into ethnopharmacological practices and spice seeds' therapeutic potential, respectively.

This is followed by an in-depth exploration of dill's pharmacological benefits in Chapter 7. Chapter 8 unveils the multifaceted properties of coriander, and Chapter 9 highlights the medicinal value of Aframomum melegueta. Finally, Chapter 10 concludes our journey with a detailed examination of Australian macadamia nuts, emphasizing their nutritional and skincare benefits.

Each chapter contributes a unique perspective, blending traditional wisdom with modern scientific insights. We believe that by delving into this rich tapestry of ethnobotanical knowledge, we not only expand our understanding of nature's healing potential but also pave the way for innovative applications and sustainable practices.

We invite you to join us in this exhilarating exploration of medicinal marvels hidden within our botanical heritage. Whether you're an expert in the field or a curious enthusiast, this book promises to captivate your intellect and inspire further exploration into the captivating world of medicinal plants.

Sara Musaddiq
The Women University Multan, Pakistan

Imama Fayyaz
The Women University Multan, Pakistan

Kiran Mustafa
The Women University Multan, Pakistan & Government Graduate College (W), Khanewal, Pakistan

Chapter 1
Antioxidant Activity and Phytochemical Composition of the Selected Fruits

D. Jancy Rani
Dr. N.G.P. Arts and Science College, Coimbatore, India

ABSTRACT

The human body's overproduction of free radicals can throw things out of balance and destroy the macromolecules like proteins, DNA, and lipids through oxidative damage. This damage is responsible for the pathogenesis of several human diseases, including cardiovascular diseases, cancers, and aging. Antioxidant phytochemicals can therefore be crucial in the management and prevention of chronic illnesses. Studies on humans and in vitro have shown that phytochemicals possess antioxidant properties. Many phytochemicals, especially polyphenols such anthocyanins, flavonoids, phenolic acids, and tannins, are known to have antioxidant and free radical scavenging properties. This chapter summarizes the phytochemical and antioxidant activity of the red coloured fruits such as grapes (Vitis vinifera), red guava (Psidium guajava), and watermelon (Citrullus lanatus), and this review highlights the recent progress on antioxidant phytochemicals, potential mechanisms in the prevention and treatment of cancer.

INTRODUCTION

Fruits are rich sources of nutritional and bioactive compounds with immense health benefits. A large number of antioxidants, both nutritive and non-nutritive occur in fruits and these antioxidants from fruits play a pivotal role in human health and

DOI: 10.4018/979-8-3693-6105-4.ch001

even pertinence of life is more or less dependent on them. These constituents when included in our daily intake reduce the risk of human diseases. Fruits which come in a wide variety of colours, flavours and textures offer a wide range of health benefits. They are important sources of minerals, fibre and vitamins which provides essential nutrients for human health. Diets rich in fruits are believed to play an important role in preventing diseases but human choices of diet are driven by necessity and economy. Since many fruits are seasonal and perishable, they must be processed to improve the keeping quality. Processing of fruits into value added products is the best option to control the huge losses. Most fruits are processed before being consumed and induce significant changes in chemical composition, influencing the concentration and retention of bioactive components in fruits. The consumption trend of fresh tropical fruits and their products is increasing steadily due to consumer's education on their exotic flavours, nutritive value, phytochemical and antioxidant content with potential health benefits (Boeing *et al.,* 2012).

Non-nutritive plant compounds with anti-inflammatory or disease-preventive qualities are known as phytochemicals. These compounds are produced by plants as a means of defense, and it is also thought that they shield people from some illnesses. (Edeoga *et al.,* 2005)

Phytochemicals, another name for phytonutrients, are non-nutritive plant compounds with protective or disease preventive properties. Most foods contain these intricate compounds, but fruits contain the most of them. Many epidemiological studies have demonstrated a strong correlation between eating fruits and vegetables and a lower risk of developing a number of diseases. Although it was traditionally thought that vitamins and minerals were important for preventing disease, new research suggests that phytochemicals can actually protect disease far more effectively than vitamins or other nutrients. Certain chronic diseases, including as cancer, diabetes, osteoporosis, and visual disorders which are particularly severe in Western nations—are linked to the protection of phytochemicals. A diet rich in fat and sugar, insufficient exercise, smoking, and other bad habits are the main causes of these diseases (Chen *et al.,* 2007). Phytochemicals are the chemicals that present naturally in plants. Now- a-days these phytochemicals become more popular due to their countless medicinal uses. They play a vital role against number of diseases such as asthma, arthritis, cancer etc. unlike pharmaceutical chemicals these phytochemicals do not have any side effects. Since the phytochemicals cure diseases without causing any harm to human beings these can also be considered as "man- friendly medicines" (Banu and Catherine, 2015). Fruits and leafy vegetables are believed to occupy a modest place as a source of trace elements due to their high water content. In addition to meeting nutrient intake levels, greater consumption of fruits and vegetables is associated with reduced risk of cardiovascular disease, stroke, and cancers of the mouth, pharynx, esophagus, lungs, stomach, and colon (Aberoumand and Deokule, 2010).

I. GRAPES (*VITIS VINIFERA*)

1.1. Grapes (*Vitis vinifera*): An Overview

One of the most well-known and extensively grown fruits in the world is the grape (Vitus). There are over sixty different species of grapes, the majority of which are located in the Northern Hemisphere's temperate zones, roughly evenly split between Asia and America (Yang et al., 2013). Based on historical records, grapes have been grown in India for approximately 4,000 years. In India, commercial grape growing dates back approximately 70 years, and it is currently regarded as the most lucrative fruit crop in the nation. Indian viticulture is distinctive in that it is practiced in nearly every climate zone in the nation from tropical to temperate across several states. Numerous foreign grape varietals were brought to the nation in the 1930s (Dhananjay and Gawande, 2021). Grape is one of the world's major fruits and has very old history. It has also been praised in "The Holy Quran". The European grapevine is believed to have originated in the area between Black and Caspian seas, where it still grows wild. It is believed that it was introduced/spreaded throughout the Europe and later by explorer to all continents (Mukhtar *et al.,* 2011).

Grapes are a type of fruit that grow in clusters of 15 to 300, and can be crimson, black, dark blue, yellow, green, orange, and pink. "White" grapes are actually green in color and are evolutionarily derived from the purple grape. Mutations in two regulatory genes of white grapes turn off production of anthocyanin, which are responsible for the color of purple grapes. Anthocyanins and other pigment chemicals of the larger family of polyphenol in purple grapes are responsible for the varying shades of purple in red wines Grapes are typically an ellipsoid shape resembling a prolate spheroid (Shah *et al.,* 2019).

1.2 Chemical Composition of Grapes

Bioactive components present in green and black grapes (*Vitis vinifera*) were compared through extraction of grapes which is done by maceration process with methanol. Phytochemical screening was performed to detect the presence of alkaloids, flavonoids, anthocyanin, anthraquinones, saponins, steroids &terpenoids, carbohydrates, glycosides, phenol compound, protein and amino acids respectively. Black grape was found to be phytochemically richer as compared to green grapes. Results reveal that green grape extract contained alkaloids, flavonoids, anthraquinones, saponins, carbohydrates and phenol compound. Steroids and glycosides were not found in preliminary (qualitative) analysis of black grapes (Sharma and Bais, 2015). The grape\skin is concentrated with *resveratrol*, anthocyanins and catechins. The Notron grape skin contains 215.6 mg phenolic compounds per gram of the extract (Zhang *et al.*, 2011).

The study was carried out by Hanaa *et al.*, (2015) to assess the proximate composition, phytochemical screening, total phenolic compounds, total flavonoids and antioxidant activities of selected solvent extracts of grape seeds. The results showed that grape seeds content was 38.2% fibers, 15.8% total lipids, 10.7% proteins, 2.58% ash, 22.37% carbohydrates and 10.4% moisture. Phytochemical analysis of grape seed extract revealed the presence of steroids, terpenoids, anthocyanins, emodins, glycosides, flavonoides and phenols in acetone (70%), ethanol (70%) and methanol. Both steroids and terpenoids were absent in water extract. Saponins were absent in methanol and water extracts. In vitro, antioxidant activity was estimated as IC50 value.

Grape juices were analyzed for total phenolic content, colour and antioxidant activity. The commercial juices had the highest average values for total monomeric anthocyanins and total phenolics. There was a strong positive correlation (R = 0.9566) between the antioxidant activity and total phenolic content for the commercial juice. In addition, the Principle Components Analysis showed a strong positive correlation between the red colour and total monomeric anthocyanins. However, the total monomeric anthocyanis and polymeric anthocyanins showed a negative correlation (Burin *et al.*, 2010).

The chemical composition of Grape fibre is known to vary depending on the Grape cultivar, growth climates, and processing conditions. The obliged characteristics of the fibre product are: total dietary fibre content above 50%, moisture lower than 9%, low content of lipids, a low energy value and neutral flavour and taste. Grape pomace represents a rich source of various high-value products such as ethanol, tartrates and malates, citric acid, Grape seed oil, hydrocolloids and dietary fibre. Used commercial Grape fibres have as a main characteristic, the high content of total dietary fibre. Amount of total dietary fibre depends on the variety of Grapes. Total dietary fibre content (TDF) in our samples of Grape fibre varied from 56.8% to 83.6%. There were also determined low contents of moisture (below 9%). In the samples of Grape fibre were determined higher amount of protein (8.6 - 10.8%), mineral (1.3 - 3.8%) and fat (2.8 - 8.6%) (Karovičová *et al.*, 2015)

1.3 *In-vitro* Antioxidant Activities

The antioxidant activity and ascorbic acid was analyzed in red wine and fruits from guava and tomatoes (raw and cooked), The ascorbic acid content was analyzed using the AOAC official titrimetry method while the carotenoid assay was analyzed using the modified method of Shigeaki Takagi as described in the determination of green leaf carotenoids. The levels of ascorbic acid in the samples ranged from <0.0001 to 216.05 mg/100g with guava having the highest value of 216.05±9.64 and red wine with the least value of <0.0001 mg/100g. Beta – carotene content in the samples had the highest value of 0.43±0.10mg/100g in raw tomatoes and the least value of 1.78×E-7mg/100g in honey samples. Similarly, lycopene content was highest (5.10±0.85mg/100g) in guava samples and least in red wine (3.12×E6mg/100g). From the results obtained, abundance of lycopene and ascorbic acid in guava could suggest their relevance in boosting the immune system as well as in cancer prevention. Effect of heat on ascorbic acid was evident in their levels in cooked and raw samples of tomatoes, as heat treatment gave marked reduction in observed values (Nwaichi *et al.,* 2015).

Dietary intake of foods rich in antioxidant properties is suggested to be cancer protective. Foods rich in antioxidant properties include grape (*Vitis vinifera*), one of the world's largest fruit crops and most commonly consumed fruits in the world. Grape skin and seed extracts exert strong free radical scavenging and chelating activities and inhibit lipid oxidation in various food and cell models *in vitro*. The use of grape antioxidants are promising against a broad range of cancer cells by targeting epidermal growth factor receptor (EGFR) and its downstream pathways, inhibiting over-expression of COX-2 and prostaglandin E2 receptors, or modifying estrogen receptor pathways, resulting in cell cycle arrest and apoptosis. Interestingly, some of these activities were also demonstrated in animal models. However, *in vivo* studies have demonstrated inconsistent antioxidant efficacy. Nonetheless, a growing body of evidence from human clinical trials has demonstrated that consumption of grape, wine and grape juice exerts many health-promoting and possible anti-cancer effects (Gollucke *et al.,* 2012).

Grape seed phenolics have been shown to protect low- density lipoprotein (LDL) against oxidation brought about by CU^{2+}, oxygen- centered radical-generating AAPH, or peroxynitrite-generating SIN-I in-vitro systems (Zhou and Raffoul, 2012) .As anti-oxidant, grape seed procyanidines can induce arrest in the phase GI and decrease iROSformation, Further interestingly, GSPE enhances the viability and growth of normal human gastric mucosal cells and murine macrophage cells (Wang *et al.,* 2012).

Praphasawat *et al.,* (2011) reported that GSPE does not induce apoptosis in neutrophils because GSPE, as a mitochondrial anti- oxidant, had no effects on neutrophils showing that GSPE induced apoptosis may be cancer specific in AML 14.3D10 cells. Oligomeric pro-anthocyanidines from Thai grape seeds were also fond to possess the anti-oxidant and anti-mutogenic effects on DNA damage induced by H_2O_2 in TK6 cells at concentrations of 100, 250, 500, and 1,000 μg/ml by 18.7, 36.4, 30.6 and 60.1%, respectively.

Three of the major antioxidant components found in grapes – resveratrol, catechin, quercetin, and grape seed extract, containing pro-anthocyanidine B-2-Gallete given typically and/or systematically strongly inhibited 7, 12- dimethylbenz anthracene inhibited epidermal hyperplasia, proliferation and inflammation. The hydroxylation of 2-deoxyguanosine was markedly inhibited by topical and dietary administration of test variables, i.e., by approximately 40-70% (Hanausek *et al.,* 2017). GSE exhibited the classical morphological characteristics of cell death under the electron microscope including cell atrophy, increased vacuoles crumbled nuclear membrane, chromosome aggregation and apoptosis of PC- 3 cells.

The effects of administration of hydro ethanol extracts of grape seed Brgund Mare variety (BM) along with DOX (30 min before drug administration) in normal (Hfl-1) and tumor cell lines (HepG2 and MIs) which showed in normal cells, the product statistically decreased cyto-toxicitty and markedly inhibited lipid peroxides and protein carbonyl formation, in a dose-depended relationship. Contrary to this, in tumor cells, the same treatment resulted in a reversed effect, cell death, melanodialdehyde and increase in protein carbonyl contents with BM dose enhancement (Postescu *et al.,* 2012).

2. RED GUAVA (*PSIDIUM GUAJAVA*)

2.1 Red Guava (*Psidium Guajava*): An Overview

Psidiumguajava L. known as Guava is a medicinal plant belongs to the family Myrtaceae. *P. guajava*is a well-known traditional medicinal plant used in various indigenous systems of medicine. It is widely distributed throughout India. *Psidium guajava* (Linn.) is popularly known as 'poor man's apple of the tropics', has a long history of traditional use for a wide range of diseases. The fruit as well as its juice is freely consumed for its great taste and nutritional benefits. Much of the traditional uses have been validated by scientific research (Asha *et al.,* 2015).

Member of the Myrtaceae, *Psidium guajava Linn.* Species (guava), plant native to tropical America has its widespread medicinal use among the world's populations. Being a plant of tropical and subtropical regions, can be found on plantations, in backyards of homes, or naturally in other areas, and could even be considered invasive species (Gull et al., 2012).

Red guava (*Psidium guajava*) is a small medicinal tree that is native to South America. It is popularly known as guava and has been used traditionally as a medicinal plant throughout the world for a number of ailments. There are two common varieties of guava: the red (*P.guajavavar.pomifera*) and the white (*P.guajava var. pyrifera*) (Wu *et al.,* 2012).

All parts of this tree, including fruits, leaves, bark, and roots, have been used for treating stomach ache and diarrhea in many countries. Guava plants have many benefits. The leaves are used traditionally by many people to overcome a wide range of diseases, while the guava fruits are believed to overcome several of health problems including and a source of antioxidants (Sandra et al., 2012). Red guava fruit is also believed to help cure dengue hemorrhagic fever (DHF). Guava fruit contains vitamin C, two times higher than other fruits such as orange. Vitamin C is an important compound that has an antioxidant activity (Espino *et al.,* 2011).

Leaves, pulp and seeds are used to treat respiratory and gastrointestinal disorders, and as an antispasmodic, anti-inflammatory, as a cough sedative, anti-diarrheic, in the management of hypertension, obesity and in the control of diabetes mellitus. It also possesses anticancer properties (Ibe *et al.,* 2014).

2.2 Chemical Composition of Red Guava

The physico chemical Characteristics of the Taiwan pink type of guava (Psidium guajava L.) was investigated. The ripe and fresh guavas were chosen so that their chemical and physical makeup could be assessed. According to the results, the guava fruit has a green colour and a pink pulp due to its high anthocyanin content (143.06 mg/100 g) and high vitamin C content (181.59 mg/100g). After additional chemical composition analysis, the following findings were obtained: protein 2.56%, fat 0.87%, moisture content 89.28%, and carbohydrate 12.4%. The guava juice's other characteristics, like its pH and acidity, were found to be 3.25 and 0.539%, respectively (Dani, 2023).

Fruit components of red guava (skin, pulp, and seeds) were assessed for their phytochemical composition, antioxidant capacity, and lycopene content. The goal of the current study was to evaluate watermelon samples that were fresh and dried. To detect the secondary metabolites such as alkaloids, flavonoids, saponins, phenols, glycosides, carbohydrates, tannins, steroids, and terpenoids contained in the metholic extracts of fresh and dried Red Guava, a preliminary qualitative phytochemical analysis was conducted using established protocols. The test results were tabulated and presented in duplicate.

Quantitative phytochemical tests were conducted on total phenol, flavonoids, and alkaloids. Lycopene analysis was conducted using UV spectroscopy. This study indicates that there are several phytochemicals present in fresh red guava. Fresh red guavas have less lycopene than pulp from dried red guavas (Rani & Vijayanchali, 2017).

2.3. *In-vitro* Antioxidant Activities

Antioxidant activity in guava and the major constituents identified in white and red guavas were ascorbic acid, gallic acid, catechin equivalents, cinnamyl alcohol, ethyl benzoate, ß-caryophyllene, (E)-3-hexenyl acetate and α-bisabolene. The antioxidant properties of the guava pulp can be related to anti-cancer effect (Bontempo *et al.,* 2011).

Lyophilized pulp of *P. guajava* in diabetic rats induces to significant hypoglycemic effects probably due to its antioxidant activity of compounds present in the pulp (Huang *et al.,* 2011).

3. WATERMELON (*CITRULLUS LANATUS*)

3.1 Watermelon (*Citrullus lanatus*): An Overview

Watermelon (*Citrullus lanatus*), botanically considered as a fruit, belong to the family Cucurbitaceae It is native to Kalahari desert of Africa but nowadays, it is also cultivated in tropical regions of the world. In the pages of history, its first harvest was documented 5000 years ago in Egypt that later spread to other parts of the world. Presently, China is the top producer followed by Turkey, United States and Iran. *Citrullus lanatus* (Watermelon) is a tropical fruit which grows in almost all part of Africa and South East Asia, belong to the family *Cucurbitacea (*Knockaert *et al.,* 2013).

Watermelon (*Citrulluslanatus*) is cultivated throughout the globe in 96 countries. There are around 1200 varieties of watermelon found in different parts of the world. Total global production of watermelon was 108.9 million tons whereas India's production was 0.4 million tons in 2013 (FAO, 2016).

The commercial quality of watermelon fruits is determined by many factors such as fruit size and shape, rind color and thickness, flesh color and texture, sugar content, aroma, flavor and nutrient composition. During the development process, the fruits of cultivated and wild watermelon undergo highly different biochemical and physiological changes such as sugar and pigment accumulation, fruit softening, and changes of flavor and aromatic volatile contents, all of which are caused by developmentally and physiologically changes in gene expression profiles (Mehra et al., 2015).

3.2. Chemical Composition of Watermelon

Watermelon is a valued source of natural antioxidants with special reference to lycopene, ascorbic acid and citruline. These functional ingredients act as protection against chronic health problems like cancer insurgence and cardiovascular disorders Lycopene is characterized by its distinctive red color in fruits and vegetable Water melons (*Citrullus lanatus*) are one of the leading worldwide and popular staple summer fruit in the world and consumed frequently as a dessert, snack, fruit salad, breakfast food, picnic food edible plate garnish in drinks and also used in many creative ways. It is a natural source of nature's most powerful antioxidants. American Heart Association identified watermelon as a constitute of a sensible low sodium, low saturated fat and low cholesterol diet and a rich source of nutrients. This property makes it a best choice for the diet conscious people. The health benefits of watermelon have been known about for a long time (Mutanen and Pajari, 2010).

Watermelon is the third most popular fruit in the world containing good quantity of nutrients. Pigment extracted from watermelon acts as functional ingredient and can be incorporated into breakfast cereals, frozen dairy desserts, yoghurts, spreads, candy, carbonated beverages, confectionary, sauces and soups, etc. (Zhao *et al.*, 2013).

Fruit of this plant is major consumed portion and variations in growth characteristics determine its end use quality. Considering the nutritional profile, consumption of 100 g watermelon provides 30 kcal. It contains almost 92% water and 7.55% of carbohydrates out of which 6.2% are sugars and 0.4% dietary fiber. It is enriched with carotenoid, vitamin C, citrulline, carotenoids and flavonoids and fat and cholesterol free, thus considered as low caloric fruit. Additionally, watermelon is rich source of β-carotene acts as an antioxidant and precursor of vitamin A. Besides the presence of lycopene, it is a source of B vitamins, especially B1 and B6 as well as minerals such as potassium and magnesium (Huh *et al.*, 2018).

Watermelon juice is an excellent source of Vitamin A, chlorophyll, Vitamin B1 and B6, potassium, iodine, magnesium, zinc and citrulline but low in calories. Watermelon also contains important nutritional compounds, including sugars, lycopene and cardiovascular health–promoting amino acids, such as arginine and glutathione. Although the seed of watermelon is often discarded as waste; it contains various amounts of carbohydrate, phenol, flavonoids, protein, fibre, phosphorus and iron (Altaş *et al.*, 2011).

The high water content (92%) of watermelon makes the juice an excellent diuretic and it is among the best dietary source of lycopene, an antioxidant whose protective effects work wonder on the human body due to its cancer fighting ability C. *lanatus*fruit have highlighted antioxidant activities, anti-inflammatory effects and several essential nutrients. Intake of *C. lanatus*fruit has been associated with a reduced incidence of many diseases. Watermelon extracts helped reduce hypertension and lower blood pressure in obese adults (Naz *et al.,* 2014).

The effect of concentration (7-50°Brix) and temperature (5-50°C) on viscosity of watermelon juice showed that apparent viscosity increased with increase in total soluble solids content and decreased with temperature. Results revealed that apparent viscosity of watermelon juice concentrate decreased with increase in spindle speed at selected temperature indicating its non-Newtonian pseudo plastic flow behavior. The value of coefficient of determination for observed and computed values was 0.955 which indicated that exponential model well explained the relationship within the limits (Sogi *et al.,* 2010).

Lycopene content in *Diploid (2X)*, *Triploid (3X)* and *Tetraploid (4X)* types of red fleshed watermelon was determined using petroleum ether, ethyl alcohol and methanol to extract lycopene from watermelon. The lycopene content of *diploid* watermelon was 3.3-5.5 mg/100g fruit where as *triploid* fruit contained 4.1-6.2 mg/100g and *tetraploid* fruit contained 3.8-5.9 mg/100g. Studies on changes in pigment content during fruit development revealed that lycopene content was low in young fruit, high in fully mature fruit and decreased in over mature fruit (Liu *et al.,* 2012).

3.3 Invitro Antioxidant activities

The biochemical composition and antioxidant activity of four watermelon cultivars was estimated during development and ripening. The ascorbic acid values were in the range of 0.6-6.2 mg/100g at different stages of ripening of watermelon (*Suman 235*) fruit. Total, reducing and non-reducing sugars were in the range of 1.98-8.14, 0.08-2.87 and 1.90-5.27% respectively in *Beauty* watermelon cultivar which were higher than other cultivars. The highest lycopene content was estimated in the *Beauty* watermelon cultivar during different stages of ripening. A strong positive correlation was observed between total polyphenols and lycopene with antioxidant activity (Soumya *et al.,* 2014).

Three varieties of watermelon seeds (Charleston gray, Crimson sweet and Black diamond) were analyzed for their proximate, minerals, phytochemicals, total phenols content and antioxidant activity. The proximate analysis and phytochemicals screening were performed using standard procedures whereas minerals content was determined by atomic absorption spectrophotometry. DPPH free radical scavenging activity and Folin-ciocalteau assays were used to determine antioxidant activity and total phenol content, respectively. The results indicated that the watermelon seeds had moisture content in the range of 7.40 - 8.50%; fat, 26.50 - 27.83%; protein, 16.33 - 17.75%; fibre, 39.09 - 43.28%; ash, 2.00 - 3.00%; carbohydrate, 9.55 - 15.32% and energy value of 354.05 - 369.11 kcal/100g. The seeds also contained appreciable minerals (Ca, P, Mg, Na, K and Zn) with K (3.40-3.5 mg/100g) being the highest while Na (0.07 - 0.08 mg/100g) was the least. DPPH% inhibition varied between 59.88-94.46% inhibitions with trolox equivalent of 82.59-130.29 μM/g depending on the variety. Saponins, tannins, triterpenoids glycosides and alkaloids were present in all samples. Crimson sweet seeds had the highest total phenol content (5416 mg GAE/100g), followed by Black diamond (3949 mg GAE/100g) and the least, Charleston gray (1494 mg GAE/100g). Similarly, Crimson sweet had the highest antioxidant activity, followed by Black diamond and lastly, Charleston gray. The present findings suggest watermelon seeds as considerable source of nutrients in the diet and may have health and economic benefits due to its fibre, minerals, phenolics content and antioxidant activity (Tabiri *et al.,* 2016).

Watermelon has various healing properties for which the pulp, the rind or the seeds are used. It is suitable to treat stroke and hypertension, it is among other diseases. The therapeutic effect of watermelon has been reported and has several antioxidant compounds. Among them, citrullin and lycopene have been demonstrated to play a prominent role in the treatment and management of ailments such as cancer and cardiovascular diseases (Oseni *et al.,* 2013), Watermelon acts as a coolant, thirst quencher, detoxifier, diuretic, febrifuge and aphrodisiac. It reduces the likelihood of weight gain, decreased plasma concentrations and attenuated atherosclerosis development. Watermelon seeds possess potent antioxidant, anti-inflammatory and analgesic properties (Seidu et al., 2014). A cup of watermelon provides 24.3% of the daily value of vitamin C, and through its beta carotene, 11.1% of the daily value of vitamin A. The antioxidant function of lycopene is its ability to help protect cells and other structures in the body from oxygen damage and has been linked in human research to prevention of heart disease. Protection of DNA inside white blood cells has also been shown to be an antioxidant role of lycopene (Edwards *et al.,* 2015).

Nutritive contents, free radical scavenging activities and phytochemical components of *C. lanatus* fruit were evaluated. The extract of the fruit was subjected to *in vitro* antioxidant assessment using 1,1-Diphenylpicryl- Hydrazyl radical (DPPH) and hydrogen peroxide radical scavenging assays. The proximate and phytochemical analyses were conducted using standard procedures. The results of this study showed that *C. lanatus* fruit had very high moisture content and its crude protein, crude fat, crude fibre and ash content were all in traceable amounts. The sugar content was considerably high in comparison with other nutritive contents. Lycopene and β-carotene contents of *C. lanatus* fruit were estimated to be 4537.83 and 308.71 μg/100g respectively. The gross energy evaluation showed a value of 0.335 Kcal/g. The fruit extract exhibited significant ($p < 0.05$) DPPH (IC50 of 0.10 mg/ml) and hydrogen peroxide radicals scavenging activity (IC50 of 0.62 mg/ml) in comparison with the positive control Butylated Hydroxytoluene (BTH). This study therefore recommends that *C. lanatus* fruit could be an excellent source of antioxidants which may prevent diseases whose pathogenesis involves oxidative stress (Adetutu *et al.,* 2015).

Watermelon fruit is one of the few excellent natural sources, richest and abundant in an antoxidant, lycopene and β-carotene. The amount of lycopene in watermelon ranges from 35 to 125 mg per kg of the edible red fleshed portion and it ranks 5th among the major contributors of lycopene in the US diet. Lycopene in fresh watermelon is as much as 40% more bioavailable than the fresh tomatoes Lycopene and vitamin C present in watermelon due to their antoxidant activity has shown very important role in human nutrition and preventing various cancers particularly prostrate, lung, colon, breast, oral cavity, pancreas and oesophageal cancers and control scurvy illness due to formation of collagen as well (Veazie *et al.,* 2012). Watermelon is a proven distinctive source of lycopene claiming new era vitamin due to health promising properties. Diversified nature of watermelon lycopene has attained core attention of researchers in terms of bioavailability and absorption. Safe and sound extraction procedure with minimal loss is desirable at industrial scale. Finally, available information from numerous cell culture and animal models have illuminated the therapeutic role of lycopene against life threatening metabolic syndromes like oxidative stress, cancer, CVD, diabetes etc.

Earlier, only tomato and its products were considered as potential sources of lycopene but now there are proven facts that watermelon also contains appreciable amount of cis-configured lycopene. Thus consumer is gradually shifting towards watermelon and its allied products for their health concerns. Nevertheless, the quantity of lycopene varies depending upon the variety and growing conditions Overall, lycopene ranges from 2.30-7.20 mg/100 g fresh weight bases, present in crystalline form in. More interestingly, lycopene contents of red fleshed watermelon are almost 40% higher than tomato i.e. 4.81 and 3.03 mg/100 g, respectively. However, yellow orange and yellow colored fleshed have relatively less lycopene content i.e. 3.68 and 2.51 mg/100 g, respectively. In tomato, lycopene is available in relatively higher quantity after heat treatment due to break down of protein-carotenoid complex. In contrast, lycopene from watermelon is available directly to the human body just after consumption cell (Tarzana *et al.,* 2017).

The radio protective effects of watermelon (*Citrullus lanatus* (Thunb.) Matsum. and Nakai) juice against oxidative damage induced by low dose X-ray exposure in mice was investigated in a study in which twelve adult male ICR mice were randomly divided into two groups consisting of radiation (Rx) and supplementation (Tx) groups. Rx received filtered tap water, while Tx was supplemented with 50% (v/v) watermelon juice for 28 days *ad libitum* prior to total body irradiation by 100 μGy X-ray on day 29. Brain, lung, and liver tissues were assessed for the levels of malondialdehyde (MDA), apurinic/apyrimidinic (AP) sites, glutathione (GSH), and superoxide dismutase (SOD) inhibition activities. Results showed significant reduction of MDA levels and AP sites formation of Tx compared to Rx ($P< 0.05$). Mice supplemented with 50% watermelon juice restore the intracellular antioxidant activities by significantly increased SOD inhibition activities and GSH levels compared to Rx. These findings may postulate that supplementation of 50% watermelon (C*itrullus lanatus(T*hunb.) Matsum. And Nakai) juice could modulate oxidative damage induced by low dose X-ray exposure (Mohammad et *al.,* 2014).

The antioxidant capacity of aqueous and organic extracts of lycopene rich foods was estimated in ketchup, fresh tomatoes, tomato paste, tomato sauce, tomato soup, vegetable juice, canned tomatoes and watermelon. The value of lycopene in watermelon was 3.7 mg/100g which was higher than obtained in tomatoes (2.3 mg/100g), tomato juice (2.9 mg/100g) and tomato soup (2.0 mg/100g). The TEAC of food extracts was greater in the aqueous versus organic fractions except for watermelon and tomato sauce where levels were similar in the two fractions. The watermelon TEAC was significantly higher than that of tomato sauce or ketchup (Djuric e*t al.,* 2001).

The hydrophilic and lypophilic antioxidant activities in watermelon cultivars harvested at four different stages was estimated by trolox equivalent antioxidant capacity (TEAC) assay. In watermelon varieties Crimson sweet, Giza, P403 and P503, the highest hydrophilic antioxidant activity was detected at red-ripe stage with values ranging from 97.4 to 546.5 μM Trolox/100g on fresh weight basis. A peak in lypophilic antioxidant activity was reached at the red ripe stage of maturity in Crimson sweet, Giza and P503 cultivars with high value of 467.9 μM Trolox/100g fresh weight basis in Crimson sweet variety. Considering the data from all watermelon cultivars good and significant correlations between lypophilic TEAC values and lycopene ($r= 0.649$) were obtained during ripening (Tlili et al., 2011).

The total antioxidant activity was estimated by DPPH method of four icebox cultivars of watermelon fruit during their development and ripening. The total antioxidant activity of F1 Arun and Beauty watermelon cultivars increased with different stages of ripening (young, pre-mature, mature, pre-ripened and ripened) and values were in the range of 48.8-53.3% and 49.8-60.3% respectively where as decreasing trend in antioxidant activity was observed with ripening in Karina King and Suman 235 cultivars. (Soumya and Rao, 2001).

REFERENCES

Aberoumand, A., & Deokule, S. S. (2010). Total Phenolic Contents of Some Plant Foods as a Antioxidant Compound. *Journal of Food Technology*, *8*(3), 131–133. doi:10.3923/jftech.2010.131.133

Adetutu, A., Olorunnisola, O. S., & Owoade, O. A. (2015). Nutritive Values and Antioxidant Activity of *Citrulluslanatus* Fruit Extract. *Food and Nutrition Sciences*, *6*(11), 1056–1064. doi:10.4236/fns.2015.611109

Altaş, S., Kızıl, G., Kızıl, M., Ketani, A., & Haris, P. I. (2011). Protective effect of Diyarbakır watermelon juice on carbon tetrachloride-induced toxicity in rats. *Food and Chemical Toxicology*, *49*(9), 2433–2438. doi:10.1016/j.fct.2011.06.064 PMID:21740951

Asha, S., Pawan, S., & Annu, D. (2015). Determination of total phenolic content and total protein in guava (*Psidiumguajava*) and lemon (*Citrus limon*). *Int J Biol Pharm Res*, *4*(11), 827–834.

Banu, K. S., & Cathrine, L. (2015). General Techniques Involved in Phytochemical Analysis. *International Journal of Advanced Research in Chemical Science*, *2*(4), 25–32.

Boeing, H., Bechthold, A., Bub, A., Ellinger, S., Haller, D., Kroke, A., Leschik-Bonnet, E., Muller, M. J., Oberritter, H., Schulze, M., Stehle, P., & Watzl, B. (2012). Critical review: Vegetables and fruit in the prevention of chronic diseases. *European Journal of Nutrition*, *51*(6), 637–663. doi:10.1007/s00394-012-0380-y PMID:22684631

Bontempo, P., Doto, A., Miceli, M., Mita, L., Benedetti, R., Nebbioso, A., & Altucci, L. (2011). Psidiumguajava L. anti-neoplastic effects: Induction of apoptosis and cell differentiation. *Cell Proliferation*, *45*(1), 22–31. doi:10.1111/j.1365-2184.2011.00797.x PMID:22172154

Burin, V. M., Falcão, L. D., Gonzaga, L. V., Fett, R., Rosier, J. P., & Bordignon-Luiz, M. T. (2010). Colour, Phenolic Content and Antioxidant Activity of Grape Juice. *Food Science and Technology (Campinas)*, *30*(4), 1027–1032. doi:10.1590/S0101-20612010000400030

Chen, L. (2007). Clément Vigneault, GS Vijaya Raghavan, Stan Kubow, 2007. Importance of the phytochemical content of fruits and vegetables to human health. *Stewart Postharvest Review*, *3*, 2.

Dani, S. (2023). A study on physical and chemical characteristics of guava (Psidium guajava L.) Taiwan pink variety. *The Pharma Innovation, 12*(5), 2023–2025. doi:10.22271/tpi.2023.v12.i5y.20175

Dhananjay. (2021). A retrospection of Indian grape varieties. *The Pharma Innovation Journal, SP-10*(11), 408–414.

Djuric, Z., & Powell, L. (2001). Antioxidant capacity of lycopene-containing foods. *International Journal of Food Sciences and Nutrition, 52*(2), 143–149. doi:10.1080/713671775 PMID:11303462

Edeoga, H. O., Okwu, D. E., & Mbaebie, B. O. (2005). Phytochemical Constiuents ofsome Nigerian medicinal plants. *African Journal ofBiotechnology, 4*(7), 685–688. doi:10.5897/AJB2005.000-3127

Edwards, A. J., Vimyard, B. T., & Willy, E. R. (2015). Consumption of Watermelon Juice increases plasma concentration of lycopene and beta carotene in humans. *The Journal ofNutrition, 133*(4), 1043–1050. doi:10.1093/jn/133.4.1043 PMID:12672916

FAO. (2016). Food and Agriculture Organisation of United Nation. http://faostat.fao.org /site /567/default.aspx#ancor

Gollucke, A., & Ribeiro, D.(2012). Use of Grape Polyphenols for Promoting Human Health: A Review of Patents. *Recent Patents on Food, Nutrition & Agriculture, 4*(1), 26–30. doi:10.2174/2212798411204010026 PMID:22316271

Gull, J., Sultana, B., Anwar, F., Naseer, R., Ashraf, M., & Ashrafuzzaman, M. (2012). Variation in Antioxidant Attributes at Three Ripening Stages of Guava (Psidiumguajava L.) Fruit from Different Geographical Regions of Pakistan. *Molecules (Basel, Switzerland), 17*(3), 3165–3180. doi:10.3390/molecules17033165 PMID:22418924

Hanaa M. A., Elshafie M. A., Ismail H. A., Mahmoud M. E., & Ibrahim H.M (2015). Chemical Studies And Phytochemicalscreening Of Grape Seeds (Vitis Vinifera L.). *Minia J. of Agric. Res. & Develop., 35*(2), 313-325.

Hanausek, M., Spears, E., Walazek, Z., Kowalczyk, M. C., Wendel, C., & Slaga, T. J. (2017). Inhibition of murine skin carcinogenesis by freeze dried grape powder and other grape derived major anto-oxidants. *Nutrition and Cancer, 63*(1), 28–38. PMID:21108125

Huang, C.-S., Yin, M.-C., & Chiu, L.-C. (2011). Antihyperglycemic and antioxidative potential of Psidiumguajava fruit in streptozotocin-induced diabetic rats. *Food and Chemical Toxicology, 49*(9), 2189–2195. doi:.2011.05.032 doi:10.1016/j.fct

Huh, Y. C., Solmaz, I., & Sari, N. (2018). Morphological characterization of Korean and Turkish watermelon germplasm. *Proceedings of the IXth Eucarpia meeting on genetics and breeding of Cucurbitaceae*, 327-33.

Ibe, C., Jacobs, C., Imo, C., Osuocha, K. U., & Okoronkwo, M. U. (2014). Evaluation of the antioxidant activities of *Psidiumguajava*and*Aloe vera*. *Journal of Pharmaceutical Research International*, *4*(3), 271–278.

Karovičová, O., Kohajdová, Z., Minarovičová, L., & Kuchtová, V. (2015). The chemical composition of Grape fibre. *Potravinárstvo*, *9*(1), 53–57. doi:10.5219/428

Knockaert, G., Pulissery, S. K., Lemmens, L., Van Buggenhout, S., Hendrickx, M., & Van Loey, A. (2013). Isomerisation of carrot β-carotene in presence of oil during thermal and combined thermal/high pressure processing. *Food Chemistry*, *138*(2-3), 1515–1520. doi:10.1016/j.foodchem.2012.10.080 PMID:23411275

Liu, Y., Hu, X., Zhao, X., & Song, H. (2012). Combined effect of high pressure carbon dioxide and mild heat treatment on overall quality parameters of watermelon juice. *Innovative Food Science & Emerging Technologies*, *13*, 112–119. doi:10.1016/j.ifset.2011.11.001

Mehra, M., Pasricha, V., & Gupta, R. K. (2015). Estimation of nutritional, phytochemical and antioxidant activity of seeds of musk melon (*Cucumismelo*) and water melon (*Citrulluslanatus*) and nutritional analysis of their respective oils. *Food Chemistry*, *3*(6), 98–102.

Mohammad, M. K. A., Mohamed, M. I., Zakaria, A. M., Abdul Razak, H. R., & Saad, W. M. M. (2014). Watermelon (Citrulluslanatus (Thunb.) Matsum. And Nakai) Juice Modulates Oxidative Damage Induced by Low Dose X-Ray in Mice. *BioMed Research International*, *2014*, 1–6. doi:10.1155/2014/512834 PMID:24877107

Monárrez-Espino, J., López-Alarcón, M., & Greiner, T. (2011). Randomized Placebo-Controlled Trial of Guava Juice as a Source of Ascorbic Acid to Reduce Iron Deficiency in Tarahumara Indigenous Schoolchildren of Northern Mexico. *Journal of the American College of Nutrition*, *30*(3), 191–200. doi:10.1080/07315724.2011.10719960 PMID:21896877

Mukhtar, U., Mukammil, S., Khalil, U. R., Alam, R., & Mohammad, A. R. (2011). Evaluation of Local and Exotic Grapes Germ plasm at mingora, swat. – sarhad. *Nongxue Xuebao*, *27*(4), 553–556.

Mutanen, M., & Pajari, A.-M. (2010). Vegetables, whole grains, and their derivatives in cancer Prevention. Dordrecht: Springer.

Naz, A., Butt, M. S., Sultan, M. T., Qayyum, M. M. N., & Niaz, R. S. (2014). Watermelon lycopene and allied health claims. *EXCLI Journal, 13*. PMID:26417290

Nwaichi, E. O., Chuku, L. C., & Oyibo, N. J. (2015). Original Research Article Profile of Ascorbic Acid, Beta-Carotene and Lycopene in Guava, Tomatoes, Honey and Red Wine. *International Journal of Current Microbiology and Applied Sciences, 4*(2), 39–43.

Oseni, O. A., & Okoye, V. I. (2013). Studies of Phytochemical and Antioxidant properties of the Fruit of Watermelon (*Citrulluslanatus*). *Journal of Pharmaceutical and Biomedical Sciences, 27*(27), 508–514.

Perkins-Veazie, P., Davis, A., & Collins, J. K. (2012). Watermelon: From dessert to functional food. *Israel Journal of Plant Sciences, 60*(4), 395–402. doi:10.1560/IJPS.60.1.402

Postescu, I. D., Chereches, G., Tatomir, C., Daicoviciu, D., & Filip, G. A. (2012). Modulation of Doxorubicin-Induced Oxidative Stress by a Grape (Vitisvinifera L.) Seed Extract in Normal and Tumor Cells. *Journal of Medicinal Food, 15*(7), 639–645. doi:10.1089/jmf.2011.0291 PMID:22519870

Praphasawat, R., Klundsupya, P., Muangman, T., Laovithayanggoon, S., Arunpairojana, V., & Himakoun, L. (2011). Anti-mutagenicity and anti-oxidative DNA damage properties of oligomeric pro-anthocyanidines from Thai grape seed in TK6 cells. *Asian Pacific Journal of Cancer Prevention, 12*(5), 1317–1321. PMID:21875289

Rani & Vijayanchali. (2017). Phytochemical, Antioxidant Activity and Lycopene Analysis of Red Guava Fruits, Journal of Research. *Extension and Development, 6*(2), 25–30.

Sandra, M., Barbalho, S. M., Farinazzi-Machado, F. M. V., Goulart, R. A., Brunnati, A. C. S., Ottoboni, A. M. M. B., & Nicolau, C. C. T. (2012). *PsidiumGuajava* (Guava): A Plant of Multipurpose Medicinal Applications. *Medicinal & Aromatic Plants, 1*(4), 1–6.

Seidu, K. T., & Otutu, O. L. (2016). Phytochemical composition and radical scavenging activities of watermelon (*Citrulluslanatus*) seed constituents. *Croatian Journal of Food Science and Technology, 8*(2), 83–89. doi:10.17508/CJFST.2016.8.2.07

Shah, M., Abbas, M., Ullah, W., & Imtiaz, M., Islam, B., Said, F., Jalal, F., & Hussain, S. (2019). Pheno-Physiological Assessment of Grapes (Vitisvinifera) Germplasm. *Acta Scientific Agriculture, 3*(3), 37–42.

Sharma, M., & Bais, R. T. (2015). Preliminary Phytochemical Screening of Methanolic Extracts of Green Grapes and Black Grapes. *International Journal of Scientific Research*, *4*(10), 2015.

Sogi, D. S., Oberoi, D. P. S., & Malik, S. (2010). Effect of Particle Size, Temperature, and Total Soluble Solids on the Rheological Properties of Watermelon Juice: A Response Surface Approach. *International Journal of Food Properties*, *13*(6), 1207–1214. doi:10.1080/10942910903013373

Soumya, V., & Ramana-Rao, T. V. (2014). Nutritional quality evaluation of four icebox cultivars of watermelon fruit during their development and ripening. *International Food Research Journal*, *21*(2).

Soumya, V., & Ramana Rao, T. V. (2014). Nutritional quality evaluation of four icebox cultivars of watermelon fruit during their development and ripening. *International Food Research Journal*, *21*, 631–639.

Tabiri, B., Agbenorhevi, J. K., Wireko-Manu, F. D., & Ompouma, E. I. (2016). Watermelon Seeds as Food: Nutrient Composition, Phytochemicals and Antioxidant Activity. *International Journal of Nutrition and Food Sciences*, *5*(2), 139–144. doi:10.11648/j.ijnfs.20160502.18

Tarazona-D'ıaz, J., Viegas, J., Moldao-Martins, M., & Aguayo, E. (2017). Bioactive Compounds from Flesh and by-product of fresh-cut Watermelon cultivars. *Journal of the Science of Food and Agriculture*, *91*(5), 805–812. doi:10.1002/jsfa.4250 PMID:21384347

Tlili, I., Hdider, C., Lenucci, M. S., Illahy, R., Jebari, H., & Dalessandro, G. (2011). Bioactive compounds and antioxidant activities during fruit ripening of watermelon cultivars. *Journal of Food Composition and Analysis*, *24*(7), 923–928. doi:10.1016/j.jfca.2011.03.016

Wang, M., Wang, L., Pan, X.-J., & Zhang, H. (2012). Monocytic differentiation of K562 cells induced by proanthocyanidins from grape seeds. *Archives of Pharmacal Research*, *35*(1), 129–135. doi:10.1007/s12272-012-0114-y PMID:22297751

Wu, J., Wan, Z., Yi, J., Wu, Y., Peng, W., & Wu, J. (2012). Investigation of the Extracts from Bidenspilosa Linn. var. radiata Sch. Bip. for Antioxidant activities and Cytotoxicity against human Tumor cells. *Journal of Natural Medicines*, *67*(1), 17–26. doi:10.1007/s11418-012-0639-x PMID:22382861

Yang, J., & Xiao, Y.-Y. (2013). Grape Phytochemicals and Associated Health Benefits. *Critical Reviews in Food Science and Nutrition*, *53*(11), 1202–1225. doi:10.1080/10408398.2012.692408 PMID:24007424

Zhang, L., Hogan, S., Li, J., Sun, S., Canning, C., Zheng, S. J., & Zhou, K. (2011). Grape skin Extract inhibits Mammalian Intestinal α-glucosidase Activity and Suppresses postprandial Glycemic response in Streptozocin-treated mice. *Food Chemistry*, *126*(2), 466–471. doi:10.1016/j.foodchem.2010.11.016

Zhao, W., Lv, P., & Gu, H. (2013). Studies on carotenoids in watermelon flesh. *Agricultural Sciences*, *4*(7), 13–20. doi:10.4236/as.2013.47A003

Zhou, K., & Raffoul, J. J. (2012). Potential Anticancer Properties of Grape Antioxidants. *Journal of Oncology*, *2012*, 1–8. doi:10.1155/2012/803294 PMID:22919383

Chapter 2
Beneath the Poppy's Veil:
Exploring Ethnobotanical Treasures, Healing Power, and Biogenic Alchemy of Poppy Seeds

Khansa Akram
The Women University Multan, Pakistan

Imama Fayyaz
The Women University Multan, Pakistan

Aleena Munir
The Women University Multan, Pakistan

Kiran Mustafa
The Women University Multan, Pakistan & Government Graduate College (W), Khanewal, Pakistan

Sara Musaddiq
https://orcid.org/0000-0002-3922-8614
The Women University Multan, Pakistan

Sadia Iqbal
The Women University Multan, Pakistan

ABSTRACT

Plants are one of the essential assets of medicine. Papaver somniferum (scientific name) is native to Southeastern Europe and Asia. The poppy plant is a common herb that is known for its nutritive purposes. Papaver somniferum, which is generally called as the opium poppy, is a source of drugs that have historically been used to help those suffering from pain. It contains several phytochemicals, mainly isolated from seeds. It consists of a lot of alkaloids such as morphine, codeine, narcotine, etc. It can be used in a variety of food products and to make soaps, paints, and other products of industry. This chapter covers cultivation aspects, production, pharmaceutical, food, cosmetic, and ethnobotanical uses. It also consists of biogenic synthesis from poppy seeds.

DOI: 10.4018/979-8-3693-6105-4.ch002

1. INTRODUCTION

Medicine and human civilization have a long history of working together to benefit the world through their various applications (Haider, Batool, & Haleem, 2012). Researchers are becoming increasingly interested in learning about the ability of various drugs to maintain health and cure diseases. It revealed the remarkable nature of herbal medicines for disease treatment. Herbal drugs are made from plants and their parts, such as the bark, stem, roots, flowers, leaves, fruits, and seeds (Mamedov & Craker, 2012). According to the World Health Organization (WHO), conventional medicine is used to treat 80% of the world's population (Shankar & Liao, 2004).

The poppy plant is a traditional plant that is used for therapeutic, nutritional, and systematic purposes (Marciano, Panicker, Liddil, Lindgren, & Sweder, 2018a) known colloquially as "Opium Poppy" (Haber, Pergolizzi, & LeQuang, 2019). The scientific name for *Papaver somniferum* is Ranunculates, the family Papaveraceae, and the genus Papaver specie is *P.somniferum*. Various varieties of Papaver somniferum plant are found all over the world (S. Chaudhary, Aslam, & Siddiqui, 2019) . The species name, somniferum, is derived from the Latin word for "sleep inducing." Opium is a liquid derived from the poppy plant. Opiate is a substance derived from opium, whereas opioids are not derived directly from poppy plants but have morphine-like effects. Opium alkaloid's pharmacological properties have been known since ancient times. Man has been using opium poppy extract for pain relief for at least 3500 years. Turkey, India, Australia, Hungary, Spain, France, and China are among the major opium poppy cultivating countries. Other countries that grow opium poppy include Afghanistan, Burma, Mexico, Pakistan, and India. Opium poppy has been used and cultured in Central and Western Europe since 4600-3800 BC, according to archeological data. Hippocrates, the father of medicine, recommended white poppy sap (Mishra et al., 2013). Poppies have long been associated with honor. Women cultivated opium poppy plants for medicinal purposes in the second century. In the sixth century, Arabian and Islamic countries used opium as a treatment and narcotic. Asians used the poppy plant as a drug in the 17th century. In the nineteenth century, Europeans began trading drugs, and opium wars erupted, with China gaining control of the industry (Mishra et al., 2013). Poppy seeds were used as a dressing for bakery products by Greeks and Europeans in the second century. Around the 7th century A.D., the opium poppy probably spread to China. In China, the opium poppy was known as the "cure all," and people relied on it for the treatment of all diseases. Chinese Despite Chinese emperors' efforts to prohibit opium use for smoking, the infamous "opium wars" of 1839-1842 and 1856-1860 with the British Empire failed, giving rise to a few influence rivalries alongside the opium agriculture moving to China

(A. G. Wheeler, 2019). This resulted in the incorrect consumption and smoking of opium all over the world, beginning in China. Because China was unable to manage national opium production and consumption, the world was subjected to widespread addiction. Opium poppy has pain relieving properties. During the Civil War, morphine was frequently utilized as an analgesic for injured troops because of the widespread availability of opium produced in the United States during the nineteenth century. Opium poppy extract was used in ancient times to alleviate the pain of surgery (Canton-Alvarez, 2019).

Opium poppy is a perennial plant. It grows from a small seed, flowers, and bears fruit "pod" all at once (Kassem & Jacquin, 2001). The entire plant grows in about 120 days. The tree stands between 1 and 5 metres (3 and 16 feet) tall. The leaves are simple, not divided into leaflets, but they can be lobed or unlobed. The leaf arrangement is alternate, with one leaf per node along the stem. On the edge of the leaf blade, there are either lobes or teeth, or both (Marciano, Panicker, Liddil, Lindgren, & Sweder, 2018b). It has large showy flowers that spread symmetrically. Flowers are typically bluish white with a white purple or purple base and range in length from 7.6 to 17.8cm (Marciano et al., 2018b). The majority of petals and sepals have four petals. The petals and sepals are not fused together. There are at least 13 stamens. Opium poppy seeds are dried, small, grey or grayish black to dark blue or white in color, and about 1mm in length (Dyer et al., 2012). They have a faint nut-like odor and a mild nutty flavor that is preferred in bakery goods and breads. The poppy species is farmed for its oil, which is used in cooking, cosmetics, and even as a salad dressing. In addition to its usage as cooking oil, this oil is also a key ingredient in the production of margarine and salad dressings (Blahovec, 2002). Paints and varnishes contain poppy seed oil because it dries quickly. It is also found in ointments, emulsions, and some soaps. Poppy seeds can also be used to make tea. Tea has a bitter taste that can be mask with coffee, honey, or lemon juice. Cultivators extract extracted from the pods while they are yet on the plant. After racking opium, the pods are removed from the plant and then dried. The pods are then opened and the seeds are extracted before drying in the sun. Because opium poppy seeds are highly nutritious, they contain up to 24% protein and other beneficial minerals. In some parts of the world, the leaves are eaten as vegetables. Opium alkaloids like morphine, codeine, and heroine are sourced from the white latex of the half-grown seed capsule. Kidney-shaped, ripe seeds that can be any shade of blue are used in bakery and other edible products. Poppy seeds contain 44-50% fixed oil, with the main components being oleic and linoleic acids. Only the unripe seeds of poppy seeds exhibit narcotic properties (Garg, Hitawala, & Agarwal, 2018). It contains proteins, sugars, plant wax, fats, and other substances, as well as some major alkaloids such as noscapine 4-8%, codeine 1-3%, papaverine 1-3%, morphine 10-15%, and thebaine 1-2%. Opium alkaloids are divided into two chemical groups: benzylisoquinolines and phenanthrenes. Codeine, morphine, and thebaine are phenanthrene alkaloids, whereas noscapine and papaverine are benzylisoquinoline alkaloid (Peter J. Facchini & Bird, 1998). Opium's pharmacologically active principles are found in its alkaloids, the most important of which is morphine, which accounts for about 10% of the weight of unprocessed opium (Bird, Franceschi, & Facchini, 2003). Papaverine and codeine are two other active alkaloids found in trace amounts. Morphine, thebaine, and codeine are all narcotics that act on the central nervous system and are analgesic. Papaverine, noscapine, and other alkaloids only act to relax involuntary smooth muscles. Opiates such as thebaine, codeine, and morphine have major effects on the spinal cord and brain. Their main goal is to relieve pain (Chain et al., 2018). Drugs also reduce anxiety, induce sedation, drowsiness, and relaxation, and enhance mood. Opioids have the following primary physiologic effects: they slow heartbeat and respiration, suppress the cough response, and relax the smooth muscles of the gastrointestinal tract. Opiates are addictive drugs that cause physical weakness. Because of their analgesic effects, opiates fall under the category of narcotics (P. J. Facchini & De Luca, 1995). Opium use causes mental and physical failure as well as shortens one's life. Dry opium has traditionally been used as an antispasmodic, expectorant, diaphoretic, astringent, hypnotic, sedative, and narcotic (Masihuddin, Jafri, Aisha, & Shahid). Morphine exhibits its activity by mimicking natural opioid peptides such as endorphins, enkephalins, and dynorphins, as it is a completely in CNS agonist at receptor. These opiates or alkaloids reconcile their pharmacological effect via four classes of receptors known as, and, the G-Protein coupled receptors. Noscapine is an antitussive agent, while the other is an anti-mitotic agent or antineoplastic; by targeting tubulin, it inhibits tumor cell growth, resulting in abnormal mitotic spindle grouping.

2. NATIVE APPLICATIONS

Herbal therapies are becoming increasingly popular due to their effectiveness and preference for a milder aftereffect. The use of medicinal herbs in traditional medicine is not only a curative strategy, but also a source of medication (Vickers, Zollman, & Lee, 2001). Humans constantly desire to eat healthy meals, and poppy seeds are the main food item with several applications. Spices have been defined as plant material of native or alien origin, fragrant or with a strong flavor, used to enhance the flavor of food. The poppy plant bears brightly colored blooms and is planted for its beauty or ornamentation. Opium poppy is widely used all over the world for its twin purpose of pharmacological "alkaloids" and food "seed." Both the seeds and the oil are considered to have a variety of health benefits (Lo

Figure 1. Poppy plant parts

& Chua, 1992). Poppy seeds, like most seeds, are high in plant lipids, fiber, and a variety of other nutrients. Manganese, a trace component beneficial to blood coagulation and bone health, is abundant in poppy seeds (Li & Yang, 2018). This mineral can also help your body gain fat and amino acids. Copper, a mineral essential to create connective tissue and transport iron, is abundant. Inflammation and sickness are caused by a high intake of poppy seed oil or an omega-6 to omega-3 ratio in the diet (Schwingshackl & Hoffmann, 2014). Poppy seeds and oils are a good source of polyphenols, an antioxidant that reduces the risk of heart disease (Mozaffarian, Micha, & Wallace, 2010). The poppy plant contains alkaloids such as codeine and morphine, which are good for pain relief, soothing, and inducing sleep. Poppy seed oil is high in mono and polyunsaturated fats (Saini & Keum, 2018). According to research, diets high in unsaturated fats reduce the incidence of heart attack and stroke by up to 17%, and poppy oil aids in wound healing and the prevention of scaly skin lesions (Lin, Zhong, & Santiago, 2017). Poppy seed is taken orally to treat diarrhea and constipation caused by infection. A high intake of dietary fiber and manganese aids in the prevention and treatment of diabetes (Macfarlane, Macfarlane, & Cummings, 2006; Yang, Wang, Zhou, & Xu, 2012). Poppy seeds include a variety of plant-derived chemical compounds that have anti-inflammatory, antioxidant, and health-promoting qualities. The seeds' antioxidants protect your body from cellular damage. According to some research, women who have products derived from poppy seed oil pumped over their fallopian tubes may boost their fertility (Dreyer et al., 2017; Wang et al., 2019). Poppy seeds are commonly used in cough and asthma syrups. The seeds are ground with milk to make a paste that helps to reduce dandruff. Poppy seed oil is used in the production of paint, soap, and varnishes. Poppy seeds are useful in the treatment of mouth ulcers because they have a cooling impact on the body (Sawarkar, Deshmukh, & Verma, 2016). Poppy seeds contain oleic acid, which is beneficial in lowering blood pressure. The seeds are a good source of zinc, and studies show that minerals can help with vision health and potentially prevent a dangerous ocular illness known as macular degeneration. Noscapine has been shown to combat tumors and even lower the number of prostate and breast cancer cells. It can also suppress cancer cell proliferation, and cancerous ulcers can be treated with poppy plant tincture (Aruna & Sivaramakrishnan, 1990). Seeds can also be used to cure kidney stones. According to the University of Maryland Medical Center, people with hypercalcemia should limit their intake of poppy seeds, which are high in oxalic acid. Many countries throughout the world utilize iodinated poppy seed oil to treat iodine deficiency (Afzali et al., 2015). Because of the high concentration of linolenic acid in seeds, it is beneficial in treating eczema. A poppy seed scrub can aid in skin cleansing (Gloor, Kägi, & Wüthrich, 1995). Poppy seeds include a high concentration of dietary fibre, which improve digestive health. Inhaling opium vapors through pipe smoking. This appears to be a common application in various countries these days. Opium is used to treat significant melancholy, sleeplessness, or socioeconomic problems that result in major worry or anxiety (Powers, Erickson, & Swortwood, 2018). The synthetically manufactured poppy alkaloid papaverine is utilized as a smooth muscle relaxant. Poppy seeds are used to flavor rolls, breads, pastries, and cakes in Middle Eastern and European cuisine. In Turkey, they are continuously trodden and used in sweets. In India, the seeds are ground and used to thicken sauces. The seeds are often used in fish, noodles, and vegetable dishes in Germany (Concheiro et al., 2015; Zentai, Sali, Szeitzné-Szabó, Szabó, & Ambrus, 2012). The seeds are high in B-complex vitamins like folic acid, pantothenic acid, and thiamin. Most vitamins work as cofactors in substrate metabolism, namely carbohydrates and lipids within the human body (Concheiro et al., 2015). Oil can also be found in ointments, soaps, and emulsions. Morphine derived from opium poppy is a strong hypnotic and tranquilizer. Morphine performs its functions through altering endorphin receptors. Morphine is a potent tonic that can induce exhilaration and remove concerns and inhibitions. Morphine consumption typically leads to addiction.

3. PHYTOCHEMICALS

Phytochemicals naturally exist in countless foods, these are plant chemicals that are non-nutritive and possess disease inhibitory or defense properties. These are referred as non-essential nutrients as they are not required by the human body for maintaining life. These chemicals are produced by plants to protect themselves but current researches have shown that it also shields human from diseases. Food containing phytochemicals are so far part of our everyday diet (Probst, Guan, & Kent, 2017).

Poppy seeds are used for dressing in cooking purposes and are nutritious oil seeds. Even though the seeds obtained from the dried fruits (pods) of the poppy plant (opium poppy) and are free of the negative effects of other products of poppy plant such as opium poppy (Regula, Cerba, Suliburska, & Tinkov, 2018).

The chemical of interest in opium poppy are alkaloids. Opium poppy also carries the blender of fats, proteins, sugars, colouring substances, wax, resins, rubber, salts such as sulphates, mucilaginous materials and water. Opium alkaloids are merged with several simple organic acids such as rare meconic acid, fumaric acid, and lactic acids (Dey et al., 2020).

The high protein content in these seeds constitutes the major part being a globulin which is almost 55% of the whole nitrogen. The seed holds riboflavin 46, pantothenic acid, thiamine 420 and folic acid and niacin. The seed oil contains palmitic acid, oleic acid, α-tocopherol,linoleic acid, β-tocopherol, stearic acid and ϒ-tocopherol respectively (S. S. Chaudhary, Tariq, Zaman, & Imtiyaz, 2013).

3.1 Poppy Seed

- **Thiamine**

Thiamine structure consists of Methylene Bridge linking aminopyrimidine and a thiazolium ring. Methyl and hydroxyethyl chains replaced the thiazole. Thiamine is

Table 1. Uses of poppy plant

Sr. No	Plant Part	Chemical Uses
1.	Flower	Poppy flower are important because they are sedative, aphrodisiac, decongestant, hypnotic, painkiller, narcotic, muscle relaxant.
2.	Capsule	The medicinal importance of capsule is painkiller, stimulant, euphoria, evaporating anxieties, tensions, fears and inhibitions
3.	Leave	Fresh leaves are used as vegetables.
4.	Seeds	Some alkaloids are present, and they have pharmaceutical importance.

considered as "anti-stress" vitamin because it helps to improve the immune system (Whitfield et al., 2018).

- **Pantothenic acid**

Pantothenic acid is also named as B5 and the class to which it belongs is secondary alcohols. It is the water soluble vitamin. In the poppy seeds, pantothenic acid is the water soluble vitamin that was found at the excessive amount than niacin and thiamin (Tahiliani & Beinlich, 1991).

- **Niacin**

Niacin also called as B3. The term niacin is also used to indicate the nicotinamide, nicotinic acid and nicotinamide. It involves in the cholesterol synthesis, DNA synthesis and metabolism of fats. It is also important in regulation of glucose. Most important source of B3 includes seeds (Bassler, 1992).

3.2 Poppy Seed Oil

- **Linoleic acid**

It is polyunsaturated omega-6 containing 18 carbon long fatty acid having two C-C double bond at the 9 and 12 positions. Linoleic acid has been discovered to be linked with various diseases known as thyroid cancer, hypertension and cirrhosis (Lehnen, da Silva, Camacho, Marcadenti, & Lehnen, 2015). Poppy seed oil contains unsaturated fatty acids which is 86-90%, and 63-75%, the high quantity fatty acid is linoleic (Lančaričová, Havrlentova, Muchová, & Bednárová, 2016).

- **Oleic acid**

Oleic acid is also named as oleate. The aliphatic tail is present in this fatty acid that present between 13 and 21 carbon atoms. It have been described as anti-inflammatory fat, it plays a role in immune system. Poppy seed oil contains 13% oleic acid as a major fatty acid (Carrillo, Cavia Mdel, & Alonso-Torre, 2012).

- **Stearic acid**

Stearic acid is long chain fatty acids. Poppy seed oil contains stearic acid and it is used in shampoo, soaps, detergents (Vanderveen, 1994).

- **Palmitic acid**

Poppy seed oil contains 8-10% of the palmitic acid. It is also known as hexadecanoic acid (Carta, Murru, Banni, & Manca, 2017).

- **α-tocopherol**

Some poppy seed oils have high levels of tocopherols and sterols. Oil from some poppy seeds was found to contain sterols and tocopherols. These oils have an average α-tocopherol content of 33.03mgkg^{-1} (Erinç, Tekin, & Özcan, 2009).

- **β-tocopherol**

β-tocopherol also known as 5,8-dimethyltocol. These are the derivatives of Vitamin E. They are having trimethyltridecyl chain which is saturated and isattached to a benzopyran ring at C-6. It is (2R)-2,5,8-trimethyl-2-[(4R,8R)-4,8,12-trimethyltridecyl]-3,4-dihydrochromen-6-ol. Poppy seed oil contains β-tocopherol (309.5 ppm-567.3 ppm) (Grzegorz, Sylwester, & Iwona, 2020).

- **ϒ-tocopherol**

It is the form of naturally occurring fat soluble vitamin E. It is present in numerous nuts and seeds. It has capability of antioxidant activity. In the ϒ-tocopherol, the methyl groups substitute the chroman-6-ol core at position 7 and 8 (Reiter, Jiang, & Christen, 2007).

Poppy seed contain small amount of alkaloids such as morphine, codeine, papaverine, narcotine and thebaine (Fritschi & Prescott, 1985). The seeds themselves barely holds any opium alkaloids but they can be defiled with the opium alkaloids from the latex because of bad harvesting practices, insect damages. Here are some major alkaloid and their uses are given below in the table (Schmidt, Boettcher, Kuhnt, Kutchan, & Zenk, 2007).

3.3 Benzoisoquinoline Alkaloids

Benzoisoquinoline nucleus is present in these compounds such as Noscapine, Papaverine etc (Beaudoin & Facchini, 2014).

- **Papaverine**

Table 2. Major alkaloids found in Papaver somniferum

Poppy Alkaloid	Pharmaceutical Uses	Description
Morphine	Narcotic analgesic	Opium's primary alkaloid; also a banned substance.
Codeine	Analgesic and Antitussive	When codeine is taken orally, it is metabolised into morphine by a variety of liver enzymes, most notably CYP2D6. As a result, codeine is a precursor of morphine. It is also considered a Schedule I drug.
Thebaine	Precursor of opioids	The pharmaceutical industry uses the morphine biosynthetic intermediary thebaine in the production of oxycodone and naloxone. It is also considered an addictive Schedule drug.
Noscapine	Antitussive	There is hope that this chemical can fight cancer. In other words, it is not a banned substance.
Papaverine	Antispasmodic	As an added bonus, it is not a banned substance.

It is a benzoisoquinoline alkaloid that is isoquinoline replaced by methoxy groups at 6 and 7 positions and 3, 4-dimethoxybenzyl group at position 1. 1-(3, 4-dimethoxybenzyl)-6,7-dimethoxyisoquinoline.

Pharmaceutical industries, food industries for the decoration use these alkaloids widely. It plays vital role as a antispasmodic drug and as a vasodilator agent. It is also used in the treatment of acute mesenteric ischemia and erectile dysfunction (H. M. Liu & Tu, 2002).

- **Noscapine**

For many decades these alkaloids obtained from opium poppy are used as a cough suppressant with no side effects (Bulduk & Taktak, 2013). Recent researches have shown that noscapine also have antitumor properties (Chougule, Patel, Sachdeva, Jackson, & Singh, 2011).

3.4 Phenanthrene Alkaloids (Morphinane Alkaloids)

These group of compounds have phenanthrene nucleus such as Morphine, Codeine, Thebaine and other alkaloids (Aynehchi & Jaffarian, 1973).

- **Morphine**

Poppy seeds carrying morphine at much better and lower concentrations were themed to steam treatments and dry heat, baking and water washing. Codeine and morphine are the narcotics, obtained from *Papaver somniferum*. To scale back pain

Figure 2. Structure of A) thiamin, B) riboflavin, C) folic acid, D) pantothenic, E) niacin, F) linoleic acid, G) oleic acid, H) stearic acid, I) palmitic acid, J) alpha-tocopherol, K) beta-tocopherol, L) gamma-tocopherol, M) papaverine, N) noscapine, O) morphine, P) codeine, Q) thebaine, R) porphyroxine

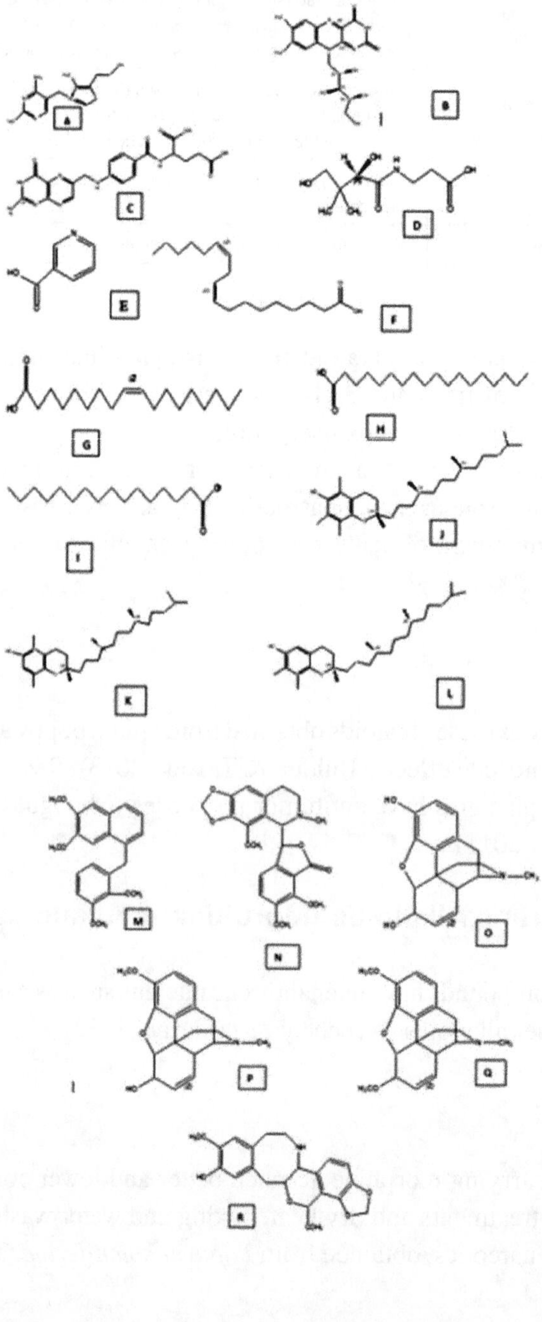

it acts directly on central nervous. Morphine offers a high prospective for addiction (Kwiecień-Obara et al., 2016).

- **Codeine**

Codeine is a naturally phenanthrene alkaloid. It is used as antidiarrheal and as a cough suppresent (Struempler, 1987). Codeine is lighter analgesic than morphine. Codeine is used in manufacturing acetyldihydrocodeine and dihydrocodeine.

- **Porphyroxine**

It is isolated from opium, which is a monophenolic alkaloid. It has three asymmetric centers and it contains a methyl acetal group.

4 BIOLOGICAL/PHARMACOLOGICAL PROPERTIES

4.1 Antioxidant Activity

Dangerous concentration of oxidative stress can be inhibited or prevented by the natural oxidants present in the spices and herbs. Free radical scavengers such as phenolic compounds and polyphenols are present in the spices and herbs. We estimated Poppy's potential to scavenge free radicals in the paper under consideration. This plant exhibited potent anticancer, hepatoprotective, antiviral, and other properties. These properties may be due to its antioxidant activity. Antioxidants are of great importance because of their positive properties on health of humans. The great oil content enables the high utilization of the poppy seed. Solid antioxidative properties are exhibited by poppy seeds due to tocopherols (Ishtiaque, Naz, Siddiqi, Jabeen, & Ahmed, 2015).

4.2 Antibacterial Activity

The antibacterial activities of botanical extracts and phytochemicals are of great medicinal importance. The antibacterial properties of plant spices were investigated against Gram -ive and Gram +ive microorganisms like S. aureus and E. coli. A total of 188 bacterial isolates representing 11 distinct genera of Gramme -ve and Gramme +ve microorganisms were collected from the mouths of seemingly healthy people, and the antibacterial activity of aqueous decoctions and a part of poppy seed (Papaver somniferum L., Papaveraceae) was investigated against them. The technique of disc diffusion was used to examine antibiotic efficacy. Antibacterial activity can be found in 14.4% of poppy seed (Cowan, 1999). Antimicrobial alkaloids produced by Papaveraceae family plants are evaluated as promising therapeutic agents for infectious syndromes (Q. Liu et al., 2017).

4.3 Narcotic Activity

An opiate alkaloid, morphine is sequestered from the Papaver somniferum plant and made by synthesis. Specific opiate receptors (mu,delta and kappa) are muddled and initiated by morphine, every single of which are jumbled in monitoring dissimilar functioning of brain. In the gastrointestinal scheme and central nervous, this agent displays widespread properties comprising anxiolysis, analgesia, sedation, euphoria, gastrointestinal system smooth muscle contraction and respiratory depression (Asadikaram et al., 2018).

4.4 Bronchodilator Activity

Respiratory disease is common and important caused of illness and death everywhere in the world. Respiratory disorders are the most common reasons for hospital stays surrounded by children. Herbal remedies for action of respiratory disorders are common training in the countless parts of the world.

4.5 Anti-Anxiety Activity

Because of its calming and relaxing effect on the central nervous system (CNS), poppy seed wash is useful for alleviating anxiety. Because of how quickly it takes effect, it can be used all during an attack of panic or other intense bout of anxiety. Like most poppies, the alkaloids in the several species of Eschscholzia have analgesic and sedative effects. Among a group of 264 patients with mild to moderate nervousness, a medication containing anchored quantities of Crataegus oxyacantha, E. californica, and magnesium was found to be safe and more effective than the control over a 3-month period. This study was randomized, double-blind, and placebo-controlled (Hanus, Lafon, & Mathieu, 2004).

4.6 Antidepressant Activity

Depression, with rigorous magnitudes to human fitness is a common mental disorder. Severe faults exist for paramount of synthetic antidepressants such as easy reappearance, contracted spectrum of antidepressants, extraordinary drug amount and adverse reactions (Belmaker & Agam, 2008). The cause of depression is complex, and some biological variations are found in numerous dissimilar categories of mental health. This hypothesis is supported by several herbal medicines, including herbal medicines that regulate these neurotransmitters by precisely inhibiting their reuptake in the brain, a mechanism that is common among different drugs (Lanni, Govoni, Lucchelli, & Boselli, 2009). For the time being, Some herbal drugs act as antidepressants by sensitizing serotonin receptors or inhibiting monoamine oxidase. (López-Muñoz & Alamo, 2009).

4.7 Analgesic Activity

When dealing with excruciating pain, morphine is a must-have analgesic reagent. When taken orally, morphine is rapidly broken down by the liver into inactive metabolites, limiting the duration of its analgesic effects unless a sustained-release formulation is used. Morphine's negative effects include respiratory depression, drowsiness, lightheadedness, nausea, vomiting, dizziness, itching, and constipation. A major issue is the brain's response to depression caused by breathing (Oudejans, van Velzen, & Dahan, 2016).

4.8 Anticarcinogenic Effect

Researchers have looked at the cancer-fighting properties of common spices and leafy greens. Benzo[a]pyrene (B[a]P) and 3'-methyl-4-dimethylaminoazobenzene (3'MeDAB) were used to induce squamous cell carcinoma in the stomachs of Swiss mice. The nine herbal remedies tested showed a significant decrease. In contrast to asafoetida, turmeric, poppy leaves, Solanum leaves, and Alternanthera leaves, only B[a]P was sufficient to induce neoplasia in poppy seeds. These results suggest that basil leaves, poppy seeds, and cumin seeds are the most often used spices in cooking and might be studied further for their potential anticancer effects. These seeds boost the activity of Glutathione S-transferase, an enzyme that helps rid the body of carcinogens, by more than 78% in the liver and oesophagus.

4.9 Hypoglycemia Activity

Aqueous extracts of seeds showed significant hypoglycemic effects in diabetic rats with glucose and alloxan (Najafipour & Beik, 2016).

4.10 Antidiarrheal Activity

Medics relies on peragoric, an opium camphor solution, to treat diarrhea by soothing the digestive system (Zarghami, 2015).

4.11 Antispasmodic Activity

Papaverine is used as an antispasmodic to treat vascular spasms (especially those involving the heart, brain, and intestines) and to treat visceral spasms.. Used to treat the symptoms of irritable bowel syndrome (Kimoto, Kessler, & Constantinou, 1990).

5. BIOGENIC SYNTHESIS

Biogenic synthesis of nanoparticles by use of medicinal plants has fewer risky outcomes than the nanoparticles synthesized by chemical methods (Thema, Beukes, Gurib-Fakim, & Maaza, 2015; Thovhogi, Diallo, Gurib-Fakim, & Maaza, 2015).

6. CONCLUSION

This chapter highlighted variety of different applications of poppy plant. We have demonstrated knowledge about habitat and phytochemistry of *Papaver somniferum*. Most of the parts of plant especially poppy seeds have showed diverse applications in Textile industry and alkaloids have many pharmaceutical properties such as antitussive, antioxidant and so on. It also plays major role in biogenic synthesis of Zinc oxide nanoparticles (ZnO-NPs) and Green synthesis of Gold nanoparticles (Au-NPs). Further and more imaginative approaches of poppy plant applications will appear in the literature over the coming decade and beyond.

REFERENCES

Afzali, M., Ghaeli, P., Khanavi, M., Parsa, M., Montazeri, H., Ghahremani, M. H., & Ostad, S. N. (2015). Non-addictive opium alkaloids selectively induce apoptosis in cancer cells compared to normal cells. *Daru : Journal of Faculty of Pharmacy, Tehran University of Medical Sciences*, *23*(1), 16. doi:10.1186/s40199-015-0101-1 PMID:25890335

Aruna, K., & Sivaramakrishnan, V. M. (1990). Plant products as protective agents against cancer. *Indian Journal of Experimental Biology*, *28*(11), 1008–1011. PMID:2283166

Asadikaram, G., Akbari, H., Vakili, S., Asiabanha, M., Shahrokhi, N., & Savardashtaki, A. (2018). The effects of opium addiction on thyroid and sex hormones in diabetic and non-diabetic male and female rats. *Acta endocrinologica (Bucharest, Romania: 2005), 14*(4), 466-472. doi:10.4183/aeb.2018.466

Aynehchi, Y., & Jaffarian, S. (1973). Determination of thebaine in various parts of Papaver bracteatum Lindl. During the growing season. *Lloydia, 36*(4), 427–429. PMID:4773272

Bassler, T. J. (1992). Niacin benefits. *The Western Journal of Medicine, 156*(2), 209–209. PMID:1536086

Beaudoin, G. A., & Facchini, P. J. (2014). Benzylisoquinoline alkaloid biosynthesis in opium poppy. *Planta, 240*(1), 19–32. doi:10.1007/s00425-014-2056-8 PMID:24671624

Belmaker, R. H., & Agam, G. (2008). Major depressive disorder. *The New England Journal of Medicine, 358*(1), 55–68. doi:10.1056/NEJMra073096 PMID:18172175

Bird, D. A., Franceschi, V. R., & Facchini, P. J. (2003). A tale of three cell types: Alkaloid biosynthesis is localized to sieve elements in opium poppy. *The Plant Cell, 15*(11), 2626–2635. doi:10.1105/tpc.015396 PMID:14508000

Blahovec, J. (2002). Mechanics of poppy seeds. *International Journal of Food Properties, 5*(2), 277–287. doi:10.1081/JFP-120005785

Bulduk, I., & Taktak, F. (2013). Isolation and Characterization of Antitumor Alkaloid from Poppy Capsules (Papaver somniferum). *Journal of Chemistry, 493870*, 1–4. Advance online publication. doi:10.1155/2013/493870

Canton-Alvarez, J. A. (2019). A Gift from the Buddhist Monastery: The Role of Buddhist Medical Practices in the Assimilation of the Opium Poppy in Chinese Medicine during the Song Dynasty (960-1279). *Medical History, 63*(4), 475–493. doi:10.1017/mdh.2019.45 PMID:31571697

Carrillo, C., Cavia Mdel, M., & Alonso-Torre, S. (2012). Role of oleic acid in immune system; mechanism of action; a review. *Nutrición Hospitalaria, 27*(4), 978–990. doi:10.3305/nh.2012.27.4.5783 PMID:23165533

Carta, G., Murru, E., Banni, S., & Manca, C. (2017). Palmitic Acid: Physiological Role, Metabolism and Nutritional Implications. *Frontiers in Physiology, 8*, 902. doi:10.3389/fphys.2017.00902 PMID:29167646

Chain, E. (2018). Update of the Scientific Opinion on opium alkaloids in poppy seeds. *EFSA journal. EFSA Journal, 16*(5), e05243–e05243. doi:10.2903/j.efsa.2018.5243

Chaudhary, S., Aslam, M., & Siddiqui, A. (2019). Tukhm Khashkhaash (Poppy Seeds): A Unani drug of multitudinous potential. *Journal of Drug Delivery and Therapeutics*, *9*(2), 528–533. doi:10.22270/jddt.v9i2.2560

Chaudhary, S. S., Tariq, M., Zaman, R., & Imtiyaz, S. (2013). The In vitro anti-acne activity of two unani drugs. *Ancient Science of Life*, *33*(1), 35–38. doi:10.4103/0257-7941.134594 PMID:25161328

Chougule, M., Patel, A. R., Sachdeva, P., Jackson, T., & Singh, M. (2011). Anticancer activity of Noscapine, an opioid alkaloid in combination with Cisplatin in human non-small cell lung cancer. *Lung Cancer (Amsterdam, Netherlands)*, *71*(3), 271–282. doi:10.1016/j.lungcan.2010.06.002 PMID:20674069

Concheiro, M., Newmeyer, M. N., da Costa, J. L., Flegel, R., Gorelick, D. A., & Huestis, M. A. (2015). Morphine and codeine in oral fluid after controlled poppy seed administration. *Drug Testing and Analysis*, *7*(7), 586–591. doi:10.1002/dta.1742 PMID:25345619

Cowan, M. M. (1999). Plant products as antimicrobial agents. *Clinical Microbiology Reviews*, *12*(4), 564–582. doi:10.1128/CMR.12.4.564 PMID:10515903

Dey, P., Kundu, A., Kumar, A., Gupta, M., Lee, B. M., Bhakta, T., . . . Kim, H. S. (2020). Analysis of alkaloids (indole alkaloids, isoquinoline alkaloids, tropane alkaloids). *Recent Advances in Natural Products Analysis*, 505-567. doi:10.1016/B978-0-12-816455-6.00015-9

Dreyer, K., van Rijswijk, J., Mijatovic, V., Goddijn, M., Verhoeve, H. R., van Rooij, I. A. J., Hoek, A., Bourdrez, P., Nap, A. W., Rijnsaardt-Lukassen, H. G. M., Timmerman, C. C. M., Kaplan, M., Hooker, A. B., Gijsen, A. P., van Golde, R., van Heteren, C. F., Sluijmer, A. V., de Bruin, J.-P., Smeenk, J. M. J., ... Mol, B. W. J. (2017). Oil-Based or Water-Based Contrast for Hysterosalpingography in Infertile Women. *The New England Journal of Medicine*, *376*(21), 2043–2052. doi:10.1056/NEJMoa1612337 PMID:28520519

Dyer, A. G., Boyd-Gerny, S., McLoughlin, S., Rosa, M. G., Simonov, V., & Wong, B. B. (2012). Parallel evolution of angiosperm colour signals: common evolutionary pressures linked to hymenopteran vision. *Proc Biol Sci*, *279*(1742), 3606-3615. doi:10.1098/rspb.2012.0827

Erinç, H., Tekin, A., & Özcan, M. (2009). Determination of fatty acid, tocopherol and phytosterol contents of the oils of various poppy (Papaver somniferum L.) seeds. *Grasas y Aceites*, *60*(4), 375–381. Advance online publication. doi:10.3989/gya.129508

Facchini, P. J., & Bird, D. A. (1998). Developmental regulation of benzylisoquinoline alkaloid biosynthesis in opium poppy plants and tissue cultures. *In Vitro Cellular & Developmental Biology. Plant*, *34*(1), 69–79. doi:10.1007/BF02823126

Facchini, P. J., & De Luca, V. (1995). Phloem-Specific Expression of Tyrosine/Dopa Decarboxylase Genes and the Biosynthesis of Isoquinoline Alkaloids in Opium Poppy. *The Plant Cell*, *7*(11), 1811–1821. doi:10.2307/3870189 PMID:12242361

Fritschi, G., & Prescott, W. R. Jr. (1985). Morphine levels in urine subsequent to poppy seed consumption. *Forensic Science International*, *27*(2), 111–117. doi:10.1016/0379-0738(85)90173-2 PMID:3979930

Garg, P., Hitawala, A. A., & Agarwal, M. (2018). Cardiotoxic Effects of Raw Opium. *Indian Journal of Critical Care Medicine : Peer-Reviewed, Official Publication of Indian Society of Critical Care Medicine*, *22*(1), 46–48. doi:10.4103/ijccm.IJCCM_427_17 PMID:29422734

Gloor, M., Kägi, M., & Wüthrich, B. (1995). Poppyseed anaphylaxis. *Schweizerische Medizinische Wochenschrift*, *125*(30), 1434–1437. PMID:7676232

Grzegorz, D., Sylwester, C., & Iwona, K. (2020). Composition and quality of poppy (papaver somniferum l.) Seed oil depending on the extraction method. *Lebensmittel-Wissenschaft + Technologie*, *110167*. Advance online publication. doi:10.1016/j.lwt.2020.110167

Haber, I., Pergolizzi, J. Jr, & LeQuang, J. A. (2019). Poppy Seed Tea: A Short Review and Case Study. *Pain and Therapy*, *8*(1), 151–155. doi:10.1007/s40122-019-0113-5 PMID:30719627

Haider, S., Batool, Z., & Haleem, D. J. (2012). Nootropic and hypophagic effects following long term intake of almonds (Prunus amygdalus) in rats. *Nutrición Hospitalaria*, *27*(6), 2109–2115. doi:10.3305/nh.2012.27.6.6140 PMID:23588464

Hanus, M., Lafon, J., & Mathieu, M. (2004). Double-blind, randomised, placebo-controlled study to evaluate the efficacy and safety of a fixed combination containing two plant extracts (Crataegus oxyacantha and Eschscholtzia californica) and magnesium in mild-to-moderate anxiety disorders. *Current Medical Research and Opinion*, *20*(1), 63–71. doi:10.1185/030079903125002603 PMID:14741074

Ishtiaque, S., Naz, S., Siddiqi, R., Jabeen, S., & Ahmed, J. (2015). Antioxidant Activity and Phenolic Contents of Ajwain, Mustard, Fenugreek and Poppy Seed. *Recent Innovations in Chemical Engineering*, *7*(2), 119–127. doi:10.2174/2405520407666150425004125

Kassem, M. A., & Jacquin, A. (2001). Somatic Embryogenesis, Rhizogenesis, and Morphinan Alkaloids Production in Two Species of Opium Poppy. *Journal of Biomedicine & Biotechnology*, *1*(2), 70–78. doi:10.1155/S1110724301000237 PMID:12488612

Kimoto, Y., Kessler, R., & Constantinou, C. E. (1990). Relaxation mechanisms of antispasmodics papaverine and thiphenamil on the human corpus cavernosum. *The Journal of Urology*, *144*(6), 1497–1499. doi:10.1016/S0022-5347(17)39785-9 PMID:2231955

Kwiecień-Obara, E., Szponar, J., Krajewska, A., Witkowska, A., Radoniewicz, A., & Szponar, M. (2016). Morphine (obtained from poppy seeds) and dextrometorfan poisoning– a case report. *Przegląd Lekarski*, *73*(8), 596–598. PMID:29677437

Lančaričová, A., Havrlentova, M., Muchová, D., & Bednárová, A. (2016). Oil content and fatty acids composition of poppy seeds cultivated in two localities of Slovakia. *Agriculture*, *62*(1), 19–27. doi:10.1515/agri-2016-0003

Lanni, C., Govoni, S., Lucchelli, A., & Boselli, C. (2009). Depression and antidepressants: Molecular and cellular aspects. *Cellular and Molecular Life Sciences*, *66*(18), 2985–3008. doi:10.1007/s00018-009-0055-x PMID:19521663

Lehnen, T. E., da Silva, M. R., Camacho, A., Marcadenti, A., & Lehnen, A. M. (2015). A review on effects of conjugated linoleic fatty acid (CLA) upon body composition and energetic metabolism. *Journal of the International Society of Sports Nutrition*, *12*(1), 36–36. doi:10.1186/s12970-015-0097-4 PMID:26388708

Li, L., & Yang, X. (2018). The Essential Element Manganese, Oxidative Stress, and Metabolic Diseases: Links and Interactions. *Oxidative Medicine and Cellular Longevity*, *7580707*, 1–11. Advance online publication. doi:10.1155/2018/7580707 PMID:29849912

Lin, T.-K., Zhong, L., & Santiago, J. L. (2017). Anti-Inflammatory and Skin Barrier Repair Effects of Topical Application of Some Plant Oils. *International Journal of Molecular Sciences*, *19*(1), 70. doi:10.3390/ijms19010070 PMID:29280987

Liu, H. M., & Tu, Y. K. (2002). The efficacy of papaverine administration by different routes for the treatment of experimental acute cerebral vasospasm. *Journal of Clinical Neuroscience*, *9*(5), 561–565. doi:10.1054/jocn.2001.1036 PMID:12383416

Liu, Q., Meng, X., Li, Y., Zhao, C.-N., Tang, G.-Y., & Li, H.-B. (2017). Antibacterial and Antifungal Activities of Spices. *International Journal of Molecular Sciences*, *18*(6), 1283. doi:10.3390/ijms18061283 PMID:28621716

Lo, D. S., & Chua, T. H. (1992). Poppy seeds: Implications of consumption. *Medicine, Science, and the Law*, *32*(4), 296–302. doi:10.1177/002580249203200403 PMID:1435125

López-Muñoz, F., & Alamo, C. (2009). Monoaminergic neurotransmission: The history of the discovery of antidepressants from 1950s until today. *Current Pharmaceutical Design*, *15*(14), 1563–1586. doi:10.2174/138161209788168001 PMID:19442174

Macfarlane, S., Macfarlane, G. T., & Cummings, J. H. (2006). Review article: Prebiotics in the gastrointestinal tract. *Alimentary Pharmacology & Therapeutics*, *24*(5), 701–714. doi:10.1111/j.1365-2036.2006.03042.x PMID:16918875

Mamedov, N., & Craker, L. (2012). Man and medicinal plants: A short review. *Acta Horticulturae*, (964), 181–190. doi:10.17660/ActaHortic.2012.964.22

Marciano, M. A., Panicker, S. X., Liddil, G. D., Lindgren, D., & Sweder, K. S. (2018). Development of a Method to Extract Opium Poppy (Papaver somniferum L.) DNA from Heroin. *Scientific Reports*, *8*(1), 2590–2590. doi:10.1038/s41598-018-20996-9 PMID:29416103

Masihuddin, M., Jafri, M. A., Aisha, S., & Shahid, C. (n.d.). Traditional Uses, Phytochemistry and Pharmacological Activities of Papaver Somniferum with Special Reference of Unani Medicine an Updated Review. doi:10.22270/jddt.v8i5-s.2069

Mishra, B., Rastogi, D. A., Siddiqui, A., Srivastava, M., Verma, N., Pandey, R., . . . Shukl, S. (2013). Opium Poppy: Genetic Upgradation Through Intervention of Plant Breeding Techniques. Academic Press.

Mozaffarian, D., Micha, R., & Wallace, S. (2010). Effects on coronary heart disease of increasing polyunsaturated fat in place of saturated fat: A systematic review and meta-analysis of randomized controlled trials. *PLoS Medicine*, *7*(3), e1000252. doi:10.1371/journal.pmed.1000252 PMID:20351774

Najafipour, H., & Beik, A. (2016). The Impact of Opium Consumption on Blood Glucose, Serum Lipids and Blood Pressure, and Related Mechanisms. *Frontiers in Physiology*, *7*, 436–436. doi:10.3389/fphys.2016.00436 PMID:27790151

Oudejans, L. C. J., van Velzen, M., & Dahan, A. (2016). Ketamine Analgesia. In V. R. Preedy (Ed.), *Neuropathology of Drug Addictions and Substance Misuse* (pp. 541–550). Academic Press. doi:10.1016/B978-0-12-800212-4.00050-9

Powers, D., Erickson, S., & Swortwood, M. J. (2018). Quantification of Morphine, Codeine, and Thebaine in Home-Brewed Poppy Seed Tea by LC-MS/MS. *Journal of Forensic Sciences*, *63*(4), 1229–1235. doi:10.1111/1556-4029.13664 PMID:28973785

Probst, Y. C., Guan, V. X., & Kent, K. (2017). Dietary phytochemical intake from foods and health outcomes: A systematic review protocol and preliminary scoping. *BMJ Open*, *7*(2), e013337–e013337. doi:10.1136/bmjopen-2016-013337 PMID:28202499

Regula, J., Cerba, A., Suliburska, J., & Tinkov, A. A. (2018). In Vitro Bioavailability of Calcium, Magnesium, Iron, Zinc, and Copper from Gluten-Free Breads Supplemented with Natural Additives. *Biological Trace Element Research*, *182*(1), 140–146. doi:10.1007/s12011-017-1065-4 PMID:28580502

Reiter, E., Jiang, Q., & Christen, S. (2007). Anti-inflammatory properties of alpha- and gamma-tocopherol. *Molecular Aspects of Medicine*, *28*(5-6), 668–691. doi:10.1016/j.mam.2007.01.003 PMID:17316780

Saini, R. K., & Keum, Y. S. (2018). Omega-3 and omega-6 polyunsaturated fatty acids: Dietary sources, metabolism, and significance - A review. *Life Sciences*, *203*, 255–267. doi:10.1016/j.lfs.2018.04.049 PMID:29715470

Sawarkar, S., Deshmukh, V., & Verma, H. (2016). Nigella sativa seed, a novel beauty care ingredient: A review. *International Journal of Pharmaceutical Sciences and Research*, *7*(8), 3185–3196. doi:10.13040/IJPSR.0975-8232.7(8).3185-96

Schmidt, J., Boettcher, C., Kuhnt, C., Kutchan, T. M., & Zenk, M. H. (2007). Poppy alkaloid profiling by electrospray tandem mass spectrometry and electrospray FT-ICR mass spectrometry after [ring-13C6]-tyramine feeding. *Phytochemistry*, *68*(2), 189–202. doi:10.1016/j.phytochem.2006.10.003 PMID:17113612

Schwingshackl, L., & Hoffmann, G. (2014). Monounsaturated fatty acids, olive oil and health status: A systematic review and meta-analysis of cohort studies. *Lipids in Health and Disease*, *13*(1), 154–154. doi:10.1186/1476-511X-13-154 PMID:25274026

Shankar, K., & Liao, L. P. (2004). Traditional systems of medicine. *Phys Med Rehabil Clin N Am, 15*(4), 725-747. doi:10.1016/j.pmr.2004.03.006

Struempler, R. E. (1987). Excretion of Codeine and Morphine Following Ingestion of Poppy Seeds. *Journal of Analytical Toxicology*, *11*(3), 97–99. doi:10.1093/jat/11.3.97 PMID:3599922

Tahiliani, A. G., & Beinlich, C. J. (1991). Pantothenic acid in health and disease. *Vitamins and Hormones*, *46*, 165–228. doi:10.1016/S0083-6729(08)60684-6 PMID:1746161

Thema, F. T., Beukes, P., Gurib-Fakim, A., & Maaza, M. (2015). Green synthesis of Monteponite CdO nanoparticles by Agathosma betulina natural extract. *Journal of Alloys and Compounds*, *646*, 1043–1048. doi:10.1016/j.jallcom.2015.05.279

Thovhogi, N., Diallo, A., Gurib-Fakim, A., & Maaza, M. (2015). Nanoparticles green synthesis by Hibiscus Sabdariffa flower extract: Main physical properties. *Journal of Alloys and Compounds*, *647*, 392–396. doi:10.1016/j.jallcom.2015.06.076

Vanderveen, J. E. (1994). Regulatory history for stearic acid. *The American Journal of Clinical Nutrition*, *60*(6, Suppl), 983s–985s. doi:10.1093/ajcn/60.6.983S PMID:7977156

Vickers, A., Zollman, C., & Lee, R. (2001). Herbal medicine. *The Western Journal of Medicine*, *175*(2), 125–128. doi:10.1136/ewjm.175.2.125 PMID:11483560

Wang, R., van Welie, N., van Rijswijk, J., Johnson, N. P., Norman, R. J., Dreyer, K., Mijatovic, V., & Mol, B. W. (2019). Effectiveness on fertility outcome of tubal flushing with different contrast media: Systematic review and network meta-analysis. *Ultrasound in Obstetrics & Gynecology*, *54*(2), 172–181. doi:10.1002/uog.20238 PMID:30740799

Wheeler, A. G., J. (2019). Bugs and Poppies: Heteroptera Associated with Developing and Mature Capsules (Fruits) of <i>Papaver somniferum</i> L. (Papaveraceae). *Proceedings of the Entomological Society of Washington, 121*(3), 511-524. 10.4289/0013-8797.121.3.511

Whitfield, K. C., Bourassa, M. W., Adamolekun, B., Bergeron, G., Bettendorff, L., Brown, K. H., Cox, L., Fattal-Valevski, A., Fischer, P. R., Frank, E. L., Hiffler, L., Hlaing, L. M., Jefferds, M. E., Kapner, H., Kounnavong, S., Mousavi, M. P. S., Roth, D. E., Tsaloglou, M.-N., Wieringa, F., & Combs, G. F. Jr. (2018). Thiamine deficiency disorders: Diagnosis, prevalence, and a roadmap for global control programs. *Annals of the New York Academy of Sciences*, *1430*(1), 3–43. doi:10.1111/nyas.13919 PMID:30151974

Yang, J., Wang, H. P., Zhou, L., & Xu, C. F. (2012). Effect of dietary fiber on constipation: A meta analysis. *World Journal of Gastroenterology*, *18*(48), 7378–7383. doi:10.3748/wjg.v18.i48.7378 PMID:23326148

Zarghami, M. (2015). Iranian Common Attitude Toward Opium Consumption. *Iranian Journal of Psychiatry and Behavioral Sciences*, *9*(2), e2074–e2074. doi:10.5812/ijpbs.2074v3 PMID:26288642

Zentai, A., Sali, J., Szeitzné-Szabó, M., Szabó, I. J., & Ambrus, Á. (2012). Exposure of consumers to morphine from poppy seeds in Hungary. *Food Additives & Contaminants. Part A, Chemistry, Analysis, Control, Exposure & Risk Assessment*, *29*(3), 403–414. doi:10.1080/19440049.2011.636762 PMID:22165856

Chapter 3
Comparative Study of Phytochemical Profile and In Vitro Antioxidant Potentials of Three Common Medicinal Herbs

G. Shobana
Srimad Andavan Arts and Science College, Trichy, India

N. Agnel Arul John
Srimad Andavan Arts and Science College, Trichy, India

Hari Haran Govindasamy
Srimad Andavan Arts and Science College, Trichy, India

ABSTRACT

The study aimed at evaluating the phytochemical profile and invitro antioxidant potentials of a three medicinal herbs including Calendula officinalis L, Lantana camara L., Desmodium gangeticum L. The aqueous extract of the selected plants were evaluated for phytochemical screening and antioxidant activities using DPPH, reducing power, hrdroxy radical, and nitric oxide radical scavenging assays. The present study exhibited higher yields of phytochemical constituents and noticeable antioxidative properties.

INTRODUCTION

Plants have long been acknowledged as vital natural sources of secondary metabolites. Thus, a lot of work has gone into researching and developing

DOI: 10.4018/979-8-3693-6105-4.ch003

phytomedicines, which include tannins, terpenoids, alkaloids, vitamins, and flavonoids, as essential therapeutic agents for treating a variety of metabolic disorders. The biological activities of phytomedicines, such as the detoxification of toxic enzymes, the inhibition of cellular damage, the regulation of gene expression, and the antimicrobial and anti-inflammatory qualities, have been connected to these diseases (Rice-Evans, Miller, Bramley, Pridham, 1985; Gengatharan, Dykes, Choo, Betalains, 2015).

Reactive oxygen species (ROS) include free radicals like hydroxyl radical, nitric oxide radical, hydrogen peroxide, superoxide anion radical, hypochlorite radical, lipid peroxides, and various singlet oxygen molecules (Hazra, Biswas, Mandal, 2008). Our body naturally creates certain substances known as free radicals during metabolic processes (Pham-Huy, He, 2008). The body produces less free radicals during normal physiological processes; however, aberrant functioning or low antioxidant levels cause oxidative stress, which in turn causes the body to produce large amounts of these free radicals (Valko, Jomova, 2011). These free radicals cause a wide range of degenerative and chronic conditions, such as Parkinson's disease, ischemic heart disease, arthritis, cancer, stroke, Alzheimer's disease, immune suppression, atherosclerosis, aging, diabetes mellitus, and chronic inflammatory diseases (Devasagayam, Tilak, Boloor, Sane, Ghaskadbi, Lele, 2004).

Antioxidants, incredibly significant chemical compounds, limit and protect us from the harmful effects of free radicals on the body's normal cells. Certain synthetic antioxidants, like butylated hydroxyl-toluene and butylated hydroxyl-anisole, are less stable and extremely dangerous, while naturally occurring antioxidants are risk-free and have few adverse effects. Antioxidants have traditionally been sourced from natural products for safety reasons (Knight, 1998). To lessen the harmful effects and damage caused by free radicals, one can use natural food sources such as fruits, vegetables, seeds, herbs, sprouts, edible mushrooms, and cereals as an efficient source of antioxidants (Kaur, Kapoor, 2001).

Three common plants, including Desmodium gangeticum Linn. (Fabaceae family), Lantana camara L. (Verbanaceae family), and Calendulo officinalis L. (Asteraceae family), are chosen for study based on the literature review.

Member of the Asteraceae family: Calendula officinalis L. It is incredibly effective at treating burns, eczema, diaper rash, sore, cracked nipples, and other itchy, inflamed skin conditions. Anti-inflammatory, antiseptic, and hemorrhagic properties are associated with Calendula officinalis L. (Harris, 1990). The verbanaceae family includes Lantana camara L. Numerous applications exist for it, including antimicrobial, fungicidal, insecticidal, and anticancerAs stated in (Himeshsoni, 2012). The Fabaceae family includes Desmodiumgangeticum L. Within Indian medicine, it is used to treat inflammatory disorders, digestive issues, febrifuges, and bitter tonics (Ross, 1999). So the present study was evaluated for phytochemical profile and invitro antioxidant activities of selected medicinal herbs.

MATERIALS AND METHODS:

Plant Identification and Authentification

Calendula officinalis L., Lantana camara L., and Desmodium gangeticum L. are the plant sources chosen for this study. The plants were gathered, recognized using the Flora of the Presidency of Madras (Gamble, 1957), and verified using a herbarium specimen kept at the RAPINAT herbarium of St. Joseph's College in Trichy, Tamilnadu, India.

Preparation of the Aqueous Poly Herbal Formulation

Using an electrical blender, the aerial parts of Calendula officinalis L., Lantana camara L., and Desmodium gangeticum L. plants were shade dried and coarsely powdered. 1.2 liters of water were combined with 200 grams of each plant material. It was then boiled until it was reduced to a third and put through a filter. The filterate was dried out by evaporation. The obtained extract was put through pre-clinical screening in paste form.

Chemical Studies

Physicochemical Constants (Anonymous, 2002)

Determination of loss on drying, Moisture Content, Total Ash, Acid Insoluble Ash and Water Soluble Ash.

Determination of Extractive Values

Determination of Hexane extractive value, Chloroform extractive value and Ethyl acetate extractive value

Determination of SOLUBILITY

i. Alcohol ii. Water

Phytochemical Analysis

Preliminary Phytochemical Screening of various extracts (Brindha, Sasikala, Purushothaman, 1981)

Preliminary phytochemical screening of plant extracts and drug powder will be carried out as per the standard textual procedure.

Quantitative Analysis of Important Phyto Constituents

Estimation of total Alkaloids (Ferguson, 1956) ·total Flavonoids (Panovska, Kulevanova, Stefova, 2005),total Phenol (Malick, Singh, 1980) and Total Tannins Price, Van Socoyoc, Butter, 1978).

In-Vitro Antioxidant Studies

DPPH radical scavenging activity (Brand-Williams, Cuvelier, Berset, 1995), Reducing power assay (Singhal, Paul, Singh, Dubey, Gaur, 2011), Hydrogen peroxide scavenging activity (Ruch, Cheng, Klaunig, 1984), Nitric oxide scavenging activity (Sreejayan, 1996).

RESULT AND DISCUSSION

The foreign matter and loss on drying of the selected plants are given in **Table 1a.**The loss on drying of *Desmodium gangeticum* Linn. was found to be lower (2.3%) than *Calendula officinalis* Linn. (4.86%) and *Lantana camara* Linn.(9%).

The extractive values and solubility of selected plants are given in Table 1b. The hexane and chloroform extractive values of *Lantana camara* Linn. was found to be higher (4% and 6%) than *Calendula officinalis* Linn.(2% and 2%) and *Desmodium gangeticum* Linn.(1.8% and 3.55%). And the water solubility of selected three plants were found to be higher than the alcohol solubility. Hexane, chloroform, and ethyl acetate extractive values suggest the presence of a significant amount of water-soluble

Table 1a. Physico chemical standards

S.No.	Plant Name	Loss on drying Value % W/W	Foreign matter Value % W/W
1	***Calendula officinalis* Linn.**	**4.86%**	**2%**
2	***Lantana camara* Linn.**	**9%**	**3%**
3	***Desmodium gangeticum* Linn.**	**2.3%**	**2%**

Table 1b. Extractive values

S.No	Parameters	Value % W/W		
		Calendula officinalis **L.**	*Lantana camara* **L.**	*Desmodium gangeticum* **Linn.**
1	Hexane	2%	4%	1.8%
2	Chloroform	2%	6%	3.55%
3	Ethyl acetate	6%	2%	4.09%
4	Ethanol	10%	8%	5.67%
5	Water	16%	26%	13.17%

constituents like sugar, tannin, glycosides, and phenols, as well as the potential for a significant amount of polar compounds.

In present study the preliminary phytochemical screening of aqueous extract and drug powder of *Calendula officinalis* Linn. revealed the presence of alkaliods, quinine, sterols, coumarin and saponin etc. These secondary metabolites having various biological activity.

In present study the preliminary phytochemical screening of aqueous extract and drug powder of *Lantana camara* Linn. revealed the sterols, glycosides, phenol, flavonoids etc., Among the various phytoconstituents, flavonoids have beneficial effects in the inflammatory conditions, that has been reported in the earlier literature.

Table 2a. Preliminary phytochemical screening of Calendula officinalis Linn.

S.No	Test	Observation	
		Plant Powder	Extract
1	Terpenoids	-ve	-ve
2	Flavonids	-ve	-ve
3	Steroids	+ve	+ve
4	Glycosides	-ve	-ve
5	Alkaloids	+ve	+ve
6	Quinine	+ve	+ve
7	Phenol	-ve	-ve
8	Tannins	+ve	+ve
9	Saponins	+ve	+ve
10	Coumarin	+ve	+ve
11	Lignin	-ve	-ve

-ve indicates absent
+ve indicates present

Table 2b. Preliminary phytochemical screening of Lantana camara Linn.

S.No	Test	Observation	
		Plant Powder	Extract
1	Terpenoids	-ve	-ve
2	Flavonids	+ve	+ve
3	Steroids	+ve	+ve
4	Glycosides	+ve	+ve
5	Alkaloids	+ve	+ve
6	Quinine	+ve	+ve
7	Phenol	+ve	+ve
8	Tannins	-ve	-ve
9	Saponins	+ve	-ve
10	Coumarin	+ve	+ve
11	Lignin	-ve	-ve

-ve indicates absent
+ve indicates present

Table 2c. Preliminary phytochemical screening of Desmodium gangeticum Linn.

S.No	Test	Observation	
		Plant Powder	Extract
1	Terpenoids	-ve	-ve
2	Flavonids	-ve	-ve
3	Steroids	+ve	+ve
4	Glycosides	+ve	+ve
5	Alkaloids	+ve	+ve
6	Quinine	-ve	-ve
7	Phenol	+ve	+ve
8	Tannins	-ve	-ve
9	Saponins	+ve	+ve
10	Coumarin	-ve	-ve
11	Lignin	-ve	-ve

-ve indicates absent
+ve indicates present

In present study the preliminary phytochemical screening of aqueous extract and drug powder of *Desmodium gangeticum* Linn. revealed the presence of alkaliods, phenol, sterols, glycosides and saponin etc. Understanding the chemical components of plants is important since it will help with the synthesis of complex compounds and the detection of biological activity. With a broad range of biological activities including anti-apoptosis, anti-aging, anti-carcinogen, anti-inflammation, anti-atherosclerosis, cardiovascular protection, and improved endothelial function, in addition to inhibition of angiogenesis and cell proliferation activities, phenolic and flavanoids are widely distributed secondary metabolites in plants that exhibit anti-oxidant activity.

In the present study, major phyto constituents of selected plants were quantitatively analysed and the results were given in the **Table 4**.The alkaloid level of *Calendula officinalis* Linn. was found to be higher than *Lantana camara* L. and *Desmodium gangeticum* Linn. Tannin was well known antioxidant, antimicrobial and astringent. The flavonoid has anti-inflammatory, antitumour, antiviral, antiallergic and antioxidant properties.

In Vitro Antioxidant Activity

The results of DPPH radical scavenging activity of aqueous extract of *Calendula officinalis* Linn., *Desmodium gangeticum* Linn. and *Lantana camara* Linn.are reported in Figure 1. The extract exhibited high DPPH radical scavenging activity with Ic50 value 419.4, 227.8 and 50.16μg/ml which is compared with ascorbic acid as a standard.

The most popular technique for examining the antioxidant potential of plant extracts is the DPPH radical scavenging assay. It leads to the production of stable free radicals, which are detectable using a standard spectrophotometric method. Reduced absorbance indicates that the extract has more effective antioxidant activity in terms of its ability to donate hydrogen atoms. Because it measures the inhibition of reactive species (free radicals) generated in the reaction mixture and because the type of reactive species used determines the assay's results, it is more indirect in nature (Padamanabhan, Manimeekali, Vasthi, Nirmala, Jagajothi, 2014). A dose-dependent response was observed in the concentration of plant extracts

Table 3. Quantitative analysis of major secondary metabolites

S.No.	Parameters	Value (mg/kg)		
		Calendula officinalis **Linn.**	*Lantana camara* **Linn.**	*Desmodium gangeticum* **Linn.**
1	Alkaloids	3.85	0.46	0.395
2	Flavonoids	20	4	2
3	Phenol	2.482	2.275	1.517
4	Tannins	4.28	5.714	17.85

Figure 1. DPPH scavenging activity of aqueous extract of Calendula officinalis Linn., Lantana camara Linn., and Desmodium gangeticum Linn.

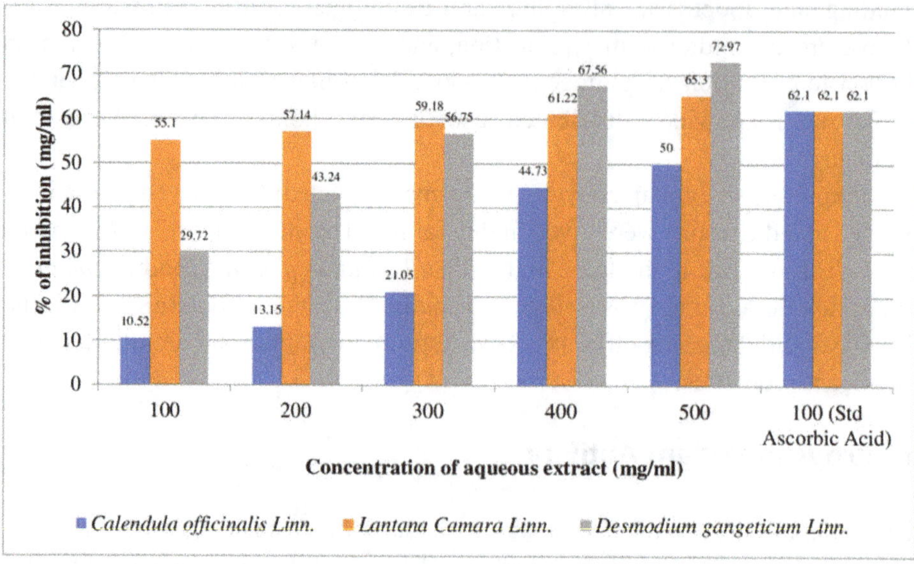

needed to scavenge DPPH. The ability of each extract to scavenge (DPPH) free radicals was used to measure its antioxidant activity, and ascorbic acid was used as the reference. Testing a compound's or plant extract's capacity to function as a free radical scavenger has frequently been done using this kind of reactivity. From the results, it may be postulated that the three plant extracts have hydrogen donors thus scavenging the free radical DPPH.

By measuring the Fe 3+ - Fe 2+ conversion by donating electron, the reducing capacity of an aqueous extract of *Calendula officinalis* Linn., *Desmodium gangeticum* Linn., and *Lantana camara* Linn. Using the ferric ion reducing power, the antioxidant properties of the extracts were evaluated based on their ability to reduce ferricyanide complex to Perl's Prussian blue ferrocyanide complex (Adamu, Naidoo, Eloff, 2014). The Fe3+–Fe2+ transformation was investigated by the assay in the presence of the extracts. The extracts' potential antioxidant activity may be indicated by their capacity to reduce (Olivera, Sousa, Ferreira, Bento, Estevinho, Pereira, 2008). The test solution's yellow color in this assay changes to various shades of green and blue ferrous ions based on the test specimen's reducing power. The presence of reducetones has been shown to have an impart antioxidant action, whereby the donation of a hydrogen atom dissolves the chain of free radicals. Antioxidants found in the sample extracts might help reduce the Fe3+/Ferric cyanide complex to its ferrous form, which spectrophotometry can detect at 700 nm. (Johnson, Janakiraman,

Figure 2. Reducing power assay of aqueous extract of Calendula officinalis Linn., Lantana camara Linn., and Desmodium gangeticum Linn.

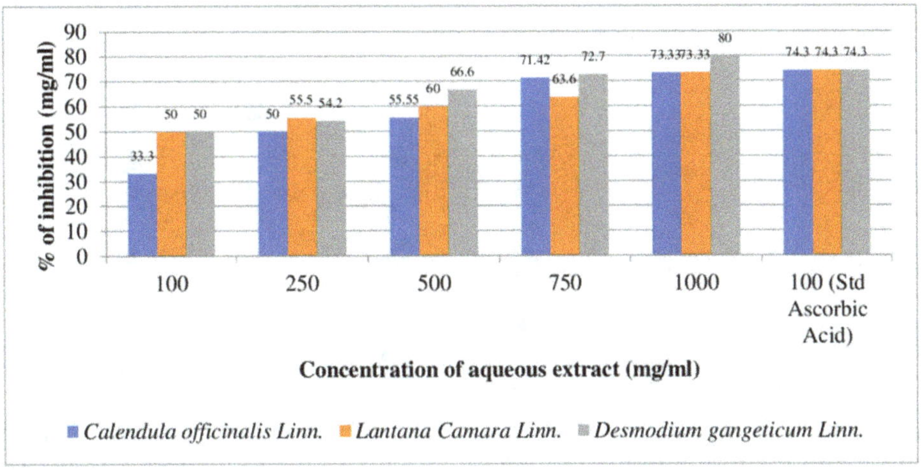

2015). Absorption increased with the intensity of the blue color, indicating strong antioxidant qualities, according to Figure 2, which shows trends of reducing potential for all of the selected plant extracts. Comparable to DPPH scavenging activity, the aqueous extract of the selected plants in this experiment demonstrated the strongest iron-reducing power activity.

Those selected plants had shown good reducing power that was comparable with standard ascorbic acid. The maximum reducing power activity was found to be 73.33, 73.30 and 80% at high concentration 1000µg/ml. The Ic50 value of these three plant extracts were found to be 267.7, 98.8 and 120.8 µg /ml.

Figure 3 display the results of the nitric oxide radical scavenging assay for *Calendula officinalis* Linn., *Desmodium gangeticum* Linn., and *Lantana camara* Linn. Depending on its concentration, nitric oxide can either protect against or induce oxidative stress in different tissues. It is also believed to be involved in the instruction of the oxidation-reduction potential of various cells. An increasing body of research suggests that some diseases are linked to either excessive or inadequate NO production (Shivatare, Musale, Bhutale, Kewatkar, Tare, 2022). There are several biological systems that contain this soluble free radical. Its roles as an effector molecule include vasodilatation, antimicrobial and antitumor effects, and neuronal communication (Hagerman, Riedl, Jones, Sovik, Ritchard, Hartzfield, Riechel, 1998). Oxygen combines with free radicals to form nitrite and peroxy nitrite anions. In present study, the nitric oxide was generated from sodium nitro prusside react with oxygen to formed nitrite which reacts with griss reagent at physiological P^H was incubated with different concentration of plant extracts, form chromophore

Figure 3. Nitric oxide inhibition activity of aqueous extract of Calendula officinalis Linn., Lantana camara Linn., and Desmodium gangeticum Linn.

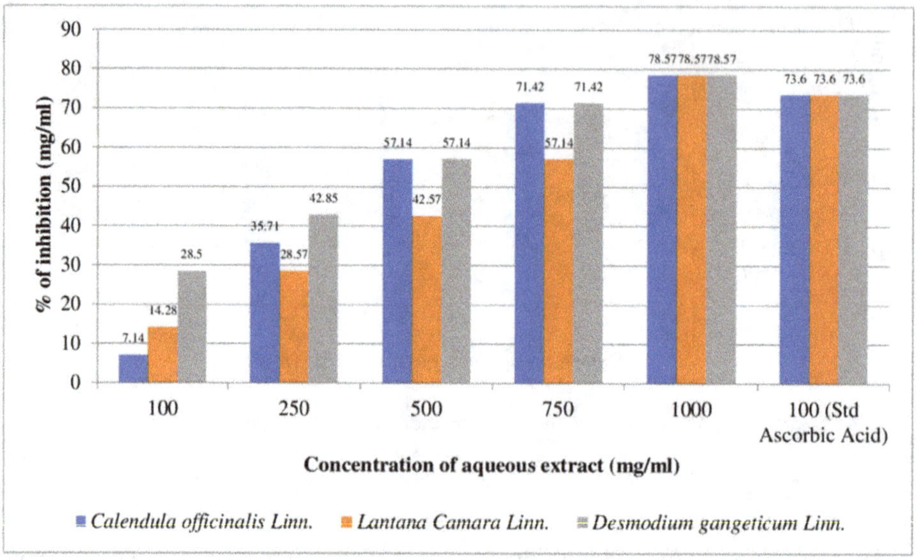

which was read at 546nm.The maximum activity of aqueous extract of these plants were found to be 77.57% at high concentration. The Ic50 value is 405, 526.7 and 520.8μg/ml,it was compared to that of the standard ascorbic acid.

Figure 4 presents the results of the plant extract's hydrogen peroxide radical scavenging assay, and illustrates the activities at various concentrations. Oxidative processes in tissues can result in the formation of hydrogen peroxide (H2O2), a biologically significant non-radical oxidizing species. Lipid peroxidation is initiated and spreads by hydrogen peroxide (H2O2), which in turn produces hydroxyl radicals (•OH) (Dama, Tare, Gore, Shende, Deore, 2010). The significance of hydrogen peroxide radical stems from its capacity to permeate biological membranes. Although hydrogen peroxide radical is a weak oxidizing agent and not very reactive, it can occasionally be harmful to cells because it can cause hydroxyl radicals to form inside the cells. Rapid cell membrane crossing is one of its many toxic effects. Once inside the cell, the hydrogen peroxide radical can likely react with Fe 2+ and possibly Cu2+ ions to form hydroxyl radical. Some enzymes can be directly inactivated by it, usually through the oxidation of essential thiol(-SH)groups.

Each of these three plant extracts demonstrated 12.25%, 15.62%, and 11.12% at the lowest concentration (100μg/ml). At 1000μg/ml, those extracts demonstrated 53.60%, 65.52%, and 60.30% at high concentrations. The plants' aqueous extracts'

Figure 4. Hydrogen peroxide assay of aqueous extract of Calendula officinalis Linn., Lantana camara Linn., and Desmodium gangeticum Linn.

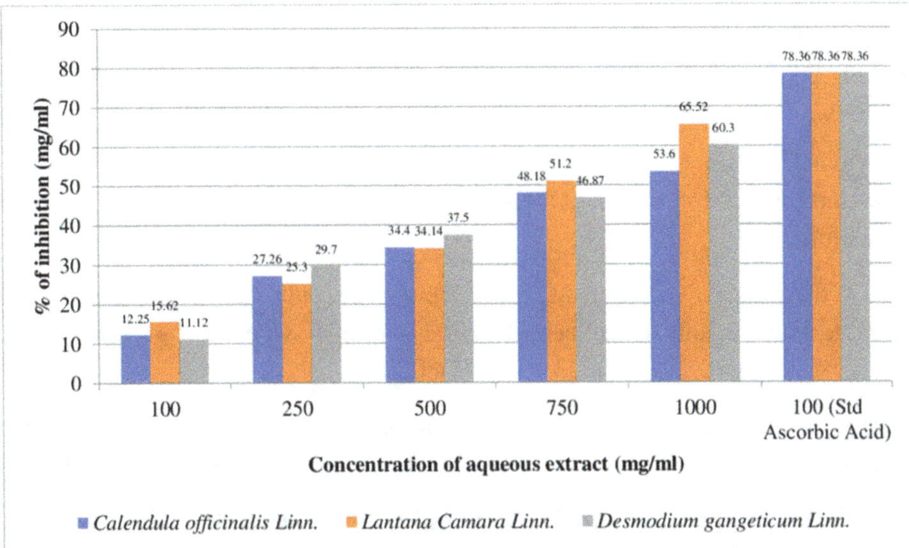

capacity to quench OH-appears to be directly tied to their ability to prevent lipid peroxidation and to act as a moderate scavenger of active oxygen species, which lowers the rate at which chain reactions occur.

CONCLUSION

Because of their high concentration of phytoconstituents (flavanoids, phenol and alkaloids, tannins, etc.), the aqueous extracts of *Calendula officinalis* Linn., *Desmodium gangeticum* Linn., and *Lantana camara* Linn. have higher free radical scavenging activities, according to the investigation's findings. These studies conducted in vitro suggest that this plant extract is a significant natural antioxidant source that could help prevent the onset of various oxidative stressors. It will take more investigation to separate and pinpoint the antioxidant compounds present in the plant extract. The in vivo antioxidant activity of this extract needs to be assessed prior to clinical application.

REFERENCES

Adamu, M., Naidoo, V., & Eloff, J. N. (2014). The Antibacterial Activity, Antioxidant Activity and Selectivity Index of Leaf Extracts of Thirteen South African Tree Species Used in ethnoveterinary Medicine to Treat helminth Infections. *BMC Veterinary Research, 10*(1), 52. doi:10.1186/1746-6148-10-52 PMID:24589020

Anonymous. (2002). Quality Control Methods for Medicinal plants materials. (An authorized publication of W.H.O. Geneva). New Delhi: A.I.T.B.S. Publishers & Distributors.

Brand-Williams, W., Cuvelier, M. E., & Berset, C. (1995). Use of a free radical method to evaluate antioxidant activity. *Lebensmittel-Wissenschaft + Technologie, 28*(1), 25–30. doi:10.1016/S0023-6438(95)80008-5

Brindha, P., Sasikala, B., & Purushothaman, K. K. (1981). Pharmacognostic studies on Merugan kizhangu. *Bulletin of Medio-Ethno Botanical Research, 3*, 84–86.

Dama, G. Y., Tare, H. L., Gore, M. S., Shende, V. S., Deore, S. R., & … . (2010). Comparative cardiotonic activity of Aegle marmelos juice with digoxin on isolated frog heart. *Int J Drug Dev Res, 2*(4), 1–8.

Devasagayam, T. P. A., Tilak, J. C., Boloor, K. K., Sane, K. S., Ghaskadbi, S. S., & Lele, R. D. (2004). Free radicals and antioxidants in human health: Current status and future prospects. *The Journal of the Association of Physicians of India, 52*(4), 794–804. PMID:15909857

Ferguson, N. M. (1956). A Textbook of Pharmacognosy. New Delhi: Mcmillan Company.

Gamble, J. S. (1957). Flora of the presidency of Madras. Sree Saraswaty Press Ltd.

Gandía-Herrero, F., Escribano, J., & García-Carmona, F. (2016). Biological activities of plant pigments betalains. *Critical Reviews in Food Science and Nutrition, 56*(6), 937–945. doi:10.1080/10408398.2012.740103 PMID:25118005

Gengatharan, A., Dykes, G. A., & Choo, W. S. (2015). Betalains: Natural plant pigments with potential application in functional foods. *Lebensmittel-Wissenschaft + Technologie, 64*(2), 645–649. doi:10.1016/j.lwt.2015.06.052

Hagerman, A. E., Riedl, K. M., Jones, G. A., Sovik, K. N., Ritchard, N. T., Hartzfeld, P. W., & Riechel, T. L. (1998). High molecular weight plant polyphenolics (tannins) as biological antioxidants. *Journal of Agricultural and Food Chemistry, 46*(5), 1887–1892. doi:10.1021/jf970975b PMID:29072454

Hazra, B., Biswas, S., & Mandal, N. (2008). Antioxidant and free radical scavenging activity of Spondias pinnata. *BMC Complementary and Alternative Medicine*, *8*(1), 122–127. doi:10.1186/1472-6882-8-63 PMID:19068130

Himeshsoni. (2012). A recent update of botanicals for wound healing activity. International Research Journal of Pharmacy, 3(7).

Johnson, M. A., & Janakiraman, N. M. (2015). Studies on Antioxidant and Phytochemical Profiles of Leptochloa Uniflora Hoechst. *International Journal of Botany*, *1*, 4.

Kadifkova Panovska, T., Kulevanova, S., & Stefova, M. (2005). In vitro antioxidant activity of some Teucrium species (Lamiaceae). *Acta Pharmaceutica (Zagreb, Croatia)*, *55*(2), 207–214. PMID:16179134

Kaur, C., & Kapoor, H. C. (2001). Antioxidants in fruits and vegetables, the millennium's health. *International Journal of Food Science & Technology*, *36*(7), 703–725.

Knight, J. A. (1998). Free radicals: Their history and current status in aging and disease. *Annals of Clinical and Laboratory Science*, *28*(6), 331–346. PMID:9846200

Malick, C. P., & Singh, M. B. (1980). Estimation of Phenol. In *Plant Enzymology and histoenzymology* (p. 286). Kalyani Publishers.

Oliveira, I., Sousa, A., Ferreira, I. C. F. R., Bento, A., Estevinho, L., & Pereira, J. A. (2008). Total Phenols, Antioxidant Potential and Antimicrobial Activity of Walnut (Juglans Regia L.) Green Husks. *Food and Chemical Toxicology*, *46*(7), 2326–2331. doi:10.1016/j.fct.2008.03.017 PMID:18448225

Padamanabhan, V., Manimekalai, G., Vasthi, K. E., Nirmala, A., & Jagajothi, A. (2014). Phytochemical Screening and Antioxidant Activity of Extracts of the Leaf and Bark of Albizzia Lebbeck (Benth). *Acad. J. Med. Plant*, 2(2), 26–31.

Pham-Huy, L. A., He, H., & Pham-Huy, C. (2008). Free radicals, antioxidants in disease and health. *International Journal of Biomedical Science; IJBS*, *4*(2), 89–96. doi:10.59566/IJBS.2008.4089 PMID:23675073

Price, M. L., Van Socoyoc, S., & Butter, L. G. (1978). A critical evaluation of the vanillin reaction as an assay for tannin in sorghum grain. *Journal of Agricultural and Food Chemistry*, *26*(5), 1214–1218. doi:10.1021/jf60219a031

Rice-Evans, C. A., Miller, N. J., Bolwell, P. G., Bramley, P. M., & Pridham, J. B. (1995). The relative antioxidant activities of plant-derived polyphenolic flavonoids. *Free Radical Research*, *22*(4), 375–383. doi:10.3109/10715769509145649 PMID:7633567

Ross. (1999). Medicinal plants of the world. Chemical constituents, traditional, and modern medical uses. Human Press Inc.

Ruch, R. T., Cheng, S. J., & Klaunig, J. E. (1984). Spin trapping of superoxide and hydroxyl radicals. J Methods in Enzymology, 105, 198-209.

Shivatare, R. S., Musale, R., Bhutale, N. K., Kewatkar, S. M., Tare, H. L., Chatur, V., Khatawakar, A. N., & Suryavanshi, D. S. (2022). Phytochemicals: A novel approach for the Management of Coronavirus Disease 2019. *Indian Journal of Pharmaceutical Sciences*, *84*(3), 519–531. doi:10.36468/pharmaceutical-sciences.946

Singhal, M., Paul, A., Singh, H. P., Dubey, S. K., & Gaur, K. (2011). Article. *Journal of Chemical and Pharmaceutical Research*, *3*(3), 639–645.

Sreejayan, N., & Rao, M. N. A. (1996). Free radical scavenging activity of curcuminoids. *Journal of Drug Research*, *46*, 169. PMID:8720307

Valko, M., & Jomova, K. (2011). Advances in metal-induced oxidative stress and human disease. *Toxicology*, *283*(2), 65–87. PMID:21414382

Chapter 4
Ethnobotanical and Pharmacological Significance of Celery Plants With Biogenic Synthesis of Nanomaterials

Javaria Kanwal
The Women University Multan, Pakistan

Mehwish Nawaz
The Women University Multan, Pakistan

Wajiha Noor
The Women University Multan, Pakistan

Mahrukh Khan
The Women University Multan, Pakistan

Hafsa Javaid
The Women University Multan, Pakistan

Abida Ashraf
Bahauddin Zakariya University, Pakistan

Nusrat Shafiq
https://orcid.org/0000-0002-3270-4227
Government College Women University, Faisalabad, Pakistan

Adnan Ali Qureshi
University of Engineering and Technology, Lahore, Pakistan

ABSTRACT

Celery, a biennial plant in the Apiaceae family, is a popular source of bioactive compounds like flavonoids, glycosides, and bergapten. These compounds have various biological activities, including anticancer, antihypertensive, antifungal, antimicrobial, and antioxidant properties. The plant has also been used for the synthesis of nanomaterials, an eco-friendly, energy-efficient, and cost-effective approach. This chapter highlights the phytochemicals, ethnobotanical importance, and pharmacological uses of celery.

DOI: 10.4018/979-8-3693-6105-4.ch004

1. INTRODUCTION

Celery i.e., *Apium graveolens* L. is a widely grown plant that belongs to the *Apiaceae* family, which was formally known as the *Umbelliferae* family (Ambrose, Manickavasagan, & Naik, 2016). The genus *Apium* contains about 20 species of *Apiaceae* family and its several varieties are found native to Eurasia. It is mostly grown in coastal regions due to favorable environmental conditions. Celery is widely cultivated in temperate zones as an important garden crop and as a popular vegetable due to bleached leaf stalks (Satyavati, Raina, & Sharma, 1987). This herb develops on a biennial basis and can be found in different forms such as celery flakes, oleo resins and seeds etc. The plant indicates diversity in both the structure and chemical composition of leaves, stems and flowers. Celery undergoes cross-pollination but it is not compatible in self-pollination (Ram P Rastogi & Mehrotra, 1990). The plant *Apium graveolens* is recognized by different names across the different global regions. The Arabic name is "Alkarafs" and in German, it is known as "Sellerie". It is commonly referred to as "celery" in English, while in Persian, it is known as "Karafs" and Spanish refers to it as "Apio". In Thailand, it is acknowledged as "Khuen chaai"(Kolarovic, Popovic, Mikov, Mitic, & Gvozdenovic, 2009). In Urdu, it is recognized as "Tukhme karafs". In Hindi, the plant is typically called "Ajmod" and the fruits are often termed celery seeds (Nadkarni, 1954). The *Apium graveolens* usually recognized as celery, is likely the most familiar form of the plant; nonetheless, this plant consists of three different types and cultivated forms, each indicating varieties in both flavor and color.

The three most important botanical cultivations of celery plant are *Apium graveolens* var. dulce i.e., "pascal celery or stalk celery", which is cultivated in West of Europe, *Apium graveolens* var. rapaceum i.e., "celeriac", which is found in abundance all over central Europe, and *Apium graveolens* var. secalinum i.e., "leaf or golden celery", which is additionally utilized as a spice (Rożek, 2013). Celery leaves have a very different arrangement, and the petiole generates greater numbers of leaves with relatively less thick and more fragile petioles. The flowers' shades may be different from green to light green and from light green to yellow (Rozek, 2007). The aesthetic and medicinal aspects of this plant render it beneficial for the folk healthcare system. This plant is utilized to cure blood pressure; thus, its commercial worth is increasing on a global level. The most widely recognized plant is celeriac because it's an essential part of gourmet American cuisine (Hamza & Amin, 2007). Agricultural aspects like fertilization, forming method and planting approach as well as genomic, autogenetic, and environmental factors, all have an important effect on the chemistry and chemical structure of the volatile oil of celery along with other species (Powanda, Whitehouse, & Rainsford, 2015). Oil made from celery seeds is produced in France, India and California, and is mainly utilized in juices, especially in savory vegetables, soups, pickles, and other foods. Tiny amounts of it are used in perfumes as well. Leaves of this plant can also be utilized to make seed oleoresins, leaf oil, and extracts from the roots or seeds. Comparing the extract of celery seed to the oil of leaves indicates that it has fewer monoterpenoids along with more sesquiterpenoids. The leaves consist of the bulk of the essential oil, while celery seeds have a small amount of oil (Sowbhagya, Srinivas, & Krishnamurthy, 2010).

Celery is a conventional kitchen vegetable along with therapeutic herbs. Celery has been extensively utilized in ancient medication to cure different diseases. It has a nice and distinctive odor (Kooti, Ghasemi-boroon, et al., 2015). Celery has a distinctive compound called 3NB (3, n-butylphthalide) that gives it its unique smell (Yaser, Muneer, Abdelhafid, & Fawzia, 2014). Different parts of the plants i.e., roots, leaves, stem and seeds shown in Fig. 1. are used to cure diseases. *Apium graveolens* has the status of living in a defense state. In addition to functioning as a good option for minerals, leafy vegetables have medicinal properties. Celery seeds have numerous medicinal applications, involving stimulants, cardiotonic, flatus-relieving, menstrual stimulants and antibiotics, and are also used for the treatment of liver, spleen, and bronchitis (Ram Nath Chopra & Chopra, 1969). In botanical medicine, *A. graveolens* finds applications as an anti-arthritic, antispasmodic, anti-rheumatic, antihypertensive, diuretic and anti-parasitic agents. Comparative analysis has been conducted on the chemical composition and anti-proliferative activities of both aerial components and roots of *Apium graveolens*. Celery-derived biogenic nanoparticles have been investigated for therapeutic applications. Research on hysteria, nervousness, and insomnia indicated that celery has calming effects (Asif et al., 2011; S. S. Fazal & R. K. Singla, 2012). As a significant functional species in the *Apiaceae* family, various extracts of *A. graveolens* have been documented to indicate multiple nutraceutical properties involving cytotoxicity, anti-hyperlipidemic, anti-diabetic, anti-platelet aggregation, antiseptic, anti-spasmodic, gastroprotective, liver tonic, antioxidative, anti-inflammatory, cardio-friendly and relaxing effects as well (Aćimović, 2019; Khairullah et al., 2021; H. Sowbhagya, 2014). The presence of different chemical metabolites, like phenolic acids, tannins, flavonoids, coumarins, and isobenzofuranoids i.e., phthalides, phenols, terpenoids, amino acids, fatty acids, organic acids, polyalcohols, and polysaccharides, is primarily responsible for the wide variety of health benefits related to *Apium graveolens*. These metabolites have been mainly recognized from the plant's seeds, petioles, and leaves (Aćimović, 2019; Al-Asmari, Athar, & Kadasah, 2017).

On the contrary, less attention has been conducted to the roots of celery, with only a few studies, reporting the presence of certain polyacetylenes and phenolic derivatives in this plant (Al-Asmari et al., 2017; Zidorn et al., 2005). Intriguingly, most of the above-mentioned types of metabolites, like phthalides, terpenoids, polyacetylenes and flavonoids, known for their diverse anticancer and chemopreventive properties, have been identified. Consequently, the regular use of celery has been proposed to offer protection against certain kinds of cancers (A. Atta & A. Alkofahi, 1998; Ingallina et al., 2020). This study is conducted to explore the Ethnobotanical, pharmacological, and traditional perspectives of *A. graveolens.*

2. ETHNOBOTANICAL SIGNIFICANCE OF CELERY PLANT

Apium is a generic name that comes from the Latin word "apis" which translate to a "bee", considering that bees are attracted to its tiny white-hue flower. "*Graveolens*" is a species name that translates to "heavy scented". The English term "celery" is derived from the Latin word "celer" meaning "swift" indicating its significance in terms of quick remedy (Khalil, Nawaz, Ben Ghnia Delgay, Rehman, & Nadeem, 2015).

2.1 Origin of Celery Plant

A. graveolens is originated in the Mediterranean regions and is derived from the wild *A. graveolens* that cultivates in mountainous regions of South Asia as well as in the wetlands of Europe and North Africa (Kochhar, 1986). Early in the 1600s, it has been cultivated as a food plant (Kooti, Ali Akbari, Asadi-Samani, Ghadery, & Ashtary-Larky, 2015). Celery is also grown in India, including, western utter Pradesh Punjab, and the Himalayas at the Northwestern side (Ram P. Rastogi et al., 1990). Nonetheless, its center of origin was much larger. Stalk celery, scientifically termed *Apium graveolens* var. dulce, is the extensively renowned variety of celery today. It was opted for cultivation in the 16th and 17th centuries. Likewise, another type, *Apium graveolens* var. rapaceum, was also grown in that tenure. This specific variety of celery continues to be grown in South Asia (Vaughan & Geissler, 2009). The two *A. graveolens* cultivars that were brought to North America from France in 1887 were the green cultivar termed "Giant Pascal" or "Pascal" and the self-blanching "Paris Golden Yellow Self Blanching" or "White Plume" cultivar. The

Figure 1. Represents (A) roots, (B) leaves, (C) stem, and (D) seeds

aforementioned techniques are interconnected and originated from older French breeds termed “solid golden white celery”. Consequently, the plant indicated little genetic variations (de Vilmorin, 1950).

2.2 Morphology of Celery Plant

Apium graveolens is characterized by its distinct morphology. It is a branching herb with a biennial growth pattern, having an approximate height of 100cm. Its branches and stems are characterized by a robust and fleshy structure. Typically, each umbrella carries an average of four to twelve branches. The leaves disclose a variety of shapes, including diamonds, triangular and spears with lengths ranging from 5 to 50 mm. The leaf’s margins are lobed or saw teeth shaped (Lim, 2012). Its fruit is hard, measuring 1 to 1.5mm in length and1.0mm in diameter, enclosing a tiny brown seed. The seeds have a slightly bitter taste, and a 10-gram portion consists of approximately 26,450 seeds. The Seeds indicate an oval shape, with a width ranging from 1.5 to 2.0 mm. The brown-colored seeds are ribbed, ovoid in shape, with a modest length of about 1.3 mm. Each carpel consists of a single seed and two carpels fused to form a single fruit (Choudhury, 2006). The plant has a superficial taproot system and a humid, ridged, branching stem. The fruit lack wings, and has a brown color with a black line on the outer surface (Amirghofran, 2010). The calyx is characterized by distinct teeth, and the small-sized flowers are white, with a greenish white hue. The tiny white flowers of this plant are arranged in umbels. The flowers possess 5 petals with an oval shape and pointed tips. The ribbed are distinct and primary, while the carpals are nearly pentagonal. The fruit has 2 mericarps, is sub-orbicular, schizocarps, and is fragrant with a diameter of 1.0 to 2.0mm (Ram P. Rastogi et al., 1990). The spice has a strong taste, and delightful and distinct odor. Celery can be employed in various varieties to add flavor to foods i.e., stalk, leaves, fresh herbs, seeds and oleoresins (H. B. Sowbhagya, 2014b).

2.3 Environmental Factors Impacting Celery Growth

Celery grows best in a temperature range of 12°C to 15°C and 22°C to 25°C. The optimal condition for its growth involves dry, cold temperatures and long seasons. Celery nourishes in cool and overcast conditions and indicates minimal tolerance to high temperatures. March and April are considered optimal months for sowing seeds, with may being the best time for seed transplantation and November considered an ideal month for fruiting. The celery can be cultivated in different kinds of soil i.e., water-logged and loamy soil is considered ideal, but saline and alkaline are unfavorable. The plant is extremely susceptible to soil reactions, and the perfect pH level for the soil is 5-7 (S. Fazal & R. K. Singla, 2012).

2.4 Utilization and Culinary Applications of Celery Plant

The leafy stem, regularly accompanied by a leaf blade, is utilized in making stews, salads sauces, vegetable juices and soups, etc. Chinese celery, distinguished by its slender stem and robust taste, is seldom employed in its raw form. However, it is usually utilized in the preparation of soups and stir-fries. Turnip-rooted celery also known as Celeriac, is primarily utilized as a cooked veggie in soups and stews, but its popularity is enhancing as a grated raw salad. Smallage, another term for leaf celery is finely chopped and utilized as a culinary ingredient, whether in its fresh or powdered state (Krishnamurthy, 2008). In Europe, the seed of celery is utilized as a food seasoning; in India, it serves as a spice. The seeds of celery can be utilized as whole or ground and added into salt to make celery salt, or they can be utilized as a spice and flavoring agent. The root extract can also be utilized to prepare celery salt. Pharmaceutical and scent industries utilized volatile oil collected from the seeds. The developing plant inhibits insects, like the cabbage white butterfly, which makes it advantageous for brassicas (Krishnamurthy, 2008).

2.5 Chemical Composition of Celery Plant

The celery plant is renowned for its distinct chemical composition. The plant has an abundant source of β-carotene, fibers, folic acid, chlorophyll potassium, magnesium, and silica. It has a significant amount of sodium, with a source of minerals like magnesium, calcium and potassium. Chopped celery leaves, in a cup, consist of 100mg of sodium. The celery essential oil is composed of phytonutrients i.e., limonene, salience and sesquiterpenes, and has a distinct odor (R. N. Chopra & De, 1929). Celery has a significant calorie and fat content. It is also recognized for being an excellent source of different other minerals and Vitamin C. Seeds of this plant consist of volatile oil, fixed oil, crude fibers, starch, proteins, starch, ash and carbohydrates (Keller & Matile, 1989), while oil derived from seed is known as fixed oil that consist of fatty acids i.e., petroselenic acid, oleic acid, stearic acid, linoleic acid, linolenic acid, and palmitic acid (Destaillats & Angers, 2002).

The plant smells good due to the presence of essential oils that are derived from leaves. The aromatic volatile oils extracted from leaves are predominantly consists of phytonutrients i.e., anhydrides, terpenes and phenol. Moreover, plants also have other phytochemicals like steroids and alkaloids. The extract of seed also consists of phytonutrients i.e., steroids, and flavonoids. In addition to this, celery extract possesses several Furocoumarins, celereoside, apiumetin, celerin, apigravrin, isopimpinellin, apiumoside, isopimpinellin, isoimperatorin, bergapten and celereoside. Phenols like tannins, isoquercitrin, phytic acid, graveobioside and apigenin are found in celery as well (Ramezani, Nasri, & Yassa, 2009).

2.6 Phytochemicals of Celery Plant

One of the most notable secondary metabolic products in celery is flavonoids, and the majority of flavonoids are produced with the formation of glycosides. Flavonoids are considered to be one of the important phytochemicals and several types of flavonoids i.e., luteolin, quercetin, isorhamnetin, apigenin and kaempferol are obtained from celery (Nijveldt et al., 2001; Romanova, Vachálková, Cipák, Krajcovicova, & Rauko, 2001). Analysis of the biological potential of flavonoids indicated that this kind of phytonutrient has anti-oxidant, antibacterial, antitumor, anti-inflammatory, and anti-activities (Kim, Son, Chang, & Kang, 2004; Marzocchella et al., 2011). Flavonoids derived from both the leaves and roots of celery are effective scavengers of 1,1-diphenyl-2-picrylhydrazyl (DPPH) and hydroxyl (OH) radicals and can lower the degree of lipid peroxidation in liposomes (Popovic, Kaurinovic, Trivic, Mimica-Dukic, & Bursać, 2006). It is also observed that apigenin disclosed remarkably reduced impacts on lipid peroxidation caused by metals (Sugihara, Arakawa, Ohnishi, & Furuno, 2000). Research promoting apigenin anti-inflammatory activities proposed that it hinders AP-1 factors and proinflammatory cytokines expression caused by PMA (Patil et al., 2015). Another study demonstrated that apigenin plays a significant role in the relaxation of blood vessels and has antiproliferative actions on the contraction of the vascular ring (Zhang et al., 2001). Quercetin as a bioactive compound protects against injury caused by heart ischemia-reperfusion by retarding iNOS, NOX2, and eNOS expression levels to preserve vascular relaxation (Li Li et al., 2009). Utilizing GC-mass spectrometry (GC-MS) and gas chromatography (GC), essential oil from celery seeds was characterized, indicating that 18 compounds were found in it. The main constituents of the essential oil are beta-selinene, limonene and p-cresyl isovalerate with percentages of 22.28, 31.15, 14.94 respectively (Lu, Li, & Wang, 2011). Analysis of the bioactive compounds in celery seeds revealed that three compounds, senkyunolide-N, sedanolide, and senkyunolide-J were found to result in complete mortality of nematode Panagrellus redivivus (Momin & Nair, 2001). Furthermore, sedanolide exhibited the retarding actions of the development of Candida parapsilasis and Candida albicans (Helaly, Mady, Eldekashy, & Craker, 2015; Momin & Nair, 2001).

3. PHARMACOLOGICAL SIGNIFICANCE OF CELERY PLANT

Celery has been acknowledged for its pharmacological significance attributed to the existence of bioactive compounds that contribute to diverse health-related effects. Some aspects of its pharmacological significance are described here. Celery has been utilized conventionally to cure multiple diseases. It is widely acknowledged that each part of celery can be used to cure one or more illness. Some of the health-related effects of celery plants are described here.

3.1 Neural Treatment

Celery is considered to be significant to treat neural diseases. Plant derived essential oil exerts a soothing and relaxing impact on the central nervous system. Certain components within it indicates muscle relaxing, calming and anti-seizure activities and are used to control high blood pressure. (Chevallier, 1996).

3.2 Treatment for Rheumatism

Launert suggested that celery herb is beneficial to cure rheumatism. Extracts collected from its seeds find application in the treatment of gout and rheumatoid arthritis. The essential oil is used by adding lukewarm water to soak the aching, gout-affected region of the feet. Moreover, tinctures from the root and fresh juice collected from the entire plant are employed in the treatment of arthritis (Krishnamurthy, 2008).

3.3 Antidiuretic Activity

In an investigation, celery diuretic quality has been employed for the formulation of herbal medicine (Houghton, 1995). The herbs, mature seeds, and roots are considered to be important as they have diuretic properties, among them, seeds are chiefly employed for diuretic functions. Seeds can also facilitate in purging toxins from the body, when uric acid crystals gather in the joints result in gout. Tinctures from the roots have been utilized for the treatment of urinary diseases, like urinary stones. Furthermore, they have been utilized as a kidney toner and purifier. Fresh juice from entire plant is employed as a remedy for inflammation of the urinary tract and urethra. This herb is employed for the treatment of kidney-related issues (Houghton, 1995).

3.4 Digestive Treatments

Consuming fresh stem is advantageous for breastfeeding mothers; however, wild celery is more effective. Wild celery alleviates digestive discomfort. The celery seeds also serve as a stimulant for digestion. The celery roots of plants serve as a digestive treatment and a stimulant for the liver (Bown & Herb Society of, 1995).

3.5 Antimicrobial Activity

The volatile oil extracted from *Apium graveolens* exhibited antimicrobial activity against bacteria, fungi and viruses. It proved to be active against multiple bacteria like *Salmonella typhi, Pseudomonas Solanacearum, Staphylococcus albus, Streptococcus pyogenes, Staphylococcus aureus, Shiglla dysenteriae* and *Streptococcus faecalis* (A. H. Atta & A. Alkofahi, 1998). The celery extract of roots indicated significant activity against *Enterococcus faecalis* and *B. cereus* (Sipailiene, Venskutonis, Sarkinas, & Cypiene, 2005). Sesquiterpene lactones derived from celery exhibited potential against *Proteus vulgaris* and *Bacillus subtilis*. It was also evaluated against fungi (Duraipandiyan, Al-Harbi, Ignacimuthu, & Muthukumar, 2012) (Jawad et al., 1985). Celery's essential oil is used to suppress the growth of *Rhodotorula Listeria monocytogenes, Staphylococcus aureus, Geotrichum, Escherichia coli* O:157:H7, *Pseudomonas aeruginosa, Yersinia enterocolitica, Pseudomonas aeruginosa, Lactobacillus plantarum,* and *Aspergillus niger* (Elgayyar, Draughon, Golden, & Mount, 2001).

3.6 Antispasmodic Effects

Seeds are well-known to possess α-limonene and β-selinene that are linked with the anti-spasmodic potential of the celery plant. It has been observed that, this plant is a natural reservoir of organic sodium which is a crucial mineral in the bloodstream. Moreover, it proves advantageous for joints, supporting the lining of the stomach and is advantageous for gastric ailments. Its oil is utilized to relieve vomiting, flatulence, colic and GI tract discomfort (Khalil et al., 2015).

3.7 Antioxidant Activity

The extract of celery plant is considered to be significant to have anti-oxidant potential as well. Extract in methanol indicated greater anti-oxidant activity in comparison to ethanolic extract, while both alcoholic extracts of celery have bioactive components i.e., phenolics which account for this anti-oxidant activity. The reducing power of this extract develops with concentration. Every bioactive component's capacity to offer electrons is associated with its antioxidant property (Sameh, Bouftira, Mahmoud, Boukef, & Boughattas, 2011).

3.8 Anti-Inflammatory Effects

Celery oils indicated anti-inflammatory actions due to the presence of multiple bioactive compounds like caffeic acid, apigenin, magnesium, cinilide, gentisic acid, coumarine, ferulic acid, bergapten, scopoletin, isopimpinellin, rutin, luteolin, mannitol, protocatechuic acid, thymol, xanthotoxin, ascorbic acid, α-pinene, chlorogenic acid, eugenol, linoleic acid, umbelliferone and myristicin etc. Seeds of celery have been used in the treatment of gout and arthritis (Cseke et al., 2016). It reduces pain and inflammation around the joints and is regularly taken to carry on healthy joints. Its seeds assist in the detoxication of our bodies and improve the flow of blood to our joints and muscles. Celery has been effectively utilized to cure rheumatoid arthritis and slow down the aging process of joints because of its sedative and nerve-persuading properties. The seeds oil is consumed to reduce pain in the neck muscles and sacrum as they possess coumarins, which are known to have muscle-relaxing properties. Moreover, seed oil is also known for phthalides as bioactive components, which are used as anti-rheumatic agents. Celery assists in mending the connective tissues, because it consists of multiple minerals like sodium, phosphorous, iron, magnesium, calcium, and zinc (Guenther, 1950).

3.9 Hepato-Protective Potential

The methanolic extracts of *Apium graveolens* in methanol exhibited hepato-protective ability in comparison to common medication drug i.e., silymarin (Gauri, Ali, & Khan, 2015).

3.10 Anticancer Effect

Celery oil is well known for its bioactive compound i.e., phthalide that exhibited health benefits because it demonstrates defense against high blood pressure, cholesterol level and cancer. The most active compound of the phthalide compound is believed to be significant to have sedanolide which is responsible for tumors in cancer patients. The plant seed oil consists of two main components i.e., sedanolide and 3-n-butyl phthalide. These active components prove to be efficient in stimulating the cleansing or purifying enzyme glutathione S-transferase (GST) in the specific tumor tissues that are being targeted (Zheng, Kenney, Zhang, & Lam, 1993). By eliminating the free radicals found in the destroyed cells, this celery plant also reduces the probability that a cell will undergo mutation and finally change into a cancerous cell. The active component "coumarins" present in celery exhibited the activity, as celery has a high content of Na and K. When these juices are drunk after working, they can function as a powerful replacement for electrolytes. Following the improvement in detoxification, the plant additionally reduces the cholesterol and inhibits cancer (H. B. Sowbhagya, 2014a).

4. BIOGENIC SYNTHESIS OF NANOMATERIALS USING CELERY PLANT

Green synthesis of nanomaterials is a one-step, rapid, low-cost, and environmentally friendly alternative to well-known physical and chemical methods that involve expensive equipment and end-product hazards, thus, the field of nanotechnology is increasing. In this study, the celery plant is opted for the synthesis of nanomaterials as different parts of the plant consist of bioactive components that are responsible for biological activities. Plant extract acts as a reducing agent and maintains the particles in the medium. Synthesized nanomaterials are used in different applications.

In a study, silver nanoparticles were produced through a one-step green synthesis method using celery leaf extract in combination with silver nitrate for reduction. The reaction mixture underwent an incubation process until a light brown color emerged. Following centrifugation, the resulting nano-silver pellet was dried in a desiccator. The obtained nanoparticles exhibited a predominantly spherical shape with an average diameter of 25 nm. XRD analysis confirmed the crystalline nature of the particles, while FTIR spectroscopy identified the organic molecules from the extract responsible for reduction and capping. The antifungal activity of these biogenic silver nanoparticles was investigated against two pathogenic fungi, *Aspergillus niger* and *Aspergillus wentii,* demonstrating significant fungicidal properties against the tested strains (Roy, Sarkar, & Ghosh, 2015).

Al-Ogaili and his colleagues suggested the synthesis of nano-sized ZnO particles with a bio-based origin, using celery leaf plant extract as a reducing and stabilizing agent, and zinc chloride metal salt as the precursor for metal ions that underwent a chemical transformation into metal oxide nanoparticles. NaOH aqueous solution served as the precipitating agent, employing a "bottom-up" approach through the co-precipitation method. The reaction, (shown in Equation.1.) occurring at room temperature with mechanical agitation, led to the formation of ZnO nanoparticles. Subsequent calcination of the nanopowder ensured proper crystal alignment and atomic periodicity order. To characterize the synthesized ZnO nanoparticles, four distinct tests were conducted: XRD, FE-SEM, EDS, and AFM. The analysis revealed an average particle size of 35 nm. Morphology assessment indicated the formation of hexagonal wurtzite nanocrystals. The ZnO nanoparticles produced through the green synthesis process are suitable for direct use as an eco-friendly additive in various industrial applications.(Al-Ogaili & Almahdawi, 2023).

Equation 1. represents the synthesis of ZnO-NPs

$ZnCL_2$+2NaOH → $Zn(OH)_2$+2NaCl

$Zn(OH)_2$+Plant extract → ZnO+H_2)

$$ZnO + H_2O \xrightarrow[\text{Calcination}]{\text{Drying}} Zno - NPs$$

In a research, cobalt oxide nanoparticles were produced through an eco-friendly method using extracts from *Camellia sinensi* leaves and *Apium **graveolens***, ensuring a synthesis process that avoids the introduction of harmful substances into the environment and is cost-effective. In the synthesis of Co_3O_4-NPs, cobalt nitrate was mixed with plant extracts, resulting in a color change from pink to grey, indicative of cobalt oxide formation. Characterization analysis revealed a particle size ranging from 21 to 55 nm. The synthesized cobalt oxide nanoparticles demonstrated the ability to inhibit the growth of *S. aureus* and *P. aeruginosa*. Consequently, these nanoparticles hold potential for use in the treatment of diseases and infections caused by these organisms (Urabe & Aziz, 2019) .

Fayyadh and his fellows proposed the synthesis of Zero-valent iron nanoparticles (C-nZVFe), utilizing celery leaf extract and employed as a catalyst in the Fenton reaction for the decolorization of aqueous solutions containing orange (OG) dyes. The synthesized C-nZVFe exhibited a size and surface area in the range of 40 to 55 nm. The concentration of dye in the aqueous solution was monitored using ultraviolet-visible (UV–vis) spectroscopy. In the decolorization process, the removal percentage of Orange dye (100 mg L^{-1}) reached 86% under specific conditions (35 mg. L^{-1}, 60 min, pH 4) (Fayyadh & Tahrim, 2020).

In an investigation, nanocomposites comprising silver nanoparticles (Ag-NPs) on the surface of reduced graphene oxide (rGO) nanosheets were synthesized through microfluidization, utilizing celery seeds extract as a green reducing agent. Characterization of the rGO/Ag nanocomposites was conducted using various techniques such as SEM, XRD, TEM, HRTEM, XPS, Zeta potential, EDX, AFM, and UV–Vis spectroscopy. The average particle sizes of the Ag-NPs were determined to be approximately 36.8 nm, 28.4 nm, and 20.9 nm through the stirred heater, ultrasonication, and microfluidization methods, respectively. The celery seeds extract played a dual role as a green reducing agent and a capping agent. Furthermore, the rGO/Ag nanocomposites demonstrated outstanding antibacterial activity against both typical Gram-positive bacteria (Staphylococcus aureus and Bacillus cereus) and Gram-negative bacteria (*Klebsiella pneumonia* and *Escherichia coli*). This sustainable and eco-friendly fabrication method via microfluidization holds promise for the development of rGO/Ag nanocomposites as effective antibacterial agents (Soleymani, Rafigh, & Hekmati, 2020).

In a study, nanostructured TiO_2/ZnO heterojunction was synthesized through a straightforward sol–gel and hydrolysis process, utilizing celery stalk as a biological template. The resulting nanostructured photocatalyst comprised anatase TiO_2 and wurtzite ZnO nanoparticles, achieving an enhanced specific surface area of 55.5 m^2/g. Crystal structure, morphology, and surface composition were thoroughly investigated using electron microscopy, and XRD. The prepared TiO_2/ZnO heterostructure was evaluated for its efficacy in the photo reduction of CO_2 into CH_4. The photocatalytic yield exhibited a remarkable fivefold increase, rising from 0.55 to 2.56 $\mu mol\ h^{-1}g^{-1}$ for the nanocomposite compared to pure TiO_2. This enhanced performance is attributed to efficient charge transfer and inhibition of electron–hole pair recombination, facilitated by the optimized band positions of ZnO and TiO_2. The study underscores the potential of employing biotemplates in designing effective photocatalysts for the conversion of CO_2 into valuable solar fuels (Wu et al., 2018).

5. CONCLUSION

Celery i.e., *Apium graveolens* is a biennial herb, that belong to the family *Apiaceae* or *Umbelliferae*. It has been grown all over the world for thousands of years and used for culinary seasoning, essential oil usage, and conventional medicinal practices. Celery plants have been consumed to cure multiple disorders like neural, digestive, diuretic, cardiovascular, liver, gout and rheumatism in conventional medicine for a long time, as parts of this plant i.e., stalk, roots, leaves and seed are enriched with several bioactive compounds like flavonoids, limonene, polyphenols, coumarin, vitamin C, and vitamin A. This study placed a focus on the history, morphology, distribution, phytochemical composition, habitat-cultivation conditions, pharmacological activities or health effects of *Apium graveolens*. Moreover, it also highlights the use of different celery parts in the synthesis of nanomaterials like silver, cobalt, zinc NPs and their nanocomposites. Nanomaterials of celery prove to be more active towards antifungal, antiproliferative and antibacterial activities.

REFERENCES

Aćimović, M. (2019). Nutraceutical potential of Apiaceae. *Reference Series in Phytochemistry*, 1311-1341.

Al-Asmari, A. K., Athar, M. T., & Kadasah, S. G. (2017). An updated phytopharmacological review on medicinal plant of Arab region: Apium graveolens linn. *Pharmacognosy Reviews*, *11*(21), 13. doi:10.4103/phrev.phrev_35_16

Al-Ogaili, F. H., & Almahdawi, F. H. (2023). Green Fabrication and Characterization of Zinc Oxide Nanoparticles Using Celery Leaf Extract. *Journal of Survey in Fisheries Sciences*, *10*(4S), 2938–2945.

Ambrose, D. C., Manickavasagan, A., & Naik, R. (2016). *Leafy medicinal herbs: botany, chemistry, postharvest technology and uses*. CABI. doi:10.1079/9781780645599.0000

Amirghofran, Z. (2010). Medicinal Plants as Immunosuppressive Agents in Traditional Iranian Medicine. *Iranian Journal of Immunology*, *7*, 65–73.

Asif, H., Akram, M., Usmanghani, K., Akhtar, N., Shah, P. A., Uzair, M., ... Rehman, R. (2011). Monograph of apium graveolens Linn. *Journal of Medicinal Plants Research*, *5*(8), 1494–1496.

Atta, A., & Alkofahi, A. (1998). Anti-nociceptive and anti-inflammatory effects of some Jordanian medicinal plant extracts. *Journal of Ethnopharmacology*, *60*(2), 117–124. doi:10.1016/S0378-8741(97)00137-2

Bown, D., & Herb Society of, A. (1995). *Encyclopedia of herbs & their uses*. Retrieved from https://archive.org/details/encyclopediaofhe0000bown

Chevallier, A. (1996). The encyclopedia of medicinal plants. Academic Press.

Chopra, R. N., & Chopra, R. (1969). Supplement to glossary of Indian medicinal plants. Academic Press.

Chopra, R. N., & De, P. (1929). Saussurea lappa in pharmacology and therapeutics. *The Indian Journal of Medical Research*, *17*(2), 351–359.

Choudhury, B. (2006). *Vegetables*. National Book Trust.

Cseke, L. J., Kirakosyan, A., Kaufman, P. B., Warber, S., Duke, J. A., & Brielmann, H. L. (2016). *Natural products from plants*. CRC Press. doi:10.1201/9781420004472

de Vilmorin, R. L. (1950). Pascal celery and its origin. *JNY Bot Gard*, *51*, 39–41.

Destaillats, F., & Angers, P. (2002). Base-catalyzed derivatization methodology for FA analysis. Application to milk fat and celery seed lipid TAG. *Lipids*, *37*(5), 527–532. doi:10.1007/s11745-002-0928-9

Duraipandiyan, V., Al-Harbi, N. A., Ignacimuthu, S., & Muthukumar, C. (2012). Antimicrobial activity of sesquiterpene lactones isolated from traditional medicinal plant, Costus speciosus (Koen ex.Retz.) Sm. *BMC Complementary and Alternative Medicine*, *12*(1), 13. doi:10.1186/1472-6882-12-13

Elgayyar, M., Draughon, F., Golden, D., & Mount, J. (2001). Antimicrobial Activity of Essential Oils from Plants against Selected Pathogenic and Saprophytic Microorganisms. *Journal of Food Protection*, *64*(7), 1019–1024. doi:10.4315/0362-028X-64.7.1019

Fayyadh, S. N., & Tahrim, N. A. (2020). Green nanoparticles investigation to remove water pollutants by Fenton reaction using celery leaves extract. *International Journal of Design & Nature and Ecodynamics*, *15*(3), 309–314. doi:10.18280/ijdne.150304

Fazal, S., & Singla, R. K. (2012). Review on the Pharmacognostical & Pharmacological Characterization of Apium Graveolens Linn. *Indo Global Journal of Pharmaceutical Sciences*, *2*(1), 36–42. doi:10.35652/IGJPS.2012.03

Gauri, M. K., Ali, S. J., & Khan, M. S. (2015). *A Review of Apium graveolens (Karafs) special reference to Unani Medicine*. Academic Press.

Guenther, E. (1950). *The Essential Oils: Individual essential oils of the plant families: Gramineae, Lauraceae, Burseraceae, Myrtaceae, Umbelliferae, and Geraniaceae*. D. Van Nostrand Company.

Hamza, A. A., & Amin, A. (2007). Apium graveolens modulates sodium valproate-induced reproductive toxicity in rats. *Journal of Experimental Zoology. Part A, Ecological Genetics and Physiology*, *307*(4), 199–206. doi:10.1002/jez.357

Helaly, A., Mady, E., Eldekashy, M. H. Z., & Craker, L. (2015). Phytochemical Analysis of Some Celery Accessions. *Journal of Medicinally Active Plants*, *1*, 1–7. doi:10.7275/R5542KJF

Houghton, P. (1995). Herbal products. 9. Bearberry, dandelion and celery. *The Pharmaceutical Journal*, *255*(6855), 272–273.

Ingallina, C., Capitani, D., Mannina, L., Carradori, S., Locatelli, M., Di Sotto, A., ... Valletta, A. (2020). Phytochemical and biological characterization of Italian "sedano bianco di Sperlonga" Protected Geographical Indication celery ecotype: A multimethodological approach. *Food Chemistry*, *309*, 125649. doi:10.1016/j.foodchem.2019.125649

Keller, F., & Matile, P. (1989). Storage of sugars and mannitol in petioles of celery leaves. *The New Phytologist*, *113*(3), 291–299. doi:10.1111/j.1469-8137.1989.tb02406.x

Khairullah, A. R., Solikhah, T. I., Ansori, A. N. M., Hidayatullah, A. R., Hartadi, E. B., Ramandinianto, S. C., & Fadholly, A. (2021). Review on the Pharmacological and Health Aspects of Apium Graveolens or Celery: An Update. *Systematic Reviews in Pharmacy*, *12*(2).

Khalil, A., Nawaz, H., Ben Ghnia Delgay, J., Rehman, R., & Nadeem, F. (2015). Value Added Products, Chemical Constituents and Medicinal Uses of Celery (Apium graveolens L.). *RE:view*, 40–48.

Kim, H., Son, K., Chang, H., & Kang, S. (2004). Anti-inflammatory Plant Flavonoids and Cellular Action Mechanism. *Journal of Pharmacological Sciences*, *96*(3), 229–245. doi:10.1254/jphs.CRJ04003X

Kochhar, S. L. (1986). *Tropical crops: a textbook of economic botany*. Macmillan London.

Kolarovic, J., Popovic, M., Mikov, M., Mitic, R., & Gvozdenovic, L. (2009). Protective effects of celery juice in treatments with doxorubicin. *Molecules (Basel, Switzerland)*, *14*(4), 1627–1638. doi:10.3390/molecules14041627

Kooti, W., Ali Akbari, S., Asadi-Samani, M., Ghadery, H., & Ashtary-Larky, D. (2015). A review on medicinal plant of Apium graveolens. *Advanced Herbal Medicine*, *1*, 48–59.

Kooti, W., Ghasemi-boroon, M., Ghafourian, M., Asadi-Samani, M., Harizi, M., Sharafi-Ahvazi, N., & Afrisham, R. (2015). The effects of celery leave extract on male hormones in rats. *Journal of HerbMed Pharmacology*, *4*(2), 56–60.

Krishnamurthy, K. (2008). Celery. In *Chemistry of spices* (pp. 401–412). CABI Wallingford UK. doi:10.1079/9781845934057.0401

Li Li, W., Jiahong, X., Duoyun, Y., Jinping, L., Jun, C., & Gang, W. (2009). Effects of Quercetin on Gene and Protein Expression of NOX and NOS after Myocardial Ischemia and Reperfusion in Rabbit. *Cardiovascular Therapeutics*, *27*(1), 28–33. doi:10.1111/j.1755-5922.2009.00071.x

Lim, T. K. (2012). *Edible medicinal and non-medicinal plants. Volume 3, Fruits*. Retrieved from https://site.ebrary.com/id/10532978 doi:10.1007/978-94-007-2534-8

Lu, Z., Li, W., & Wang, P. (2011). Chemical Composition and Ability of Scavenging DPPH Radical of Essential Oil and Residue from The Celery Seed. *Advanced Materials Research*, *183-185*, 18–21. doi:10.4028/www.scientific.net/AMR.183-185.18

Marzocchella, L., Fantini, M., Benvenuto, M., Masuelli, L., Tresoldi, I., Modesti, A., & Bei, R. (2011). Dietary Flavonoids: Molecular Mechanisms of Action as Anti-Inflammatory Agents. *Recent Patents on Inflammation & Allergy Drug Discovery*, *5*(3), 200–220. doi:10.2174/187221311797264937

Momin, R., & Nair, M. (2001). Mosquitocidal, Nematicidal, and Antifungal Compounds from Apium graveolens L. Seeds. *Journal of Agricultural and Food Chemistry*, *49*(1), 142–145. doi:10.1021/jf001052a

Nadkarni, A. (1954). *Nadkarni's Indian Materia Medica*. Nadkarni's Indian Materia Medica.

Nijveldt, R., van Nood, E., Hoorn, D., Boelens, P., Norren, K., & Leeuwen, P. (2001). Flavonoids: A review of probable mechanisms of action and potential applications. *The American Journal of Clinical Nutrition*, *74*(4), 418–425. doi:10.1093/ajcn/74.4.418

Patil, R., & Lamani, B. (2015). Apigenin inhibits PMA-induced expression of pro-inflammatory cytokines and AP-1 factors in A549 cells. *Molecular and Cellular Biochemistry*, *403*(1-2), 95–106. Advance online publication. doi:10.1007/s11010-015-2340-3

Popovic, M., Kaurinovic, B., Trivic, S., Mimica-Dukic, N., & Bursać, M. (2006). Effect of celery (Apium graveolens) extracts on some biochemical parameters of oxidative Stress in mice treated with carbon tetrachloride. *Phytotherapy Research*, *20*(7), 531–537. doi:10.1002/ptr.1871

Powanda, M. C., Whitehouse, M. W., & Rainsford, K. (2015). Celery seed and related extracts with antiarthritic, antiulcer, and antimicrobial activities. *Novel natural products: therapeutic effects in pain, arthritis and gastro-intestinal diseases*, 133-153.

Ramezani, M., Nasri, S., & Yassa, N. (2009). Antinociceptive and anti-inflammatory effects of isolated fractions from Apium graveolens seeds in mice. *Pharmaceutical Biology*, *47*(8), 740–743. doi:10.1080/13880200902939283

Rastogi, R. P., & Mehrotra, B. (1990). *Compendium of Indian medicinal plants* (Vol. 1). CSIR.

Rastogi, R. P., Mehrotra, B. N., Sinha, S., Pant, P., & Seth, R. (1990). Compendium of Indian medicinal plants. Lucknow: Central Drug Research Institute and Publications & Information Directorate, New Delhi Lucknow.

Romanova, D., Vachálková, A., Cipák, L., Krajcovicova, Z., & Rauko, P. (2001). Study of antioxidant effect of apigenin, luteolin and quercetin by DNA protective method. *Neoplasma*, *48*, 104–107.

Roy, K., Sarkar, C., & Ghosh, C. (2015). Apium graveolens leaf extract-mediated synthesis of silver nanoparticles and its activity on pathogenic fungi. *Digest Journal of Nanomaterials and Biostructures, 10*(2), 393–400.

Rozek, E. (2007). Growth and yielding of leaf celery [Apium graveolens L. var. secalinum Alef.] cultivated for two-cut harvest. *Herba Polonica, 3*(53).

Rożek, E. (2013). YIeldIng of leaf celery Apium graveolens l. var. secAlinum alef. dependIng on the nuMber of harvests and IrrIgatIon. *Modern Phytomorphology, 3*, 83–86.

Sameh, B., Bouftira, I., Mahmoud, A., Boukef, K., & Boughattas, N. (2011). Antioxidant Activity of Apium graveolens Extracts. *Journal of Biologically Active Products from Nature, 1*(5-6), 340–343. Advance online publication. doi:10.1080/22311866.2011.10719102

Satyavati, G., Raina, M., & Sharma, M. (1987). *Medicinal plants of India* (Vol. 2). Indian Council of Medical Research.

Sipailiene, A., Venskutonis, R., Sarkinas, A., & Cypiene, V. (2005). Composition and antimicrobial activity of celery (Apium graveolens) leaf and root extracts obtained with liquid carbon dioxide. *Acta Horticulturae*, (677), 71–77. doi:10.17660/ActaHortic.2005.677.9

Soleymani, A. R., Rafigh, S. M., & Hekmati, M. (2020). Green synthesis of RGO/Ag: As evidence for the production of uniform mono-dispersed nanospheres using microfluidization. *Applied Surface Science, 518*, 146264. doi:10.1016/j.apsusc.2020.146264

Sowbhagya, H. (2014). Chemistry, technology, and nutraceutical functions of celery (Apium graveolens L.): An overview. *Critical Reviews in Food Science and Nutrition, 54*(3), 389–398. doi:10.1080/10408398.2011.586740

Sowbhagya, H., Srinivas, P., & Krishnamurthy, N. (2010). Effect of enzymes on extraction of volatiles from celery seeds. *Food Chemistry, 120*(1), 230–234. doi:10.1016/j.foodchem.2009.10.013

Sowbhagya, H. B. (2014a). Chemistry, Technology, and Nutraceutical Functions of Celery (Apium graveolens L.): An Overview. *Critical Reviews in Food Science and Nutrition, 54*(3), 389–398. doi:10.1080/10408398.2011.586740

Sowbhagya, H. B. (2014b). Chemistry, technology, and nutraceutical functions of celery (Apium graveolens L.): An overview. *Critical Reviews in Food Science and Nutrition, 54*(3), 389–398. doi:10.1080/10408398.2011.586740

Sugihara, N., Arakawa, T., Ohnishi, M., & Furuno, K. (2000). Anti and pro-oxidative effects of flavonoids on metal induced lipid hydroperoxide dependent lipid peroxidation in cultured hepatocyte loaded with a-Linolenic acid. *Free Radical Biology & Medicine*, *27*(11-12), 1313–1323. doi:10.1016/S0891-5849(99)00167-7

Urabe, A. A., & Aziz, W. J. (2019). Biosynthesis of cobalt oxide (Co3O4) nanoparticles using plant extract of Camellia sinensis (L.) Kuntze and Apium graveolens L. as the antibacterial application. *World News of Natural Sciences*, (24), 356–364.

Vaughan, J., & Geissler, C. (2009). *The New Oxford Book of Food Plants*. OUP Oxford.

Wu, K., Dong, X., Zhu, J., Wu, P., Liu, C., Wang, Y., Wu, J., Hou, J., Liu, Z., & Guo, X. (2018). Designing biomimetic porous celery: TiO2/ZnO nanocomposite for enhanced CO2 photoreduction. *Journal of Materials Science*, *53*(16), 11595–11606. doi:10.1007/s10853-018-2397-y

Yaser, A., Muneer, A., Abdelhafid, B., & Fawzia, A. (2014). Chemical composition, phytochemical constituents and antioxidant activities of the seeds extract of Apiumgraveoleus L from Yemen. *Research Journal of Pharmaceutical, Biological and Chemical Sciences*, *5*(1), 1074–1087.

Zhang, Y.-H., Park, Y.-S., Kim, T.-J., Fang, L.-H., Ahn, H.-Y., Hong, J., Kim, Y., Lee, C.-K., & Yun, Y. (2001). Endothelium-dependent vasorelaxant and antiproliferative effects of apigenin. *General Pharmacology*, *35*(6), 341–347. doi:10.1016/S0306-3623(02)00113-1

Zheng, G. Q., Kenney, P. M., Zhang, J., & Lam, L. K. (1993). Chemoprevention of benzo[a]pyrene-induced forestomach cancer in mice by natural phthalides from celery seed oil. *Nutrition and Cancer*, *19*(1), 77–86. doi:10.1080/01635589309514238

Zidorn, C., Jöhrer, K., Ganzera, M., Schubert, B., Sigmund, E. M., Mader, J., Greil, R., Ellmerer, E. P., & Stuppner, H. (2005). Polyacetylenes from the Apiaceae vegetables carrot, celery, fennel, parsley, and parsnip and their cytotoxic activities. *Journal of Agricultural and Food Chemistry*, *53*(7), 2518–2523. doi:10.1021/jf048041s

Chapter 5
Ethnobotanical Insights Into Diabetes Treatment:
A Focus on Morocco's Traditional Medicinal Plants

Khawla Esmaoui
Faculty of Sciences Ben M'Sick, University Hassan II, Casablanca, Morocco

Hind Bouzoubaa
Faculty of Sciences Ben M'Sick, University Hassan II, Casablanca, Morocco

Houriya Mestaghanmi
Faculty of Sciences Ben M'Sick, University Hassan II, Casablanca, Morocco

Mohammed El Khasmi
Faculty of Sciences Ben M'Sick, University Hassan II, Casablanca, Morocco

ABSTRACT

Diabetes prevalence is rising globally, including in Morocco, where traditional medicine plays a crucial role. The review catalogs 310 plant species from 84 families used in diabetes treatment. The authors analyze preparation methods and plant parts utilized, highlighting Morocco's ethnopharmacological landscape. Between June 2022 and January 2023, they surveyed literature from Elsevier and PubMed (1980-2022). The results reveal diverse plant species with antidiabetic potential. However, further research is essential to identify active compounds and validate efficacy for integration into modern healthcare practices.

DOI: 10.4018/979-8-3693-6105-4.ch005

1. INTRODUCTION

Diabetes is a chronic disease that can occur when the pancreas does not produce sufficient insulin or when the body becomes unable to effectively use the insulin produced. High blood sugar levels, also known as hyperglycemia, is a common consequence of uncontrolled diabetes and over time it causes serious damage to many of the body's systems, including the nerves and blood vessels (Roglik et al., 2013). The common symptoms of diabetes include itching, excessive thirst, weight loss, increased hunger, wasting, blurry vision, increased urination, rapid heartbeat, and low blood pressure (Barkaoui et al., 2017).

Diabetes is a major public health problem which affects millions of people in the world. In addition, it has been predicted that the number of people affected by diabetes will increase to 300 million by 2025. In Morocco, Diabetes is considered as a real social scourge with severe consequences in terms of morbidity and mortality (OMS,2021); it causes more than 24,000 deaths each year; the prevalence of diabetes has increased from 108 million in 1980 to 425 million in 2017 (Kehailou et al., 2019). According to the International Diabetes Federation (IDF), the prevalence of type II diabetes was 463 million in 2019, and it is projected to reach approximately 700 million by the year 2045 (Kehailou et al., 2021).

Nowadays, despite advances in modern medicine, there is a growing interest regarding the consumption of medicinal plants. According to the WHO more than 80% of the population relies on herbal medicines for their health needs especially in the underdeveloped countries (Abouri et al., 2012). These remedies may have therapeutic effects and can be an affordable and accessible option, but also come with potential toxic side-effects (Kee et al., 1994).

Medicinal plants are widely used in the treatment of numerous pathologies and are included in the composition of several medications (Farh et al., 2018); In Morocco, the use of medicinal plants to treat chronic diseases, such as diabetes mellitus is part of an ancient tradition. Several authors have emphasized this practice, and studies carried out in different regions have shown that the use of plants to treat these chronic diseases is very widespread (Ziyyat et al., 1997).

Morocco is well known for its rich vegetation, medicinal plants biodiversity, geographical and climatic conditions (Msanda et al., 2005; Ouhaddou et al., 2014) making it a prime location for the study of traditional pharmacopoeia. As diabetes imposes a significant economic burden, there is a growing interest in the use of medicinal plants for its treatment, especially as there is no evidence of resistance to whole-plant extracts (Roglik et al., 2013).

It is essential to document the medicinal component of Morocco's flora for conservation and sustainable use (Abouri et al., 2012), where the disease is a growing public health problem, and the plant biodiversity provides a wealth of potential therapeutic options (Giannenas et al., 2020).

This review focuses on the traditional use of medicinal plants known for their antidiabetic effect in Morocco by the Moroccan society and aims to present a comprehensive overview of ethnobotanical field works published between 1991 and 2015. The objective is to synthesize separate ethnobotanical studies and provide a document that reflects the profile of ethnobotany, including the background of traditional plant use and the underlying principles and practical applications of the field.

2. MATERIAL AND METHODS

Study Area

The study was conducted in Morocco, a North African country with diverse geographic regions, including mountain chains, coastal areas along the Mediterranean Sea and Atlantic Ocean, as well as arid desert regions. The cultural diversity of Morocco, which includes Arab and Berber populations, contributes to a rich tradition of medicinal plant use.

Data Collection

A comprehensive search for relevant literature was carried out between June 2022 and January 2023 using several databases, including Elsevier, Web of Science, Medline, Google Scholar, and PubMed. The search encompassed articles published between 1980 and 2022 and utilized both English and French keywords. The key search terms included "medicinal plants," "ethnobotany," "ethnopharmacology," "ethnomedicine," and "Morocco."

Inclusion Criteria

The literature reviewed included ethnobotanical articles and pharmacological papers related to the use of medicinal plants in the treatment of diabetes. Only the most frequently mentioned plants used for diabetes treatment in Morocco were selected for detailed analysis.

Data Analysis

Data were extracted from selected articles, focusing on plant species, parts used, preparation methods, and documented antidiabetic effects. The collected data were organized and categorized for analysis.

3. RESULTS

A total of 310 medicinal plant species belonging to 84 families have been reported as being used for diabetes treatment [Table1]. The form of preparation and parts used vary between species:

4. DISCUSSION

Within plant families, The family of Asteraceae represent the highest number of species used as source of antidiabetic medicinal plant followed by Fabaceae, Lamiaceae, Apiaceae, Poaceae, Rosacae, Brassicaceae, Rutaceae and Apocynaceae. Asteraceae was the most frequently mentioned plant family, aligning with the predominance of this family in the findings of various studies conducted in other countries (Giovannini et al.,2016; Hamza et al., 2019).

Asteraceae, also known as Compositae and commonly referred to the 'daisy' family, constitutes a vast group of flowering plants with approximately 32,000 species. This family has a global presence, growing in diverse regions worldwide, excluding Antarctica. Over the ages, these plants have been utilized for medicinal purposes and have been integral components of the human diet. (Mohanta et al.,2023).

It is important to note that not all these plants have undergone comprehensive validation by the scientific community (Odeyemi et al.,2018). Within this family, a considerable number of plants have shown diverse effects on the body, comprising the reduction of blood sugar levels, stimulation of insulin production, restoration of damaged pancreatic β-cells, inhibition of carbohydrate-digesting enzymes, and protection against oxidative stress. Most of the identified anti-diabetic compounds from these plants belong to the categories of saponins, flavanones, tannins, and flavonoids (aglycones) (Mohanta et al.,2023).

Various preparation methods are employed by the population, including decoction, infusion, powdering, maceration, raw application, poultice, and cooking. Among these, infusion is the most common; it involves steeping herbs in hot water, providing an effective means of extracting oils and flavors. This method is particularly popular for consuming the medicinal properties of herbs due to its convenience (Studzińska-

Table 1. Plants used in the treatment of diabetes in Morocco

Family	Plant species	Local Name	English name	Parts Used	Preparation	Number of citation	Réf
Aloaceae	*Aloe succotrina Lam.*	Sebar	Aloe	Whole plant	Powder, Infusion	2	(El Hachlafi et al., 2020; Kharchoufa et al., 2018).
	Aloe vera [L.] Burm.f.	Sebar	Aloe vera	Leaves	Decoction	1	(Belhaj et al., 2020).
Amaranthaceae	*Beta Vulgaris L.*	Selk	Leaves beets	seeds	-	1	(Kharchoufa et al., 2018).
	Hammada scoparia [Pomel]	Assay	-	seeds, Leaves	Decoction	3	(Barkaoui et al., 2017; Boufous et al., 2017; Abouri et al., 2012).
Amaryllidaceae	*Allium sativum L.*	Touma	Garlic	Bulb	Raw, cooked, Decoction, Scorched	20	(Belhaj et al.; 2021; Jouad et al., 2001; Barkaoui et al., 2017; Eddouks et al., 2022;Tahraoui et al., 2007; Orch et al., 2015; Gourri et al., 2013; Laadim et al., 2017; Hachi et al., 2016; Benkhnigue et al., 2014; Ziyyat et al., 1997; Eddouks et al., 2002; El Hachlafi et al., 2020; Benali et al., 2017; Belhaj et al., 2020; Ziyyat et al.,1997; Laadim et al., 2017; Boufous et al., 2017; Mechchate et al., 2020; Eddouks, 2018).
	Allium porrum L.	Borro	Leek	Bulb, Stem	-	2	(Gourri et al., 2013; Skalli et al., 2019).
	Allium cepa L.	Basla	onion	Bulb;	Raw	16	(Jouad et al., 2001, Tahraoui et al., 2007, Orch et al., 2015, Gourri et al., 2013, Laadim et al., 2017,Hachi et al., 2016; Bousta et al., 2014; Skalli et al., 2019, Benkhnigue et al., 2014; El Hachlafi et al., 2020; Eddouks et al., 2002; Belhaj et al., 2020; Boufous et al., 2017; Naceiri et al., 2021; Mechchate et al., 2020,Eddouks, 2018).
Anacardiaceae	*Pistacia atlantica Desf.*	Btem/Igg/Drou	Atlas pistachio	Fruit, Leaves, bark	Decoction	3	(Barkaoui et al., 2017); Lahsissene et al., 2009; Hachi et al., 2016).
	Pistacia lentiscus L.	Drou	Mastic tree	Leaves, Bark	Decoction, Infusion	5	(Orch et al., 2015; Lahsissene et al., 2009; Gourri et al., 2013; Hachi et al., 2016; Belhaj et al., 2020).
	Rhus albida schousboe	Zewayâ, ânaffis	-	Fruit	-	1	(Gourri et al., 2013).
	Rhus tripartita	Jdâri	Sumac	Leaves	Powder	1	(Gourri et al., 2013).
Apiaceae	*Ammi. visnaga Lam.*	Bachnikha	Bishop's weed	Fruit, seeds, stem	Decoction	14	(Jouad et al., 2001; El-Hilaly et al., 2003; Tahraoui et al., 2007; Tahri et al., 2012; Orch et al., 2015; Hachi et al., 2016; Bousta et al., 2014; Ziyyat et al., 1997; Eddouks et al., 2002; El Hachlafi et al., 2020; Benali et al., 2017; Belhaj et al., 2020; Naceiri et al., 2021; Mechchate et al., 2020).

continued on following page

Table 1. Continued

Family	Plant species	Local Name	English name	Parts Used	Preparation	Number of citation	Réf
Apiaceae	*Ammodaucus leucotrichus Coss.*	Kamoun soufi	Heary cumin	seeds	Powder, Decoction, Infusion	5	(Orch et al., 2015, Gourri et al., 2013; Laadim et al., 2017 ;
	Heracleum sphondylium	-	Hogweed	Arial Parts	-	1	(Eddouks, 2018).
	Apium graveolens	Krafess	Leaves celery	seeds	Infusion	3	(Laadim et al., 2017; El Hachlafi et al., 2020; Mechchate et al., 2020).
	Coriandrum sativum L	Kezbour	Coriander	seeds, Leaves	Infusion, Poultice, Decoction, Powder Friction	11	(Belhaj et al., 2021; Tahraoui et al., 2007; Gourri et al., 2013; Laadim et al., 2017; Bousta et al., 2014; Skalli et al., 2019; Benkhnigue et al., 2014; Eddouks et al., 2002; Belhaj et al., 2020; Mechchate et al., 2020; Eddouks, 2018).
	Cuminum cyminum L.	Kamoun	Cumin	-	-	3	(Tahri et al., 2012; Gourri et al., 2013; Benali et al., 2017).
	Daucus carota	Khizzu	Carrot	seeds, Root	Decoction	2	(Gourri et al., 2013, Hachi et al., 2016).
	Foeniculum. vulgare Gaertn	Nafaa	Sweet fennel	Fruit, seeds, Root, Leaves	Decoction, infusion, Powder, Poultice	18	(Belhaj et al., 2021; Jouad et al., 2001; El-Hilaly et al., 2003; Barkaoui et al., 2017; Eddouks et al., 2022; Tahraoui et al., 2007 ;Orch et al., 2015; Jamila et al., 2014; Gourri et al., 2013; Kachmar et al., 2021; Laadim et al., 2017; Eddouks et al., 2002; Benali et al., 2017; Belhaj et al., 2020; Boufous et al., 2017; Naceiri et al., 2021,Eddouks, 2018; Abouri et al., 2012).
	Ptychotis verticillata Dub.	Nounkha	Ammi ptychotis	Arial Parts	Infusion	1	(Hachi et al., 2016).
	Ridolfia segetum [L.] Moris	Slilo	false fennel	Leaves	Cooked	1	Chaachouay et al., 2019).
	Carum carvi L.	el-quarwiya	Caraway	seeds, Fruit	Decoction, maceration, powder, infusion	10	(El-Hilaly et al., 2003; Barkaoui et al., 2017; Eddouks et al., 2022; Tahraoui et al., 2007; Jamila et al., 2014; Gourri et al., 2013; Eddouks et al., 2002; El Hachlafi et al., 2020; Belhaj et al., 2020; Naceiri et al., 2021).
	Echinops spinosus L.	Taskra – Bongar	Tesskra	Root, flowers, aerial parts	Decoction	1	(Abouri et al., 2012).
	Pimpinella anisum L.	Habbat Hlawa	Aniseeds	seeds	Decoction, Raw	11	(El-Hilaly et al., 2003; Barkaoui et al., 2017; Tahraoui et al., 2007; Gourri et al., 2013, Kachmar et al., 2021; Laadim et al., 2017; Skalli et al., 2019; El Hachlafi et al., 2020; Belhaj et al., 2020; Mechchate et al., 2020; Eddouks, 2018).

continued on following page

Table 1. Continued

Family	Plant species	Local Name	English name	Parts Used	Preparation	Number of citation	Réf
Apiaceae	*Eryngium ilicifolium Lam.*	Tasnant/Iglifin	-	Stem and Leaves	Decoction and powder	1	(Barkaoui et al., 2017)
	Pastinaca sativa L.	Left lmahfour	Wild parsnip	Root	Raw	1	(Barkaoui et al., 2017)
	Petroselinum crispum Mill.	Ma^adnouss	Parsley	seeds, Leaves	Decoction	7	(Tahraoui et al., 2007; Orch et al., 2015; Gourri et al., 2013; Ziyyat et al., 1997; El Hachlafi et al., 2020; Mechchate et al., 2020; Eddouks, 2018).
Apocynaceae	*Caralluma europaea*	Aknariy nouzrou/Oukan/ Daghmous/ Oukan iddan	Caralluma europaea	Whole plant	Maceration, Decoction, Infusion	7	(Belhaj et al., 2021; Laadim et al., 2017; Hachi et al., 2016; Katiri et al., 2017; Mechchate et al., 2020; Barkaoui et al., 2017; Rhattas et al., 2021).
	Nerium oleander L.	Defla	Oleander	Leaves	Decoction, fumigation, Infusion, Poultice	22	Jouad et al., 2001; El-Hilaly et al., 2003; Barkaoui et al., 2017; Eddouks et al., 2022; Tahraoui et al., 2007; Orch et al., 2015; Lahsissene et al., 2009; Jamila et al., 2014; Beblamdini et al., 2014; Gourri et al., 2013; Kachmar et al., 2021; Laadim et al., 2017; Hachi et al., 2016; Ziyyat et al., 1997; Katiri et al., 2017; Benkhnigue et al., 2014; Eddouks et al., 2002; Rhattas et al., 2016: Eddouks et al., 2022; Belhaj et al., 2020; Naceiri et al., 2021; Abouri et al., 2012).
	Vinca minor L	-	Common Periwinkle	Whole plant	-	1	(Eddouks, 2018).
	Periploca angustifolia	Asllif	-	Fruit	Decoction	1	(Barkaoui et al., 2017).
	Borago officinalis L.	lis¯an at-t¯ur	Borage	Leaves, flower	Decoction	1	(Tahraoui et al., 2007).
	Ptychotis verticillata L.	Nu^nkha	-	Arial parts	-	1	(Ziyyat et al., 1997).
Arecaceae	*Chamaerops humilis L.*	Ddoum	Dwarf fan palm	Roots, Leaves, Fruit	Cooked, Infusion	5	(Belhaj et al., 2021; El-Hilaly et al., 2003; Alaoui et al., 2017; Hachi et al., 2016; Belhaj et al., 2020).
	Phoenix dactylifera	Tmar	Date palm	Fruits	Direct consumption, Decoction	9	Tahraoui et al., 2007; Gourri et al., 2013; Benkhnigue et al., 2014; Laadim et al., 2017; Hachi et al., 2016; Ziyyat et al., 1997; Belhaj et al., 2020; Boufous et al., 2017; Abouri et al., 2012).
Aristolochiaceae	*Aristolochialonga L.*	Barztam	Bereztem	Rhizome	Powder	4	Jamila et al., 2014; Gourri et al., 2013. Katiri et al., 2017; El Hachlafi et al., 2020).

continued on following page

Table 1. Continued

Family	Plant species	Local Name	English name	Parts Used	Preparation	Number of citation	Réf
Asparagaceae	*Asparagus altissimus Munby/Asparagus albus L.*	Skkoum	Asperge blanche	Leaves	Decoction	5	Idm'hand et al., 2020; Jamila et al., 2014; Hachi et al., 2016; El Hachlafi et al., 2020; Benali et al., 2017).
Asphodelaceae	*Asphodelus tennuifolius Cav.*	Asphodèle	Tâzya	Tubers	Pwder	1	(Gourri et al., 2013).
Asteraceae	*Artemisia absinthium*	Chiba	Wormwood	Whole plant, Arial parts	Poultice, Decoction, Infusion	12	(Belhaj et al., 2021; Tahri et al., 2012; Orch et al., 2015; Lahsissene et al., 2009; Jamila et al., 2014; Laadim et al., 2017; El Hachlafi et al., 2020; Benali et al., 2017; Belhaj et al., 2020; Boufous et al., 2017; Mechchate et al., 2020; Kharchoufa et al., 2018).
	Artemisia huguetii [Caball.]	-	-	Leaves	Decoction	1	(Boufous et al., 2017).
	Cotula cinerea [Delile]	-	Saharan camomile	Arial parts, leaves	Decoction, Infusion	1	(Boufous et al., 2017).
	Warionia saharae benth and Coss	Afessas	-	Leaves	Decoction, Infusion	3	(Barkaoui et al., 2017; Gourri et al., 2013; Boufous et al., 2017).
	Anvillearadiata coss &Dur	Nakd	Nogde	Root, Whole plant, Leaves	Decoction, Powder	2	(Jamila et al., 2014; Gourri et al., 2013)
	Achillea Lipiophylla M.Bieb.	El-qorte	Achille	Leaves	Decoction, Infusion	1	(Belhaj et al., 2020).
	Artemisia mesatlantica Maire	Chih elkhryassi	Artemisia	Whole plant	Decoction	3	(Belhaj et al.; 2021; Idm'hand et al., 2020; Benkhnigue et al., 2014).
	Artemisia herba-alba	Chih	White mugwort	Leaves, stems, roots, Arial parts	Decoction, infusion, powder	24	(Belhaj et al.; 2021; Jouad et al., 2001; Barkaoui et al., 2017; Eddouks et al., 2022; Tahraoui et al., 2007; Orch et al., 2015; Jamila et al., 2014; Benlamdini et al.,2014; Gourri et al., 2013; Benkhnigue et al., 2014; Laadim et al., 2017; Hachi et al., 2016; Bousta et al., 2014; Ziyyat et al., 1997; Skalli et al., 2019; Katiri et al., 2017; El Hachlafi et al., 2020; Eddouks et al., 2002,Benali et al., 2017; Belhaj et al., 2020; Naceiri et al., 2021; Mechchate et al., 2020; Kharchoufa et al., 2018; Abouri et al., 2012).
	Artemisia abrotanum L.	-	southern wormwood	Arial parts	Decoction	1	(Benkhnigue et al., 2014).

continued on following page

Table 1. Continued

Family	Plant species	Local Name	English name	Parts Used	Preparation	Number of citation	Réf
Asteraceae	*Brocchia cinera*	Gartofa	-	Leaves, Flower	Decoction, Powder	1	(Jamila et al., 2014).
	Cichorium intybus	Bû-aggad	Chicorée sauvage	Root	Infusion	1	(Gourri et al., 2013).
	Cynara cardunculus L	Khorchouf	Artichoke	Roots, Leaves, Whole plant	Powder, Decoction, infusion	7	(Belhaj et al., 2021; Jamila et al., 2014; Gourri et al., 2013; Benkhnigue et al., 2014; Laadim et al., 2017; El Hachlafi et al., 2020; Belhaj et al., 2020).
	Cynara scolymus L	El qoq	Artichoke	Fruits, leaves	Decoction	1	(Belhaj et al., 2021).
	Cladanthus scariosus	Arzgi/irzgi	-	Flower	Decoction	1	(Katiri et al., 2017).
	Cynara arabicus	Tafsse	Cladanthus	Arial Parts	Infusion	1	(Belhaj et al., 2020).
	Cynara humilis L.	Tagemmit Taymant	Petit artichaut	Root	Decoction, Powder	1	(Belhaj et al., 2020).
	Dittrichia viscosa	Bagramane	Aunée visqueuse	Leaves	Raw, Infusion	5	(Gourri et al., 2013; Hachi et al., 2016; Benali et al., 2017; Belhaj et al., 2020; Mechchate et al., 2020).
	Inula viscosa L.	Terrahla	-	Leaves, Whole plant	Decoction, Infusion, Powder	4	(Orch et al., 2015; Jamila et al., 2014; Katiri et al., 2017; Eddouks, 2018).
	Inula graveolens [L.] Desf	-	-	Arial parts, Leaves	Powder, Decoction	1	(Boufous et al., 2017).
	Lactuca sativa L.	Lkhoss	Lettuce	Leaves, Arial parts	Infusion, raw	5	(Belhaj et al., 2021; Rhattas et al., 2016; Jouad et al., 2001; Eddouks et al., 2002; Eddouks, 2018).
	Launea arborescens [Batt] Murb.	oum lbeyna	-	Stem, Root	Decoction	4	(Gourri et al., 2013; Boufous et al., 2017; Kharchoufa et al., 2018; Abouri et al., 2012).
	Matricaria nrecutita/ Chamaemelum nobile/ Anthemis nobilis	Babounj	-	Aerial parts	Infusion	6	(Laadim et al., 2017; El Hachlafi et al., 2020; Belhaj et al., 2020; Naceiri et al., 2021; Mechchate et al., 2020; Eddouks, 2018).
	Scolymus hispanicus L.	Garnina	Golden thistle	Roots, flowers	Decoction, infusion	4	(Belhaj et al., 2021; Hachi et al., 2016; Katiri et al., 2017; Naceiri et al., 2021).

continued on following page

Table 1. Continued

Family	Plant species	Local Name	English name	Parts Used	Preparation	Number of citation	Réf
Asteraceae	*Sonchus oleraceus [L.]* / *L./Sonchus tenerrimus L*	Tifaf	Sow thistle	Leaves	Decoction	2	(Belhaj et al., 2021; Gourri et al., 2013).
	Matricaria chamomilla L.	Mansania	Chamomile	Leaves, Flower	Decoction, Infusion	2	(Orch et al., 2015; Boufous et al., 2017).
	Tanacetum vulgare L.	Lbalssem	-	Leaves	Infusion	1	(Chaachouay et al., 2019).
	Calendula arvensis M.Bieb.	Jemra, Azwiwel	-	Flower	Infusion	3	(El Hachlafi et al., 2020; Chaachouay et al., 2019; Naceiri et al., 2021).
	Echinops spinosissimus Turra	Taskra	-	Flower	Decoction	4	(Barkaoui et al., 2017; Katiri et al., 2017; Boufous et al., 2017; Kharchoufa et al., 2018).
	Launaea arborescens [Batt.]	Iferskel/ Moulbina	-	Stem, Leaves and flower	Decoction and infusion	2	(Barkaoui et al., 2017; Jamila et al., 2014).
	Scorzonera undulata Vahl	Tamtla	-	Flower	Raw	1	(Barkaoui et al., 2017).
	Anacyclus pyrethrum L.	Iguntas / Tagundecht	Pyrètre d'afrique	Root	Infusion, Powder	1	(Hachi et al., 2016).
	Ormenis mixta [L.] Dumort.	Hella	-	Flower	Infusion	1	(El Hachlafi et al., 2020).
	Chrysanthemum coronarium L.	Hmessou	Chrysanthème à couronnes	Flower	Infusion, Decoction	1	(Belhaj et al., 2020).
	Silybum marianum [L.]Gaertn.	Tawra, Chouklahmir	-	seeds, Leaves	Decoction	1	(El Hachlafi et al., 2020).
	A. absinthium L.	Chiba	Wormwood	Arial Parts, Leaves	Decoction	7	(Jouad et al., 2001; Eddouks et al., 2022; Tahraoui et al., 2007; Bousta et al., 2014; Ziyyat et al., 1997; Eddouks et al., 2002; Eddouks, 2018)
	Lactuca scariola L.	-	-	Leaves	-	1	(Eddouks, 2018).
	Helianthus annuus L	-	-	seeds	-	1	(Eddouks, 2018).
	Cynara cardunculus L.	Khorchouf	Artichoke thistle	Roots	Decoction	2	(Jouad et al., 2001; Tahraoui et al., 2007).

continued on following page

Table 1. Continued

Family	Plant species	Local Name	English name	Parts Used	Preparation	Number of citation	Réf
Asteraceae	*Cichorium intybus L.*	El okif	-	Whole plant	Decoction	1	(El-Hilaly et al., 2003)
Brassicaceae	*Lepidium sativum L.*	Hab rchad	Garden Cress	seeds	Decoction maceration, maceration	14	(Belhaj et al., 2021; Jouad et al., 2001; Eddouks et al., 2022; Tahraoui et al., 2007; Orch et al., 2015; Gourri et al., 2013; Laadim et al., 2017; Skalli et al., 2019; El Hachlafi et al., 2020; Eddouks et al., 2002; Belhaj et al., 2020; Boufous et al., 2017; Naceiri et al., 2021; Mechchate et al., 2020).
	Anastatica hierochuntica L.	Lkemsa	Rose of Jericho	Stem	Decoction	2	(Gourri et al., 2013; Hachi et al., 2016).
	Alyssum spinosum L.	Aguerbaz	Thorny Alysson	Stem, Leaves	Decoction	1	(Hachi et al., 2016).
	Anvillea radiata Coss. &Durieu	Nagd – Ojrg	-	Whole plant	Infusion	1	(Abouri et al., 2012).
	Brassicarapa L./ Brassica napus L./ Brassica Rapa	Left	Vegetable turnip	Leaves, Rhizom	Decoction	5	(Alaoui et al., 2017; Gourri et al., 2013; Hachi et al., 2016; Belhaj et al., 2020; Boufous et al., 2017).
	Raphanus sativus L./ Raphanusraphanistrum subsp. Sativus	Lefjel	Radish	seeds, roots, Rhizome	Decoction, raw	8	(Belhaj et al.; 2021; Jouad et al., 2001; Barkaoui et al., 2017; Gourri et al., 2013; Laadim et al., 2017; Belhaj et al., 2020; Mechchate et al., 2020; Kharchoufa et al., 2018).
	Brassica oleracea L.	seflûr, Krumb, Melfûf	Cabbage	Fruit	Raw, Maceration	4	Gourri et al., 2013; Laadim et al., 2017; Hachi et al., 2016; Belhaj et al., 2020).
	Diplotaxis pitardiana Maire	Kerkâz	-	Whole plant	-	1	(Gourri et al., 2013).
	Nasturtium officinale	Gernûnes	-	stem	-	1	(Gourri et al., 2013).
Burseraceae	*Boswellia carterii Birdw*	Salabane	Carteri	Fruits	Decoction, infusion	3	(Belhaj et al., 2021; Benali et al., 2017; Belhaj et al., 2020).
	Boswellia sacra Flueck.	Loubane Dacare	-	seeds	Infusion	1	(Belhaj et al., 2020).
	Commiphora myrrha	Lmorra	-	-	-	1	(Mechchate et al., 2020).
Buxaceae	*Buxus balearica Lam.*	Azazer / Ibakous	Balearic boxwood	Leaves	Decoction	1	(Hachi et al., 2016).

continued on following page

Table 1. Continued

Family	Plant species	Local Name	English name	Parts Used	Preparation	Number of citation	Réf
Cactaceae	*Opuntia ficus indica Mill.*	Hendiya, Kermous	-	Flowers, stem, Fruit	Decoction and powder	9	(Jouad et al., 2001; Barkaoui et al., 2017; Tahraoui et al., 2007; Laadim et al., 2017; Hachi et al., 2016; Bousta et al., 2014; Boufous et al., 2017; Naceiri et al., 2021; Mechchate et al., 2020).
	Opunitia ficus barbarica	aknari	Prickly pear	Flower, Fruit, Leaves	Powder	1	(Gourri et al., 2013).
Capparaceae	*Capparis spinosa L.*	Lkabbar	Caper	Fruits, leaves, Flower	Decoction, Powder infusion	16	(Belhaj et al., 2021; Jouad et al., 2001; Eddouks et al., 2022; Tahraoui et al., 2007; Orch et al., 2015; Jamila et al., 2014; Gourri et al., 2013; Kachmar et al., 2021; Hachi et al., 2016; Bousta et al., 2014; Ziyyat et al., 1997; Katiri et al., 2017; Eddouks et al., 2002; Benali et al., 2017; Belhaj et al., 2020; Naceiri et al., 2021).
	Capparis decidua [Forssk] Edgew	Ignin	-	Fruit	Powder	1	(Gourri et al., 2013).
	Maerua crassifolia forssk	Âtîl, Ssadra l khadra	-	Leaves	Powder	1	(Gourri et al., 2013).
Caprifoliaceae	*Sambucus nigra L.*	Sembouqa	-	Leaves, Fruit	Decoction, Infusion	1	(Belhaj et al., 2020).
Caryophyllaceae	*Herniaria glabra L.*	Hrasset lehjer	-	Arial Parts, Leaves, Whole plant	Decoction, Infusion	4	(Jouad et al., 2001; Tahraoui et al., 2007; Jamila et al., 2014; Gourri et al., 2013).
	S. purpurea Pers. G. Don fil	Zahrat arrimal	-	Arial Parts	-	1	(Jouad et al., 2001).
	Atriplex halimus L.	Legtef	Arroche sauvage	Leaves	Powder, Decoction	2	(Idm'hand et al., 2020; Gourri et al., 2013).
	Corrigiola telephiifolia Pourr.	Sarghina	Telephium-leaved corrigiole	-	Powder	1	(Hachi et al., 2016).
	Paronychia argentea Lam	Tahidourt n'imksaoum	-	Leaves, Stem	Infusion	1	(Katiri et al., 2017).

continued on following page

Table 1. Continued

Family	Plant species	Local Name	English name	Parts Used	Preparation	Number of citation	Réf
Chenopodiaceae	*Chenopodium ambrosioides L.*	Mkhinza	Mexican tea	Flowers, leaves	Infusion, Fresh juice	13	(Jouad et al., 2001; Tahraoui et al., 2007; Orch et al., 2015; Gourri et al., 2013; Laadim et al., 2017; Hachi et al., 2016; Bousta et al., 2014; Ziyyat et al., 1997; Rhattas et al., 2016; Eddouks et al., 2022; Naceiri et al., 2021; Mechchate et al., 2020; Eddouks, 2018).
	Anabasis aretiodes	Sallaa	-	Arial Parts	-	2	(Eddouks et al., 2022; Eddouks et al., 2002)
	Bassia muricata [L.] Asch	-	-	Arial parts, Leaves	Infusion	1	(Boufous et al., 2017).
	Fredolia aretioides	Sellî'	cauliflower of bou hamama	Stem	Decoction	1	(Gourri et al., 2013).
	Salsola tetragona	Li'rad	-	Leaves	Powder, Decoction,	1	(Gourri et al., 2013).
Cistaceae	*Cistus monspeliensis*	Meril-Stevia-Markil-Amzil -shtapa	Cistus of Montpellier	-	poultice, Infusion	1	(Belhaj et al., 2021).
	Cistus albidus L.	Tuzzala	White cicstus	Leaves	Decoction	2	(Lahsissene et al., 2009; Hachi et al., 2016).
	Cistus creticus L.	Irgel	-	Leaves	Decoction	2	(Barkaoui et al., 2017; Katiri et al., 2017).
	Cistus salviifolius L.	Tirgelt	-	Leaves	Decoction	1	(Barkaoui et al., 2017).
	Cistus Ladanifer.L	Touzalt	Cistus ladaniferous	seeds	Powder, Infusion	1	(Belhaj et al., 2020).
	Cistus Laurifolius L.	Agullid	Poplar-leaved cistus	seeds	Powder, Decoction,	1	(Belhaj et al., 2020).
Cruciferae	*B. oleracea L.*	Kroumb	-	Arial parts	-	1	(Jouad et al., 2001).
	Raphanus sativus sp.	jirjir	-	seeds	-	1	(Jouad et al., 2001)
Cucurbitaceae	*Bryonia dioica Jacq.*	Terbouna	Red bryony	Stems, fruits	Decoction	3	(Belhaj et al., 2021; Gourri et al., 2013; Belhaj et al., 2020).
	Citrullus colocynthis Schrad.	Hdejja, Hantel, Taferzizte	Bitter apple, Colocynth	Fruits, seeds	Infusion, Decoction, Powder, Maceration	18	Idm'hand et al.,2020; Jouad et al., 2001; Barkaoui et al., 2017; Eddouks et al., 2022; Tahraoui et al., 2007; Jamila et al., 2014; Gourri et al., 2013; Benkhnigue et al., 2014; Laadim et al., 2017; Ziyyat et al., 1997; Skalli et al., 2019; Katiri et al., 2017; El Hachlafi et al., 2020; Benali et al., 2017; Chaachouay et al., 2019; Boufous et al., 2017; Eddouks, 2018; Abouri et al., 2012).

continued on following page

Table 1. Continued

Family	Plant species	Local Name	English name	Parts Used	Preparation	Number of citation	Réf
Cucurbitaceae	*Cucumis. sativus L.*	Khiyar	Cucumber	Fruits, seeds	Raw, Powder	8	(Belhaj et al., 2021; Jouad et al., 2001,2; Barkaoui et al., 2017; Gourri et al., 2013; Hachi et al., 2016; Belhaj et al., 2020; Mechchate et al., 2020).
	Citrullus vulgaris Schard.	Dellah	Watermelon	Leaves	Infusion, maceration	2	(Belhaj et al., 2021; Idm'hand et al., 2020).
	Cucurbita pepo L.	Garaa Khedra	-	Fruit	Cooked	4	(Orch et al., 2015; Gourri et al., 2013; Hachi et al., 2016; Chaachouay et al., 2019)
	Cucurbita maxima	Lgar'a hamra	pumpkin	Leaves	Decoction	2	(Gourri et al., 2013; Boufous et al., 2017).
Cupressaceae	*Juniperus oxycedrus L.*	L arâar chrini	Prickly juniper	Leaves	Maceration, Powder	3	(Belhaj et al., 2021; Idm'hand et al., 2020; Jouad et al., 2001).
	Juniperus phoenicea L.	Arar Finiqi	Red juniper	Leaves	Decocotion	5	(Chaachouay et al., 2019; Lahsissene et al., 2009; Benkhnigue et al., 2014; Hachi et al., 2016; El Hachlafi et al., 2020).
	Juniperus thurifera L.	Tawayt	Thuriferous juniper	Leaves	Decoction	1	(Hachi et al., 2016).
	Tetraclinis articulata Benth.	Aarar	Berber thuya, Arar tree	Arial Parts	Maceration, Powder	11	(Eddouks et al., 2022; Tahraoui et al., 2007; Orch et al., 2015; Lahsissene et al., 2009; Kachmar et al., 2021; Laadim et al., 2017; Hachi et al., 2016; Ziyyat et al., 1997; Eddouks et al., 2002; Benali et al., 2017; Naceiri et al., 2021).
Cuscutaceae	*Cuscuta epithymum Mur.*	Zîtra	-	-	-	1	(Jouad et al., 2001).
Cyperaceae	*Cyperus rotundus L*	Tara	Sochet rond	-	-	1	(Gourri et al., 2013).
Cynomoriaceae	*Cynomorium coccineum*	Tertût	Maltese mushroom	Stem	Powder	1	(Gourri et al., 2013).
Dracaenaceae	*Dracaena draco subsp.*	Ajgal	-	Stem and Leaves	Decoction	1	(Barkaoui et al., 2017).
Ericaceae	*Arbutus unedo L.*	Sasnou	Arbutus	Leaves, Roots	Powder, Maceration	5	(Jouad et al., 2001; Orch et al., 2015; Hachi et al., 2016; Ziyyat et al., 1997; Naceiri et al., 2021).
Ephedraceae	*Ephedra altissima Desf.*	Tougel argan	-	Stem, Leaves and wholeplant	Decoction	1	(Barkaoui et al., 2017)

continued on following page

Table 1. Continued

Family	Plant species	Local Name	English name	Parts Used	Preparation	Number of citation	Réf
Ephedraceae	*Ephedra alata Dec/ Ephedra nebrodensis Guss.*	šš̆dîda	Ephedra	Stem	Maceration, Decoction	5	(Gourri et al., 2013; Hachi et al., 2016; El Hachlafi et al., 2020; Boufous et al., 2017; Abouri et al., 2012).
	Ephedra fragilis desf	Amater	-	Leaves, Stem	Decoction	2	(Katiri et al., 2017; Kharchoufa et al., 2018).
Euphorbiaceae	*Euphorbia echinus Hook.f. & Coss.*	Tikiwt/ Daghmous	Euphorbia echinus	Stems, leaves	Powder, maceration, infusion	6	Barkaoui et al., 2017; Gourri et al., 2013; Bousta et al., 2014; El Hachlafi et al., 2020; Benali et al., 2017; Abouri et al., 2012).
	Euphorbia resinifera O.Berg	Zeggoum	Gum Euphorbium	Flowers, latex	Infusion, latex water	6	(Belhaj et al.2021; Hachi et al., 2016; Katiri et al., 2017; Belhaj et al., 2020; Kharchoufa et al., 2018; Abouri et al., 2012).
	Mercurialis annua L	L-hurriga l melsaa	Annual Mercurial	Stem, Whole plant	Decoction, Infusin	2	(Gourri et al., 2013; Idm'hand et al.,2020).
	Euphorbia officinarum [L.]	-	-	Arial parts	Powder	1	(Boufous et al., 2017).
Fabaceae	*Acacia gummifera Willd.*	Telh	-	Roots	Decoction	1	(Belhaj et al.; 2021)
	Cassia Senna L.	Sana	-	Leaves, whole plant	Decoction, Infusion	2	Jamila et al., 2014; El Hachlafi et al., 2020).
	Cassia absus L.	El habba sawdae	Obtuse-leaved Cassia	seeds	Powder, Decoction	1	(Belhaj et al., 2020).
	Acacia arabica	âmûr, sllâhâ	red gum tree	_	Powder	1	(Gourri et al., 2013)
	Cytisus battandieri Maire	Akhamelel	-	Leaves	Decoction	1	(Naceiri et al., 2021).
	Acacia senegal L.	Al'elk	Senegal gum tree	_	Powder	1	(Gourri et al., 2013).
	Acacia tortilis	Telh	Acacia saharien	Leaves, stem, bark	Powder, Decoction	1	(Gourri et al., 2013).

Table 1. Continued

Family	Plant species	Local Name	English name	Parts Used	Preparation	Number of citation	Réf
Fabaceae	*Trigonella foenum-graecum L.*	Lhelba	fenugreek	seeds	Powder, Decoction, Fusion	18	(Belhaj et al., ;2021,3; Barkaoui et al., 2017; Tahri et al., 2012; Orch et al., 2015; Lahsissene et al., 2009; Jamila et al., 2014; Gourri et al., 2013,Benkhnigue et al., 2014; Kachmar et al., 2021; Laadim et al., 2017; Hachi et al., 2016; Bousta et al., 2014; Skalli et al., 2019; Katiri et al., 2017; El Hachlafi et al., 2020; Benali et al., 2017; Naceiri et al., 2021; Mechchate et al., 2020).
	Anagyris foetida L.	Foul gnawa	Mediterranean stinkbush	Leaves	Infusion	*4*	(Belhaj et al.; 2021; Gourri et al., 2013; Belhaj et al., 2020; Mechchate et al., 2020).
	Arachis hypogaea	L-gerta, Kawkaw	Arachide	-	Powder	*1*	(Gourri et al., 2013).
	Ceratonia siliqua L.	Lkharoub	Carob tree	Leaves, seeds	Infusion, powder, Decoction	*8*	(Belhaj et al.,2021; Barkaoui et al., 2017; Gourri et al., 2013; Laadim et al., 2017; Skalli et al., 2019; Belhaj et al., 2020; Naceiri et al., 2021; Mechchate et al., 2020).
	Cicer arietinum L.	Lhemmes	-	seeds	Decoction	4	(Idm'hand et al., 2020; Gourri et al., 2013; Belhaj et al., 2020; Mechchate et al., 2020)
	Retama monosperma [L.] Boiss.	Rtam	Retama	Roots, leaves	Decoction, infusion	2	(Belhaj et al., 2021; El Hachlafi et al., 2020).
	Lupinus albus L.	termes	Lupinus	seeds	Powder, infusion, seeds	8	(Belhaj et al., 2021; Jamila et al., 2014; Gourri et al., 2013; Benkhnigue et al., 2014; El Azzouzi et al., 2015; Bousta et al., 2014; Belhaj et al., 2020; Naceiri et al., 2021).
	Lupinus luteus L.	Ssemkala	-	seeds	Decoction	2	(Belhaj et al., 2021; Rhattass et al.,2016)
	Lupinus pilosus L.	Rjel Djaja	-	seeds	Infusion	1	(Chaachouay et al., 2019).
	Medicago sativa L.	Fessa	luzerne	Leaves, seeds	Infusion	5	(Gourri et al., 2013; Rhattass et al.,2016; Hachi et al., 2016; Belhaj et al., 2020; Kharchoufa et al., 2018).
	Viciafaba L.	Foul	-	seeds, Leaves	Decoction, Infusion	4	(Jamila et al., 2014; Gourri et al., 2013; Mechchate et al., 2020; Kharchoufa et al., 2018).
	Glycine max [L.] Merr.	Soja	-	seeds	Maceration, Raw	5	(Barkaoui et al., 2017; Rhattas et al.,2016; Naceiri et al., 2021; Mechchate et al., 2020; Abouri et al., 2012).
	Lupinus angustifolius L.	Ibawn dekouk	-	seeds	Powder, Infusion	2	(Barkaoui et al., 2017; Katiri et al., 2017).
	Ononis natrix L.	Hennet reg	yellow ononis	Leaves	-	1	(Gourri et al., 2013).
	Ononis tournefortii	Afezdad	Ononis	Leaves	-	1	(Belhaj et al., 2021).

continued on following page

Table 1. Continued

Family	Plant species	Local Name	English name	Parts Used	Preparation	Number of citation	Réf
Fabaceae	*Phaseolus aureus Roxb*	Soja	Soy	seeds	-	1	(Belhaj et al., 2020).
	Phaseolus vulgaris L	lubiya	Bean	-	Powder	4	(Gourri et al., 2013; Hachi et al., 2016; Bousta et al., 2014; Belhaj et al., 2020).
	Vicia sativa	Ain larneb	-	seeds	Powder	1	(Belhaj et al.; 2021)
	Glycyrrhiza glabra L.	Ark sous	-	Bark	Infusion	3	(Skalli et al., 2019; Idm'hand et al.,2020; Kharchoufa et al., 2018).
Fagaceae	*Quercus suber L.*	D'bbagh	-	Bark	Powder	1	(Orch et al., 2015).
	Quercus ilex L.	Lbalout	-	Bark	Decoction	1	(Jamila et al., 2014).
	Quercus faginea Lam.	Lasfa	-	Leaves	-	1	(Kharchoufa et al., 2018).
Fumariaceae	*Fumaria officinalis L.*	Lwarda D'lard	Fumitory	Roots	Decoction, Poultice, Inf$usion	2	(Belhaj et al., 2020; Boufous et al., 2017).
Gentianaceae	*Centaurium Erythraea Rafn*	Gosset l'hayya	-	Arial parts	Decoction, Infusion	9	(Jouad et al., 2001; Orch et al., 2015; Jamila et al., 2014; Kachmar et al., 2021; Hachi et al., 2016; El Hachlafi et al., 2020; Benali et al., 2017; Belhaj et al., 2020; Naceiri et al., 2021).
	Pelargonium roseum Willd.	Laattercha	-	Stem	Infusion	2	(Hachi et al., 2016; Idm'hand et al.,2020).
Globulariaceae	*Globularia alypum L.*	Ain larneb	-	Leaves, Whole plant	Decoction, Powder, Infusion	15	(Jouad et al., 2001; Eddouks et al., 2022; Orch et al., 2015; Jamila et al., 2014; Gourri et al., 2013; Benkhnigue et al., 2014; Laadim et al., 2017; Hachi et al., 2016; Bousta et al., 2014; Ziyyat et al., 1997; Katiri et al., 2017; Eddouks et al., 2002; Benali et al., 2017; Eddouks, 2018; Abouri et al., 2012).
Graminaceae	*Phlaris paradoxa L.*	Zuan	Hood canarygrass	seeds	Powder	2	(El-Hilaly et al., 2003; Tahraoui et al., 2007).
	Cynodon dactylon [L.] Pers.	njem	Bermudagrass	Arial Parts, Root	Decoction	2	(Tahraoui et al., 2007; Eddouks, 2018).
	Zea mays L.	dra	Maize, Corn	seeds	Dried/ Scorched	1	(Tahraoui et al., 2007).
	Pennisetum typhoides [stapf]		-	seeds	Decoction	1	(Boufous et al., 2017).
	Sorghum 6ulgare L.	Bachna	-	-	-	1	(Ziyyat et al., 1997)

continued on following page

Table 1. Continued

Family	Plant species	Local Name	English name	Parts Used	Preparation	Number of citation	Réf
Iridaceae	*Crocus sativus L.*	Zafrane lhor	Saffron	Flower	-	3	(Gourri et al., 2013; Ziyyat et al., 1997; Boufous et al., 2017).
Juglandaceae	*Juglans regia L.*	Lgargaa/ Swak	Common walnut	Barks, leaves, Fruit	Decoction, infusion	10	(Belhaj et al., 2021; Jouad et al., 2001; Tahraoui et al., 2007; Gourri et al., 2013; Benkhnigue et al., 2014; Hachi et al., 2016; Ziyyat et al., 1997; Katiri et al., 2017; El Hachlafi et al., 2020; Belhaj et al., 2020).
Juncaceae	*Juncus maritimus Lank.*	Smâr	-	Stem	-	2	(Gourri et al., 2013; Ziyyat et al.,1997).
Lamiaceae	*Ajuga iva [L.] Schreb.*	Chendgura	-	Leaves, stem, Arial parts	Decoction, Powder	19	(Idm'hand et al.,2020; Jouad et al., 2001; El-Hilaly et al., 2003; Barkaoui et al., 2017; Tahraoui et al., 2007; Tahri et al., 2012; Orch et al., 2015; Lahsissene et al., 2009; Gourri et al., 2013; Benkhnigue et al., 2014; Kachmar et al., 2021; Laadim et al., 2017; Hachi et al., 2016; Ziyyat et al., 1997; Katiri et al., 2017; El Hachlafi et al., 2020; Belhaj et al., 2020; Boufous et al., 2017; Naceiri et al., 2021).
	Ballota hirsuta Benth.	Merrouwt	-	Leavesy, stems	Decoction, Infusion	1	(Belhaj et al., 2020).
	Calamintha officinalis Moench	Manta	-	Arial parts	Decoction, infusion	5	(Orch et al., 2015; Skalli et al., 2019; Belhaj et al., 2020; Naceiri et al., 2021; Mechchate et al., 2020).
	Lavandula dentata L.	Khzama	Fringed lavender	Flower, Whole plant, Stem, Leaves	Poultice, infusion, Decoction, Raw	13	(Belhaj et al., 2021; Jouad et al., 2001; Barkaoui et al., 2017; Eddouks et al., 2022; Tahraoui et al., 2007; Orch et al., 2015; Ziyyat et al., 1997; Eddouks et al., 2002; El Hachlafi et al., 2020; Belhaj et al., 2020; Boufous et al., 2017; Mechchate et al., 2020,Eddouks, 2018).
	*Lavandula maroccana Murbeck***	Lguiz	-	Arial parts	Decoction	1	(Katiri et al., 2017).
	Marrubium vulgare L.	Merriwa, Merriwta	white horehound	Arial parts, Leaves, stem	Decoction, Infusion	20	Jouad et al., 2001; Eddouks et al., 2002; Tahraoui et al., 2007; Orch et al., 2015; Gourri et al., 2013; Benkhnigue et al., 2014; Kachmar et al., 2021; Laadim et al., 2017; Bousta et al., 2014; Ziyyat et al., 1997; Skalli et al., 2019; Eddouks et al., 2002; El Hachlafi et al., 2020; Benali et al., 2017; Belhaj et al., 2020; Boufous et al., 2017 ;Naceiri et al., 2021,Mechchate et al., 2020; Eddouks, 2018; Abouri et al., 2012).
	Mentha pulegium L.	Flayou	Mint	Arial parts, Leaves, stem	-	13	Jouad et al., 2001; Tahraoui et al., 2007; Orch et al., 2015; Gourri et al., 2013; Laadim et al., 2017; Hachi et al., 2016; Ziyyat et al., 1997; Skalli et al., 2019; El Hachlafi et al., 2020; Benali et al., 2017; Boufous et al., 2017; Mechchate et al., 2020).

continued on following page

Table 1. Continued

Family	Plant species	Local Name	English name	Parts Used	Preparation	Number of citation	Réf
Lamiaceae	*Mentha viridis L.*	Naanaa	Green mint	Leaves, Stem	Raw	3	(Gourri et al., 2013; Eddouks et al., 2002; Belhaj et al., 2020).
	Origanum compactum Benth.	Zaatar	Oregano	Leaves, Flower	Infusion, Infusion, Powder	15	(Belhaj et al.; 2021; Jouad et al., 2001; Eddouks et al., 2022; Tahraoui et al., 2007; Tahri et al., 2012; Orch et al., 2015; Jamila et al., 2014; Gourri et al., 2013; Benkhnigue et al., 2014; Hachi et al., 2016; Ziyyat et al., 1997; Skalli et al., 2019; El Hachlafi et al., 2020; Belhaj et al., 2020; Mechchate et al., 2020).
	Origanum majorana L.	Merdedouch	Marjoram	Leaves	Powder	3	(Gourri et al., 2013; El Hachlafi et al., 2020; Belhaj et al., 2020).
	Ocimum basilicum L.	Lahbak	Basil	Arial parts	Decoction, infusion	4	(Orch et al., 2015; Hachi et al., 2016; El Hachlafi et al., 2020; Belhaj et al., 2020).
	Rosmarus officinalis L.	Azir	Rosemary	Arial parts, Leaves, Flower	Decoction, infusion	12	(Jouad et al., 2001; Chaachouay et al., 2019; Orch et al., 2015; Gourri et al., 2013; Laadim et al., 2017; Hachi et al., 2016; Ziyyat et al., 1997; Skalli et al., 2019; El Hachlafi et al., 2020; Benali et al., 2017; Naceiri et al., 2021,Mechchate et al., 2020).
	Salvia officinalis L.	Salmia	Garden sage	Leaves, Flower	Infusion, Decoctio, maceratio	24	(Belhaj et al., 2021; Idm'hand et al.,2020; Jouad et al., 2001; Barkaoui et al., 2017; Tahraoui et al., 2007; Tahri et al., 2012,Orch et al., 2015; Jamila et al., 2014; Gourri et al., 2013; Benkhnigue et al., 2014; El Azzouzi et al.,2015; Kachmar et al., 2021; Hachi et al., 2016; Bousta et al., 2014; Ziyyat et al., 1997; Skalli et al., 2019; Katiri et al., 2017; El Hachlafi et al., 2020; Benali et al., 2017; Belhaj et al., 2020; Boufous et al., 2017; Naceiri et al., 2021; Mechchate et al., 2020; Eddouks, 2018).
	Lavandula multifida L.	kohayla	Lavender harvests	–	Decoction, infusion	2	(Belhaj et al.; 2021. Gourri et al., 2013
	Lavandula stoechas L.	halhal	-	Leaves	Decoction	7	(Idm'hand et al.,2020; Barkaoui et al., 2017; Tahri et al., 2012. Gourri et al., 2013. Kachmar et al., 2021; Laadim et al., 2017; Hachi et al., 2016).
	Lavandula stoechas L.	Al'Halhal	-	Flower	Decoction, infusion	4	(Orch et al., 2015; Lahsissene et al., 2009; Jamila et al., 2014; Belhaj et al., 2020).
	Rosmarinus officinalis L.	Azir	Common Rosemary	Leaves	Decoction, infusion	7	(Belhaj et al.; 2021; Eddouks et al., 2022; Tahraoui et al., 2007; Tahri et al., 2012; Lahsissene et al., 2009; Belhaj et al., 2020; Boufous et al., 2017).

continued on following page

Table 1. Continued

Family	Plant species	Local Name	English name	Parts Used	Preparation	Number of citation	Réf
Lamiaceae	*Teucrium polium L./ Mentha pulegium L.*	Jiida/flyou	Shrubby germander	Whole plant, Leaves	Decoction, powder	5	(Belhaj et al.; 2021; Barkaoui et al., 2017; Eddouks et al., 2022; Katiri et al., 2017; Belhaj et al., 2020).
	Lavandula stoechas L.	Halhal	French lavender	Flower, Leaves	-	1	(Tahraoui et al., 2007).
	Mentha spicata L.	Naânaâ	Spearmint	Leaves, Stem	-	2	(Tahraoui et al., 2007; Eddouks, 2018).
	Mentha suaveolens Ehr	Marceta	-	Leaves, Whole plan	Decoction, Infusion	2	(Jamila et al., 2014; Skalli et al., 2019).
	Origanum majorana L.	Merdad	Sweet marjoram	Leaves	Infusion	1	(Tahraoui et al., 2007).
	Thymus vulgaris L.	Zeitra	Thyme	Leaves, Arial parts, Flowers	Infusion, Powder, Decoction, Maceration	16	(Belhaj et al., 2021; Idm'hand et al., 2020; Barkaoui et al., 2017; Tahraoui et al., 2007; Orch et al., 2015,Gourri et al., 2013; Benkhnigue et al., 2014; Laadim et al., 2017; Hachi et al., 2016; Skalli et al., 2019; El Hachlafi et al., 2020; Benali et al., 2017; Belhaj et al., 2020; Boufous et al., 2017; Naceiri et al., 2021,Mechchate et al., 2020).
	Mentha suaveolens L.	Marseta	Round-leaved mint	Leaves	Poultice	2	(Lahsissene et al., 2009; Skalli et al., 2019).
Lauraceae	*Laurus nobilis L.M. pulegium L.*	Ourak moussa	Laurel	Leaves	Decoction, infusion	4	(Belhaj et al.; 2021; Rhattas et al., 2016; Ziyyat et al., 1997; Belhaj et al., 2020).
	Cinnamomum cassia [L.] J.Presl/ Cinnamomum verum J. Presl	Lqrfa	Soybean	Bark, Arial parts, seeds, fruit	Powder, Fried seeds	13	(Idm'hand et al., 2020; Eddouks et al., 2022; Tahraoui et al., 2007; Jamila et al., 2014; Gourri et al., 2013; Kachmar et al., 2021; Laadim et al., 2017; Hachi et al., 2016; Skalli et al., 2019; Eddouks et al., 2002; Belhaj et al., 2020; Mechchate et al., 2020; Eddouks, 2018).
	Persea gratissima Gaertn/Persea americana Mill.	Avocat	Avocado	-	Powder	4	(Gourri et al., 2013; Bousta et al., 2014; Skalli et al., 2019; El Hachlafi et al., 2020).
Leguminosae	*Hispida maxim. Glycine soja sichet Zucc.*	Soja	-	Fruits	-	1	(Jouad et al., 2001)
	Lupinus albus L.	Foul mesri	-	seeds	-	4	(Jouad et al., 2001; Eddouks et al., 2022; Eddouks et al., 2002; Eddouks, 2018).

continued on following page

Table 1. Continued

Family	Plant species	Local Name	English name	Parts Used	Preparation	Number of citation	Réf
Leguminosae	*Trigonella foeniculum-graecum L.*	Halba	-	seeds	Maceration, Maceration, Infusion	8	(Jouad et al., 2001; El-Hilaly et al., 2003; Eddouks et al., 2022; Tahraoui et al., 2007; Ziyyat et al., 1997; Eddouks et al., 2002; Boufous et al., 2017; Eddouks, 2018).
	Retama sphaerocarpa [Boiss.] L.	rtem	Common retama	Root	Decoction	1	(Tahraoui et al., 2007).
Liliaceae	*Aloe succotrina Lam/ Aloe vera Burm.*	Sibr	Aloevera	Arial parts, Leaves	-	5	(Jouad et al., 2001; Eddouks et al., 2022; Tahraoui et al., 2007; Gourri et al., 2013; Eddouks et al., 2002).
	Asphodelus tenuifolius [Cav.]		-	Leaves, Fruit	Decoction	1	(Boufous et al., 2017).
	Illicium verum Hook.f.	Badiane	-	seeds	Powder, Decoction	1	(El Hachlafi et al., 2020).
	Urginea maritima [L.] Baker	Bssallansa	-	Leaves	Decoction	1	(Naceiri et al., 2021).
Linaceae	*Linum usitatissimum L.*	Zeriaa ketane	Flax	seeds	Decoction, infusion	12	(Belhaj et al., 2021; Eddouks et al., 2022; Gourri et al., 2013; Laadim et al., 2017; Hachi et al., 2016; Skalli et al., 2019; Eddouks et al., 2002; Benali et al., 2017; Belhaj et al., 2020; Mechchate et al., 2020; Eddouks, 2018; Kharchoufa et al., 2018).
Lythraceae	*Punica granatum L.*	Rommane	Grenadier	Bark	Powder, Decoction, Infusion	5	(Belhaj et al., 2021; Barkaoui et al., 2017; Hachi et al., 2016; El Hachlafi et al., 2020; Mechchate et al., 2020).
	Lawsonia inermis	Hanna	-	Leaves	-	3	(Eddouks et al., 2022; Eddouks et al., 2002,Eddouks, 2018).
Malvacea	*Adansonia digitata L.*	Tajmakht	-	Fruit	Infusion	1	(Idm'hand et al., 2020).
	Abelmoschus esculentus [L.] Moench	Lmloukhia	Gombo	Fruit	Maceration	2	(Hachi et al., 2016; Skalli et al., 2019).
	Hibiscus sabdariffa L.	Bissam, karkadi	Pink Tea	Flower	-	3	(Gourri et al., 2013; Katiri et al., 2017; Belhaj et al., 2020).
Moraceae	*Ficus carica L.*	Karmouss	Fig	Leaves, Fruits	Maceration, infusion, Decoction, Powder	13	(Belhaj et al., 2021; Jouad et al., 2001,5; Barkaoui et al., 2017; Tahraoui et al., 2007,Gourri et al., 2013,Hachi et al., 2016, Ziyyat et al., 1997,Eddouks et al., 2002; Belhaj et al., 2020; Boufous et al., 2017; Naceiri et al., 2021,Mechchate et al., 2020; Abouri et al., 2012).
	Morus alba L.	Ettout	-	Leaves	Infusion	3	(Chaachouay et al., 2019; Hachi et al., 2016; Eddouks et al., 2002).

continued on following page

Table 1. Continued

Family	Plant species	Local Name	English name	Parts Used	Preparation	Number of citation	Réf
Molluginaceae	*Opophytum theurkauffii Maire L.*	âfzû	-	Leaves, Fruit	Decoction	1	(Gourri et al., 2013).
Musaceae	*Musa paradisiaca L.*	Banane	Banana tree	Leaves	Decoction	1	(Gourri et al., 2013).
Myrtaceae	*Myrtus communis L.*	Rayhane	Common Myrtle	Leaves, Fruits	Infusion, Decoction	9	(Belhaj et al., 2021; Jouad et al., 2001; Orch et al., 2015; Jamila et al., 2014; Laadim et al., 2017; El Hachlafi et al., 2020; Mechchate et al., 2020; Eddouks, 2018; ; Kharchoufa et al., 2018).
	Eugenia caryophyllata thumb./Syzygium aromaticum	Qronfel	Clove	Flower	Decoction, Infusion, maceration, powder	11	Jouad et al., 2001; Tahraoui et al., 2007; Orch et al., 2015; Gourri et al., 2013; Laadim et al., 2017; Ziyyat et al., 1997; Skalli et al., 2019; Benali et al., 2017; Belhaj et al., 2020; Naceiri et al., 2021; Eddouks, 2018).
	Eucalyptus globulus Labill [sp.]	Kalitous	Eucalyptus	Leaves, Flowers	Decoction, Infusion	12	(Jouad et al., 2001; Eddouks et al., 2022; Tahraoui et al., 2007; Orch et al., 2015; Gourri et al., 2013; Laadim et al., 2017; Hachi et al., 2016; Ziyyat et al., 1997; Eddouks et al., 2002; Belhaj et al., 2020; Naceiri et al., 2021 ;Mechchate et al., 2020).
	Jasminum fruticans L.	Yasmin	Jasmine	Leaves, flowers	Maceration, infusion	2	(Belhaj et al., 2021; Idm'hand et al.,2020).
	Myrtus communis L.	Raihane	Myrtle	Leaves, Fruits	Decoction, Infusion	7	(Eddouks et al., 2022; Tahraoui et al., 2007; Hachi et al., 2016; Ziyyat et al., 1997; Eddouks et al., 2002; Belhaj et al., 2020; Naceiri et al., 2021).
Myristicacea	*Myristica fragrans Houtt.*	Lgouza	Mace	seeds	Powder	2	(Barkaoui et al., 2017; Naceiri et al., 2021).
Nitrariaceae	*Peganum harmala L.*	Lharmal	Harmal	seeds	Infusion, powder	2	(Belhaj et al., 2021; Rhattas et al., 2016).
Oleaceae	*Olea europaea L.*	Zaytoun	Olive	Fruits, leaves	Raw, decoction, infusion, maceration, Powder	26	(Belhaj et al.; 2021; Rhattas et al.,2016; Idm'hand al., 2020; El-Hilaly et al., 2003; Barkaoui et al., 2017; Eddouks et al., 2022; Tahraoui et al., 2007; Orch et al., 2015; Jamila et al., 2014; Gourri et al., 2013; Benkhnigue et al., 2014; Laadim et al., 2017; Hachi et al., 2016; Bousta et al., 2014; Ziyyat et al., 1997; Skalli et al., 2019; Katiri et al., 2017; Eddouks et al., 2002; El Hachlafi et al., 2020; Benali et al., 2017; Belhaj et al., 2020; Boufous et al., 2017; Naceiri et al., 2021; Mechchate et al., 2020; Eddouks, 2018; Abouri et al., 2012).
	Papaver rhoeas L.	Belaaman	-	seeds	Powder	1	(Barkaoui et al., 2017).
	Fraxinus augustifolia Vahl.	Touzalt	-	Leaves	Infusion	2	(Hachi et al., 2016; Ziyyat et al., 1997).

continued on following page

Table 1. Continued

Family	Plant species	Local Name	English name	Parts Used	Preparation	Number of citation	Réf
Oxalidaceae	*Hyphaene thebaica L.*	Dûm, Kârôr	-	-	-	1	(Gourri et al., 2013).
Papaveraceae	*Papaver rhoeas L*	Gawche / Belaamen	-	seeds	Decoction	3	(Katiri et al., 2017; Naceiri et al., 2021; Mechchate et al., 2020).
	Fumaria officinalis L.	Lwarda D'lard	Fumitory	Roots	Decoction, Poultice, Infusion	1	(Belhaj et al., 2020).
Pedaliaceae	*Sesamum indicum Dc.*	Jenjlan	-	seeds	Infusion, Powder, Decotion	9	(Jouad et al., 2001; Eddouks et al., 2022; Orch et al., 2015; Gourri et al., 2013; Laadim et al., 2017; Eddouks et al., 2002; Belhaj et al., 2020; Mechchate et al., 2020; Eddouks, 2018).
	Pennisetum typhoides	Illan	-	seeds	Infusion	1	(Laadim et al., 2017).
Plantaginaceae	*Plantago ovata Forssk*	Katouna	-	seeds	Infusion	1	(Naceiri et al., 2021).
Plumbaginaceae	*Limonium sinuatum [L.] Mill.*	Lgarsa	-	Leaves	Decoction	2	(Idm'hand et al.2020; Gourri et al., 2013).
Poaceae	*Avena sativa L./Avena sterilis L*	Khourtal	Oat	seeds	Infusion, powder, decoction	4	(Belhaj et al., 2021; Hachi et al., 2016; Katiri et al., 2017,Benali et al., 2017).
	Hordeum vulgare L.	Cheir	Barly	seeds, whole plant	Decoction, Powder	6	(Belhaj et al., 2021; Jamila et al., 2014; Gourri et al., 2013; Laadim et al., 2017; Belhaj et al., 2020; Boufous et al., 2017).
	Cynodon dactylon L.Pers	Njem	Quackgrass	Root	Decoction	2	(Gourri et al., 2013; Benali et al., 2017).
	Panicum miliaceum L.	Tafsout	Millet	seeds	Decoction	2	(Belhaj et al., 2021; Idm'hand et al., 2020).
	Panicum turgidum Forssk.	ûmm - rekba	Panic	Stem	Deccoction	1	(Gourri et al., 2013).
	Pennisetum thyphoides	Illane	Mil chandelles	seeds	-	1	(Belhaj et al., 2020).
	Phalaris canariensis L./Phalaris paradoxa L./Lolium perenne L./	Zwan abiyad	-	seeds	Powder, infusion, Maceration	10	(Belhaj et al., 2021; Orch et al., 2015; Gourri et al., 2013; Laadim et al., 2017; Hachi et al., 2016; Bousta et al., 2014; Katiri et al., 2017; El Hachlafi et al., 2020; Belhaj et al., 2020; Mechchate et al., 2020).
	Polypogon monspeliensis	Tûgga	Polypogon de Montpellier	Fruit	-	1	(Gourri et al., 2013).

continued on following page

Table 1. Continued

Family	Plant species	Local Name	English name	Parts Used	Preparation	Number of citation	Réf
Poaceae	*Stipatenacissima L.*	Halfa	-	Whole plant, Leaves	Decoction, Infusion	1	(Jamila et al., 2014).
	Sorghum vulgare Pers.	Bachna	Sorgho	seeds	Infusion, Decoction	1	(Hachi et al., 2016).
	Triticum durum Desf./ Triticumaestivum L.	Lkamh	Wheat	seeds	Infusion	3	(Belhaj et al., 2021; Jamila et al., 2014; Belhaj et al., 2020).
	Zea mays L.	Asngar/Lkbal	-	seeds	Decoction	1	(Katiri et al., 2017).
Polygonaceae	*Emex spinosa*	Lhenzab	-	Leaves, Bulb	Powder	1	(Gourri et al., 2013).
	Rumex palustris Sm.	Homaida	-	Leaves	Decoction	1	Gourri et al., 2013).
Portulacaceae	*Portulaca oleracea*	Rejla	-	Aerial parts	Decoction	2	(Laadim et al., 2017; Katiri et al., 2017).
Punicaceae	*Punica granatum L.*	Reman	Pomegranate	Peel of fruit	Crushed, Decoction, Powder, Infusion	8	(El-Hilaly et al., 2003; Eddouks et al., 2022; Tahraoui et al., 2007; Jamila et al., 2014; Gourri et al., 2013; Eddouks et al., 2002; Eddouks, 2018; Abouri et al., 2012).
Ranunculaceae	*Nigella sativa L.*	Sanouje	Black seeds	seeds	Powder, infusion, Decoction	19	(Belhaj et al., 2021; Jouad et al., 2001; Eddouks et al., 2022; Tahraoui et al., 2007; Orch et al., 2015; Gourri et al., 2013; Benkhnigue et al., 2014; Laadim et al., 2017; Hachi et al., 2016; Bousta et al., 2014; Ziyyat et al., 1997; Skalli et al., 2019; Eddouks et al., 2002; El Hachlafi et al., 2020; Benali et al., 2017; Belhaj et al., 2020; Boufous et al., 2017; Naceiri et al., 2021,Mechchate et al., 2020).
Rhamnaceae	*Ziziphus lotus [L.] Lam.*	Ssedra, Azougar,nbegue	Wild jujube	Leaves, roots, fruits, Flower	Infusion, decoction, powder	14	(Belhaj et al.,2021; Idm'hand et al.,2020; Jouad et al., 2001; El-Hilaly et al., 2003; Barkaoui et al., 2017; Tahraoui et al., 2007; Gourri et al., 2013; Ziyyat et al., 1997; Katiri et al., 2017; Benali et al., 2017; Belhaj et al., 2020; Boufous et al., 2017; Naceiri et al., 2021; Abouri et al., 2012).
	Ziziphus jujuba Mil	Zafzouf	-	Leaves	Decoction	1	(Belhaj et al., 2021).
Rosaceae	*Prunus amygdalus Batsch./Prunus dulcis*	Loz lharr	Bitter almond tree	Leaves, seeds	Decoction, Infusion	16	(Belhaj et al., 2021; Jouad et al., 2001; Eddouks et al., 2022; Tahraoui et al., 2007; Orch et al., 2015; Gourri et al., 2013; Laadim et al., 2017; Ziyyat et al., 1997; Eddouks et al., 2002; El Hachlafi et al., 2020; Belhaj et al., 2020; Boufous et al., 2017; Naceiri et al., 2021; Mechchate et al., 2020; Eddouks, 2018; Skalli et al., 2019).

continued on following page

Table 1. Continued

Family	Plant species	Local Name	English name	Parts Used	Preparation	Number of citation	Réf
Rosaceae	*Prunus armeniaca L.*	Imchmach	Apricot tree	Fruits	Powder	1	(Belhaj et al., 2020).
	Eriobotrya japonica Lindl.	Mzah	-	Leaves	-	2	(Ziyyat et al., 1997; El Hachlafi et al., 2020).
	Prunus amygdalus Stokes	Loz	-	seeds, flower	Powder, Oil	3	(Jamila et al., 2014; Gourri et al., 2013; Skalli et al., 2019).
	C. vulgaris Pers/ Cydonia oblonga Mill.	Sferjel	Quince	Fruit	Fruit	3	(Jouad et al., 2001; Tahraoui et al., 2007; Gourri et al., 2013).
	Fragaria vesca L.	Fraiz berri	Woodland strawberry	Fruit	-	1	(Jouad et al., 2001).
	R. idaeus L.	Toute berri	Red raspberry	Fruit	-	1	(Jouad et al., 2001).
	R. fructicosus L.	Toute chaouki	-	Leaves, Fruit	Decoction, Infusion	2	(Jouad et al., 2001; Orch et al., 2015).
	Malus domestica Borkh./Malus communis	Tûffah	Apple	Fruit	Raw	6	(Jouad et al., 2001; Orch et al., 2015; Laadim et al., 2017; Katiri et al., 2017; Chaachouay et al., 2019; Boufous et al., 2017).
	Eriobotria japonica [Thunb.]	Mzah	loquat	Leaves	Decoction	1	(El-Hilaly et al., 2003).
	Prunus dulcis [Mill.] D. A.	Louz imrzig	Sweet Almond	seeds	Raw	2	(Barkaoui et al., 2017; Katiri et al., 2017).
	Rubus ulmifolius	Tanajelt	Leaves	-	Decoction	2	(Katiri et al., 2017; Id'mhand et al.,2020).
Rubiaceae	*Ruta montana L.*	Fijel, awermi	Common rue	Arial parts, Flower	Decoction, Infusion, Powder	3	(Jouad et al., 2001; Tahraoui et al., 2007; Orch et al., 2015).
	Rubia tinctorium L.	Al'Fouwa	Garance	Root	Decoction	2	(Orch et al., 2015; Gourri et al., 2013).
Rutaceae	*Citrus aurantium L./ Citrus bigaradia Riss.*	Larange	Orange flower	Flowers, fruits	Poultice, juice, infusion	11	(Belhaj et al., 2021; Jouad et al., 2001; Eddouks et al., 2022; Tahraoui et al., 2007; Orch et al., 2015; Gourri et al., 2013; Laadim et al., 2017; Bousta et al., 2014; Eddouks et al., 2002; Belhaj et al., 2020; Naceiri et al., 2021).
	Citrus paradisi Macfad.	Pamplemouss	Grapefruit	Fruits	Juice	1	(Belhaj et al., 2021).
	Ruta chalepensis [L.]	-	Fringed rue	Arial parts	Decoction	1	(Boufous et al., 2017).

continued on following page

Table 1. Continued

Family	Plant species	Local Name	English name	Parts Used	Preparation	Number of citation	Réf
Rutaceae	*Citrus sinensis*	Laymoune	Oranger	Fruit	-	2	(Gourri et al., 2013; Rhattas et al.,2016).
	Citrus amara L.	-	Bitter orange	Fruit	Juice	1	(Belhaj et al., 2021).
	Citrus limette	El hamed lbldi	Sweet lime	Flower	Infusion	1	(Belhaj et al., 2020).
	Haplophyllum vermiculare	l-fijel	-		Powder	2	(Gourri et al., 2013; Benali et al., 2017).
	Ruta montana [L.] L.	Awermi	-	-	Decoction	2	(Belhaj et al., 2021; Ziyyat et al.,1997).
	Ruta graveolen	Rûta	Rue	Stem	-	2	(Gourri et al., 2013; Idm'hand et al., 2020).
Salicaceae	*Salix alba L.*	Ud el-mâ	White willow	Leaves	Decoction	2	(Jamila et al., 2014; Naceiri et al., 2021).
Solanaceae	*Capsicum annuum L.*	Felfla	Pepper	Fruits	Raw	2	(Belhaj et al., 2021; Orch et al., 2015).
	Capsicum frutescens	Felfel Hârr	Chilli pepper	Fruits	-	1	(Hachi et al., 2016).
	Mandragora autumnalis Bertol.	Baydh Al'Ghoul	Mandrake	seeds, Root, Flower	Decoction, Powder	2	(Orch et al., 2015; Belhaj et al., 2021).
	Nicotiana tabacum L.	Nefha	Tobacco	Leaves	Decoction	1	(Orch et al., 2015).
	Solanum lycopersicum L./Lycopersicum esculentum	Maticha	Tomato	Fruits	Raw	2	(Belhaj et al., 2021; Gourri et al., 2013).
	Datura stramonium L. [T]	krenk	Jimson weed	seeds	Maceration	2	(Tahraoui et al., 2007; Gourri et al., 2013).
Sapotaceae	*Argania spinosa [L.] Skeels*	Argan	Argan	seeds	Raw	9	Barkaoui et al., 2017; Gourri et al., 2013; Laadim et al., 2017; Hachi et al., 2016; Bousta et al., 2014; Skalli et al., 2019; Katiri et al., 2017; El Hachlafi et al., 2020; Abouri et al., 2012).
Tamaricaceae	*Tamarix articulata Vahl*	aeris	Athel tamarisk	Leaves	Decoction	1	(Tahraoui et al., 2007).
Taxaceae	*Taxus baccata L.*	Igen	-	Leaves Roots	Decoction	1	(Belhaj et al., 2020).
Theaceae	*Camellia sinensis [L.] Kuntze*	Attay	Tea	Leaves	Infusion	4	(Barkaoui et al., 2017; Rhattas et al., 2016; Gourri et al., 2013; El Hachlafi et al., 2020).

continued on following page

Table 1. Continued

Family	Plant species	Local Name	English name	Parts Used	Preparation	Number of citation	Réf
Thymelaeacea	*Thymelaea hirsuta [L.] Endl./Daphne gnidium L.*	Lmetnan	-	Leaves, stems	Infusion, powder	5	(Belhaj et al., 2021; Ziyyat et al., 1997; Benali et al., 2017; Belhaj et al., 2020; Kharchoufa et al., 2018).
	Thymelaea tartonraira [L.] All	Talazazt	A scrub shrub	Leaves	Decoction	1	(Tahraoui et al., 2007).
Urticaceae	*Urtica. pilulifera L./ Urtica doica*	Harrigua	Stinging nettle	Leaves, seeds, Arial parts	Decoction, Infusion	10	(Jouad et al., 2001; Barkaoui et al., 2017; Tahraoui et al., 2007; Orch et al., 2015; Jamila et al., 2014; Hachi et al., 2016; Ziyyat et al., 1997; Katiri et al., 2017; Benali et al., 2017; Naceiri et al., 2021).
	Urtica urens L.	Tikzint	Annual nettle	Flower, Leaves	Poultice, Infusion, Powder	1	(Belhaj et al., 2020).
	Forskahlea tenacissima	Lessig	-	Leaves	Powder	1	Gourri et al., 2013).
Verbenaceae	*Aloysia citriodora/ Verbena officinalis L.*	Louiza	Lemon verbena	Leaves	Decoction, Infusion	9	(Tahraoui et al., 2007; Orch et al., 2015; Gourri et al., 2013; Kachmar et al., 2021; Bousta et al., 2014; Ziyyat et al., 1997; Skalli et al., 2019; El Hachlafi et al., 2020; Belhaj et al., 2020).
	Vitex agnus-castus [L.]	-	Chaste tree	Leaves, Fruits	Infusion	1	(Boufous et al., 2017).
Vitaceae	*Vitis vinifera L.*	Al-ainab	Grapes	Young leaves	Decoction	5	(Belhaj et al.; 2021; Tahraoui et al., 2007; Gourri et al., 2013; Belhaj et al., 2020; Eddouks, 2018).
Xanthorrhoeaceae	*Asphodelus microcarpus Salzm. & Viv.*	Blaluz	-	Tubers	Decoction, raw	1	(Belhaj et al., 2020).
Zingiberaceae	*Zingiber officinal Rosc.*	Sekinjbir	Ginger	Rhizome	Decoction, Powder, maceration	7	(Orch et al., 2015; Gourri et al., 2013; Skalli et al., 2019; El Hachlafi et al., 2020; Benali et al., 2017; Belhaj et al., 2020; Mechchate et al., 2020).
	Curcuma longa L.	Al-kharkoum	Tumeric	Rhizome	Infusion	1	(Skalli et al., 2019).
	Bassia muricata [L.] Asch.	Tasnant	-	Aerial parts	Infusion	1	(Abouri et al., 2012)
Zygophyllaceae	*Zygophyllum gaetulum Emb. & Maire*	Aaggaya	-	Leaves, Stem	Infusion, Decoction	12	(Belhaj et al.; 2021; Jouad et al., 2001; Eddouks et al., 2022; Tahraoui et al., 2007; Gourri et al., 2013; Hachi et al., 2016; Eddouks et al., 2002; Ziyyat et al.,1997; Boufous et al., 2017; Mechchate et al., 2020; Eddouks, 2018; Abouri et al., 2012).
	Peganum harmala L.	Harmal	Syrian rue	Sedds	Infusion, Powder	8	(Eddouks et al., 2022; Tahraoui et al., 2007; Jamila et al., 2014; Hachi et al., 2016; Ziyyat et al., 1997; Eddouks et al., 2002; Eddouks, 2018

Sroka et al.,2021). On the other hand, decoction, another prevalent technique, entails boiling plant material to extract constituents. It is preferred for tougher plant parts, such as roots, bark, and seeds, facilitating the release of active compounds that might be challenging to extract through alternative methods (Abubakar et al., 2020).

When it comes to the parts used, our evaluation of literature showed that people use different plant parts (leaves, aerial part, fruit, stem, seed, root, bark, flower, stem, rhizome, tuber) otherwise there is a strong preference for the arial parts which is in accordance with the literature for various purposes. The Aerial parts, such as the leave and flowers, often contain a high concentration in bioactive compounds as well as secondary metabolites that have medicinal properties (Okhuarobo et al., 2014).

4.1 Amaryllidaceae

Allium Cepa L.

Regarding the plants listed in this chapter we can figure out that the family of Amaryllidaceae contains only 3 species:

Allium cepa L belongs to the family Liliaceae also known as Onion, is one of the most widely cultivated species of the Allium family (Brahimi et al.,2022). It is native to Southwestern Asia and is now grown globally, including in Morocco where it is the most cultivated species (Ozougwu et al.2011).

Eldin and coworkers (Eldin et al., 2010) confirmed that administration of *Allium cepa* crude hydroalcoholic extract in animal models [alloxan-induced diabetic rats] induce a significant hypoglycemia and good health effects due to the regeneration of pancreatic beta-ll.

In another study, the antidiabetic effect of the juice of Allium Cepa bulbs was investigated using normal and STZ-induced diabetic then verified in STZ-induced diabetic male Wistar rats the results showed that it reduced the fasting blood glucose levels of diabetic rats by 50% (Ojieh et al.,2015) This is in line with the results found by Augustine (Augustine et al.,2020) regarding the Allium *Cepa* juice which induces significantly lowered fasting blood sugar; comparison was made between untreated and treated groups with different dosages.

Allium sativum L

The consumption of garlic in the culinary field has increased because several studies have revealed the effects of garlic on health (Szychowski et al., 2018). The study made by Mirunalini et al on the administration of raw garlic in diabetic patients has shown a significant decrease of blood glucose and TBARS levels (Mirunalini et al., 2011).Another study was conducted on 2002 by Rahmy et al wich results is reducing by 29%, 68% and 57%the serum glucose levels of the garlic-treated diabetic rats in comparison with control diabetic rats, The authors attribute this effect to the formation of colloidal suspension in the intestines and stomach and when the mucilaginous type fiber of garlic is hydrated it affects the gastro-intestinal transit by slowing the glucose absorption (Thakran, et al.,2004).

4.2 Apiaceae

Ammi. visnaga Lam

Ammi visnaga Lam., named as "Bachnikha" locally (Ahmed et al.,2021); distributed in Eastern Mediterranean region, southwestern Asia, North Africa, Atlantic Islands, Europe and North America; known for its health benefits it's used as a treatment for many disease including diabetes; An aqueous extract of A. visnaga given to both normal and streptozotocin diabetic rats has shown to possess a significant hypoglycemic effect more pronounced in STZ than normal rats (Jouad et al.,2002). Additionally, a decoction prepared from the fruits of the A.visnaga has reduced the blood glucose levels by 51% in normoglycemic rats in comparison to an oral hypoglycemic agent [tolbutamide] (Bnouham et al., 2002).

Coriandrum sativum L.

Coriandrum sativum L. Known as coriander is an species cultivated globally; It's seeds have been used traditionally to treat various disease related to rheumatism, indigestion, diabetes (Eidi et al.,2009) The powdered part, aqueous extract and alcoholic extract of Coriandrum sativum produced a significant reduce in the concentration of plasma glucose two weeks after administration, this was more marked in patients who had a history of inadequate control and were taking oral hypoglycemic agents (Waheed et al.,2006) the leaf extract of Coriandrum sativum L. has shown a significant decrease of blood glucose ;It was predicted that the extract has the ability to regenerate β-cells after the destruction by alloxan from where the antidiabetic effect (Sreelatha et al.,2012).The administration of coriander ethanol extract resulted in a significant in fasted healthy animals; In streptozotocin-induced diabetic fasted animals The administration of coriander seed ethanol also decrease significantly serum glucose levels (Eidi et al.,2009).

Foeniculum vulgare Mill.

Foeniculum vulgare Mill. known as fennel is an aromatic plant found in the mediterranean area that has been used for centuries as a medicinal plant (Abou

El-Soud et al., 2011). The leaves aqueous extract of Foeniculum vulgare were evaluated by oral administration in streptozotocin [STZ]-induced and normal diabetic rats;after the fifteenth day of treatment blood glucose levels were decreased in both normal [$p<0.05$] and STZ diabetic rats [$p<0.0001$] (El-Ouady et al., 2020). In another study; Administration of the aqueous extract of Foeniculum vulgare to diabetic rats was effective in reducing the hyperglycaemia from 339.3+0.48 mg/dl,in Streptozotocin-induced diabetic rats Administration of the extract to diabetic rats resulted in correction of the hyperglycaemia from 339.3+0.48 mg/dl to 101.4+ 0.34 mg/dl (Anitha et al.,2014).

4.3 Apocynaceae

Nerium oleander L.

Nerium oleander L. known commonly as oleander is found along the entire Mediterranean coast stretching from southern Portugal in the west to Syria and streambeds of the Western and Southern Anatolia (Bas et al., 2012). The antidiabetic activity of the chloroform and ethanolic extract of Nerium oleander L was studied by Sikarwar et al.in alloxan induced diabetic albino rats and compared to glibenclamide; These extract has shown a significant antidiabetic activity and also prevented body weight loss in diabetic rats (Sikarwar et al ., 2009); A study was conducted in streptozotocin- induced diabetic rats to compare the effect of Nerium oleander extract [250 mg/ kg bw/day for 4 weeks] with sulfonylurea drug [glimepiride] on insulin, glucose levels and some liver enzymes activities, Compared to controls, diabetic rats showed hypoinsulinemia and hyperglycemia . The findings indicate that both glimepiride and the plant extract improved insulin and glucose levels (Mwafy et al.,2011).

4.4 Arecaceae

Phoenix dactylifera

Phoenix dactylifera known as date palm is one of the oldest trees cultivated by human,Recently there has been a surge of interest in various health benefits of dates leading to various in vitro and animal studies as well as the identification and quantification of different types of phytochemicals (Vayalil et al., 2012). The administration of date seed extract has been shown to have a significant antidiabetic effect (El fouhil et al.,2013). A study made by halabi et al demonstrated that the oral administration of date pit fortified bread at 10% and 15% to diabetic rats for 45 days results in a significant reduction in blood glucose level compared to the alloxanized control group, (Halaby et al., 2014) Tohamil et al found that a dose of 400mg/kg resulted in antidiabetic effects from the 6th days onwards,while a dose of 200mg/kg showed similar effect from the 10th day onwards (Tohamy et al.,2013). Additionally it was found that in streptozotocin diabetic rats,combining 10mL dose of date seed extract with insulin leads to a boost in endogenous insulin secretion, potentially contributing to its hypoglycemic effect (El fouhil et al.,2013).

4.5 Asteraceae

Artemisia absinthium L.

Artemisia absinthium L. commonly known as wormwood,is a widely distributed and abundant species withis the artemisia genus. It has been found that it has several beneficial effect on the body.In diabetic individuals,the use of wormwood capsules for 30days decrease the glucose level in comparison to levels on day 0.It has been demonstrated that 1 g dose of Artemisia absinthium led to reductions of 32% glucose wich is very significant (Li et al., 2015). It was observed in another study a that after 10 treatment period significant time-dependent hypoglycemic activity in the medium dose [500 mg/kg BW, $P < 0.01$] and high dose [1000 mg/kg BW, $P < 0.001$] (Li et al., 2015).

Artemisia herba-alba

Artemisia herba-alba commonly known as wormwood,it can be found in various regions,including the steppes and deserts of north africa[Morocco,Tunisia,Al geria..],the middle East [Turkey,jordan,saudi arabia...],Asia....It'is known for its therapeutic and medicinal properties (Houti et al., 2023) .The potential anti-diabetich effect of Artemisia herba-alba was explored on diabetic rats induced by alloxan.by administration 0,39g/kg of an aqueous solution of Artemisia herba-alba wich reduced significantly blood glucose levels similar to regular insulin and repaglinide (Tastekin et al.,2006). This results are in line with what was found by Iriadam et al.(Iriadam et al.,2006). Another study demonstrated that the administration of aqueous extract at a dose of 100mg/KG of body weight, for 60 days in male albino rats induced a significant reduction in glucose levels [$p <0.01$] (Mohamed et al., 2010).

4.6 Brassicaceae

Lepidium sativum L.

Lepidium sativum also known as Garden cress, has significant economic importance due to its roots, leaves and seeds, although it is primarily cultivated for the seeds. The plant has been traditionally used in medicine for its various biological activities. [84]A single administration of Lepidium sativum extract caused a significant decrease in blood glucose levels in normal rats from the first hour after administration,becoming more significant at the sixth hour. Same results for in STZ rats after two hours of oral administration (Eddouks et al.,2005) The methanol extract of lepidium sativum seeds was administered in alloxan-induced diabetic male rats which resulted in decreasing blood glucose levels in diabetic groups (Attia et al.,2019).

4.7 Capparaceae

Capparis spinosa L.

Capparis spinosa commonly known as caper a species of saharo-Arabian and mediterranean origins, characterized by food, ecological, socio economi and ornamental importance. In addition to its uses caper possesses a range of biological activities including antidiabetic, antibacterial, anti-inflammatory (Ammari et al., 2022; Sun et al.,2022). The antidiabetic effect was examined by administration root extract of capparis spinosa; then it was found that effectively it lowered blood glucose levels to the same extent as glibenclamide-treated group (Kazemian et al.,2015). Another study where the caper fruit was examined in type 2 diabetic patients;the results demonstrated a significant reduction in fasting blood glucose levels compared to the control group at the end of the study (Huseini et al.,2013). Eddouks et al studied the hypoglycemia effect of Capparis sminosa by oral administration of its aqueous extract in MLDS diabetic mice. In the control group, blood glucose levels decreased between the first and fourth hour (Eddouks et al., 2017) These findings demonstrate the valuable hypoglycemia potential of c.spinosa .

4.8 Cucurbitaceae

Citrullus colocynthis [C. colocynthis]

Citrullus colocynthis [C. colocynthis] Schrad commonly known as colocynth able to grow in high temperature conditions and found typically in the desert regions of North Africa, The Middle East, and Western Asia valued for its therapeutic properties

(Rahimi et al.,2012; Shahabi et al.,2008; Da Silva et al., 2017). Sebbagh et al conducted a study with seed extract of c of C.colocynthis oil compared to sunflower and olive oils in rats with streptozotocin [STZ]-induced diabete Their finding revealed that a diet rich in C.colocynthis oil resulted in a significantly greater hypoglycemic effect in rats treated with or without insulin compared to diets containing other oils. (Sebbagh et al.,2009). Another study conducted by Agarwal et al. assessing the therapeutic effects of aqueous, chloroform and ethanolic extracts of C. colocynthis root in normal and alloxan-induced diabetic rats; the results showed that the aqueous extract presents a significant hypoglycemia effect compared to the chloroform and ethanol (Agarwal et al.,2012). Another study made by Aghanouri et al in Wistar rats with streptozotocin-induced diabetes which findings are a significant antihyperglycemic effect after 10 days of administration, indicating that C.colocynthis is a promising herbal medicine for the management of diabetic conditions (Aghanouri et al.,2009).

4.9 Lamiaceae

Marrubium vulgare L.

Marrubium vulgare originally from the region between the mediterranean sea and central Asia, is a widely distributed species on all continents.Numerous studies have shown that it has antioxidant properties that make it a potential treatment for cancer, diabetes mellitus and liver disease (Knöss,1999).

In a study by Chakir et al, it was found that oral administration of methanolic extracts of Marrubium vulgare resolute in a significant reductions in blood glucose levels, serum urea, uric acid, creatinine as well improvement in lipid profiles compared to streptozotocin-induced diabetic rats; this results suggest that the antidiabetic effect may be due to extrapancreatic mechanisms, involving the modulation of glycogen synthesis and inhibition of intestinal glucose absorption (Chakir et al.,2015). In another study, the effects of different Marrubium vulgare extracts [Methanol, water and butanol] on autoimmune diabetes mellitus induced by cyclosporine and streptozotocin were investigated; the study found that the extracts caused a significant decrease in blood glucose levels when compared to diabetic mice (Elmhdwi et al., 2015). Hella et al. conducted a series of in vivo experiments on alloxan-induced diabetes during which the aqueous extract of *M. vulgare was administered to* experimental animals for 15 days; The results showed a significant dose-dependent antidiabetic and antihyperlipidemic effects (Boudjelal et al.,2012).

Salvia officinalis

Salvia officinalis also known as sage is a plant originally from the Middle East and Mediterranean regions, but has since become widely distributed throughout the world .Sage has a long history of use in traditional medicine for treating various conditions, such as inflammation, dizziness, tremor, paralysis, diarrhea ulcers, gout, rheumatism,, and hyperglycemia (Ghorbani et al., 2017). Behradmanesh et al made a study to investigate the hypoglycemic effect of salvia officinalisblood glucose, Glycosylated hemoglobin [HbA1c], lipid profile, liver and kidney function tests in patients with type II diabetes. The results showed a significant decrease in blood sugar and cholesterol levels in the salvia officinalis group compared to the control group, while there were no significant changes in glycosylated hemoglobin and fasting sugar levels between the two groups (Behradmanesh et al., 2013) .An other study was carried by Eidi et al to evaluated the oral administration of sage extract and glibenclamide on the serum glucose, lipids, liver and kidney function markers in normal and diabetic rats, it was showed that it reduced serum glucose,lipids and improved insulin levels in streptozotocin-induced diabetic rats; The effects were comparable to those of glibenclamide (Eidi et al., 2009); Another study by Eidi et al where they examined the effects of the methanolic extract of Salvia officinalis on serum glucose levels in fasted rats wich findings showed that Intraperitoneal administration of sage methanolic extract significantly reduces serum glucose concentration only in streptozotocin-induced diabetic faste rats but not in healthy fasted rats without affecting insulin release from the pancreas. (Eiddi et al., 2005).

4.10 Oleaceae

Olea europaea L.

Olea Europaea is commonly known as olive tree grown in various regions, but most of its cultivation, about 98%, takes place in the mediterranean (Somova et al.,2003). Olive oil is the primary fat in the Mediterranean diet, which has been linked to numerous health benefits such as anti-inflammatory, antiviral, hypoglycemia… These benefits can be attributed to the high levels of monounsaturated fatty acids and bioactive phytochemicals such as phenolic compounds, carotenoids, tocopherols, and phospholipids present in olive oil (Talhaoui et al., 2018). Al-Azzawie and Alhamdani studied the hypoglycemic and antioxidant effects of Olea europaea in alloxan-diabetic rabbits which results in a significant decrease in blood glucose levels (Al-Azzawie et al., 2006). Similar Komaki et al found that olive-leaf extract, specially luteolin and oleanolic acid, had inhibitory effects on postprandial blood glucose increase in diabetic rats and humans (Komaki et al.,2003). . In another study

Eidi et al studied the alcoholic extract of olive leaves in diabetic rats; the results indicate that its extract produced a significant hypoglycemia ee=ffect and affected insulin release from the pancreas (Eidi et al.,2009).

5. CONCLUSION

This comprehensive ethnobotanical study sheds light on the rich tradition of medicinal plant use in Morocco for the management of diabetes. Our analysis revealed a diverse array of plant species from various botanical families that have been employed for their antidiabetic properties. These findings underscore the importance of Morocco's ethnopharmacological heritage and the potential therapeutic value of traditional medicinal plants in the context of diabetes treatment. With the increasing global burden of diabetes, exploring and harnessing the knowledge of these plants offers a promising avenue for affordable and accessible healthcare solutions, particularly in resource-limited regions. Furthermore, this study highlights the need for continued research, conservation, and sustainable use of Morocco's medicinal plant biodiversity. As diabetes poses a significant public health challenge in Morocco and beyond, preserving this valuable ethnobotanical knowledge is essential for future generations. Collaborative efforts between ethnobotanists, pharmacologists, and traditional healers can further explore the bioactive compounds and mechanisms of action of these plants, potentially leading to the development of novel antidiabetic agents. In this way, the fusion of traditional wisdom with modern scientific inquiry holds promise in addressing the complex and growing health concern of diabetes in Morocco and globally.

While this study has provided a comprehensive overview of the traditional use of medicinal plants in Morocco for diabetes management, it is important to acknowledge its limitations. The research is based on existing ethnobotanical literature, which may not encompass the full spectrum of plant-based remedies employed by diverse communities across the country. Additionally, while the ethnobotanical data is valuable, it is essential to bridge the gap between traditional knowledge and modern pharmacological research by conducting further studies to isolate and identify the active compounds responsible for the antidiabetic effects of these plants. Moreover, the safety and potential toxicities associated with the use of these medicinal plants need to be thoroughly investigated to ensure their responsible and effective integration into healthcare practices.

Future research in this field should focus on pharmacological and clinical studies to validate the efficacy and safety of these traditional remedies. This includes chemical profiling of active compounds, in vitro and in vivo studies, as well as randomized controlled trials involving diabetic patients.

REFERENCES

Abou El-Soud, N., El-Laithy, N., El-Saeed, G., Wahby, M. S., Khalil, M., Morsy, F., & Shaffie, N. (2011). Antidiabetic activities of Foeniculum vulgare Mill. essential oil in streptozotocin-induced diabetic rats. *Macedonian Journal of Medical Sciences*, *4*(2), 139–146.

Abouri, M., El Mousadik, A., Msanda, F., Boubaker, H., Saadi, B., & Cherifi, K. (2012). An ethnobotanical survey of medicinal plants used in the Tata Province, Morocco. *Int J Med Plants Res*, *1*(7), 99–123.

Abubakar, A. R., & Haque, M. (2020). Preparation of medicinal plants: Basic extraction and fractionation procedures for experimental purposes. *Journal of Pharmacy & Bioallied Sciences*, *12*(1), 1. doi:10.4103/jpbs.JPBS_175_19 PMID:32801594

Aćimović, M., Jeremić, K., Salaj, N., Gavarić, N., Kiprovski, B., Sikora, V., & Zeremski, T. (2020). Marrubium vulgare L.: A phytochemical and pharmacological overview. *Molecules (Basel, Switzerland)*, *25*(12), 2898. doi:10.3390/molecules25122898 PMID:32599693

Agarwal, V., Sharma, A. K., Upadhyay, A., Singh, G., & Gupta, R. (2012). Hypoglycemic effects of Citrullus colocynthis roots. *Acta Poloniae Pharmaceutica*, *69*(1), 75–79. PMID:22574509

Ahmed, S. S. T., Fahim, J., & Abdelmohsen, U. R. (2021). Chemical and biological potential of Ammi visnaga [L.] Lam. and Apium graveolens L.: A review. *Journal of advanced Biomedical and Pharmaceutical Sciences, 4*(3), 160-176.

Al-Azzawie, H. F., & Alhamdani, M. S. S. (2006). Hypoglycemic and antioxidant effect of oleuropein in alloxan-diabetic rabbits. *Life Sciences*, *78*(12), 1371–1377. doi:10.1016/j.lfs.2005.07.029 PMID:16236331

Alaoui, A., & Laaribya, S. (2017). Etude ethnobotanique et floristique dans les communes rurales Sehoul et Sidi-Abderrazak. *Nature & Technology.*

Anitha, T., Balakumar, C., Ilango, K. B., Jose, C. B., & Vetrivel, D. (2014). Antidiabetic activity of the aqueous extracts of Foeniculum vulgare on streptozotocin-induced diabetic rats. *Int J Adv Pharm Biol Chem*, *3*(2), 487–494.

Attia, E. S., Amer, A. H., & Hasanein, M. A. (2019). The hypoglycemic and antioxidant activities of garden cress [Lepidium sativum L.] seed on alloxan-induced diabetic male rats. *Natural Product Research*, *33*(6), 901–905. doi:10.1080/14786419.2017.1413564 PMID:29237302

Augusti, K. T. (1973). Studies on the effects of a hypoglycemic principle from Allium Cepa Linn. *The Indian Journal of Medical Research*. PMID:4757998

Augustine, I. (2020). Hypolipidaemic and Antidiabetic Potency of Allium cepa [Onions] Bulb in Alloxan-Induced Diabetic Rats. Acta Scientific Nutritional Health, 4, 1-8.

Baharvand-Ahmadi, B., Bahmani, M., Tajeddini, P., Naghdi, N., & Rafieian-Kopaei, M. (2016). An ethno-medicinal study of medicinal plants used for the treatment of diabetes. *Journal of Nephropathology*, *5*(1), 44–50. doi:10.15171/jnp.2016.08 PMID:27047810

Bailey, C. J., & Day, C. (1989). Traditional plant medicines as treatments for diabetes. *Diabetes Care*, *12*(8), 553–564. doi:10.2337/diacare.12.8.553 PMID:2673695

Barkaoui, M., Katiri, A., Boubaker, H., & Msanda, F. (2017). Ethnobotanical survey of medicinal plants used in the traditional treatment of diabetes in Chtouka Ait Baha and Tiznit [Western Anti-Atlas], Morocco. *Journal of Ethnopharmacology*, *198*, 338–350. doi:10.1016/j.jep.2017.01.023 PMID:28109915

Bas, A. L., Demirci, S., Yazihan, N., Uney, K., & Ermis Kaya, E. (2012). Nerium oleander distillate improves fat and glucose metabolism in high-fat diet-fed streptozotocin-induced diabetic rats. *International Journal of Endocrinology*. PMID:23251156

Behradmanesh, S., Derees, F., & Rafieian-Kopaei, M. (2013). Effect of Salvia officinalis on diabetic patients. *Journal of Renal Injury Prevention*, *2*(2), 51. PMID:25340127

Belhaj, S., Chaachouay, N., & Zidane, L. (2021). Ethnobotanical and toxicology study of medicinal plants used for the treatment of diabetes in the High Atlas Central of Morocco. *Journal of Pharmacy & Pharmacognosy Research*, *9*(5), 619–662. doi:10.56499/jppres21.1014_9.5.619

Belhaj, S., Dahmani, J., Belahbib, N., & Zidane, L. (2020). Ethnopharmacological and Ethnobotanical study of Medicinal plants in the Central High Atlas, Morocco الدراسة الإيثنوفرماكولوجية والإثنوبوتانية للنباتات الطبية في الأطلس المتوسط الكبير للمغرب. *Ethnobotany Research and Applications*, *20*, 1–40. doi:10.32859/era.20.18.1-40

Benali, T., Khabbach, A., Ennabili, A., & Hammani, K. (2017). Ethnopharmacological prospecting of medicinal plants from the Province of Guercif. *Moroccan Journal of Biology*, *14*(14), 1–14.

Benkhnigue, O., Ben Akka, F., Salhi, S., Fadli, M., Douira, A., & Zidane, L. (2014). Catalogue des plantes médicinales utilisées dans le traitement du diabète dans la région d'Al Haouz-Rhamna. *J Anim Plant Sci*, *23*(1), 3539–3568.

Benlamdini, N., Elhafian, M., Rochdi, A., & Zidane, L. (2014). Étude floristique et ethnobotanique de la flore médicinale du Haut Atlas oriental. *Journal of Applied Biosciences*, *78*(1), 6771–6787. doi:10.4314/jab.v78i1.17

Bnouham, M., Mekhfi, H., Abdelkhaleq, L., & Ziyyat, A. (2002). Ethnopharmacology Forum Medicinal Plants used in the treatment of diabetes in Morocco. *International Journal of Diabetes and Metabolism*, *10*, 33–50.

Boudjelal, A., Henchiri, C., Siracusa, L., Sari, M., & Ruberto, G. (2012). Compositional analysis and in vivo anti-diabetic activity of wild Algerian Marrubium vulgare L. infusion. *Fitoterapia*, *83*(2), 286–292. doi:10.1016/j.fitote.2011.11.005 PMID:22100836

Boufous, H., Marhoume, F., Chait, A., & Bagri, A. (2017). Ethnopharmacological survey of medicinal plants with hallucinogenic effect and used against pain, inflammatory diseases, diabetes and urinary lithiasis in Zagora "Morocco". *Journal of Intercultural Ethnopharmacology*, *6*(4), 342–350. doi:10.5455/jice.20170721062527

Bousta, D., Boukhira, S., Aafi, A., Ghanmi, M., & el Mansouri, L. (2014). Ethnopharmacological Study of anti-diabetic medicinal plants used in the Middle-Atlas region of Morocco. *International Journal of Pharma Research and Health Sciences*, *2*(1), 75–79.

Brahimi, A., El Ouardi, M., Kaouachi, A., Boudboud, A., Hajji, L., Hajjaj, H., & Mazouz, H. (2022). Characterization of the Biochemical Potential of Moroccan Onions. *International Journal of Food Sciences*, *2022*, 2022. doi:10.1155/2022/2103151 PMID:35308133

Chaachouay, N., Benkhnigue, O., Fadli, M., El Ibaoui, H., & Zidane, L. (2019). Ethnobotanical and ethnopharmacological studies of medicinal and aromatic plants used in the treatment of metabolic diseases in the Moroccan Rif. *Heliyon*, *5*(10), e02191. doi:10.1016/j.heliyon.2019.e02191 PMID:31720440

Chakir, A. R. S., Elbadaoui, K., & Alaoui, T. I. (2015). Antidiabetic activities of methanolic extracts of Marrubium vulgare leaves in rats. *International Journal of Pharmaceutical and Phytopharmacological Research*, *4*(5), 258–263.

Da Silva, J. A. T., & Hussain, A. I. (2017). Citrullus colocynthis [L.] Schrad. [colocynth]: Biotechnological perspectives. *Emirates Journal of Food and Agriculture*, *29*(2), 83–90. doi:10.9755/ejfa.2016-11-1764

Daradka, H. M., Abas, M. M., Mohammad, M. A., & Jaffar, M. M. (2014). Antidiabetic effect of Artemisia absinthium extracts on alloxan-induced diabetic rats. *Comparative Clinical Pathology*, *23*(6), 1733–1742. doi:10.1007/s00580-014-1963-1

Eddouks, M. (2017). Contribution to the study of medicinal plants used in the treatment of diabetes, obesity and hypertension in Tafilalet region. *Arabian Journal of Medicinal and Aromatic Plants*, *3*(2), 124–161.

Eddouks, M., Lemhadri, A., Hebi, M., Hidani, A. E., Zeggwagh, N. A., Bouhali, B. E., ... Burcelin, R. (2017). Capparis spinosa L. aqueous extract evokes antidiabetic effect in streptozotocin-induced diabetic mice. *Avicenna Journal of Phytomedicine*, *7*(2), 191. PMID:28348974

Eddouks, M., Maghrani, M., Lemhadri, A., Ouahidi, M. L., & Jouad, H. (2002). Ethnopharmacological survey of medicinal plants used for the treatment of diabetes mellitus, hypertension and cardiac diseases in the south-east region of Morocco. *Journal of Ethnopharmacology*, *82*(2-3), 97–103. doi:10.1016/S0378-8741(02)00164-2 PMID:12241983

Eddouks, M., Maghrani, M., Zeggwagh, N. A., & Michel, J. B. (2005). Study of the hypoglycaemic activity of Lepidium sativum L. aqueous extract in normal and diabetic rats. *Journal of Ethnopharmacology*, *97*(2), 391–395. doi:10.1016/j.jep.2004.11.030 PMID:15707780

Eidi, A., & Eidi, M. (2009). Antidiabetic effects of sage [Salvia officinalis L.] leaves in normal and streptozotocin-induced diabetic rats. *Diabetes & Metabolic Syndrome*, *3*(1), 40–44. doi:10.1016/j.dsx.2008.10.007

Eidi, A., Eidi, M., & Darzi, R. (2009). Antidiabetic effect of Olea europaea L. in normal and diabetic rats. *Phytotherapy Research*, *23*(3), 347–350. doi:10.1002/ptr.2629 PMID:18844257

Eidi, M., Eidi, A., Saeidi, A., Molanaei, S., Sadeghipour, A., Bahar, M., & Bahar, K. (2009). Effect of coriander seed [Coriandrum sativum L.] ethanol extract on insulin release from pancreatic beta cells in streptozotocin-induced diabetic rats. *Phytotherapy Research*, *23*(3), 404–406. doi:10.1002/ptr.2642 PMID:19003941

Eidi, M., Eidi, A., & Zamanizadeh, H. (2005). Effect of Salvia officinalis L. leaves on serum glucose and insulin in healthy and streptozotocin-induced diabetic rats. *Journal of Ethnopharmacology*, *100*(3), 310–313. doi:10.1016/j.jep.2005.03.008 PMID:16125023

El Azzouzi, F., & Zidane, L. (2015). La flore médicinale traditionnelle de la région de Béni-Mellal [Maroc]. *Journal of Applied Biosciences, 91*(1), 8493–8502. doi:10.4314/jab.v91i1.8

El Fouhil, A. F., Ahmed, A. M., Atteya, M., Mohamed, R. A., Moustafa, A. S., & Darwish, H. H. (2013). An extract from date seeds stimulates endogenous insulin secretion in streptozotocin-induced type I diabetic rats. *Functional Foods in Health and Disease, 3*(11), 441–446. doi:10.31989/ffhd.v3i11.33

El Fouhil, A. F., Ahmed, A. M., Darwish, H. H., Atteya, M., & Al-Roalle, A. H. (2011). An extract from date seeds having a hypoglycemic effect. Is it safe to use? *Saudi Medical Journal, 32*(8), 791–796. PMID:21858387

El Hachlafi, N., Chebat, A., Bencheikh, R. S., & Fikri-Benbrahim, K. (2020). Ethnopharmacological study of medicinal plants used for chronic diseases treatment in Rabat-Sale-Kenitra region [Morocco]. *Ethnobotany Research and Applications, 20*, 1–23. doi:10.32859/era.20.2.1-23

El-Hilaly, J., Hmammouchi, M., & Lyoussi, B. (2003). Ethnobotanical studies and economic evaluation of medicinal plants in Taounate province. *Journal of Ethnopharmacology, 86*(2-3), 149–158. doi:10.1016/S0378-8741(03)00012-6 PMID:12738079

El-Ouady, F., Lahrach, N., Ajebli, M., Haidani, A. E., & Eddouks, M. (2020). Antihyperglycemic effect of the aqueous extract of Foeniculum vulgare in normal and streptozotocin-induced diabetic rats. *Cardiovascular & Haematological Disorders-Drug Targets, 20*(1), 54-63.

Eldin, I. M. T., Ahmed, E. M., & Abd, E. H. (2010). Preliminary study of the clinical hypoglycemic effects of Allium cepa [red onion] in type 1 and type 2 diabetic patients. Environmental health insights, 4, EHI-S5540.

Elmhdwi, M. F., Muktar, M. A., & Attitalla, I. H. (2015). Hypoglycemic effects of Marrubium vulgare [Rubia] in experimentally induced autoimmune Diabetes Mellitus. *International Journal of Pharmacy & Life Sciences, 6*(4).

Fakchich, J., & Elachouri, M. (2021). An overview on ethnobotanico-pharmacological studies carried out in Morocco, from 1991 to 2015: Systematic review. *Journal of Ethnopharmacology, 267*, 113200. doi:10.1016/j.jep.2020.113200 PMID:32750461

Farh, M., Kadil, Y., Tahri, E. H., Abounasr, M., Riad, F., El Khasmi, M., & Tazi, A. (2018). Évaluation des effets anxiolytiques, antidépressifs et mnésiques de l'huile des graines de la nigelle chez le rat. *Phytothérapie, 16*(1), 5–13. doi:10.1007/s10298-017-1095-7

Ghorbani, A., & Esmaeilizadeh, M. (2017). Pharmacological properties of Salvia officinalis and its components. *Journal of Traditional and Complementary Medicine, 7*(4), 433–440. doi:10.1016/j.jtcme.2016.12.014 PMID:29034191

Ghourri, M., Zidane, L., & Douira, A. (2013). Usage des plantes médicinales dans le traitement du Diabète Au Sahara marocain. *Journal of Animal &Plant Sciences, 17*(1), 2388–2411.

Giannenas, I., Sidiropoulou, E., Bonos, E., Christaki, E., & Florou-Paneri, P. (2020). The history of herbs, medicinal and aromatic plants, and their extracts: Past, current situation and future perspectives. In *Feed additives*. Academic Press. doi:10.1016/B978-0-12-814700-9.00001-7

Giovannini, P., Howes, M. J. R., & Edwards, S. E. (2016). Medicinal plants used in the traditional management of diabetes and its sequelae in Central America: A review. *Journal of Ethnopharmacology, 184*, 58–71. doi:10.1016/j.jep.2016.02.034 PMID:26924564

Hachi, M., Ouafae, B., Hachi, T., Mohamed, E. B., Imane, B., Atmane, R., & Zidane, L. (2016). Contribution to the ethnobotanical study of antidiabetic medicinal plants of the Central Middle Atlas region. *Lazaroa, 37*, 1-11.

Halaby, M. S., Farag, M. H., & Gerges, A. H. (2014). Potential effect of date pits fortified bread on diabetic rats. *International Journal of Nutrition and Food Sciences, 3*(2), 49–59. doi:10.11648/j.ijnfs.20140302.16

Hamza, N., Berke, B., Umar, A., Cheze, C., Gin, H., & Moore, N. (2019). A review of Algerian medicinal plants used in the treatment of diabetes. *Journal of ethnopharmacology, 238*, 111841.

Huseini, H. F., Hasani-Rnjbar, S., Nayebi, N., Heshmat, R., Sigaroodi, F. K., Ahvazi, M., ... Kianbakht, S. (2013). Capparis spinosa L.[Caper] fruit extract in treatment of type 2 diabetic patients: A randomized double-blind placebo-controlled clinical trial. *Complementary Therapies in Medicine, 21*(5), 447–452. doi:10.1016/j.ctim.2013.07.003 PMID:24050578

Idm'hand, E., Msanda, F., & Cherifi, K. (2020). Ethnobotanical study and biodiversity of medicinal plants used in the Tarfaya Province, Morocco. *Acta Ecologica Sinica, 40*(2), 134–144. doi:10.1016/j.chnaes.2020.01.002

Iriadam, M., Musa, D., Gumushan, H., & Baba, F. (2006). Effects of two Turkish medicinal plants Artemisia herba-alba and Teucrium polium on blood glucose levels and other biochemical parameters in rabbits. *Journal of Cell and Molecular Biology, 5*(1), 19–24.

Jamila, F., & Mostafa, E. (2014). Ethnobotanical survey of medicinal plants used by people in Oriental Morocco to manage various ailments. *Journal of Ethnopharmacology*, *154*(1), 76–87. doi:10.1016/j.jep.2014.03.016 PMID:24685583

Jouad, H., Haloui, M., Rhiouani, H., El Hilaly, J., & Eddouks, M. (2001). Ethnobotanical survey of medicinal plants used for the treatment of diabetes, cardiac and renal diseases in the North centre region of Morocco. *Journal of Ethnopharmacology*, *77*(2-3), 175–182. doi:10.1016/S0378-8741(01)00289-6 PMID:11535361

Jouad, H., Maghrani, M., & Eddouks, M. (2002). Hypoglycemic effect of aqueous extract of Ammi visnaga in normal and streptozotocin-induced diabetic rats. *Journal of Herbal Pharmacotherapy*, *2*(4), 19–29. doi:10.1080/J157v02n04_03 PMID:15277079

Kachmar, M. R., Naceiri Mrabti, H., Bellahmar, M., Ouahbi, A., Haloui, Z., El Badaoui, K., Bouyahya, A., & Chakir, S. (2021). Traditional knowledge of medicinal plants used in the Northeastern part of Morocco. *Evidence-Based Complementary and Alternative Medicine*, *2021*, 2021. doi:10.1155/2021/6002949 PMID:34512779

Katiri, A., Barkaoui, M., Msanda, F., & Boubaker, H. (2017). Ethnobotanical survey of medicinal plants used for the treatment of diabetes in the Tizi n'Test region. *Journal of Pharmacognosy & Natural Products*, *3*(1), 2472–0992. doi:10.4172/2472-0992.1000130

Kazemian, M., & Abad, M., Reza Haeri, M., Ebrahimi, M., & Heidari, R. (2015). Anti-diabetic effect of Capparis spinosa L. root extract in diabetic rats. *Avicenna Journal of Phytomedicine*, *5*(4), 325. PMID:26445712

Keen, R. W., Deacon, A. C., Delves, H. T., Moreton, J. A., & Frost, P. G. (1994). Indian herbal remedies for diabetes as a cause of lead poisoning. *Postgraduate Medical Journal*, *70*(820), 113–114. doi:10.1136/pgmj.70.820.113 PMID:8170881

Kehailou, F. Z., Jabari, M., Labriji, A., El Khair, M. M., Bouzoubaa, H., Ouasmyne, G., ... Mestaghanmi, H. (2021). Quality of Life Assessment with EQ-5D-3L in a Moroccan Diabetic Population. *OAlib*, *8*(5), 1–14. doi:10.4236/oalib.1107357

Kehailou, F. Z., Jabari, M., Labriji, A., & Mestaghanmi, H. (2019). Prevalence de l'anemie chez une population des diabetiques, Casablanca-maroc. *Am J Innov Res Appl Sci*, *9*(1), 109–116.

Kharchoufa, L., Merrouni, I. A., Yamani, A., & Elachouri, M. (2018). Profile on medicinal plants used by the people of North Eastern Morocco: Toxicity concerns. *Toxicon*, *154*, 90–113. doi:10.1016/j.toxicon.2018.09.003 PMID:30243516

Knöss, W. (1999). Marrubium vulgare [White Horehound]: In vitro culture, and the production of diterpene marrubiin and other secondary metabolites. *Medicinal and Aromatic Plants XI*, 274-289.

Laadim, M., Ouahidi, M. L., Zidane, L., El Hessni, A., Ouichou, A., & Mesfioui, A. (2017). Ethnopharmacological survey of plants used for the treatment of diabetes in the town of Sidi Slimane. *Journal of Pharmacognosy and Phytotherapy*, *9*(6), 101–110. doi:10.5897/JPP2016.0437

Lahsissene, H., Kahouadji, A., & Hseini, S. (2009). Catalogue des plantes médicinales utilisées dans la région de Zaër. *Lejeunia, revue de botanique*.

Lee, Y. J., Thiruvengadam, M., Chung, I. M., & Nagella, P. (2013). Polyphenol composition and antioxidant activity from the vegetable plant'Artemisia absinthium' L. *Australian Journal of Crop Science*, *7*(12), 1921–1926.

Li, Y., Zheng, M. I. N., Zhai, X., Huang, Y., Khalid, A., Malik, A., ... Hou, X. (2015). Effect of Gymnema sylvestre, Citrullus colocynthis and Artemisia absinthium on blood glucose and lipid profile in diabetic human. *Acta Poloniae Pharmaceutica*, *72*(5), 981–985. PMID:26665406

Lim, T. K. (2012). *Edible medicinal and non-medicinal plants*. Springer.

Mechchate, H., Es-safi, I., Bari, A., Grafov, A., & Bousta, D. (2020). Ethnobotanical survey about the management of diabetes with medicinal plants used by diabetic patients in region of FezMeknes, Morocco. *Journal of Ethnobotany Research and Applications*.

Mirunalini, S., Krishnaveni, M., & Ambily, V. (2011). Effects of raw garlic [Allium sativum] on hyperglycemia in patients with type 2 diabetes mellitus. *Pharmacologyonline*, *2*, 968–974.

Mohamed, A. E. H. H., El-Sayed, M., Hegazy, M. E., Helaly, S. E., Esmail, A. M., & Mohamed, N. S. (2010). Chemical constituents and biological activities of Artemisia herba-alba. *Records of Natural Products*, 4.

Mohanta, Y. K., Mishra, A. K., Nongbet, A., Chakrabartty, I., Mahanta, S., Sarma, B., Panda, J., & Panda, S. K. (2023). Potential use of the Asteraceae family as a cure for diabetes: A review of ethnopharmacology to modern day drug and nutraceuticals developments. *Frontiers in Pharmacology*, *14*, 14. doi:10.3389/fphar.2023.1153600 PMID:37608892

Msanda, F., El Aboudi, A., & Peltier, J. P. (2005). Biodiversity and biogeography of Moroccan argan tree communities. *Cahiers Agricultures*, *14*(4), 357–364.

Mwafy, S., & Yassin, M. (2011). Antidiabetic activity evaluation of glimepiride and Nerium oleander extract on insulin, glucose levels and some liver enzymes activities in experimental diabetic rat model. Academic Press.

Naceiri Mrabti, H., Bouyahya, A., Naceiri Mrabti, N., Jaradat, N., Doudach, L., & Faouzi, M. E. A. (2021). Ethnobotanical survey of medicinal plants used by traditional healers to treat diabetes in the Taza region of Morocco. *Evidence-Based Complementary and Alternative Medicine*, *2021*, 2021. doi:10.1155/2021/5515634 PMID:33986815

Odeyemi, S., & Bradley, G. (2018). Medicinal plants used for the traditional management of diabetes in the Eastern Cape, South Africa: Pharmacology and toxicology. *Molecules (Basel, Switzerland)*, *23*(11), 2759. doi:10.3390/molecules23112759 PMID:30366359

Ojieh, A. E., Adegor, E. C., Okolo, A. C., Lawrence, E. O., Njoku, I. P., & Onyekpe, C. U. (2015). Hypoglycemic and hypolipidaemic effect of allium cepa in streptozotocin-induced diabetes. *International Journal of Science and Engineering*, *6*(10), 23–29.

Okhuarobo, A., Falodun, J. E., Erharuyi, O., Imieje, V., Falodun, A., & Langer, P. (2014). Harnessing the medicinal properties of Andrographis paniculata for diseases and beyond: A review of its phytochemistry and pharmacology. *Asian Pacific Journal of Tropical Disease*, *4*(3), 213–222. doi:10.1016/S2222-1808(14)60509-0

OMS. (2012).Statistiques sanitaires mondiales. OMS.

Orch, H., Douira, A., & Zidane, L. (2015). Étude ethnobotanique des plantes médicinales utilisées dans le traitement du diabète, et des maladies cardiaques dans la région d'Izarène. *Journal of Applied Biosciences*, *86*(1), 7940–7956. doi:10.4314/jab.v86i1.3

Ouhaddou, H., Boubaker, H., Msanda, F., & El Mousadik, A. (2014). An ethnobotanical study of medicinal plants of the Agadir Ida Ou Tanane province. *Journal of Applied Biosciences*, *84*(1), 7707–7722. doi:10.4314/jab.v84i1.5

Ozougwu, J. C. (2011). Anti-diabetic effects of Allium cepa [onions] aqueous extracts on alloxan-induced diabetic Rattus novergicus. *Journal of Medicinal Plants Research*, *5*(7), 1134–1139.

Painuli, S., Quispe, C., Herrera-Bravo, J., Semwal, P., Martorell, M., Almarhoon, Z. M., Seilkhan, A., Ydyrys, A., Rad, J. S., Alshehri, M. M., Daştan, S. D., Taheri, Y., Calina, D., & Cho, W. C. (2022). Nutraceutical profiling, bioactive composition, and biological applications of Lepidium sativum L. *Oxidative Medicine and Cellular Longevity*, *2022*, 2022. doi:10.1155/2022/2910411 PMID:35096265

Rahimi, R., Amin, G., & Ardekani, M. R. S. (2012). A review on Citrullus colocynthis Schrad.: From traditional Iranian medicine to modern phytotherapy. *Journal of Alternative and Complementary Medicine (New York, N.Y.), 18*(6), 551–554. doi:10.1089/acm.2011.0297 PMID:22784342

Rhattas, M., Douira, A., & Zidane, L. (2016). Étude ethnobotanique des plantes médicinales dans le Parc National de Talassemtane. *Journal of Applied Biosciences, 97*(0), 9187–9211. doi:10.4314/jab.v97i1.5

Roglic, G., & Norris, S. L. (2018). Medicines for treatment intensification in type 2 diabetes and type of insulin in type 1 and type 2 diabetes in low-resource settings: Synopsis of the World Health Organization guidelines on second-and third-line medicines and type of insulin for the control of blood glucose levels in nonpregnant adults with diabetes mellitus. *Annals of Internal Medicine, 169*(6), 394–397. doi:10.7326/M18-1149 PMID:30178023

Sebbagh, N., Cruciani-Guglielmacci, C., Ouali, F., Berthault, M. F., Rouch, C., Sari, D. C., & Magnan, C. (2009). Comparative effects of Citrullus colocynthis, sunflower and olive oil-enriched diet in streptozotocin-induced diabetes in rats. *Diabetes & Metabolism, 35*(3), 178–184. doi:10.1016/j.diabet.2008.10.005 PMID:19264524

Shahabi, S., Hassan, Z. M., Mahdavi, M., Dezfouli, M., Rahvar, M. T., Naseri, M., Jazani, N. H., & Khalkhali, H. R. (2008). Hot and Cold natures and some parameters of neuroendocrine and immune systems in traditional Iranian medicine: A preliminary study. *Journal of Alternative and Complementary Medicine (New York, N.Y.), 14*(2), 147–156. doi:10.1089/acm.2007.0693 PMID:18307392

Sikarwar, M. S., Patil, M. B., Kokate, C. K., Sharma, S., & Bhat, V. (2009). Antidiabetic activity of Nerium indicum leaf extract in alloxan-induced diabetic rats. *Journal of Young Pharmacists : JYP, 1*(4), 330. doi:10.4103/0975-1483.59323

Skalli, S., Hassikou, R., & Arahou, M. (2019). An ethnobotanical survey of medicinal plants used for diabetes treatment in Rabat, Morocco. *Heliyon, 5*(3), e01421. doi:10.1016/j.heliyon.2019.e01421 PMID:30976694

Somova, L. I., Shode, F. O., Ramnanan, P., & Nadar, A. (2003). Antihypertensive, antiatherosclerotic and antioxidant activity of triterpenoids isolated from Olea europaea, subspecies africana leaves. *Journal of Ethnopharmacology, 84*(2-3), 299–305. doi:10.1016/S0378-8741(02)00332-X PMID:12648829

Sreelatha, S., & Inbavalli, R. (2012). Antioxidant, antihyperglycemic, and antihyperlipidemic effects of Coriandrum sativum leaf and stem in alloxan-induced diabetic rats. *Journal of Food Science, 77*(7), T119–T123. doi:10.1111/j.1750-3841.2012.02755.x PMID:22671941

Studzińska-Sroka, E., Galanty, A., Gościniak, A., Wieczorek, M., Kłaput, M., Dudek-Makuch, M., & Cielecka-Piontek, J. (2021). Herbal infusions as a valuable functional food. *Nutrients*, *13*(11), 4051. doi:10.3390/nu13114051 PMID:34836310

Sun, Y., Yang, T., & Wang, C. (2022). Capparis spinosa L. as a potential source of nutrition and its health benefits in foods: A comprehensive review of its phytochemistry, bioactivities, safety, and application. *Food Chemistry*, 135258. PMID:36587515

Szychowski, K. A., Rybczynska-Tkaczyk, K., Gawel-Beben, K., Swieca, M., Karas, M., Jakuczyk, A., ... Gminski, J. (2018). Characterization of active compounds of different garlic [Allium sativum L.] cultivars. *Polish Journal of Food and Nutrition Sciences*, *68*(1), 68. doi:10.1515/pjfns-2017-0005

Tahraoui, A., El-Hilaly, J., Israili, Z. H., & Lyoussi, B. (2007). Ethnopharmacological survey of plants used in the traditional treatment of hypertension and diabetes in south-eastern Morocco. *Journal of Ethnopharmacology*, *110*(1), 105–117. doi:10.1016/j.jep.2006.09.011 PMID:17052873

Tahri, N., El Basti, A., Zidane, L., Rochdi, A., & Douira, A. (2012). Etude ethnobotanique des plantes medicinales dans la province de Settat. Kastamonu University Journal of Forestry Faculty, 12(2), 192-208.v

Talhaoui, N., Trabelsi, N., Taamalli, A., Verardo, V., Gomez-Caravaca, A. M., Fernández-Gutiérrez, A., & Arraez-Roman, D. (2018). Olea europaea as potential source of bioactive compounds for diseases prevention. *Studies in Natural Products Chemistry, 57*, 389-411.

Tastekin, D., Atasever, M., Adiguzel, G., Keles, M., & Tastekin, A. (2006). Hypoglycaemic effect of Artemisia herba-alba in experimental hyperglycaemic rats. *Bulletin of the Veterinary Institute in Pulawy*, 50.

Thakran, S., Siddiqui, M. R., & Baquer, N. Z. (2004). Trigonella foenum graecum seed powder protects against histopathological abnormalities in tissues of diabetic rats. *Molecular and Cellular Biochemistry*, *266*(1/2), 151–159. doi:10.1023/B:MCBI.0000049153.14295.0d PMID:15646037

Tohamy, A. A., Mohammed, R. S., Abdalla, M. S., Ibrahim, A. K., & Mahran, K. F. (2013). The effect of Lupinus Albus [Termis] and Hyphaene thebaica [doum] on some biochemical parameters in streptozotocin-induced diabetic rats. *The Egyptian Journal of Hospital Medicine*, *53*(1), 789–794. doi:10.12816/0001640

Vayalil, P. K. (2012). Date fruits [Phoenix dactylifera Linn]: An emerging medicinal food. *Critical Reviews in Food Science and Nutrition*, *52*(3), 249–271. doi:10.1080/10408398.2010.499824 PMID:22214443

Waheed, A., Miana, G. A., Ahmad, S. I., & Khan, M. A. (2006). Clinical investigation of hypoglycemic effect of Coriandrum sativum in type-2 [NIDDM] diabetic patients. *Pakistan Journal of Pharmacology*, *23*(1), 7–11.

Ziyyat, A., Legssyer, A., Mekhfi, H., Dassouli, A., Serhrouchni, M., & Benjelloun, W. (1997). Phytotherapy of hypertension and diabetes in oriental Morocco. *Journal of Ethnopharmacology*, *58*(1), 45–54. doi:10.1016/S0378-8741(97)00077-9 PMID:9324004

Chapter 6
Ethnobotanical Insights Into the Bioactive Properties of Commercially Important Spice Seeds

Bancha Yingngam
https://orcid.org/0000-0001-7215-9123
Ubon Ratchathani University, Thailand

ABSTRACT

Ethnobotany provides key insights from ancient traditions, particularly focusing on spice seeds. Despite their growing commercial value in food and pharmaceuticals, there is still a need for thorough research into their traditional uses and scientifically validated properties. The objective of this chapter is to compile and integrate current ethnobotanical knowledge with recent scientific findings and highlight the inherent medicinal and pharmacological properties of these plants. A comprehensive examination was conducted on a range of spice seeds. The traditional applications, methods of preparation, and cultural significance of these plants have been described, and the latest scientific findings have confirmed their therapeutic potential. In summary, numerous spice seeds are traditionally considered healing agents, and modern research has confirmed their benefits. The convergence of ethnobotanical knowledge and current scientific advancements has established the foundation for comprehensive integration into modern medicinal practices.

DOI: 10.4018/979-8-3693-6105-4.ch006

INTRODUCTION

Ethnobotany, the study of the relationship between people and plants, is increasingly relevant in contemporary times (Mutinda et al., 2023). As the world has shifted its focus toward sustainable health solutions and natural remedies, ethnobotany serves as a critical link between ancient wisdom and modern healthcare demands (Anwar et al., 2024). In an era dominated by synthetic drugs, which are often associated with numerous side effects, the ethnobotanical approach offers a more encompassing view (Daley & Badal, 2024). It integrates traditional plant-based knowledge, developed over centuries, into contemporary medicinal practices (He et al., 2024). This approach goes beyond merely identifying plant-based remedies; it explores their roles in cultural practices and traditional healing methods (Aggarwal et al., 2024). Ethnobotany provides a deep understanding of how diverse cultures utilize the power of plants for health and wellness, offering invaluable insights for current medical research and the development of natural products (Schultz & Garbe, 2023).

Spices are aromatic or pungent plant-derived substances that are typically obtained from the seeds, berries, bark, roots, or fruits of various plants. They are used to enhance the flavor, aroma, and color of food. Spice seeds, as a specific category of spice, are the dried seeds of plants that are prized for their flavors, aromas, and preservative qualities (P. Das et al., 2023). Unlike leafy herbs, which are often used for their fresh or dried foliage, spice seeds are usually dried and can be used whole or ground into powder. Spice seeds, which are fundamental to culinary traditions worldwide, also hold a crucial position at the intersection of tradition and modernity within medical practices (Varghese & Ray, 2023). In addition to their ability to enhance flavors, these seeds are valued for their medicinal properties across various cultures (Burdock, 2022). Each spice seed, ranging from black cumin to coriander, carries a narrative deeply embedded in the history of traditional medicine (Nath & Debnath, 2023). Recently, these seeds have been revisited by scientists and researchers who are exploring their potential in contemporary medicine (Maqbool et al., 2024). This renewed interest marks a fascinating juncture where ancient wisdom meets modern scientific inquiries, leading to a deeper understanding of the therapeutic potential of these seeds (Debnath et al., 2023). As sources of numerous bioactive compounds, spice seeds are now being examined through the lens of modern science, which validates and expands upon their traditional uses, opening new avenues for natural and effective treatment options (Nath & Debnath, 2023).

The significant issue at hand is the underutilization and insufficient scientific understanding of the full therapeutic potential of many spice seeds (Usmani et al., 2023). While certain plant species, such as black pepper and nutmeg, have undergone extensive study, a large number of these species have been minimally explored (Katoch et al., 2024). This gap represents a missed opportunity to discover potential natural remedies for various health conditions (Arunachalam et al., 2023). Another challenge lies in effectively integrating these ethnobotanical insights into mainstream medicine in a way that is both scientifically credible and ethically sound (Domingo-Fernández et al., 2023). Additionally, the sustainable sourcing and conservation of these spice seeds pose crucial concerns, particularly as their demand escalates (Anand et al., 2023). Despite their recognized commercial significance, achieving a balance between traditional application and the scientifically verified properties of these seeds is an expansive and intriguing domain awaiting further exploration. Addressing these challenges requires a multidisciplinary approach involving ethnobotanists, pharmacologists, chemists, conservationists, and other stakeholders.

This book chapter aims to shed light on these issues by exploring the intersection of ethnobotanical knowledge and modern research to fully utilize the potential of spice seeds in contemporary medicinal practices. The objectives of this chapter are as follows:

(1) To provide a comprehensive overview of the historical and cultural significance of spice seeds across various civilizations, highlighting their crucial role in traditional medicine and culinary arts.
(2) To conduct an in-depth examination of selected spice seeds, focusing on their unique bioactive properties and potential therapeutic applications.
(3) To bridge ancient ethnobotanical knowledge with contemporary scientific research to validate the medicinal benefits of spice seeds.
(4) To emphasize the continuing relevance of spice seeds in today's healthcare context.
(5) To identify and outline future directions of spice seed research, pinpointing areas ready for investigation, thereby establishing a clear direction for future researchers and practitioners.

The organization of this chapter is as follows: Section 2 explores the historical ethnobotanical context, tracing the deep-rooted relationship between spice seeds and human civilizations. Section 3 provides a detailed examination of the selected spice seeds, offering comprehensive insight into their significance and applications. Section 4 highlights the scientific validation and therapeutic potential of these seeds, merging traditional wisdom with modern research findings. Section 5 discusses the promising future of spice seeds in medical applications, while Section 6 focuses on the importance of collaborative efforts, emphasizing the synergy between ethnobotany and modern pharmacology. Finally, Section 7 concludes the chapter and discusses potential future research directions for the study of spice seeds.

HISTORICAL ETHNOBOTANICAL CONTEXT

The annals of human history are replete with references to plants, seeds, and herbs that have been instrumental in shaping societies, guiding religious practices, and offering medicinal solace (Waswa et al., 2024). Ethnobotany, as an interdisciplinary study, explores these interactions, tracing the journey of plant use through the ages. Among the diverse array of plants, spice seeds have etched a unique narrative, interlacing culinary, medicinal, and cultural threads (D'Ambrosio et al., 2023). This section will explore the origins, early uses, and cultural significance of these plants across civilizations, as detailed below.

Origins and Early Uses

The Middle East and Mediterranean Basin

In the Middle East and Mediterranean Basin, ancient civilizations are often hailed, and the use of spice seeds has a long and storied history (D'Ambrosio et al., 2023). This is particularly notable in the early use of seeds such as fennel (*Foeniculum vulgare* Mill.), black cumin (*Nigella sativa* L.), and mustard (*Brassica* spp.) (Jadid et al., 2023). These areas, with their profound historical roots and ancient narratives, offer unique insight into the multifaceted roles of spice seeds (Öztürk et al., 2023). For example, black cumin, a seed native to this region, has been highly esteemed for centuries, not only for its culinary value but also for its medicinal properties (Cilak et al., 2021). Ancient Egyptian civilizations, in particular, provide ample evidence of the significant role played by black cumin (Maqbool et al., 2024). In addition to its use as a spice for flavoring food, black cumin is widely utilized in various health remedies and as a key ingredient in topical ointments (Ramalingam et al., 2024). Its usage in these ancient societies was not limited by its availability; rather, its use also stemmed from a deep understanding of its beneficial properties (Máthé & Turgut, 2023). This highlights how the people of these early civilizations recognized and utilized the diverse benefits of spice seeds, integrating them into their daily lives and healthcare practices.

The reverence of spice seeds in the Middle East and Mediterranean Basin during ancient times was not a coincidence but rather a result of a profound understanding of their health benefits. These seeds are integral to both culinary and medicinal practices, signifying their importance in everyday life as well as in the field of health and wellness (Qari et al., 2024). The use of spice seeds such as black cumin in traditional medicine in this region is particularly noteworthy. They were believed to possess healing properties that could alleviate various ailments, and this belief was deeply embedded in the cultural and health practices of the time (Máthé & Turgut, 2023).

This historical use of spice seeds in the Middle East and Mediterranean Basin is a testament to their enduring legacy (Tabaza & Aburjai, 2024). This finding underscores the longstanding recognition of their therapeutic potential and the knowledge of these early societies in using natural resources for health and healing. The enduring significance of these seeds in both cuisine and traditional medicine highlights the rich heritage of understanding and utilizing nature's bounty for holistic well-being.

Indian Subcontinent

With its rich and diverse cultural heritage, the Indian subcontinent has always placed significant emphasis on the medicinal properties of various natural elements, particularly spice seeds (Maqbool et al., 2024). This is most notably observed in Ayurveda, an ancient and holistic system of Indian medicine. Ayurveda, which translates to 'the science of life', has long recognized the therapeutic benefits of numerous spice seeds and integrated them into both dietary and medicinal practices (Kulkarni et al., 2023). A prominent example of this phenomenon is fenugreek (*Trigonella foenum-graecum*). This spice seed is a fundamental ingredient in Indian cooking and is known for its distinctive aroma and flavor. However, its role extends beyond culinary uses; fenugreek has been a key component in Ayurvedic treatments for a multitude of health issues. This includes managing digestive problems, mitigating inflammation, and even treating skin conditions. The enduring use of fenugreek in the Indian subcontinent is a reflection of a broader cultural understanding that views food as an integral part of overall health and wellness (Akhtari et al., 2024; Mishra et al., 2024).

This holistic approach in Ayurveda, where food and health are deeply intertwined, is a distinctive feature of traditional Indian medicine. From this perspective, every ingredient in the diet is considered for its health benefits and potential healing properties. Spice seeds such as fenugreek fruits not only are flavor enhancers but are also valued for their health-promoting qualities (Archita et al., 2023). The principles of Ayurveda advocate for a balance in the body's fundamental elements, and diet plays a crucial role in maintaining this equilibrium (Nedungadi et al., 2023). Therefore, the consumption of spice seeds is limited not only to merely savoring their taste but also to leveraging their medicinal properties to improve health. This deep-seated belief in the curative power of natural ingredients, such as spice seeds, has passed through generations and remains an integral part of the Indian approach to health and wellness. The use of these seeds in everyday life on the Indian subcontinent is a living testament to the ancient wisdom encapsulated in Ayurveda, demonstrating a seamless blend of gastronomy and therapy (Sanker et al., 2023).

East Asia

In East Asia, particularly in ancient Chinese and Japanese cultures, the use of spice seeds has been deeply rooted in medicinal traditions, as recorded in various historical texts and practices (Li et al., 2024; Maqbool et al., 2024). Seeds such as sesame and perilla were highly valued not only for their culinary contributions but also for their perceived medicinal properties (Park et al., 2023). In these traditions, health and wellness were viewed through a lens that emphasized the balance of internal energies, a concept central to traditional East Asian medicine systems such as traditional Chinese medicine (TCM) and Kampo in Japan (Liu et al., 2023; Okayasu et al., 2024). Sesame seeds, for instance, are often used for their nourishing qualities and are believed to promote longevity and vitality (Zhang, 2021). On the other hand, Perilla seeds are commonly used for their therapeutic effects on respiratory ailments and their antiallergic effects (Cao et al., 2024). The use of these seeds went beyond mere symptom treatment; they were considered vital in maintaining the delicate balance of the body's internal energies, known as 'qi' in TCM, thereby preventing illness and promoting overall health (Ai et al., 2023).

This holistic approach to health in East Asian medicine, where dietary ingredients play a critical role, reflects a deep-seated belief in the power of natural substances to heal and maintain health. In this context, spice seeds are more than just food additives; they are integral components of a comprehensive healthcare system. From this perspective, the body is viewed as an interconnected system in which diet, lifestyle, and the environment all play crucial roles in maintaining health (Zhang et al., 2024). Accordingly, the incorporation of specific spice seeds into the diet was not only about enjoying their flavors but also about harnessing their health benefits to treat and prevent diseases. This approach represents a profound understanding of the interconnectedness of food and health, where each dietary choice is made with an awareness of its impact on the body's overall balance and well-being (Spence, 2024). The enduring legacy of using spice seeds in East Asian medicinal practices is a testament to the region's rich heritage in natural healing and the enduring relevance of this ancient wisdom in contemporary health practices (He et al., 2023).

Southeast Asia

In Southeast Asia, a region known for its rich biodiversity and cultural diversity, the use of spice seeds in both culinary and medicinal practices is deeply ingrained in local traditions (Pal & Lal, 2024). The region's warm, tropical climate provides an ideal environment for a wide variety of spice plants to flourish, leading to a rich tapestry of flavors and medicinal properties embedded in its food culture (Inprasit et al., 2024). Spices such as black or white pepper, turmeric, coriander, and cumin are not fundamental to the flavor profiles of Southeast Asian cuisine and are also known for their healing properties (Moulick et al., 2023). For instance, turmeric, with its active compound curcumin, is widely used in this region for its anti-inflammatory and antioxidant properties (Iweala et al., 2023). It has been a staple in traditional medicine systems such as Jamu in Indonesia and traditional Thai medicine for centuries (Shaik et al., 2023). Similarly, coriander and cumin are known for their digestive and antimicrobial properties (Suttisansanee et al., 2023). In Southeast Asian culture, these spice seeds are integrated into daily cooking, which provides insight into their health benefits. This region exemplifies how culinary practices are intertwined with traditional knowledge of natural remedies, with spice seeds playing a dual role in both nourishing the body and preventing or treating various ailments.

Traditional Thai medicine, for instance, is deeply rooted in the country's rich cultural and botanical heritage. Influenced by ancient Indian (Ayurveda) and Chinese medicinal practices, it contains a variety of native herbs and spices (Chansiw et al., 2024). These natural ingredients are used not only for their flavor during cooking but also for their therapeutic properties in healing and maintaining health. For example, black and white peppers, both of which are derived from the plant *Piper nigrum* L., have distinct uses in traditional Thai medicine, as indicated by their unique properties and flavors. In Thai culture, these spices not only are considered culinary staples but also integral to various medicinal practices. Black pepper is highly valued for its ability to warm plants and is often used to stimulate digestion and treat gastrointestinal issues such as indigestion, flatulence, and constipation. Due to their pungency and heating qualities, black pepper plants are believed to stimulate their appetite and enhance the absorption of other medicinal compounds. It is also used to treat colds and respiratory issues, as it acts as a natural decongestant. Additionally, black pepper is sometimes applied topically in the form of pastes or oils to alleviate muscle pain and arthritis due to its anti-inflammatory and analgesic properties (Inprasit et al., 2024). White pepper, the mature and dried fruit of the same plant, is used differently in traditional Thai medicine. It tends to be less pungent than black pepper and is often used for its more subtle warming properties. White pepper is frequently employed in the treatment of stomachache and minor digestive disorders. The stomach of pepper plants is considered to be gentler than that of black pepper plants, making it suitable for people with more sensitive digestive systems. Additionally, white pepper is used to help alleviate coughs and reduce phlegm, as well as to promote circulation and relieve pain, particularly in patients with conditions such as headaches or menstrual discomportions (Chansiw et al., 2024; Nutmakul & Chewchinda, 2023).

This holistic approach to health and wellness in Southeast Asia is further exemplified in the traditional medicinal practices of the region, where spice seeds are key components. These practices are based on a deep understanding of the natural world and beliefs about the healing power of plants. In traditional medicine, spice seeds are often used in combination with other herbs and plants, creating remedies that are tailored to treat specific ailments. This approach is characterized by a focus on restoring balance within the body, similar to what has been done in other Eastern medical philosophies. The use of spice seeds in this context not only addresses specific health issues but also maintains overall well-being and prevents illness. Knowledge of these medicinal properties of spice seeds has progressed through generations, reflecting the rich heritage of natural healing. Today, with the growing interest in natural and holistic health practices worldwide, the traditional use of spice seeds in Southeast Asia is receiving increased attention, highlighting the region's contribution to the global understanding of natural health and wellness.

Africa

The history of spice seeds in Africa is deeply intertwined with the continent's rich and diverse cultural tapestry. Unlike Europe, where many spices were imported, Africa had its own indigenous spice seeds, in addition to those introduced through trade (Mapeka et al., 2024). For example, the Ethiopian Highlands are known for their indigenous spices, such as korarima, a type of cardamom, and the long pepper. These spices are used not only for their flavor but also for their preservative qualities in a climate that often demands such properties (Hordofa & Tolossa, 2020). African spice seeds, such as grains of paradise, a West African spice with a peppery flavor, and Ethiopian berbere, a blend of several spices, are integral to traditional cuisines and are used in a variety of dishes to add depth and heat (Skeels, 2017). Intercontinental trade routes, particularly those across the Sahara Desert, facilitated the exchange of spices between North Africa and Sub-Saharan regions, enriching the culinary traditions of the continent (Lucas-Sánchez et al., 2023). Spices were traded alongside gold and salt, which were among the most coveted commodities in the ancient and medieval world. The Swahili coast, a key player in the spice trade, served as a hub where African, Arab, and later European traders exchanged goods and cultural practices, including culinary traditions (Marchant, 2021).

In terms of early uses, spice seeds in Africa were not confined solely to culinary applications but also played a significant role in social, medicinal, and religious contexts. Traditional African medicine has made extensive use of local herbs and spices (Nyirenda & Chipuwa, 2024). For instance, many spices have been used for their supposed healing properties in various local medicinal practices. These practices are often intertwined with spiritual beliefs, where spices are used in rituals and as offerings to deities or ancestors. In some African cultures, spices were integral to rites of passage, weddings, and other ceremonies, symbolizing various cultural values such as purification, protection, or fertility. The use of spice seeds in traditional African beauty and health regimens also underscores their importance beyond the kitchen. For example, certain spices were used in skincare and haircare, taking advantage of their aromatic and therapeutic properties. This multifunctional use of spices highlights their significance in the daily lives of various African communities, firmly establishing them as a cornerstone of African heritage and identity (Aworh, 2023).

Europe

Spice seeds first made their way into Europe through a complex network of trade routes, most notably the Silk Road. This ancient network of trade routes connects the East and West, facilitating not only the exchange of goods but also the sharing of knowledge and cultural practices (Wang et al., 2023). The spice trade, primarily driven by the demand for exotic flavors and medicinal properties, played a pivotal role in shaping early European cuisine and medicine. Spices such as pepper, cinnamon, cardamom, and nutmeg, which are native to the East Indies and the Indian subcontinent, were highly prized in European markets. Their rarity and the long, perilous journeys they had to undertake to reach European shores added to their value. By the Middle Ages, these spices were not only culinary ingredients but also symbols of wealth and status. They were often used by the aristocracy to demonstrate affluence, both in lavish feasts and as a form of currency in trade and dowries (Divakaran et al., 2018).

The early use of spice seeds in Europe was not limited to culinary endeavors; spice seeds also held significant medicinal value. Medieval European medicine heavily relied on the principles of the four humors, and spices were thought to possess properties that could balance these bodily fluids. For instance, pepper is believed to aid digestion and is used to treat ailments related to the stomach. Similarly, nutmeg was valued for its supposed ability to ward off plague and other diseases. Monasteries played a crucial role in the cultivation and study of these spices and herbs, as they were often the centers of learning and medicine during this period (Larsson & Foley, 2023). The use of spices in medicine has also extended to the realms of spirituality and superstition. For example, certain spices were used in rituals and were thought to have protective effects on evil spirits. This blend of practical and mystical uses of spices reflects the multifaceted nature of their impact on European society, transcending mere culinary use and highlighting their integral role in the cultural and spiritual fabric of the time.

South America

South America's relationship with spice seeds has a unique narrative that is deeply rooted in its indigenous cultures and biodiverse landscapes. Prior to European contact, several South American civilizations, such as the Incas, Mayans, and Aztecs, had a rich tradition of using native spices and herbs (Caceres Guido et al., 2015). One of the most notable of these varieties was the aji superpicante, or superhot chili pepper, which has been cultivated in this region for thousands of years (Tejeda et al., 2022). Archaeological evidence suggests that chili peppers were used in various parts of South America. These peppers not only are used for heat and flavor when cooking but also hold significant value in rituals and as medicinal plants (Roa Solís et al., 2023). Other native spices include annatto, which is derived from the seeds of the achiote tree and was used by indigenous peoples for culinary coloring, as well as for body paint and healing rituals (da Silva et al., 2023). The Amazon rainforest, a treasure trove of biodiversity, is another source of various herbs and spices, many of which are still being discovered and understood by the outside world. The indigenous peoples of South America had intricate knowledge of these plants, utilizing them in their daily lives for food, medicine, and spiritual practices (Ferreira et al., 2023).

The arrival of Europeans in South America brought about a significant exchange of spices and culinary practices, often referred to as the Columbian Exchange. This period saw the introduction of European, African, and Asian spices to the continent, which were quickly adopted and integrated into local cuisines. However, the impact was bidirectional; South American spices such as chili peppers and vanilla (originally from Mexico but also cultivated in South America) made their way to Europe and Asia, where they were enthusiastically received and became staples in various cuisines worldwide (Hancock, 2022). In terms of early uses, many South American spices were used medicinally by indigenous peoples. For instance, chili peppers were used for their analgesic properties, and coca leaves were chewed for energy and to alleviate altitude sickness. In religious contexts, certain spices and herbs play a role in rituals and ceremonies and are often believed to carry spiritual significance. The use of these spices was deeply embedded in the social and cultural fabric of the indigenous communities, reflecting a profound understanding of and respect for the natural world (Busnel & Manrique López, 2023).

North America

The story of spice seeds in North America is intrinsically linked with the continent's indigenous peoples and their deep understanding of the local flora. Long before European colonization, native American tribes were adept at utilizing a wide range of native plants and seeds for culinary, medicinal, and ceremonial purposes. One of the most significant among these was the sunflower seed, which is native to North America and used by tribes such as the Aztecs and Otomi in present-day Mexico. These seeds not only are vital food sources but are also used in oil extraction and as dyeing agents (Guzmán-Gutiérrez et al., 2023). Another notable example is the use of sassafras, a tree whose leaves and bark were used by native Americans for flavoring food and as a health remedy. The indigenous peoples had a profound knowledge of the land and the seasons, which guided their use of these plants. These plants practice sustainable harvesting methods, ensuring the preservation and continuity of these natural resources (Mariappan et al., 2023). The spice trade in North America took a different turn than Europeans did, who introduced their own spices and incorporated indigenous spices into their diets. This resulted in a fusion of culinary traditions, enriching the gastronomic landscape of the continent. In early North American societies, the use of spice seeds extended beyond culinary applications, playing a significant role in health and spiritual practices. Native American medicine, which is deeply rooted in a holistic understanding of health, utilizes a variety of local herbs and spices for healing. For instance, various tribes use spicebush berries as a remedy for colds and fever, and chili peppers are used for their pain-relieving properties. These plants were often used in conjunction with rituals and ceremonies, underscoring their spiritual significance. Spices and herbs are also integral to rites of purification and protection and are often burned in smudging ceremonies to cleanse spaces and individuals with negative energy. Early European settlers, influenced by Native American herbal knowledge, began incorporating these native spices into their medicinal practices. This blend of indigenous and European herbal medicine laid the foundation for the diverse botanical knowledge that characterizes North American ethnobotany today. Thus, the early use of spice seeds in North America reflects a deep connection with the natural world, showcasing the intricate ways in which these plants were woven into the fabric of daily life across different cultures.

Cultural Significance Across Civilizations

Religious Practices

Throughout history, spice seeds not only have been valued for their culinary and medicinal properties but also have played a significant role in various religious practices. One such seed is black cumin, which holds a special place in Islamic traditions. In Islamic scriptures, black cumin is often cited for its healing properties, which are believed to be effective in both physical and spiritual realms (Ali & Sultan, 2023). This belief is rooted in the famous saying of the Prophet Muhammad, which states that black cumin can cure every disease except death. This high regard for black cumin in Islamic culture transcends its use as a mere spice or health remedy; it becomes a symbol of divine healing and protection (Saad, 2024). The integration of black cumin in religious practices, such as in the form of oils used in anointing or in herbal preparations for internal consumption, reflects its revered status. This spiritual significance of black cumin is indicative of a broader phenomenon in which spice seeds are interwoven with religious beliefs and rituals, symbolizing not only health and nourishment but also spiritual wellness and divine blessing. The incorporation of spice seeds in religious practices extends beyond Islam and can be observed in various other religious traditions around the world. For instance, in Hinduism, certain seeds, such as sesame seeds, are used in rituals and offerings to deities, symbolizing purity, prosperity, and spiritual awakening (Sachan & Chandra, 2023). In the Christian tradition, mustard seeds are often referenced, symbolizing faith and the potential for growth from small beginnings. This cross-cultural reverence for spice seeds in religious contexts highlights their significance beyond the physical realm. These seeds serve as metaphors and physical embodiments of religious and spiritual principles, playing an integral role in the expression of faith and devotion (Panchuk, 2023). The use of spice seeds in religious ceremonies and rituals often has deep symbolic meanings, reflecting the beliefs and values of the culture. This spiritual dimension of spice seeds, in addition to their culinary and medicinal uses, underscores the multifaceted roles these natural elements play in human life, bridging the material and spiritual worlds.

Trade and Economy

The spice trade, a driving force in the historical development of trade and economy, was significantly influenced by the high demand for spice seeds. Spices such as black pepper, cardamom, cinnamon, and cloves were not only culinary luxuries but also commodities of immense value. This quest for spices played a pivotal role in shaping the economic and political landscapes of ancient and medieval times. The allure of these aromatic treasures led to the establishment of extensive trade routes that connected distant lands, from the Far East to the Mediterranean and beyond. The spice trade was not just about the exchange of goods; it represented a complex network of cultural, economic, and political interactions (Wang et al., 2023). This prompted explorers to embark on perilous voyages across unknown seas, leading to the discovery of new lands and the encounter of diverse civilizations. The lucrative nature of the spice trade attracted merchants and adventurers alike, often resulting in fierce competition and conflicts. European powers, in particular, vied for control of the spice trade routes, leading to the age of exploration and colonization. The impact of the spice trade on the global economy and politics cannot be overstated—it was a catalyst for globalization, laying the groundwork for the interconnected world we know today.

The economic significance of the spice trade has also had profound impacts on the societies involved. For producing regions, especially in the East Indies and the Indian subcontinent, spices were a crucial source of wealth and power. The cultivation and export of spice seeds became a major economic activity, shaping the social and economic structures of these societies. On the other hand, for importing regions, particularly in Europe, spices were a luxury item that was often worth more than their weight in gold. They symbolized wealth and social status, and their demand led to the development of new economic systems and trading companies, such as the Dutch East India Company and the British East India Company. These companies not only controlled the spice trade but also wielded enormous political power, often acting as quasigovernmental entities in their own right. The spice trade thus had far-reaching implications, affecting everything from daily cuisine to the course of empires. It was a key factor in the emergence of a global economy, where the exchange of goods was intertwined with cultural exchanges, the spread of ideas, and the shaping of world history (Bennett, 2023).

Cultural Festivities and Traditions

Spice seeds play a vibrant and symbolic role in cultural festivities and traditions across the world and are deeply intertwined with the social customs and heritage of various communities. In India, for instance, mustard seeds hold a significant place in many festivals and rituals. They are often used in traditional ceremonies, symbolizing fertility, protection, and good luck. During certain Hindu festivals, mustard seeds are scattered around homes or used in rituals to ward off evil spirits and bring good fortune (Sidhu, 2023). This practice not only is a mere supposition but also reflects a deep-rooted cultural belief in the power of nature and its connection to human life. Similarly, in southern parts of India, seeds such as sesame and jagged fruits are used in making sweets during festivals such as Sankranti, symbolizing the removal of bitterness and the introduction of sweetness into life. These traditions showcase how spice seeds are more than just culinary ingredients; they are integral components of cultural identity and embedded with meanings and values that transcend generations (Kulkarni et al., 2023).

In Europe, the use of spice seeds in cultural festivities is equally prominent, especially in culinary traditions. Caraway seeds, for example, are staples in European baked goods, particularly during celebrations and festive seasons. In countries such as Germany and Austria, caraway seeds are commonly used in breads and pastries served during Christmas and other celebrations, symbolizing love, protection, and health (Sarand et al., 2023). The inclusion of these seeds in festive cooking is a way of imparting not only flavor but also blessings and good wishes to family and friends. This tradition of incorporating spice seeds into celebratory foods is a testament to their enduring cultural significance. It highlights the role of food not only in nourishing the body but also in nourishing the soul, bringing people together, and preserving cultural heritage. Across different cultures, the use of spice seeds in festivities and traditions underscores their importance as carriers of symbolism and harbingers of goodwill, which are deeply embedded in the tapestry of human societal customs.

SELECTED SPICE SEEDS: A DEEPER LOOK

With their myriad flavors and therapeutic properties, the vast world of spice seeds offers an intriguing exploration of the confluence of tradition and modern science. This section provides a comprehensive overview of specific spice seeds, examining their historical applications and juxtaposing these with contemporary research.

Black Cumin

Black cumin, derived from *Nigella sativa* L. (Table 1), holds a special place in the tapestry of traditional medicine across various cultures, particularly in the Middle East, Southeast Asia, and parts of Europe (Ramalingam et al., 2024; Ramineedu et al., 2023). Traditionally, black cumin has been revered for its wide array of medicinal properties and is often referred to as a "cure for everything but death" in the Islamic literature (Saad, 2024). It has been used for centuries to treat a variety of ailments, including asthma, hypertension, diabetes, inflammation, and even certain types of cancer (Kumar et al., 2024; Maqbool et al., 2024). In traditional Indian medicine, or Ayurveda, black cumin has been utilized for its digestive and anti-inflammatory properties and is often prescribed for issues related to the stomach and liver (Hossain et al., 2024; Ramalingam et al., 2024). It has also been used as an immune system booster and is a common ingredient in cold and flu remedies (Deshmukh, 2024). Furthermore, these seeds have been applied topically in various cultures for skin conditions and to promote wound healing, demonstrating the diverse applications of black cumin in traditional healing practices (Kmail et al., 2023).

Black cumin seed essential oil is composed of a variety of volatile compounds. The most prominent component is thymoquinone, which typically constitutes approximately 30-48% of the oil. Another major constituent is p-cymene, ranging from 7-15%. Thymohydroquinone and dithymoquinone are also significant and usually constitute approximately 1-5% of the oil. Other notable constituents include α-pinene (1-3%), α-thujene (0.5-3%), and β-pinene (0.5-3%) (Figure 1). The exact composition can vary depending on factors such as the origin of the black cumin seeds, the climate and soil where they are grown, and the method of oil extraction. These volatile compounds confer distinct medicinal properties and a sharp, slightly bitter flavor to black cumin seed oil (Gholamnezhad et al., 2016; Nautiyal, 2019). In recent years, modern scientific research has begun to explore and substantiate many of the traditional claims surrounding black cumin. One of the key bioactive components of black cumin is thymoquinone, which has been the focus of much related research (Anlar & Bacanli, 2020). Studies have shown that thymoquinone has potent anti-inflammatory and antioxidant properties, contributing to its potential therapeutic effects (Khalifa et al., 2021). Research has also explored its potential in managing asthma, with findings suggesting that it may help patients relax tracheal muscles and reduce inflammation in the airways (Li et al., 2023). In the field of metabolic disorders, black cumin has been studied for its potential to improve blood sugar and cholesterol levels, thereby offering benefits for managing conditions such as diabetes and heart disease (Ramineedu et al., 2023). Furthermore, there is growing interest in the anticancer effects of black cumin, as preliminary studies have indicated that thymoquinone may induce cell death in certain types of cancer cells (Shabani et al., 2023). While additional research is needed to fully understand the underlying mechanisms and efficacy of black cumin in these areas, the findings of the present study provide a scientific basis for its traditional use and suggest its potential for incorporation into modern therapeutic strategies. The exploration of black cumin in scientific research not only validates its historical significance but also opens new avenues for its application in contemporary healthcare.

Fenugreek

Fenugreek, derived from the plant *Trigonella foenum-graecum* L. (Table 1), has a long and rich history in traditional medicine, particularly in the Indian subcontinent and Middle Eastern regions. Traditionally, fenugreek has been used for a variety of health-related purposes. In Ayurveda, the ancient Indian system of medicine, fenugreek, is valued for its digestive and anti-inflammatory properties. It has been commonly used to treat digestive ailments, such as indigestion and constipation, and has also been used as a natural remedy for conditions such as arthritis and sore throat (Ahmad et al., 2023; Akhtari et al., 2024). Apart from its medicinal uses, fenugreek has been a staple in culinary practices, particularly in Indian, Middle Eastern, and North African cuisines, where it is used both as an herb (leaves) and as a spice (seeds) (Bakhtiar et al., 2024). The seeds have a slightly bitter taste and are often cultivated in spice mixes or curries or as flavoring agents in various dishes.

The volatile compounds in fenugreek seed essential oils are distinct. The most predominant component is neryl acetate, which typically represents approximately 15-60% of the oil. Other major constituents include sotolon (also known as fenugreek lactone), which accounts for approximately 10-20%, and anethole, which ranges from 5-15%. Trigonelline, a unique alkaloid, and 3-hydroxy-4,5-dimethyl-2(5H)-furanone are also present in notable amounts, each constituting approximately 1-5%. Additional minor components include α-pinene, β-pinene, and limonene, each generally found in smaller percentages (Figure 2) (Dhull et al., 2024; Liang et al.,

Table 1. A comprehensive list of commercially important spice seeds, their botanical names, and primary regions of cultivation

Common Name	Botanical Name	Family	Primary Regions of Cultivation	Reference
Black cumin	*Nigella sativa* L.	Ranunculaceae	Middle East, South Asia, North Africa	Ramalingam et al. (2024); Ramineedu et al. (2023); Yagoub (2023)
Fenugreek	*Trigonella foenum-graecum* L.	Fabaceae	Indian subcontinent, North Africa, Middle East	Bakhtiar et al. (2024)
Mustard	*Brassica* spp.	Brassicaceae	Asia, Europe, North America	Nguyen et al. (2023)
Black pepper	*Piper nigrum* L.	Piperaceae	India, Thailand, Vietnam, Indonesia, Brazil	Hadi et al. (2023)
Dill	*Anethum graveolens* L.	Apiaceae	Europe, Asia	Cătunescu et al. (2023)
Cardamom	*Elettaria cardamomum* (L.) Maton	Zingiberaceae	India, Guatemala, Sri Lanka, Papua New Guinea	Ikechukwu et al. (2023); Khan and Aslam (2023b)
Celery	*Apium graveolens* L.	Apiaceae	Mediterranean region, North America, Europe	Hussain et al. (2023)
Nutmeg	*Myristica fragrans* Houtt.	Myristicaceae	Indonesia, Grenada, Sri Lanka, India	Khanam et al. (2023); Trifan et al. (2023)
Caraway	*Carum carvi* L.	Apiaceae	Europe, North Africa, Western Asia	Kilian et al. (2023); von Maydell et al. (2024)
Fennel	*Foeniculum vulgare* Mill.	Apiaceae	Mediterranean, Middle East, Indian subcontinent	Rafieian et al.
Coriander	*Coriandrum sativum* L.	Apiaceae	Mediterranean region, Middle East, Indian subcontinent	Machado et al. (2023)

2023). Modern scientific research has begun to corroborate many of these traditional applications of fenugreek and shed light on its potential health benefits. One of the most studied aspects of fenugreek medicine is its potential to aid in blood sugar regulation, making it a topic of interest in diabetes research. Seeds contain a type of soluble fiber that can slow carbohydrate digestion and absorption, thus helping to lower blood sugar levels (Correia et al., 2023). Additionally, fenugreek seeds have been found to contain compounds such as diosgenin and saponins, which may have anti-inflammatory and cholesterol-lowering effects. This makes fenugreek a candidate for research on managing cardiovascular health (Ahmad et al., 2023; Singh et al., 2023). There is also emerging evidence suggesting that fenugreek may have a positive effect on lactation, increasing milk production in breastfeeding mothers. Moreover, the potential use of fenugreek for improving skin health and reducing menstrual cramps is being explored (Akhtari et al., 2024). While related research is still ongoing, the findings thus far provide a scientific basis for the traditional uses

Figure 1. Selected chemical compounds found in black cumin seeds

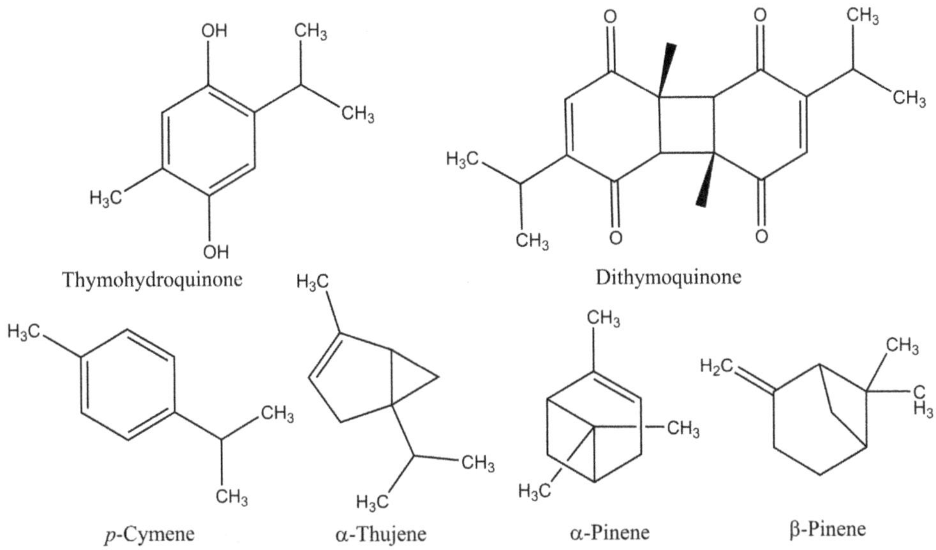

of fenugreek and open up new possibilities for its application in modern medicine. The continued study of fenugreek could lead to new, natural, and effective treatments for a range of health conditions, bridging the gap between traditional wisdom and contemporary scientific understanding.

Mustard

Mustard seeds, derived from various plants in the genus *Brassica* spp., have been used extensively in both traditional culinary and medicinal practices across various cultures, particularly in Asia and Europe (Nguyen et al., 2023). Traditionally, mustard seeds are known for their strong, pungent flavor and are used as a spice in cooking, contributing to the flavor profile of many dishes, including pickles, curries, and sauces. In Indian Ayurvedic medicine, mustard seeds have been used for their supposed therapeutic effects, such as stimulating digestion, promoting circulation, and serving as a remedy for colds and muscle pain (Maqbool et al., 2024). Seeds have been applied in the form of poultices to relieve muscle aches and pains, and the warm growth of these plants is believed to help in clearing respiratory congestion. Additionally, in many cultures, mustard seeds, which are believed to stimulate hair growth and improve scalp health, have been used as hair tonics (Ahmed et al., 2020).

Mustard seed essential oil, derived primarily from *Brassica nigra* (black mustard) or *Brassica alba* (white mustard), contains a unique composition of

Figure 2. Selected chemical compounds found in fenugreek seeds

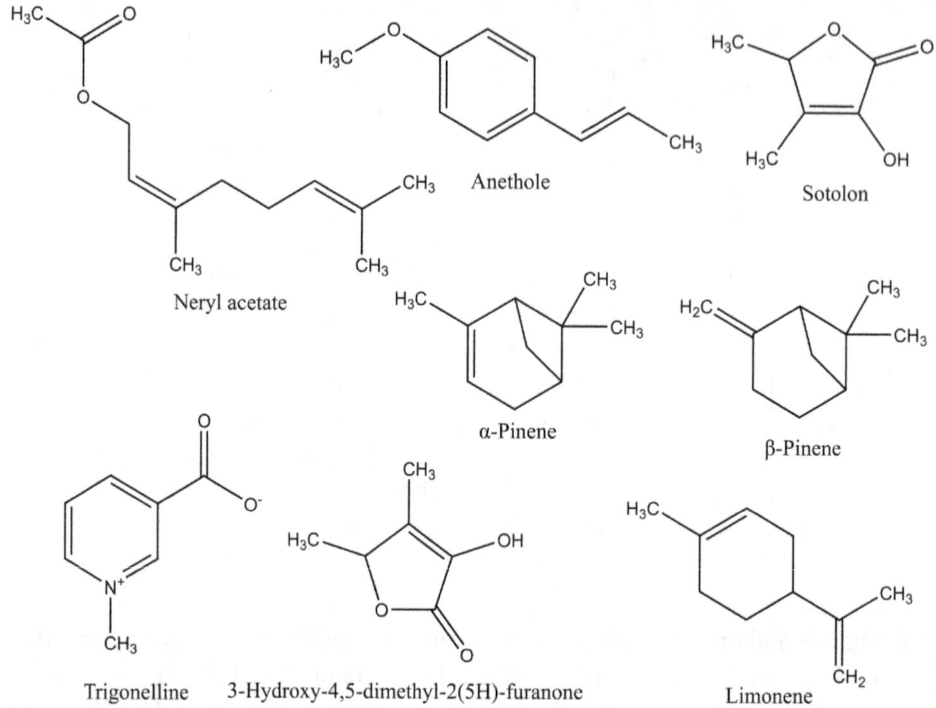

volatile compounds. The most significant component in black mustard oil is allyl isothiocyanate, which can constitute up to 90-95% of the oil. In white mustard oil, the primary compound is sinigrin, which is a glucosinolate precursor to allyl isothiocyanate, and its content can vary greatly (Figure 3). Other minor constituents include various other isothiocyanates, thiocyanates, and oxazolidinethiones, each typically present in much smaller percentages. The exact composition of these volatile compounds can vary depending on factors such as the type of mustard seed, the soil and climate where it is grown, and the method of oil extraction (Brindisi et al., 2023; Cerón et al., 2023).

Contemporary scientific studies on mustard seeds have started to confirm and further explore their traditional uses. A primary focus of this research is the significant presence of glucosinolates, which breakdown to form isothiocyanates, in mustard seeds. These compounds have been studied for their potential anticancer properties (Akhtar & Khan, 2024). Research suggests that these compounds may help inhibit the growth of cancer cells and could play a role in cancer prevention, particularly in cancers of the digestive tract (Rizwan & Masoodi, 2023). Another area of interest is the anti-inflammatory properties of mustard seeds. Studies have indicated that certain compounds in mustard seeds may help reduce inflammation, which is beneficial for treating conditions such as arthritis and other inflammatory diseases (Lang et al., 2018). Additionally, mustard seeds are rich in omega-3 fatty acids, minerals, and vitamins, which contribute to their overall health benefits, potentially improving cardiovascular health and aiding in the management of blood sugar levels (Akhtar & Khan, 2024). While additional research is needed to fully understand the extent and mechanisms of these health benefits, the present findings provide a scientific basis for many of the traditional applications of mustard seeds. Ongoing research continues to explore the potential of mustard seeds in various therapeutic applications, bridging traditional knowledge and modern science.

Black Pepper

Black pepper (Table 1), produced from the dried fruit of the *Piper nigrum* L. plant, is one of the most widely used spices in the world, both in culinary and medicinal contexts. Traditionally, black pepper plants have held a place in various cultures, particularly in Indian Ayurvedic medicine, traditional Chinese medicine, and traditional Thai medicine (Ganapathy et al., 2023; Hadi et al., 2023). In these systems, black pepper has been used not only as a flavor enhancer but also for its health benefits. It has been prescribed for its digestive stimulating properties, as it is believed to enhance the secretion of stomach acids, aiding in digestion. Additionally, black pepper has been used to treat colds and coughs due to its expectorant properties, which help clear congestion in the respiratory tract. It has also been recognized for its

Figure 3. Selected chemical compounds found in mustard seeds

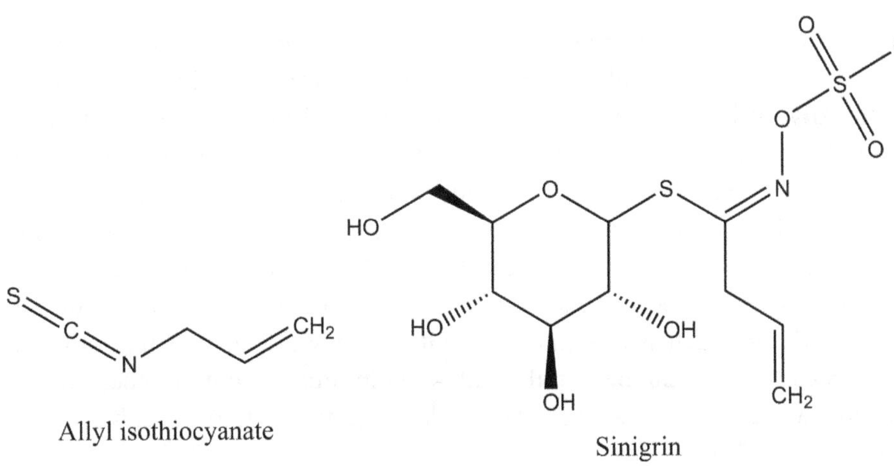

anti-inflammatory properties and has been used to alleviate pain and inflammation, particularly in joint-related conditions (Takooree et al., 2019).

Black pepper essential oils are characterized by the presence of complex blends of volatile compounds. The primary constituent is β-caryophyllene, which typically accounts for approximately 10-30% of the oil. Another major component is limonene, which comprises approximately 10-20%. Additionally, α-pinene and sabinene each make up approximately 9-14%, while β-pinene and δ-3-carene are present in the range of 5-10%. Other significant constituents include myrcene and α-phellandrene, which are usually found in smaller percentages (1-5%) (Figure 4) (Weluwanarak et al., 2023). In modern scientific research, black pepper, particularly its active compound piperine, has been the subject of extensive study. Piperine has been found to enhance the bioavailability of various nutrients and drugs, meaning that it can increase the absorption of these substances in the body. This property of piperine has significant implications, as it can potentially enhance the effectiveness of various medications and nutritional supplements (Dilrukshi et al., 2024). Furthermore, recent studies have shown that black pepper may have potent antioxidant effects that can help in combating oxidative stress and reducing the risk of chronic diseases such as cancer and heart disease. The anti-inflammatory properties of black pepper have also been validated in scientific studies, supporting its traditional use in treating inflammation (Maqbool et al., 2024). Additionally, there is growing evidence suggesting that black pepper might play a role in weight management by promoting the breakdown of fat cells (Dilrukshi et al., 2024). While additional research is needed to fully understand and confirm these benefits, the present findings provide a strong scientific basis for the traditional use of black pepper and open new avenues for its application in modern health and nutritional practices.

Dill

Dill seeds (Table 1), obtained from the herb *Anethum graveolens* L., have a storied history in traditional uses, particularly in European and Asian cultures. Traditionally, dill has been used for its culinary and medicinal properties. In the culinary world, dill seeds are known for their aromatic, slightly sweet, and citrus-like flavor and are commonly used in pickling, bread baking, and spices in various dishes. Medically, dill has been utilized in traditional herbal medicine for its carminative properties, helping in the relief of intestinal gas, bloating, and indigestion. It has also been used traditionally as a mild sedative or anxiolytic and to treat conditions such as insomnia due to its calming effects. In some cultures, dill has been used to stimulate milk production in breastfeeding mothers and to treat colic in infants, showcasing its versatile application in traditional healthcare (Sobhy et al., 2024).

Figure 4. Selected chemical compounds found in black pepper seeds

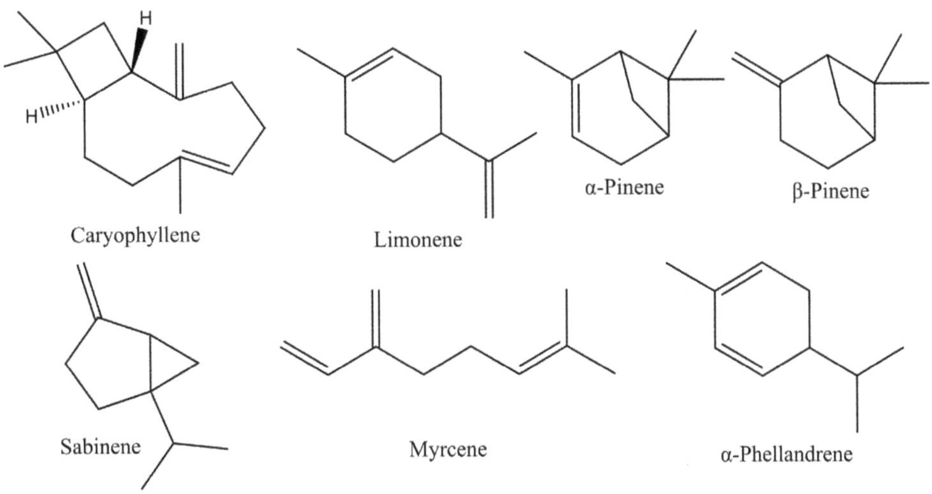

Dill seed essential oils feature a distinctive profile of volatile compounds. The principal component is carvone, which typically comprises approximately 40-60% of the oil. The drill apiole, another major constituent, usually accounts for approximately 20-40% of the muscle mass. Additionally, limonene is present in significant amounts, making up approximately 10-20% of the oil. Other notable components include α-phellandrene (1-5%) and eugenol (0.5-3%) (Figure 5) (Khaliq & Mushtaq, 2023). In terms of modern scientific research, dill seeds have been the subject of various studies examining their potential health benefits, which support many of their traditional uses. One of the significant areas of research focuses on the digestive health benefits of dill. Compounds in dill seeds, such as carvone and limonene, have been found to possess carminative and antispasmodic properties, which can help in soothing digestive ailments. This aligns with the traditional use of dill in treating digestive issues (Sobhy et al., 2024). Additionally, research has explored the potential antimicrobial properties of dill seeds, which could be beneficial in combating bacterial infections. Another area of interest is the potential of dill seeds to regulate the menstrual cycle and manage symptoms of menstruation, such as cramps, which ties back to their traditional uses in women's health (Bansal et al., 2024). Moreover, some studies have indicated that dill may have antioxidant effects, contributing to overall health by combating oxidative stress (Sobhy et al., 2024).

Figure 5. Selected chemical compounds found in dill seeds

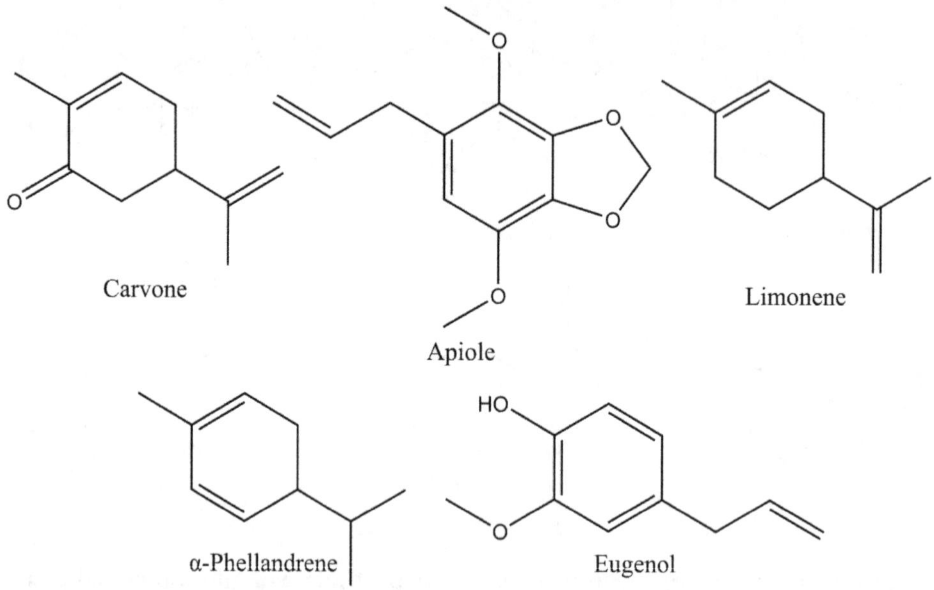

Cardamon

Cardamom (Table 1), obtained from the seeds of plants in the *Elettaria* and *Amomum* genera, is a spice with a long history of traditional use, particularly in South Asian and Middle Eastern cultures. Traditionally, cardamom is highly valued not only for its aromatic, sweet, and spicy flavor profile, which makes it a favorite in various culinary preparations, such as curries, desserts, and teas but also for its medicinal properties. In Ayurvedic medicine, cardamom has been used for centuries as a digestive aid. It is believed to help alleviate symptoms of indigestion, bloating, and gas and is often incorporated into dietary remedies for these conditions. Additionally, it has been used traditionally to treat bad breath and to improve oral health. In addition to being useful for digestive and oral health, cardamom has been used in traditional medicine systems to treat colds and flu, as it is believed to possess expectorant properties, helping to clear congestion in the respiratory tract (Maqbool et al., 2024; Pathak et al., 2024).

Cardamom seed essential oil is rich in volatile compounds. The primary constituent is 1,8-cineole, which typically accounts for approximately 20-50% of the oil. α-Terpinyl acetate is another major component and comprises approximately 20-40% of the total material. Additionally, linalool is present in significant amounts, usually ranging from 5-10%. Other important constituents include limonene (2-5%), myrcene (1-3%), and sabinene (1-3%) (Figure 6) (Sakkaravarthy et al., 2023).

Modern scientific research has begun to explore and validate the traditional uses of cardamom, with several studies highlighting its potential health benefits. One of the key areas of interest is the impact of cardamom on digestive health. Research has supported the traditional use of cardamom as a digestive aid, as compounds in cardamom may help stimulate digestion and alleviate symptoms such as gas and bloating. These studies align with the traditional role of cardamoms in helping Ayurveda people remedy digestive issues. Another significant area of research is the potential of cardamom for improving cardiovascular health. Studies suggest that cardamom may help lower blood pressure, thereby benefiting heart health. This is thought to be due to the antioxidant and diuretic properties of cardamom, which can help improve blood circulation and reduce hypertension. Additionally, the antioxidant properties of cardamom are being researched for its potential to prevent chronic diseases, including cancer and diabetes (Galal & Morsy, 2023).

Celery

Celery (Table 1), derived from the plant *Apium graveolens* L., has been utilized for centuries in various traditional medicine systems, including Ayurveda and TCM. Traditionally, celery has been valued not only for its culinary use as a vegetable but also for its therapeutic properties. In Ayurvedic practices, celery seeds have been used to treat colds, flu, water retention, and various types of arthritis. It is believed that celery has diuretic properties that help reduce fluid retention and flush toxins from the body. In TCM, celery is used to lower blood pressure and calm nerves and

Figure 6. Selected chemical compounds found in the cardamom seeds

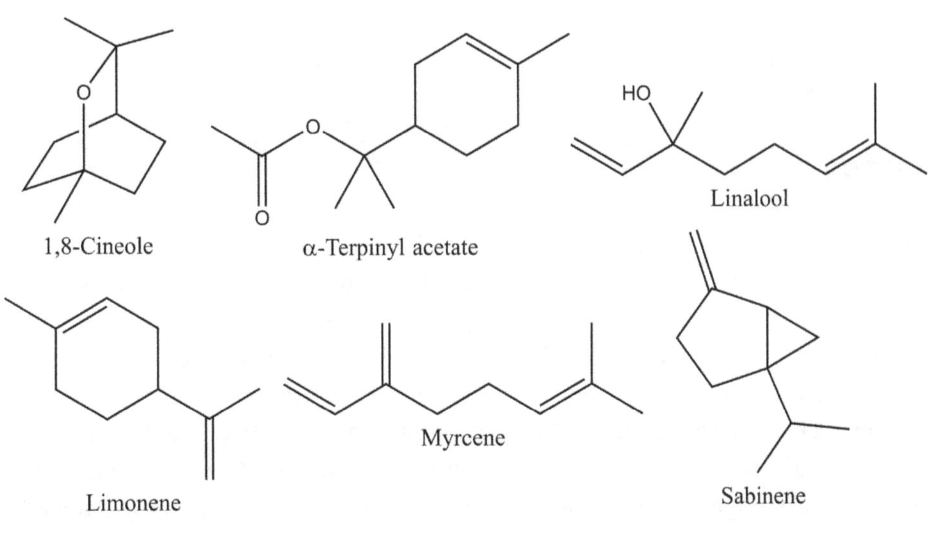

is considered beneficial for the liver and the spleen. Additionally, celery has been traditionally used for its potential to aid in digestion and for its anti-inflammatory properties, helping to alleviate conditions such as gout and rheumatoid arthritis (Shayani Rad et al., 2023).

Celery seed essential oils contain a diverse array of volatile compounds. The most abundant component is typically limonene, which can constitute approximately 60-70% of the oil. Along with limonene, sedanolide, and sedanenolide, which are phthalides unique to celery, generally make up approximately 10-17% of the total bacteria. Other significant constituents include β-selinene (5-10%) and α-selinene (2-6%). Smaller but notable percentages of compounds such as terpinolene, myrcene, and α-terpinene are also present. The exact composition of these compounds can vary depending on factors such as the celery variety, cultivation conditions, and extraction methods. These volatile compounds give celery seed oil its characteristic earthy, spicy aroma (Dąbrowska et al., 2020; Mahajan et al., 2023). Modern scientific research has started to validate many of these traditional uses of celery. One of the key areas of research is the potential cardiovascular benefits of celery. Studies have shown that compounds in celery, such as phthalides, may help lower blood pressure and cholesterol levels, thereby promoting heart health. These findings support its traditional use in TCM as a remedy for hypertension. The diuretic effects of celery have also been studied, with research indicating that celery can aid in reducing water retention and may aid in the treatment of urinary tract infections (Ouasti et al., 2023). Furthermore, research into the anti-inflammatory properties of celery has shown promise, particularly in the context of conditions such as arthritis. Celery contains luteolin, an antioxidant flavonoid that has been found to have anti-inflammatory effects and could be beneficial for reducing pain and inflammation associated with arthritis and other inflammatory diseases (G. Das et al., 2023). Additionally, studies are exploring the potential of celery for cancer prevention due to its antioxidant properties, which can aid in combating oxidative stress, a factor in the development of cancer (Shayani Rad et al., 2023).

Nutmeg

Nutmeg (Table 1), the aromatic seed of the *Myristica fragrans* Houtt. tree and has a long-standing reputation in traditional medicine and culinary arts across various cultures. Traditionally, nutmeg has been prized for more than just its warm, spicy flavor, which makes it a favorite in dishes ranging from sweet desserts to savory meat preparations. Its medicinal value has been recognized for centuries, especially in systems such as Ayurveda and TCM. In these traditional practices, nutmeg has been used for its digestive and carminative properties, helping to relieve gastrointestinal issues such as indigestion, gas, and diarrhea. Moreover, it has been used as a remedy for insomnia and anxiety due to its sedative qualities and analgesic properties for relieving pain. In some cultures, nutmeg was used to treat infections and inflammatory conditions, reflecting its broad application in traditional medicine (Sultan et al., 2023).

Modern scientific research has begun to uncover and validate many of these traditional applications of nutmeg. The spice contains several bioactive compounds, and the major components in its essential oil include sabinene (15-25%), α-pinene (10-20%), β-pinene (5-15%), myristicin (5-15%), limonene (1-8%), and camphene (1-8%) (Figure 7). Other significant constituents are terpineol, α-terpineol, and safrole, each typically found in smaller percentages. The exact composition can vary depending on factors such as the nutmeg variety, geographical origin, and extraction methods. These volatile compounds are responsible for the characteristic aromatic and flavor properties of nutmeg, and they also contribute to its medicinal and psychoactive effects (Ashokkumar et al., 2021). Myristicin, for instance, has drawn attention for its psychoactive properties and is being researched for its potential effects on the nervous system. While it offers possible therapeutic benefits, it also underscores the need for caution, as high doses can be toxic and lead to serious health issues. Eugenol, another compound found in nutmeg, is known for its anti-inflammatory and pain-relieving properties. Research has shown its potential in treating conditions such as arthritis and other inflammatory diseases (Sultan et al., 2023). Additionally, some studies suggest that nutmeg may have antimicrobial properties, making it a candidate for developing alternative treatments for bacterial infections. The antioxidant properties of nutmeg are also being explored and could have implications for preventing or managing conditions caused by oxidative stress, such as heart disease and certain types of cancer (Sultan et al., 2023).

Caraway

Caraway (Table 1), derived from the plant *Carum carvi* L., is a spice seed with a rich history in culinary and medicinal traditions across various cultures, particularly in Europe, North Africa, and Western Asia. Culinaryly, caraway seeds are known for their distinctive earthy and slightly peppery flavors and are eaten by citrus plants, making them a popular ingredient in numerous dishes. They are commonly used in European cuisine, especially in rye bread and sauerkraut, and as seasoning agents in cheeses and liqueurs. In Middle Eastern cuisines, caraway seeds are often added to savory dishes, soups, or sweets for their unique flavor. This widespread use in cooking highlights the versatility of caraway seeds as flavor enhancers in a diverse range of culinary traditions (Aly et al., 2023).

Caraway seed essential oil is notable for its unique composition of volatile compounds. The predominant component is carvone, which typically accounts for 40-60% of the oil. D-limonene is another major constituent and comprises

Figure 7. Selected chemical compounds found in nutmeg seeds

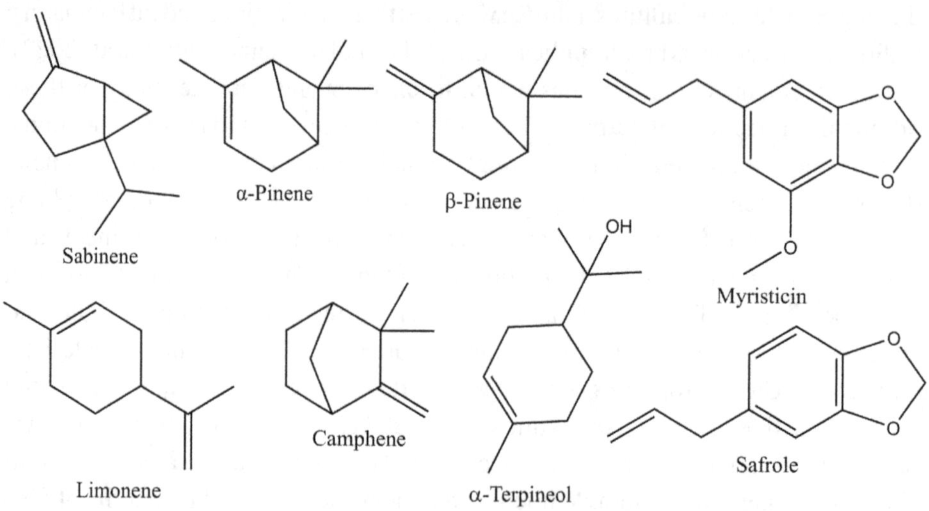

approximately 30-40% of the total material. Together, these two components form the bulk of the oil. Other minor but notable constituents include α-pinene (1-5%), β-myrcene (1-3%), and carveol (0.1-2.5%) (Malhotra, 2006). From a medicinal standpoint, caraway seeds have been utilized for centuries for their health benefits. Traditionally, they have been known for their digestive properties. They are often used as remedies for gastrointestinal issues such as bloating, gas, and indigestion. This is due to the presence of carvone and limonene, which are compounds that are believed to have carminative (gas-relieving) and antispasmodic effects (Abdallah et al., 2023). In herbal medicine, caraway seeds have also been used to treat ailments such as bronchitis and as appetite stimulants (G. Das et al., 2023). The potential antibacterial and antifungal properties of these materials have made them a subject of interest in scientific research. Recent studies have explored the potential of caraway seeds in various health-related applications, including their ability to improve digestive health, antioxidant properties, and even the possibility of having a positive impact on weight loss and cholesterol levels (G. Das et al., 2023; El-Assri et al., 2023). This scientific interest in caraway seeds not only validates their traditional uses but also suggests new therapeutic applications, which could be significant for future health-related developments.

Fennel

Fennel seeds (Table 1), derived from the *Foeniculum vulgare* Mill. plant are other spice seeds with a long history in culinary and medicinal traditions worldwide. In the kitchen, fennel seeds are known for their sweet, slightly licorice-like flavor, making them a popular ingredient in a variety of cuisines. They are commonly used in Mediterranean dishes and are often added to paste sauces, sausages, and baked goods. In Indian cuisine, fennel seeds are not only used in cooking but also often chewed as a mouth freshener and digestive aid after meals. This dual role of fennel seeds as flavor enhancers and digestive aids is indicative of their broader significance in both culinary and medicinal practices. Fennel seeds contain a rich array of volatile compounds. Its most prominent component is anethole, which typically constitutes approximately 50-80% of the oil. Following anethole, fenchone comprises approximately 5-20%, and estragole (methyl chavicol) is present in a range of 1-10%. Other significant constituents include α-pinene (1-5%), limonene (1-5%), and β-myrcene (0.5-3%). The concentrations of these components can vary based on factors such as the fennel variety, growing conditions, and extraction methods. This unique blend of volatile compounds gives fennel seed oil its distinct anise-like aroma and flavor, and it is widely used in food, pharmaceuticals, and aromatherapy for its various health benefits (Damayanti & Setyawan, 2012).

Medicinally, fennel seeds have been used historically in various cultures for their supposed health benefits. These plants are known for their carminative properties, meaning that they help relieve gas and bloating, making them common remedies for digestive issues (Zafar et al., 2023). In traditional medicine systems such as Ayurveda and traditional Chinese medicine, fennel seeds have been used to treat various ailments, from menstrual cramps to respiratory problems (Patel et al., 2023). Its use in treating colic in infants is also well documented (Gavish et al., 2023). Specifically, fennel seeds are rich in compounds such as anethole, estragole, and fenchone, which have been studied for their potential health benefits (Šunić et al., 2023). Research has shown that these compounds can have anti-inflammatory, antibacterial, and antispasmodic effects. Studies suggest that fennel seeds may help reduce symptoms of menopause and improve overall hormonal balance in women. Additionally, the potential role of these compounds in improving eyesight, regulating blood pressure, and even possessing antioxidant properties is being explored (Zafar et al., 2023).

Coriander

Coriander seeds (Table 1), derived from the *Coriandrum sativum* L. plant, play a significant role in the culinary and medicinal traditions of many cultures around the world. In the culinary field, these seeds are known for their distinct, slightly citrusy flavor and are used extensively in a variety of cuisines, ranging from Latin American to South Asian dishes. They are often used in spice blends, such as garam masala in Indian cuisine, and as a component in Latin American salsas and stews. In addition to their culinary uses, coriander seeds have a rich history of medicinal use. Traditionally, they have been used for their digestive, carminative, and anti-inflammatory properties. In Ayurvedic medicine, coriander seeds are used to treat digestive ailments such as indigestion and flatulence. They are believed to have a cooling effect on the body and are often recommended for soothing stomach ailments and for promoting overall digestive health (Agrawal et al., 2024).

From a scientific perspective, coriander seeds are interesting because they contain a variety of bioactive compounds. These include essential oils such as linalool and pinene, which have been shown to have antimicrobial and anti-inflammatory properties (Ersİn & Şenkal, 2023). The primary constituent is linalool, which typically comprises 50-70% of the oil. Other significant components include γ-terpinene (6-10%), α-pinene (1-5%), geranyl acetate (1-4%), and camphor (1-4%). Lesser, but still notable, components include limonene, β-pinene, and myrcene, each of which is generally present in smaller percentages (less than 3%). The composition of coriander can vary depending on factors such as the coriander variety, cultivation conditions, and method of oil extraction. These volatile compounds give coriander seed oil its characteristic warm, spicy, and sweet aroma (Memet et al., 2014). Research into these compounds has revealed that they may have several health benefits, such as lowering blood sugar levels, reducing anxiety, and even possessing antioxidant properties that can help prevent cellular damage caused by free radicals (Ghazanfari et al., 2023). Additionally, studies suggest that coriander seeds may improve heart health by lowering bad cholesterol (LDL) and increasing good cholesterol (HDL) levels in the blood (Hafeez et al., 2023).

SCIENTIFIC VALIDATIONS AND THERAPEUTIC POTENTIAL

The intricate tapestry of human knowledge is adorned with threads of ancient wisdom and embroidered with golden strands of modern scientific discovery. In the quest to understand the therapeutic potential of spice seeds, a harmonious dance emerges between time-tested ethnobotanical practices and cutting-edge scientific research, as detailed below.

The Convergence of Ancient Wisdom and Modern Science

The traditional use of spice seeds was often based on empirical observations, passed through generations, and intertwined with cultural and spiritual beliefs. Modern science, through its empirical methodologies, provides a lens through which these traditional practices can be validated, refined, or sometimes even refuted.

Shared Objectives

The shared objectives of traditional practices and contemporary scientific research in the field of spice seeds lie in the pursuit of understanding and harnessing their therapeutic properties for holistic well-being. Traditional systems of medicine, such as Ayurveda and TCM, and traditional Thai medicine have long embraced the medicinal qualities of various spice seeds, often categorizing them based on their warming or cooling effects on the body (Wardani et al., 2023). For instance, in these ancient systems, a spice such as cardamom is considered ‘warm’ and is used to treat ailments believed to originate from ‘cold’ conditions in the respiratory tract (Khan & Aslam, 2023a). On the other hand, coriander, known for its cooling properties, is used to alleviate ‘hot’ conditions such as inflammation (Agrawal et al., 2024). This categorization is based on centuries of empirical observations and cultural wisdom, reflecting a holistic view of health where the properties of spice seeds are integrated into a broader understanding of the body, health, and disease.

Contemporary scientific research, while distinct in methodology and language, shares the fundamental goal of traditional practices: to uncover and utilize the health benefits of spice seeds. Modern pharmacology and medical research translate the concepts of warming and cooling into more scientifically quantifiable terms such as anti-inflammatory, antioxidant, or vasodilatory effects. For example, the warming effect of coriander can be correlated with its anti-inflammatory and digestive-stimulating properties, as identified in pharmacological studies. Similarly, the cooling effect of coriander has been explored in terms of its potential to soothe inflammation and act as an antispasmodic. This scientific validation not only strengthens the credibility of traditional uses but also expands the understanding of these seeds by elucidating the biochemical mechanisms underlying their effects. The convergence of traditional wisdom and modern science thus represents a comprehensive approach, blending the insights of ancient practices with the rigors of scientific investigation. This synergy aims to fully realize the potential of spice seeds to promote health and treat diseases, bridging traditional knowledge and contemporary research for the improvement of holistic health care.

Reinforcing Credibility

The process of scientific validation plays a crucial role in reinforcing the credibility of ancient practices involving the use of spice seeds, effectively bridging the gap between traditional knowledge and modern medicine. This reinforcement is essential in propelling these age-old practices into the realm of mainstream acceptance. A prime example is the traditional use of fenugreek, a spice known for its culinary and medicinal applications across various cultures. Traditionally, fenugreek has been used for its ability to control blood sugar levels, particularly in Ayurvedic medicine (Singh et al., 2023). This traditional belief has been substantiated through scientific research, which has demonstrated that fenugreek contains soluble fiber and other compounds that can modulate insulin activity and thus help in managing blood glucose levels. Studies have shown that fenugreek can slow the absorption of sugars in the stomach and stimulate insulin, factors that are critical in controlling diabetes. This convergence of traditional wisdom and scientific evidence not only validates the traditional use of fenugreek but also enhances its acceptance and application in modern therapeutic practices, especially for managing diabetes (Madhu et al., 2023).

The impact of such scientific validations extends beyond just reinforcing the credibility of traditional practices; it also contributes to the development of new, evidence-based therapeutic applications. By understanding the mechanisms through which these spice seeds exert their effects, researchers are able to develop more targeted and effective treatments. For example, the discovery of specific compounds in fenugreek that have insulin-modulating properties has opened up possibilities for the development of new diabetes medications or dietary supplements based on these compounds. Additionally, this process of validation often leads to a deeper understanding of the interplay between diet, traditional remedies, and health, providing valuable insights into the field of nutrition and holistic health care. The scientific study of these traditional practices not only confirms their efficacy but also contributes to a more nuanced understanding of health and wellness, integrating the wisdom of the past with the scientific advancements of the present. In this way, the scientific validation of ancient practices using spice seeds plays a pivotal role in enhancing their credibility and facilitating their integration into contemporary healthcare paradigms.

Table 2 presents a comparative analysis of traditional ethnobotanical uses versus findings from modern research for various spice seeds. The common names of the spices, their historical uses, their principal bioactive compounds, the latest research findings, and references to relevant studies are listed. For instance, black cumin, traditionally used for immune and respiratory health, aligns with modern research findings of its antioxidant and anti-inflammatory properties (Kavyani et al., 2023). Similarly, the traditional role of fenugreek as a digestive aid and for lactation support is supported by modern research demonstrating its effects on blood sugar regulation and anti-inflammatory properties (Akhtari et al., 2024). Mustards, which are used traditionally for respiratory congestion and muscle pain and as digestive aids, contain glucosinolates and allyl isothiocyanate. Studies by Usman et al. (2023) and Zafar et al. (2023) indicated its antimicrobial and potential anticancer effects. The black pepper plant, known for its ability to aid in digestion and treat colds and inflammation, contains piperine. This enhances nutrient and drug bioavailability and has anti-inflammatory effects, as shown in previous research (Thahira Banu and

Aswini, 2023). Dill, which is used for digestive aid, infant colic relief, and calming, contains carvone and limonene and has antimicrobial, antispasmodic, and carminative properties (G. Das et al., 2023). Cardamom, aiding in digestive and oral health and respiratory relief, contains cineole and has been reported to assist in digestion, act as a diuretic and have expectorant properties (Khan & Aslam, 2023a). Celery, which is traditionally used as a diuretic for arthritis treatment and digestive aid, contains phthalides and apigenin, Hussain et al. (2023) which have antihypertensive, anti-inflammatory, and diuretic properties. Nutmeg, utilized for pain relief, digestion, and insomnia treatment, contains myristicin and eugenol and is reported to have antioxidant, anti-inflammatory, and psychoactive effects (Ramírez-Alarcón et al., 2023). Caraway, which aids in digestion, relieves bloating and gas and is used as a flavoring agent, contains carvone and limonene and was suggested by Sulieman et al. (2023)to possess antispasmodic, carminative, and digestive aid properties. Fennel, which is used for digestive disorders, colic, and respiratory issues, contains anethole, estragole, and fenchone and was developed by Zafar et al. (2023) to indicate antispasmodic, carminative, and digestive aid effects. Finally, coriander, a digestive aid with anti-inflammatory and culinary flavorings, contains linalool and pinene and has been identified as having antioxidant, antimicrobial, and digestive aid properties (Agrawal et al., 2024). This comparison reveals how modern research often validates and provides a scientific basis for traditional ethnobotanical uses, although it also extends our understanding of these seeds through new discoveries and potential applications. It is important to note that while there is significant overlap between traditional uses and scientific findings, the efficacy and safety of these seeds for specific medical applications still require careful clinical validation.

Recent Breakthroughs and Discoveries

As the scientific community has focused on spice seeds, numerous breakthroughs have been made, affirming their multifaceted therapeutic potential.

Bioactive Compounds

The field of modern pharmacology has made remarkable strides in identifying and isolating bioactive compounds in spice seeds through the use of advanced analytical techniques. These compounds are responsible for the medicinal properties of seeds and offer a wide array of health benefits. For example, one of the most significant discoveries in recent years has been the isolation of thymoquinone from black cumin. Thymoquinone has been extensively studied for its wide range of pharmacological properties. Research has demonstrated its potent antioxidant capabilities, which play a crucial role in protecting cells from oxidative stress, a factor implicated in various

Table 2. Comparative analysis of traditional ethnobotanical uses of spice seeds versus findings from modern research studies

Common Name	Traditional Ethnobotanical Uses	Principal Bioactive Compounds	Modern Research Findings	Reference
Black cumin	Immune booster, digestive, respiratory conditions	Thymoquinone	Antioxidant, anti-inflammatory, potential anticancer	Akhzari et al. (2023); Rashwan et al. (2023)
Fenugreek	Digestive aid, lactation support, anti-inflammatory	Diosgenin, saponins	Blood sugar regulation, anti-inflammatory properties	Ahmad et al. (2023); Akhtari et al. (2024)
Mustard	Respiratory congestion relief, muscle pain, digestive aid	Glucosinolates, allyl isothiocyanate	Antimicrobial properties, potential anticancer effects	Usman et al. (2023); Zafar et al. (2023)
Black pepper	Digestive stimulant, treatment for colds, anti-inflammatory	Piperine	Increases nutrient and drug bioavailability, anti-inflammatory effects	Thahira Banu and Aswini (2023)
Dill	Digestive aid, relief for colic in infants, calming effects	Carvone, limonene	Antimicrobial, antispasmodic, and carminative properties	G. Das et al. (2023)
Cardamom	Digestive health, oral health, respiratory relief	Cineole	Digestive aid, diuretic, expectorant properties	Khan and Aslam (2023a)
Celery	Diuretic, arthritis treatment, digestive aid	Phthalides, apigenin	Antihypertensive, anti-inflammatory, diuretic properties	Hussain et al. (2023)
Nutmeg	Pain relief, digestive aid, insomnia treatment	Myristicin, eugenol	Antioxidant, anti-inflammatory, psychoactive effects	Ramírez-Alarcón et al. (2023)
Caraway	Digestive aid, bloating and gas relief, flavoring agent	Carvone, limonene	Antispasmodic, carminative, digestive aid properties	Sulieman et al. (2023)
Fennel	Treatment for digestive disorders, colic, respiratory issues	Anethole, estragole, fenchone	Antispasmodic, carminative, digestive aid properties	Zafar et al. (2023)
Coriander	Digestive aid, anti-inflammatory, culinary flavoring	Linalool, pinene	Antioxidant, antimicrobial, digestive aid properties	Agrawal et al. (2024)

chronic diseases, including cancer and heart disease. Additionally, thymoquinone has been recognized for its anti-inflammatory properties, suggesting that it is a potential therapeutic agent for the treatment of inflammatory diseases such as arthritis (Warinhomhoun et al., 2023). Furthermore, studies have explored the anticancer properties of thymoquinone, and the available evidence suggests that this compound can induce cancer cell apoptosis (programmed cell death), suggesting that it is a promising novel anticancer agent (Salah et al., 2023). These breakthroughs not only validate the traditional uses of these spice seeds but also enhance our understanding of their therapeutic potential. Ongoing research on the isolation and understanding of bioactive compounds from spice seeds is needed to test the rich medicinal value of these natural resources, suggesting that their use in modern medicine is promising.

Mechanisms of Action

The exploration of the mechanisms of action of bioactive compounds in spice seeds is a critical area of focus in contemporary research, as it provides a deeper understanding of how these natural substances exert their therapeutic effects at the molecular and cellular levels. Advanced scientific techniques are being employed to study how these compounds interact with specific cellular receptors, enzymes, and other molecular targets within the human body (Singh et al., 2021). For example, capsaicin, the active ingredient in chili peppers, is known to bind to the transient receptor potential vanilloid 1 (TRPV1) receptor, a type of protein found in nerve cells. This binding action is what triggers the sensation of heat and pain and is also believed to be responsible for capsaicin's pain-relieving effects, as it can lead to the desensitization of pain receptors over time (Abdel-Salam & Mózsik, 2023). Understanding the mechanisms of action of these compounds not only validates traditional uses of spice seeds but also aids in the development of targeted and effective therapeutic strategies. Another aspect of this research is the study of how spice seeds can modulate gene expression (Delgadillo-Puga et al., 2023). For instance, certain compounds in spice seeds can activate or inhibit genes involved in inflammation, cell proliferation, and detoxification processes. This regulation at the genetic level can have significant implications for the prevention and treatment of various diseases. For instance, studies on sulforaphane, a compound found in cruciferous vegetables such as mustard seeds, have shown its potential to activate genes that produce detoxification enzymes, which play a role in cancer prevention (Orouji et al., 2023). Furthermore, understanding these mechanisms also helps in identifying potential side effects and interactions with other drugs, ensuring the safe and effective use of these compounds in medical treatments. Ongoing research into the mechanisms of action of spice seeds is a dynamic and continually evolving field revealing the complex ways in which these natural substances can contribute to health and well-being. This deeper insight not only furthers our understanding of medicinal plants but also paves the way for new discoveries and innovations in the realm of natural health products.

Clinical Trials

Clinical trials play an indispensable role in the scientific validation of the therapeutic properties of spice seeds. These trials, particularly those that are double-blind and placebo-controlled, offer a rigorous and objective method for evaluating the efficacy and safety of treatments derived from spice seeds. Such trials are essential for translating the traditional and preliminary scientific understanding of these seeds into concrete, evidence-based medical applications (Akhtari et al., 2024). Recent clinical studies on mustard seeds have begun to illuminate their potential health benefits. These trials focused on evaluating the cardiovascular benefits of mustard seeds and examining how their high omega-3 fatty acid content and antioxidant properties may contribute to heart health. In these studies, participants typically receive either a mustard seed supplement or a placebo, with researchers measuring indicators such as blood pressure, cholesterol levels, and markers of inflammation. The significance of these findings lies in their provision of empirical evidence supporting the traditional use of mustard seeds for promoting heart health and their potential role in preventing heart disease (Poddar et al., 2022). However, as reported by Poddar et al. (2022), the controversy surrounding mustard oil and cardiovascular health largely arises from its composition and varying international regulations. Mustard oil, popular in South Asia and low in saturated fat, contains a notable proportion of erucic acid. This fatty acid is linked to myocardial lipidosis in rodents, raising concerns about its safety for human consumption. As a result, the U.S. Food and Drug Administration (FDA) banned its use in cooking, citing these potential health risks. In contrast, regions such as Australia, New Zealand, and the European Union have established upper limits for tolerable intake of mustard oil instead of an outright ban. This variance in regulatory approaches underscores the need for further clarification, particularly in light of the extensive use of oil in Asian cooking and its increasing popularity among both Americans and Indian immigrants in the U.S., despite the FDA ban. Clinical trials thus serve as a critical bridge between traditional knowledge and modern medical practice, ensuring that the use of spice seeds in treatment is based on solid scientific evidence. These findings could help in establishing standardized protocols for the use of these natural substances in healthcare, potentially leading to new, effective, and natural treatment options for a variety of health conditions.

THE FUTURE OF SPICE SEEDS IN MEDICINE

The scope of medical research is extensive and continuously growing. As scientific exploration probes further into the lush realms of ethnobotany, the significance of spice seeds in the broad spectrum of therapeutic interventions becomes increasingly evident. Their powerful bioactive compounds and millennial-old relationships with human health position them as indispensable protagonists in the unfolding narrative of integrative medicine.

The Need for Further Research

While significant strides have been made in understanding the therapeutic potential of spice seeds, many aspects of these plants have not been explored or are underexplored. The need for further research is outlined in the following sections.

In-Depth Phytochemical Analysis

The intricate world of spice seeds, which are rich in a myriad of bioactive compounds, necessitates in-depth phytochemical analysis to fully comprehend their therapeutic potential. Each spice seed is a complex mixture of various compounds, each with unique properties and effects on human health. Systematic phytochemical analyses involve isolating and identifying these individual molecules, which is essential for understanding how they work individually and in synergy (Gupta et al., 2023). For example, black pepper contains more piperine, the most well-known compound, than any other compound (Yadav et al., 2023). It also contains several other compounds, including terpenes and alkaloids, which contribute to its overall health benefits (Vieira et al., 2022; Zeng et al., 2017). Detailed phytochemical analysis can help in identifying these lesser-known compounds, shedding light on their specific roles and potential health benefits. Such analyses not only broaden the scope of our understanding of these seeds but also provide crucial information for developing new pharmaceuticals and supplements.

Moreover, comprehensive phytochemical analysis is key to understanding the interactions between different compounds within a single seed and their combined effects on health. The therapeutic efficacy of a spice seed is often not due to a single compound but rather the result of a complex interplay of multiple constituents (Kim et al., 2022; Ren et al., 2023). Advanced analytical techniques such as mass spectrometry and nuclear magnetic resonance spectroscopy are currently available for this purpose, enabling researchers to delve into the intricate chemical makeup of these seeds (Yan et al., 2023). For instance, the full spectrum of curcuminoids in turmeric, each with slightly different properties, needs to be studied to understand the overall efficacy of turmeric (Gupta et al., 2024). Furthermore, this analysis can also help in identifying the most effective extraction and processing methods for these compounds, which is crucial for ensuring their stability and bioavailability in therapeutic formulations (Vannabhum et al., 2023). In-depth phytochemical analysis is thus a foundational step in advancing the knowledge of spice seeds and unlocking their full therapeutic potential, paving the way for new discoveries and innovations in natural health products.

Clinical Validation

The transition from traditional, anecdotal evidence to scientifically validated knowledge is a critical step in the journey of spice seeds from the kitchen to clinical application. Many traditional uses of these seeds, passed through generations and supported by cultural practices, still await rigorous validation through randomized clinical trials. These trials are the gold standard in medical research, providing unbiased and reliable evidence on the efficacy and safety of treatments (He et al., 2023). Clinical validation of spice seeds is crucial not only for enhancing their credibility in the scientific community but also for paving the way for their formal integration into medical treatment regimens. Once the benefits of a spice seed are validated through clinical trials, it can transition from being a complementary or alternative remedy to being a recognized part of conventional medicine. This progression is essential for ensuring that patients receive treatments that are both effective and safe. For example, the active compound curcumin has been subjected to numerous clinical trials to validate its anti-inflammatory and antioxidant properties. These studies are essential for determining how curcumin can be effectively used to treat conditions such as arthritis or cancer therapy. The success of these trials can lead to the development of standardized curcumin supplements or drugs, providing patients with more natural treatment options (Sorkin & Gahche, 2020). Clinical validation is, therefore, a critical step in the journey of spice seeds from age-old remedies to modern therapeutics, bridging the gap between traditional wisdom and contemporary medical practice.

Safety and Dosage

The exploration of the therapeutic potential of spice seeds must be accompanied by a thorough understanding of their safety profiles, potential interactions with other medications, and determination of optimal dosages. These are critical areas that require further research to ensure that the use of spice seeds in medical treatments is both safe and effective. While many spice seeds have been used traditionally without apparent harm, the concentrated extracts and compounds used in modern therapeutics can have different effects (Rietjens et al., 2023). For instance, cinnamon, a spice derived from the bark of trees and not from seeds, can cause digestive issues when consumed in high quantities. Additionally, excessive intake of cinnamon poses a risk of liver damage, primarily due to the presence of the compound coumarin in its bark. Despite being a bark-based spice, cinnamon serves as a relevant example in this context. The acceptable daily intake of coumarin is approximately 0.1 mg/kg body weight; however, the coumarin content in cinnamon varies significantly, from 0 to as high as 8.79 mg/g, averaging approximately 2.41 mg/g. With an average concentration of 2.4 mg/g, an adult weighing 60 kg could exceed the safe coumarin threshold by consuming just 248 mg of cinnamon (Grant et al., 2023). These findings underscore the importance of establishing safe dosage guidelines. Moreover, understanding potential drug interactions is crucial, especially for patients who are already on medication regimens. Certain compounds in spice seeds can interact with pharmaceutical drugs, either by enhancing or inhibiting their effects, possibly leading to unforeseen complications.

In addition to safety and drug interactions, the determination of optimal dosages is another area that demands attention. The effective dosage can vary significantly depending on the specific spice seed, the condition being treated, and individual patient factors such as age, weight, and overall health. Clinical trials and pharmacological studies play a vital role in this aspect, helping to establish dosing guidelines that maximize therapeutic benefits while minimizing risks (Grant et al., 2023). For instance, studies on black cumin might focus on determining the optimal dose to harness its anti-inflammatory properties without causing adverse effects (Rashwan et al., 2023). Similarly, research into the use of fenugreek for diabetes management is needed to ascertain the appropriate dosage that effectively regulates blood sugar levels (Fakhr et al., 2023). These investigations are crucial for transitioning spice seeds from traditional use to their application in modern medicine. Patients can safely incorporate these natural remedies into their treatment plans, guided by evidence-based dosage recommendations. Thus, further research on safety, drug interactions, and dosage is necessary not only for the prevention but also for the effective integration of spice seeds into healthcare.

Lesser-Known Seeds: The Untapped Reservoir

While some spice seeds, such as those of cardamom, black cumin, fenugreek, and mustard, have undergone extensive study, many others remain relatively obscure, awaiting their turn into the scientific spotlight. Below are suggestions for future research on lesser-known spice seeds.

Diverse Ethnobotanical Repositories

Indigenous communities around the globe serve as custodians of an immense wealth of ethnobotanical knowledge, particularly regarding local flora and lesser-known spice seeds. These communities have developed a deep understanding of their surrounding plant life through centuries of close interaction and observation, often forming unique and highly specialized relationships with these natural resources. This vast, yet frequently undocumented, ethnobotanical wisdom is a reservoir of potential therapeutic agents waiting to be explored. For example, in various remote parts of the Amazon rainforest, indigenous tribes use spice seeds and plants that are virtually unknown to the outside world, harnessing their properties to treat a wide array of ailments. Similarly, in the mountainous regions of the Himalayas, local communities have a rich tradition of using specific seeds and herbs for their medicinal properties, many of which have not yet been studied or recognized by modern science.

Engaging with these indigenous communities to learn about their use of lesser-known spice seeds is crucial in discovering new therapeutic agents. This engagement needs to be carried out to ensure that it does not lead to the exploitation of these communities or their resources. Collaborative efforts, where scientists work alongside indigenous knowledge holders, can be a mutually beneficial approach. This collaboration can help in scientifically validating the traditional uses of these spice seeds while also ensuring that the communities are acknowledged and benefitting from any commercial developments that might arise. For instance, a lesser-known seed used by a community for its anti-inflammatory properties could, upon scientific examination, yield novel compounds with significant pharmacological potential. These discoveries not only contribute to the global repository of medicinal knowledge but also offer new possibilities for drug development and therapeutic applications. Therefore, tapping into the diverse ethnobotanical repositories of indigenous communities is not only about uncovering new medicinal plants and seeds but also about preserving cultural heritage and fostering a more inclusive and respectful approach to scientific discovery.

Modern Biotechnological Tools

The application of modern biotechnological tools such as genomic sequencing and metabolomic profiling represents a significant advancement in the exploration of lesser-known spice seeds for potential therapeutic uses. Genomic sequencing allows for a deep dive into the genetic blueprint of plants, uncovering the genetic basis for the synthesis of bioactive compounds within these seeds. This method can reveal specific genes that are responsible for producing particular phytochemicals with medicinal properties (P. Das et al., 2023). For instance, by using genomic sequencing on a relatively unknown seed used traditionally by a local community, researchers could discover unique genetic sequences that encode enzymes responsible for synthesizing compounds with anti-inflammatory or anticancer properties. These discoveries not only add to the repository of medicinal compounds but also open the door to the development of new drugs based on these genetic insights (P. Das et al., 2023).

Metabolomic profiling complements genomic studies by providing a detailed chemical fingerprint of the seeds. This technique involves analyzing the complete set of metabolites present in a sample, which can reveal a wealth of information about the compounds that contribute to the medicinal effects of seeds. Metabolomic profiling can identify known and unknown compounds, offering a comprehensive view of the chemical composition of seeds (Djiazet et al., 2023). For example, metabolomic analysis of a lesser-known seed might uncover a complex array of antioxidants, alkaloids, or other phytochemicals with potential health benefits, some of which may be novel compounds not previously identified. These findings are invaluable in the drug discovery process, as they can guide further research into the pharmacological properties of these compounds. The application of these sophisticated biotechnological tools is crucial for revealing the full therapeutic potential of lesser-known spice seeds. This method aids not only in discovering new prospects for drug development but also in deepening our comprehension of the biological mechanisms that underpin the medicinal qualities of these plants. This sets the stage for the development of innovative treatments and therapies derived from the unexplored riches of nature.

Conservation and Sustainable Utilization

The growing recognition of the therapeutic potential of lesser-known spice seeds is likely to lead to an increase in their demand, which highlights the critical issues of conservation and sustainable utilization. As these seeds move from obscurity to prominence, it becomes imperative to ensure that their natural habitats are protected and that their harvesting is performed in ways that do not deplete or harm the environment. Overexploitation, driven by commercial interests, can lead to a significant loss of biodiversity and the degradation of ecosystems. This concern is particularly acute for seeds that come from rare or endangered plant species or those that are endemic to specific, often fragile, ecosystems (Mipeshwaree Devi et al., 2023). For instance, a sudden spike in demand for a newly discovered seed with anticancer properties could lead to unsustainable harvesting practices, threatening plant survival and the ecological balance of its habitat. Therefore, implementing conservation strategies, such as creating protected areas or reserves and promoting the cultivation of these plants under controlled conditions, is essential.

Sustainable utilization encompasses not only environmental aspects but also social and economic dimensions. Farming practices that are environmentally friendly and socially responsible are adopted to ensure that local communities, particularly those that have traditionally used these seeds, benefit from their commercialization. This can include practices such as fair trade and ethical sourcing, which ensure that the harvesters or growers of these seeds receive fair compensation for their labor and knowledge. Additionally, sustainable utilization also means exploring and implementing agricultural methods that maximize yield while minimizing environmental impact. This could involve the use of organic farming techniques, crop rotation, and cultivation methods that are suited to the specific growing conditions of these seeds. By focusing on conservation and sustainable utilization, we can ensure that the exploration and commercialization of lesser-known spice seeds contribute to the well-being of both the environment and the communities that rely on them while also providing valuable resources for medical research and development (dos Santos et al., 2023).

COLLABORATIVE APPROACHES TO HOLISTIC HEALTH

In the field of medical research, collaboration acts as a key illuminator, uncovering new and unexplored paths. With the growing global awareness of the holistic potential of natural remedies, developing a synergistic method that seamlessly integrates ethnobotanical insights with the thoroughness of pharmacology is essential. This collaborative approach to holistic health is described in detail below.

Bridging Ethnobotany and Pharmacology

Informed Exploration

The concept of informed exploration in bridging ethnobotany and pharmacology centers around utilizing the wealth of traditional knowledge about spice seeds to guide modern pharmacological research. Ethnobotany provides invaluable insights into the potential therapeutic uses of various spice seeds, knowledge that has been accumulated and refined over centuries (Ramalingam et al., 2024). For instance, traditional uses of certain seeds in indigenous medicine can offer clues about their possible pharmacological effects. This traditional wisdom acts as a starting point for pharmacologists, allowing them to focus their research on specific seeds or compounds that have been historically used for medicinal purposes (Maqbool et al., 2024). For example, the use of turmeric in Ayurvedic medicine for its anti-inflammatory properties can direct pharmacological studies to investigate its active compound, curcumin, and its derivatives for potential use in treating inflammatory diseases (Anand et al., 2023). This approach streamlines the drug discovery process, saving time and resources that would otherwise be spent on testing a wide range of compounds with unknown effects.

Moreover, informed exploration through the integration of ethnobotanical knowledge into pharmacology has accelerated the discovery of new drugs and enhanced the understanding of how these compounds work. Pharmacologists can gain insights into the underlying mechanisms of action and potential therapeutic benefits of these spice seeds by studying their historical context and traditional applications.

This approach is particularly important when the bioactive compounds in seeds interact with the body in complex ways that are not fully understood. For example, the use of black cumin in traditional medicine for a variety of ailments suggests a range of pharmacological activities that merit detailed scientific investigation. By exploring these activities in the context of traditional knowledge, researchers can develop a more nuanced understanding of the therapeutic potential of seeds. This informed exploration, therefore, not only led to the discovery of new drugs but also contributed to a deeper understanding of traditional medicines, validating and enriching this ancient wisdom with scientific evidence.

Enhanced Validation

Enhanced validation in the context of bridging ethnobotany and pharmacology refers to the process of applying scientific methods to confirm, refine, and sometimes reinterpret the traditional uses of spice seeds. Ethnobotanical practices, rooted in historical and cultural contexts, offer rich insight into how different cultures use spice seeds for medicinal purposes. However, these traditional practices often lack the empirical evidence required to meet modern scientific standards. This is where pharmacological studies come into play, providing the necessary empirical validation through rigorous testing and analysis. For example, the spice ginger is traditionally used in many cultures for its anti-nausea effects. Pharmacological research can validate this traditional use by conducting clinical trials and laboratory experiments to understand how ginger compounds interact with the human body to produce these effects. Even though ginger is not a spice seed but a rhizome, it serves as a pertinent example in this context, demonstrating how traditional uses of plant-based substances can be scientifically substantiated. Crichton et al. (2023) evaluated the impact of ginger as an adjuvant therapy for chemotherapy-induced nausea and vomiting (CINV). A double-blind trial with 103 participants was conducted to compare the effects of ginger capsules and placebo in chemotherapy-naïve adults. The study primarily measured quality of life related to chemotherapy-induced nausea, along with various secondary outcomes, such as vomiting, fatigue, nutritional status, and mental health aspects. The results showed a significant improvement in the quality of life related to nausea and overall CINV-related outcomes in those taking ginger. Specifically, ginger was found to reduce the severity of delayed nausea and fatigue and the incidence of delayed nausea and vomiting in later chemotherapy cycles. It also appeared to positively impact malnutrition rates. However, ginger did not significantly affect anticipatory or acute CINV, overall health-related quality of life, anxiety, or depression. No serious adverse events were associated with ginger use. The study concluded that ginger is a safe and effective supplement for enhancing the quality of life of patients receiving chemotherapy by reducing CINV-related issues. Further research is suggested to determine the optimal dosing regimen. This scientific validation not only confirms the efficacy of traditional practices but can also lead to the development of new, standardized medicinal products based on these spices.

Furthermore, the integration of ethnobotany and pharmacology in enhanced validation can lead to the refinement and even the discovery of new uses for spice seeds. In some cases, scientific research uncovers mechanisms of action or therapeutic effects that were not recognized in traditional practices. For instance, the traditional use of cinnamon in various cultures for its flavoring properties might overlook its potential antidiabetic effects, which pharmacological studies have revealed (Cheikhyoussef et al., 2023). Researchers can not only confirm the validity of these practices but also expand our understanding of the medicinal properties of spice seeds by subjecting traditional uses to scientific scrutiny. This process ensures that the knowledge passed through generations is not only preserved but also enhanced with scientific rigor. In this way, enhanced validation through the combination of ethnobotany and pharmacology serves as a critical bridge, connecting the wisdom of the past with the scientific advancements of the present and paving the way for the development of novel, effective, and culturally sensitive healthcare solutions.

Bioprospecting With Respect

Bioprospecting with respect is an approach that emphasizes ethical and equitable practices in the exploration of biological resources, particularly those derived from indigenous knowledge. Ethnobotanical insights, which are often sourced from indigenous communities, contain valuable information about the medicinal uses of various plants and spice seeds. These insights are the result of generations of observation, experimentation, and cultural practices. However, when such knowledge is used for commercial purposes, it is crucial to ensure that the communities providing this wisdom are respected, acknowledged, and fairly compensated. This involves not only recognizing their contribution but also actively involving them in the research and development process (Chaudhary et al., 2023). For instance, if a pharmaceutical company is developing a new drug based on the traditional use of a spice seed known to an indigenous community, this community must be consulted, informed about the research, and receive a fair share of the benefits, such as royalties or other forms of compensation. This practice not only is ethically sound but also fosters trust and collaboration, which are essential for sustainable and mutually beneficial relationships.

Moreover, bioprospecting with respect extends to ensuring that the utilization of ethnobotanical knowledge does not lead to the exploitation or depletion of natural resources. This means adopting sustainable practices for the harvesting and use of spice seeds and other biological resources, in line with the traditional practices and ecological knowledge of indigenous communities (Rani et al., 2023). For example, if a particular spice seed is identified as having significant medicinal properties, it is important to harvest it in a way that does not threaten its existence in the natural habitat. This might involve following traditional harvesting techniques or developing cultivation methods that do not harm the environment. Such practices not only preserve biodiversity but also respect the deep connections that indigenous communities often have with their natural surroundings. Bioprospecting with respect, therefore, is about more than just fair economic practices; it is about adopting an approach that is environmentally sustainable, culturally sensitive, and socially just, ensuring that the benefits of ethnobotanical research are shared by all, especially those who have been the custodians of this knowledge for centuries (Hazarika et al., 2023).

A Call for Integrated Research

In this section, the author explores the pressing need for a comprehensive and multidisciplinary approach to investigate the therapeutic potential of spice seeds, as detailed in the following subsections.

Interdisciplinary Teams

Interdisciplinary teams, especially in the context of research on spice seeds, bring together a spectrum of expertise that is crucial for comprehensive study and understanding. Such teams typically include ethnobotanists, pharmacologists, biochemists, and anthropologists, each contributing unique perspectives and skills (Srikanth & Goel, 2023). Ethnobotanists, for instance, offer insights into the historical and traditional uses of these seeds across various cultures. These tools help in understanding the cultural significance and the historical context in which these seeds were used, which is vital for identifying potential areas for pharmacological research. Pharmacologists and biochemists, on the other hand, explore the scientific analysis of these seeds. The authors studied the bioactive compounds present in the seeds, their pharmacological effects, and the molecular mechanisms underlying these effects. This scientific exploration is critical for validating the traditional uses of these seeds and for discovering new therapeutic potential. Anthropologists on these teams play a role in understanding the sociocultural dimensions of spice seed use. They examine how cultural practices, beliefs, and societal structures influence the use of these seeds and how this knowledge is transferred across generations. This understanding can be pivotal in ensuring that modern scientific research on spice seeds is respectful and inclusive of traditional knowledge and practices. For example, in the study of turmeric (*Curcuma longa*), which is widely used in South and Southeast Asian cultures, an interdisciplinary approach can reveal a wealth of information. Ethnobotanists provide historical data on the use of curcumin in traditional medicine, while biochemists might investigate the anti-inflammatory properties of curcumin, an active compound. Pharmacologists could then explore its potential in drug development, and anthropologists could offer insights into its cultural significance and the implications of its commercialization. Such an integrated approach ensures that research is not only scientifically robust but also culturally sensitive and ethically sound. This approach opens up opportunities for discoveries that are more holistic and reflective of both the scientific and cultural dimensions of spice seeds, leading to more effective and culturally appropriate applications.

Shared Knowledge Platforms

The development of shared knowledge platforms that document and make traditional ethnobotanical knowledge accessible to researchers worldwide represents a significant step forward in the field of natural product research. These platforms can serve as comprehensive databases where information about the history, culture, and medicinal uses of various plants and seeds is systematically cataloged and made available (Vallet et al., 2023). These platforms can significantly expedite the discovery and development of new therapeutics by bridging the gap between traditional wisdom and modern science. For instance, a database that compiles the traditional uses of spice seeds such as black cumin or turmeric across different cultures can provide a starting point for scientists looking to investigate their medicinal properties. Researchers can use this information to identify patterns, such as the use of a particular seed in various cultures for similar ailments, which might suggest a common bioactive compound with therapeutic potential. This approach not only accelerates the research process by providing a wealth of preexisting knowledge but also ensures that the research is grounded in real applications and cultural practices. Moreover, shared knowledge platforms play a crucial role in preserving and respecting the traditional knowledge of indigenous and local communities. Much of this knowledge, accumulated over generations, is at risk of being lost or overlooked in the modern scientific paradigm. Such platforms not only help safeguard these communities but also acknowledge and honor their contributions by documenting and sharing this wisdom. They facilitate a collaborative environment where traditional knowledge holders and modern researchers can interact and learn from each other, fostering mutual respect and understanding. For example, a shared database might reveal that certain Amazonian tribes have been using a specific seed for its anti-inflammatory properties, which could then be studied scientifically for potential drug development. This not only leads to new scientific discoveries but also ensures that the communities that have been custodians of this knowledge are recognized and potentially benefit from any commercial developments. In this way, shared knowledge platforms are not limited by scientific advancement; they are also instruments of cultural preservation and ethical research practice, ensuring that the exploration of natural products is inclusive, respectful, and beneficial for all parties involved.

Translational Research

Translational research in the context of studying spice seeds involves a critical process of converting scientific findings from laboratory research into practical therapeutic applications or treatments for patients. This type of research is pivotal in bridging the gap between laboratory discoveries and clinical practice, ensuring that the potential health benefits of spice seeds are realized in real-world medical settings (Srikanth & Goel, 2023). For example, laboratory research on turmeric has identified curcumin as a compound with significant anti-inflammatory and antioxidant properties. Therefore, translational research could apply these findings in clinical trials to test their efficacy and safety in treating conditions such as arthritis or chronic inflammation in humans. This process involves testing for therapeutic efficacy and determining the appropriate dosages, delivery methods, and potential side effects. The goal is to develop safe, effective, and accessible treatments that can be integrated into healthcare practice, thereby directly benefiting patients. An integrated approach in translational research is essential for the rapid and effective development of these treatments. This approach allows scientists, clinicians, and researchers from various disciplines to work collaboratively, ensuring a seamless transition of knowledge from the laboratory to the clinic. This multidisciplinary collaboration is crucial for addressing the complex challenges that arise in translating research into practice. For instance, researchers might discover the potential use of black cumin for treating respiratory ailments. Translational research should then involve pulmonologists, pharmacologists, and clinical trial experts to further study and develop this approach as a viable treatment option. To fully realize the therapeutic benefits of spice seeds, research should not be confined to the laboratory; rather, research should also be rigorously tested and validated in clinical settings. This approach, where translational research plays a pivotal role, aims to bridge the gap between experimental discoveries and their application in medical scenarios. By integrating ancient wisdom with modern scientific research on spice seeds, this strategy is geared toward developing effective healthcare treatments. The ultimate objective is to provide tangible health benefits to patients across the globe by seamlessly blending time-honored practices with contemporary scientific advancements for the best possible health outcomes.

CONCLUSION AND FUTURE SCOPE

Spice seeds, which are rich in bioactive compounds, bridge the gap between time-honored traditions and cutting-edge research. Their diverse roles, spanning gastronomy, culture, and medicine, resonate globally, surpassing temporal and geographical boundaries. This chapter intertwines the depth of ethnobotanical wisdom with the precision of modern pharmacology, painting a hopeful narrative. This suggests that answers to contemporary health conundrums might indeed be rooted in time-tested knowledge from past generations. Exploration of the known aspects of spice seeds reveals uncharted territories awaiting further discovery. Beyond the well-documented seeds lie countless lesser-known variants, each potentially a reservoir of therapeutic compounds. The integration of advanced technologies, such as artificial intelligence (AI), holds promise for facilitating the rapid tracking of these seeds. However, with the growing commercial interest in these seeds, ethical considerations, especially concerning indigenous knowledge, are taking center stage. A shift toward proactive healthcare emphasizes the daily embrace of these potent seeds, upheld by both historical records and scientific evidence. The way forward calls for a collaborative global effort, harnessing expertise from various disciplines. In essence, these diminutive seeds hold expansive potential. As custodians of this knowledge, there is an inherent duty to harness their potential, crafting a future imbued with health, well-being, and sustainable harmony.

ABBREVIATIONS

CINV chemotherapy-induced nausea and vomiting
FDA Food and Drug Administration
TCM traditional Chinese medicine

ACKNOWLEDGMENT

The author is deeply grateful to the expert reviewers for their insightful feedback and recommendations, which significantly enhanced the quality of this chapter.

REFERENCES

Abdallah, E. G., Amer, A. M., Omar, H. S., & Hussein, M. H. (2023). In vitro propagation and Agrobacterium-mediated genetic transformation of caraway (*Carum carvi* L.). *Egyptian Pharmaceutical Journal*, *22*(1), 94–104. doi:10.4103/epj.epj_136_22

Abdel-Salam, O. M. E., & Mózsik, G. (2023). Capsaicin, The Vanilloid Receptor TRPV1 Agonist in Neuroprotection: Mechanisms Involved and Significance. *Neurochemical Research*, *48*(11), 3296–3315. doi:10.1007/s11064-023-03983-z PMID:37493882

Aggarwal, G., Sharma, M., Singh, R., & Sharma, U. (2024). Ethnopharmacologically important highly subsidized Indian medicinal plants: Systematic review on their traditional uses, phytochemistry, pharmacology, quality control, conservation status and future prospective. *Journal of Ethnopharmacology*, *320*, 117385. doi:10.1016/j.jep.2023.117385 PMID:37951375

Agrawal, M., Singhal, M., Jasoria, Y., Chaudhary, H., & Prajapati, B. G. (2024). Pharmacological Aspects of Coriander Essential Oils. In *Pharmacological Aspects of Essential Oils* (pp. 227–242). CRC Press.

Ahmad, N., Riaz, S., Ali, A., & Riaz, T. (2023). Fenugreek (*Trigonella foenum-graecum*): An Overview of Food Uses and Health Benefits. In T. Ismail, S. Akhtar, & C. E. Lazarte (Eds.), *Neglected Plant Foods Of South Asia: Exploring and valorizing nature to feed hunger* (pp. 453–471). Springer International Publishing. doi:10.1007/978-3-031-37077-9_18

Ahmed, A. G., Hussein, U. K., Ahmed, A. E., Kim, K. M., Mahmoud, H. M., Hammouda, O., Jang, K. Y., & Bishayee, A. (2020). Mustard seed (*Brassica nigra*) extract exhibits antiproliferative effect against human lung cancer cells through differential regulation of apoptosis, cell cycle, migration, and invasion. *Molecules (Basel, Switzerland)*, *25*(9), 2069. doi:10.3390/molecules25092069 PMID:32365503

Ai, C., Zou, Y., Liu, H., Yang, Z., & Xi, J. (2023). Traditional Chinese herbal medicine for allergic diseases: A review. *The American Journal of Chinese Medicine*, *51*(4), 779–806. doi:10.1142/S0192415X23500374 PMID:37060193

Akhtar, M. J., & Khan, S. A. (2024). Chemistry and biological activity of mustard oil: Therapeutic benefits and risk to healthcare. *Revista Brasileira de Farmacognosia, 34*(1), 65–79. doi:10.1007/s43450-023-00450-2

Akhtari, E., Ram, M., Zaidi, S. M. A., Marques, A. M., Rahimi, R., & Bahramsoltani, R. (2024). Fenugreek (*Trigonella foenum-graecum* L.) in women's health: A review of clinical evidence and traditional use. *Journal of Herbal Medicine, 43*, 100816. doi:10.1016/j.hermed.2023.100816

Akhzari, M., Mohammadi, S., Jalili, S., Barazesh, M., & Noorizadeh, K. (2023). Nigella sativa, a jack of all trades plant in medicine: Pharmacological aspects in diseases treatment and prevention. *The Natural Products Journal, 13*(6), 2–24. doi:10.2174/2210315513666221005093047

Ali, T., & Sultan, H. (2023). An Islamic perspective on infection treatment and wound healing. *Religions, 14*(8), 1044. doi:10.3390/rel14081044

Aly, A., Maraei, R., Rezk, A., & Diab, A. (2023). Phytochemical constitutes and biological activities of essential oil extracted from irradiated caraway seeds (*Carum carvi* L.). *International Journal of Radiation Biology, 99*(2), 318–328. doi:10.1080/09553002.2022.2078004 PMID:35549975

Anand, A. V., Sreedevi, M. J., & Swapna, T. S. (2023). Plant Conservation Associated with Traditional Knowledge: Past and Future. In S. T. Sukumaran & K. T R (Eds.), Conservation and Sustainable Utilization of Bioresources (pp. 261-290). Springer Nature Singapore. doi:10.1007/978-981-19-5841-0_11

Anlar, H. G., & Bacanli, M. (2020). Thymoquinone: the active compound of black seed (Nigella sativa). In V. R. Preedy (Ed.), *Pathology* (pp. 369–378). Academic Press. doi:10.1016/B978-0-12-815972-9.00035-4

Anwar, T., Qureshi, H., Naeem, H., Siddiqi, E. H., Hanif, A., Anwaar, S., Noreen, Z., Iqbal, J., Ali, B., Iqbal, R., & Paray, B. A. (2024). Prospects of the sustainability of traditional and ethnic foods and their applications for basic health care in human society. *Journal of King Saud University. Science, 36*(2), 103016. doi:10.1016/j.jksus.2023.103016

Archita, K., Ved, P., Vikesh Kumar, S., & Rohit, S. (2023). Exploring the therapeutic potential of herbal plants in managing blood sugar levels: A comprehensive evaluation. *Pharmacognosy Research, 16*(1), 10–18. doi:10.5530/pres.16.1.2

Arunachalam, K., Sreeja, P. S., & Yang, X. (2023). Mechanisms and therapeutic actions of edible fruits in inflammatory bowel disease: A review of preclinical studies. *Food Chemistry Advances, 3*, 100498. doi:10.1016/j.focha.2023.100498

Ashokkumar, K., Vellaikumar, S., Muthusamy, M., Dhanya, M. K., & Aiswarya, S. (2021). Compositional variation in the leaf, mace, kernel, and seed essential oil of nutmeg (Myristica fragrans Houtt.) from the Western Ghats, India. *Natural Product Research, 36*(1), 432–435. doi:10.1080/14786419.2020.1771713 PMID:32515616

Aworh, O. C. (2023). African traditional foods and sustainable food security. *Food Control, 145*, 109393. doi:10.1016/j.foodcont.2022.109393

Bakhtiar, Z., Hassandokht, M., Naghavi, M. R., & Mirjalili, M. H. (2024). Variability in proximate composition, phytochemical traits and antioxidant properties of Iranian agro-ecotypic populations of fenugreek (*Trigonella foenum-graecum* L.). *Scientific Reports, 14*(1), 87. doi:10.1038/s41598-023-50699-9 PMID:38167975

Bansal, K., Sharma, S., & Bajpai, M. (2024). Herbal medicines-a fruitful approach to periodic illness dysmenorrhoea: Evidence-based review. *Current Pharmaceutical Biotechnology, 25*(2), 179–195. doi:10.2174/1389201024666230623161113 PMID:37357517

Bennett, M. D. (2023). Caribbean plantation economies as colonial models: The case of the English East India Company and St. Helena in the late seventeenth century. *Atlantic Studies, 20*(4), 508–539. doi:10.1080/14788810.2022.2034569

Brindisi, L. J., Lyu, W., Juliani, H. R., Wu, Q., Tepper, B. J., & Simon, J. E. (2023). Determination of glucosinolates and breakdown products in Brassicaceae baby leafy greens using UHPLC-QTOF/MS and GC/MS. *Food Chemistry Advances, 3*, 100389. doi:10.1016/j.focha.2023.100389

Burdock, G. A. (2022). Assessment of black cumin (*Nigella sativa* L.) as a food ingredient and putative therapeutic agent. *Regulatory Toxicology and Pharmacology, 128*, 105088. doi:10.1016/j.yrtph.2021.105088 PMID:34838871

Busnel, R., & Manrique López, H. (2023). The political economy of a failed drug reform: Insights from Peru's main legal coca valley. *The International Journal on Drug Policy, 117*, 104050. doi:10.1016/j.drugpo.2023.104050 PMID:37267739

Caceres Guido, P., Ribas, A., Gaioli, M., Quattrone, F., & Macchi, A. (2015). The state of the integrative medicine in Latin America: The long road to include complementary, natural, and traditional practices in formal health systems. *European Journal of Integrative Medicine, 7*(1), 5–12. doi:10.1016/j.eujim.2014.06.010

Cao, M., Zhan, M., Jing, H., Wang, Z., Wang, Y., Li, X., & Miao, M. (2024). Network pharmacology and experimental evidence: MAPK signaling pathway is involved in the anti-asthma roles of Perilla frutescens leaf. *Heliyon, 10*(1), e22971. doi:10.1016/j.heliyon.2023.e22971 PMID:38163225

Cătunescu, G. M., Bodea, I. M., David, A. P., Pop, C. R., & Rotar, A. M. (2023). Essential oils from Apiaceae family (parsley, lovage, and dill). In G. A. Nayik & M. J. Ansari (Eds.), *Essential Oils* (pp. 241–308). Academic Press. doi:10.1016/B978-0-323-91740-7.00015-3

Cerón, D. A. C., de Alencar, E. R., Faroni, L. R. D. A., Silva, M. V. D. A., & Salvador, D. V. (2023). Toxicity of allyl isothiocyanate applied in systems with or without recirculation for controlling *Sitophilus zeamais, Rhyzopertha dominica, and Tribolium castaneum* in corn grains. *Journal of the Science of Food and Agriculture, xx*(x), xxx–xxx. doi:10.1002/jsfa.12713 PMID:37205787

Chansiw, N., Chusri, P., Praman, S., Hawiset, T., Sukhorum, W., Champakam, S., & Charoensup, R. (2024). Anti-inflammatory potential of a Thai traditional remedy called Prabchompoothaweep in an animal model of acute and subacute inflammation. *Journal of Ethnopharmacology, 319*, 117380. doi:10.1016/j.jep.2023.117380 PMID:37925003

Chaudhary, N., Ghimire, S. K., & Chaudhary, R. P. (2023). Traditional Amchi Medicinal Practice in Trans-Himalaya of Nepal: Conservation and Bioprospecting. In K. Arunachalam, X. Yang, & S. Puthanpura Sasidharan (Eds.), *Bioprospecting of Tropical Medicinal Plants* (pp. 307–327). Springer Nature Switzerland. doi:10.1007/978-3-031-28780-0_11

Cheikhyoussef, A., Cheikhyoussef, N., Sharma, R., & Hussein, A. A. (2023). Health-Promoting Effects of Cardamom (*Elettaria cardamomum*). In M. F. Ramadan (Ed.), *Cardamom (Elettaria cardamomum): Production, Processing and Properties* (pp. 133–156). Springer International Publishing. doi:10.1007/978-3-031-35426-7_11

Cilak, G. O., Mujdeci, G. N., & Kabak, B. (2021). Legislation on aromatic herbs in food. In C. M. Galanakis (Ed.), *Aromatic Herbs in Food* (pp. 405–438). Academic Press. doi:10.1016/B978-0-12-822716-9.00011-1

Correia, A. G. S., Alencar, M. B., dos Santos, A. N., da Paixão, D. C. B., Sandes, F. L. F., Andrade, B., Castro, Y., & de Andrade, J. S. (2023). Effect of saffron and fenugreek on lowering blood glucose: A systematic review with meta-analysis. *Phytotherapy Research, 37*(5), 2092–2101. Advance online publication. doi:10.1002/ptr.7817 PMID:36992660

Crichton, M., Marshall, S., Isenring, E., Lohning, A., McCarthy, A. L., Molassiotis, A., ... Marx, W. (2023). Effect of a standardized ginger root powder regimen on chemotherapy-induced nausea and vomiting: A multicenter, double-blind, placebo-controlled randomized trial. *Journal of the Academy of Nutrition and Dietetics, xx*(x), xxx–xxx. doi:10.1016/j.jand.2023.09.003 PMID:37699474

D'Ambrosio, U., Pozo, C., Vallès, J., & Gras, A. (2023). East meets west: Using ethnobotany in ethnic urban markets of Barcelona metropolitan area (Catalonia) as a tool for biocultural exchange. *Journal of Ethnobiology and Ethnomedicine*, *19*(1), 63. doi:10.1186/s13002-023-00636-x PMID:38105250

da Silva, A. K. A., dos Santos, C. R., Carneiro, M. L. B., Joanitti, G. A., Luz, G. V. S., Rosa, S. S. R. F., . . . Tatmatsu-Rocha, J. C. (2023). Chapter 1 - Bixa orellana L. and its implications in human health: Perspectives and new trends. In R. Atta ur (Ed.), Studies in Natural Products Chemistry (Vol. 77, pp. 1-36). Elsevier. doi:10.1016/B978-0-323-91294-5.00001-4

Dąbrowska, J. A., Kunicka-Styczyńska, A., & Śmigielski, K. B. (2020). Biological, chemical, and aroma profiles of essential oil from waste celery seeds (Apium graveolens L.). *The Journal of Essential Oil Research*, *32*(4), 308–315. doi:10.1080/10412905.2020.1754937

Daley, D. K., & Badal, S. (2024). Plant crude drugs. In S. B. McCreath & Y. N. Clement (Eds.), *Pharmacognosy* (2nd ed., pp. 75–89). Academic Press. doi:10.1016/B978-0-443-18657-8.00002-5

Damayanti, A., & Setyawan, E. (2012). Essential oil extraction of fennel seed (*Foeniculum vulgare*) using steam distillation. *International Journal of Science and Engineering*, *3*(2), 12–14. doi:10.12777/ijse.3.2.12-14

Das, G., Das, S., Talukdar, A., Venil, C., Bose, S., Banerjee, S., ... Patra, J. (2023). Pharmacology and ethnomedicinal potential of selected plants species from Apiaceae (Umbelliferae). *Combinatorial Chemistry & High Throughput Screening*, *26*(2), 256–288. doi:10.2174/1386207325666220406110404 PMID:35388749

Das, P., Chandra, T., Negi, A., Jaiswal, S., Iquebal, M. A., Rai, A., & Kumar, D. (2023). A comprehensive review on genomic resources in medicinally and industrially important major spices for future breeding programs: Status, utility and challenges. *Current Research in Food Science*, *7*, 100579. doi:10.1016/j.crfs.2023.100579 PMID:37701635

Debnath, S., Kumar, H., & Sharma, A. (2023). *Foeniculum vulgare* Mill. from spice to pharma: Recent advances in its medicinal value, bioactivities and perspectives. *Traditional and Integrative Medicine*, *8*(2), 217–229. doi:10.18502/tim.v8i2.13087

Delgadillo-Puga, C., Torre-Villalvazo, I., Cariño-Cervantes, Y. Y., García-Luna, C., Soberanes-Chávez, P., de Gortari, P., Noriega, L. G., Bautista, C. J., & Cisneros-Zevallos, L. (2023). Cardamom (*Elettaria cardamomum* (L.) Maton) seeds intake increases energy expenditure and reduces fat mass in mice by modulating neural circuits that regulate adipose tissue lipolysis and mitochondrial oxidative metabolism in liver and skeletal muscle. *International Journal of Molecular Sciences*, *24*(4), 3909. doi:10.3390/ijms24043909 PMID:36835337

Deshmukh, P. (2024). Utilization of herbs and spices as a conventional medicine in covid-19: A study. *Multidisciplinary Science Journal*, *6*(1), 2024008–2024008. doi:10.31893/multiscience.2024008

Dhull, S. B., Bamal, P., Chandak, A., Nain, K. B., & Malik, A. (2024). Fenugreek (*Trigonella foenum-graecum*): An herb with impressive nutritional and antidiabetic properties. In M. Naeem & T. Aftab (Eds.), *Antidiabetic Medicinal Plants* (pp. 305–325). Academic Press. doi:10.1016/B978-0-323-95719-9.00012-4

Dilrukshi, E. A. C., Nishiyama, Y., Ito, K., & Nomura, S. (2024). Alleviation of acute stress response by black pepper aroma administration. *Journal of Physiological Anthropology*, *43*(1), 3. doi:10.1186/s40101-023-00352-1 PMID:38167228

Divakaran, M., Jayasree, E., Nirmal Babu, K., & Peter, K. V. (2018). Legacy of Indian Spices: Its Production and Processing. In A. B. Sharangi (Ed.), *Indian Spices: The Legacy, Production and Processing of India's Treasured Export* (pp. 13–30). Springer International Publishing. doi:10.1007/978-3-319-75016-3_2

Djiazet, S., Kavitake, D., Ravi, R., Mezajoug Kenfack, L. B., Pullakhandam, R., Prabhavathi Devi, B. L. A., Tchiégang, C., Reddy, G. B., & Shetty, P. H. (2023). Investigation of physicochemical properties, nutritional status, and volatile compounds profiling of four underutilized spices of Cameroon origin. *Food Chemistry Advances*, *3*, 100465. doi:10.1016/j.focha.2023.100465

Domingo-Fernández, D., Gadiya, Y., Mubeen, S., Bollerman, T. J., Healy, M. D., Chanana, S., Sadovsky, R. G., Healey, D., & Colluru, V. (2023). Modern drug discovery using ethnobotany: A large-scale cross-cultural analysis of traditional medicine reveals common therapeutic uses. *iScience*, *26*(9), 107729. doi:10.1016/j.isci.2023.107729 PMID:37701812

dos Santos, M. L., Chandran, D., Lejaniya, A. S., & da Silva, L. E. (2023). Conservation and Sustainable Use of Medicinal Plants. In S. Jha & M. Halder (Eds.), *Medicinal Plants: Biodiversity, Biotechnology and Conservation* (pp. 327–341). Springer Nature Singapore. doi:10.1007/978-981-19-9936-9_13

El-Assri, E.-M., Hajib, A., Choukri, H., Gharby, S., Lahkimi, A., Eloutassi, N., & Bouia, A. (2023). Nutritional quality, lipid, and mineral profiling of seven Moroccan Apiaceae seeds. *South African Journal of Botany*, *160*, 23–35. doi:10.1016/j.sajb.2023.06.042

Ersîn, E., & Coşge Şenkal, B. (2023). Impacts of essential oil and extracts obtained from coriander cultivars (Coriandrum sativum L.) on important some pathogenic bacteria. *Current Perspectives on Medicinal and Aromatic Plants*, *6*(1), 27–37. doi:10.38093/cupmap.1281520

Fakhr, L., Chehregosha, F., Zarezadeh, M., Chaboksafar, M., & Tarighat-Esfanjani, A. (2023). Effects of fenugreek supplementation on the components of metabolic syndrome: A systematic review and dose–response meta-analysis of randomized clinical trials. *Pharmacological Research*, *187*, 106594. doi:10.1016/j.phrs.2022.106594 PMID:36470549

Ferreira, P. M. P., Arcanjo, D. D. R., & Peron, A. P. (2023). Drug development, Brazilian biodiversity and political choices: Where are we heading? *Journal of Toxicology and Environmental Health. Part B, Critical Reviews*, *26*(5), 257–274. doi:10.1080/10937404.2023.2193762 PMID:36967535

Galal, S. M., & Morsy, N. F. S. (2023). Cardamom in Food Applications. In M. F. Ramadan (Ed.), *Cardamom (Elettaria cardamomum): Production, Processing and Properties* (pp. 243–253). Springer International Publishing. doi:10.1007/978-3-031-35426-7_18

Ganapathy, A. A., Haripriya, V. M., Acharya, N., Somappa, S. B., & Kumaran, A. (2023). Ethnobotanical significance of medicinal plants: Beta-amyloid and tau aggregation inhibitors against Alzheimer's disease. *Journal of Biochemical and Molecular Toxicology*, *37*(6), e23339. doi:10.1002/jbt.23339 PMID:37009720

Gavish, L., Korchia, G., Cohen, S., & Maor, Y. (2023). A real-world evaluation of an herbal treatment for infantile colic reported by 1218 parents in Israel. *Journal of Pediatric Nursing*, *69*, e39–e44. doi:10.1016/j.pedn.2022.12.002 PMID:36528454

Ghazanfari, N., Tabatabaei Yazdi, F., Mortazavi, S. A., & Mohammadi, M. (2023). Using pulsed electric field pretreatment to optimize coriander seeds essential oil extraction and evaluate antimicrobial properties, antioxidant activity, and essential oil compositions. *Lebensmittel-Wissenschaft + Technologie*, *182*, 114852. doi:10.1016/j.lwt.2023.114852

Gholamnezhad, Z., Havakhah, S., & Boskabady, M. H. (2016). Preclinical and clinical effects of Nigella sativa and its constituent, thymoquinone: A review. *Journal of Ethnopharmacology*, *190*, 372–386. doi:10.1016/j.jep.2016.06.061 PMID:27364039

Grant, J. K., Dangl, M., Ndumele, C. E., Michos, E. D., & Martin, S. S. (2023). A historical, evidence-based, and narrative review on commonly used dietary supplements in lipid-lowering. *Journal of Lipid Research*, *xx*(x), 100493. doi:10.1016/j.jlr.2023.100493 PMID:38145747

Gupta, N., Bhattacharya, S., Urbanová, K., Dutta, A., Hazra, A. K., Fernández-Cusimamani, E., & Leuner, O. (2023). Systematic analysis of antimicrobial activity, phytochemistry, and in silico molecular interaction of selected essential oils and their formulations from different Indian spices against foodborne bacteria. *Heliyon*, *9*(12), e22480. doi:10.1016/j.heliyon.2023.e22480 PMID:38107328

Gupta, S., Singh, A. P., Singh, G., Ding, X., & Sharma, A. (2024). Plant based metabolomics: A future prospective and versatile tool for metabolite databases of *Curcuma longa*. *Journal of Future Foods*, *4*(1), 37–47. doi:10.1016/j.jfutfo.2023.05.003

Guzmán-Gutiérrez, S. L., Reyes-Chilpa, R., González-Diego, L. R., Silva-Miranda, M., López-Caamal, A., García-Cruz, K. P., Jiménez-Mendoza, M. S., Arciniegas, A., & Espitia, C. (2023). Five centuries of Cirsium ehrenbergii Sch. Bip. (Asteraceae) in Mexico, from Huitzquilitl to Cardo Santo: History, ethnomedicine, pharmacology and chemistry. *Journal of Ethnopharmacology*, *301*, 115778. doi:10.1016/j.jep.2022.115778 PMID:36202165

Hadi, S., Restuhadi, F., & Lukmanasari, P. (2023). Export competitiveness analysis of pepper (Piper Nigrum L.) commodity in the international market. *Jurnal Agrinika: Jurnal Agroteknologi dan Agribisnis*, *7*(2), 134-143. doi:10.30737/agrinika.v7i2.4773

Hafeez, A., Ali, S. S., Akhtar, J., Naz, S., Alhidary, I. A., Israr, M., & Khan, R. U. (2023). Garlic (*Allium sativum*), fenugreek (*Trigonella foenum-graecum*) and coriander (*Coriandrum sativum*): Performance, nutrient digestibility and blood metabolites in broilers. *Journal of Applied Animal Research*, *51*(1), 624–629. doi:10.1080/09712119.2023.2264966

Hancock, J. F. (2022). Five Hundred Years After the Great Encounter. In J. F. Hancock (Ed.), *World Agriculture Before and After 1492: Legacy of the Columbian Exchange* (pp. 135–144). Springer International Publishing. doi:10.1007/978-3-031-15523-9_10

Hazarika, S., Borah, P., Deb, P. K., Venugopala, K. N., & Hemalatha, S. (2023). Icacinaceae plant family: A recapitulation of the ethnobotanical, phytochemical, pharmacological, and biotechnological aspects. *Current Pharmaceutical Design*, *29*(15), 1193–1217. doi:10.2174/1381612829666230502164605 PMID:37132105

He, G., Yang, S., & Wang, Y. (2023). The potential of Amomum tsao-ko as a traditional Chinese medicine: Traditional clinical applications, phytochemistry and pharmacological properties. *Arabian Journal of Chemistry, 16*(8), 104936. doi:10.1016/j.arabjc.2023.104936

He, X., Chen, X., Yang, Y., Xie, Y., & Liu, Y. (2024). Medicinal plants for epileptic seizures: Phytoconstituents, pharmacology and mechanisms revisited. *Journal of Ethnopharmacology, 320*, 117386. doi:10.1016/j.jep.2023.117386 PMID:37956914

Hordofa, T. S., & Tolossa, T. T. (2020). Cultivation and postharvest handling practices affecting yield and quality of major spices crops in Ethiopia: A review. *Cogent Food & Agriculture, 6*(1), 1788896. doi:10.1080/23311932.2020.1788896

Hossain, R., Khan, R. A., Mukhopadhyay, N., Jain, D., & Islam, M. T. (2024). Phytochemical Constituents 6 and Antidiabetic Features of Black Cumin (Nigella sativa L.). In Antidiabetic Medicinal Plants and Herbal Treatments (pp. 95-115). CRC Press.

Hussain, M., Sabri, R., Zia-Ul-Haq, M., & Riaz, M. (2023). Celery. In M. Zia-Ul-Haq, A. Abdulkreem Al-Huqail, M. Riaz, & U. Farooq Gohar (Eds.), *Essentials of Medicinal and Aromatic Crops* (pp. 1165–1190). Springer International Publishing. doi:10.1007/978-3-031-35403-8_45

Ikechukwu, M., Maroyi, A., & Cheikhyoussef, A. (2023). Cardamom Botany, Cultivars, and Genetic Diversity. In M. F. Ramadan (Ed.), *Cardamom (Elettaria cardamomum): Production, Processing and Properties* (pp. 39–50). Springer International Publishing. doi:10.1007/978-3-031-35426-7_3

Inprasit, J., Itharat, A., Ruangnoo, S., Thisayakorn, K., Sukkasem, K., Prommee, N., Khoenok, W., Sriyam, K., Pahusee, D., & Davies, N. M. (2024). Ethnopharmacological analysis based on Thai traditional medicine theory and anti-inflammatory activity of Sa-Tri-Lhung-Klod remedy as a postpartum anti-inflammatory drug. *Journal of Ethnopharmacology, 319*, 117207. doi:10.1016/j.jep.2023.117207 PMID:37739101

Iweala, E. J., Uche, M. E., Dike, E. D., Etumnu, L. R., Dokunmu, T. M., Oluwapelumi, A. E., Okoro, B. C., Dania, O. E., Adebayo, A. H., & Ugbogu, E. A. (2023). *Curcuma longa* (Turmeric): Ethnomedicinal uses, phytochemistry, pharmacological activities and toxicity profiles—A review. *Pharmacological Research. Modern Chinese Medicine, 6*, 100222. doi:10.1016/j.prmcm.2023.100222

Jadid, N., Widodo, A. F., Ermavitalini, D., Sa'adah, N. N., Gunawan, S., & Nisa, C. (2023). The medicinal Umbelliferae plant Fennel (*Foeniculum vulgare* Mill.): Cultivation, traditional uses, phytopharmacological properties, and application in animal husbandry. *Arabian Journal of Chemistry*, *16*(3), 104541. doi:10.1016/j.arabjc.2023.104541

Katoch, P., Raina, K., Sharma, R., & Chaudhary, A. (2024). Ayush Kwath: A major contribution of ayurveda in preventing COVID-19 infection. *Current Drug Therapy*, *19*(1), 60–80. doi:10.2174/1574885518666230601150338

Kavyani, Z., Musazadeh, V., Golpour-hamedani, S., Moridpour, A. H., Vajdi, M., & Askari, G. (2023). The effect of Nigella sativa (black seed) on biomarkers of inflammation and oxidative stress: An updated systematic review and meta-analysis of randomized controlled trials. *Inflammopharmacology*, *31*(3), 1149–1165. doi:10.1007/s10787-023-01213-0 PMID:37036558

Khalifa, A. A., Rashad, R. M., & El-Hadidy, W. F. (2021). Thymoquinone protects against cardiac mitochondrial DNA loss, oxidative stress, inflammation and apoptosis in isoproterenol-induced myocardial infarction in rats. *Heliyon*, *7*(7), e07561. doi:10.1016/j.heliyon.2021.e07561 PMID:34355083

Khaliq, F. A., & Mushtaq, A. (2023). The five most traded compounds worldwide: Importance, opportunities, and risks. *International Journal of Chemical and Biochemical Science*, *23*, 185–194.

Khan, M. R., & Aslam, S. (2023a). Cardamom Safety. In M. F. Ramadan (Ed.), *Cardamom (Elettaria cardamomum): Production, Processing and Properties* (pp. 227–241). Springer International Publishing. doi:10.1007/978-3-031-35426-7_17

Khan, M. R., & Aslam, S. (2023b). Plant Morphological Traits of Elettaria cardamomum. In M. F. Ramadan (Ed.), *Cardamom (Elettaria cardamomum): Production, Processing And Properties* (pp. 51–66). Springer International Publishing. doi:10.1007/978-3-031-35426-7_4

Khanam, M., Dar, A. H., Beg, F., Khan, S. A., Nayik, G. A., & Karabagias, I. K. (2023). Nutmeg essential oil. In G. A. Nayik & M. J. Ansari (Eds.), *Essential Oils* (pp. 391–399). Academic Press. doi:10.1016/B978-0-323-91740-7.00012-8

Kilian, I. C., Swenson, S. J., Mengual, X., Gemeinholzer, B., Hamm, A., Wägele, J. W., & Peters, R. S. (2023). More complex than you think: Taxonomic and temporal patterns of plant–pollinator networks of caraway (*Carum carvi* L.). *Molecular Ecology*, *32*(13), 3702–3717. doi:10.1111/mec.16943 PMID:37004150

Kim, J. H., Kim, T. I., & Ma, J.-Y. (2022). Synergistic effects of novel herbal decoctions from *Panax ginseng* and *Morus alba* on tyrosinase activity and melanogenesis in vitro. *Heliyon*, *8*(2), e08866. doi:10.1016/j.heliyon.2022.e08866 PMID:35198755

Kmail, A., Said, O., & Saad, B. (2023). How thymoquinone from *Nigella sativa* accelerates wound healing through multiple mechanisms and targets. *Current Issues in Molecular Biology*, *45*(11), 9039–9059. doi:10.3390/cimb45110567 PMID:37998744

Kulkarni, A., Nair, A., Hirode, K., Chavan, M., Jain, M., Majumdar, P., & Jamkhedkar, S. (2023). Indian Traditional Foods from an Ayurvedic Perspective. In *Ethnic Knowledge and Perspectives of Medicinal Plants* (pp. 395–424). Apple Academic Press. doi:10.1201/9781003353089-19

Kumar, V., Kumar, S., Chauhan, R. S., & Dwivedi, S. (2024). Medicinal Plants and Their Role in Mitigation of Arsenic-Induced Toxicity. In *Phytoremediation Potential of Medicinal and Aromatic Plants* (pp. 154–171). CRC Press.

Lang, S., Popp, T., Kriegs, C. S., Schmidt, A., Balszuweit, F., Menacher, G., Kehe, K., Thiermann, H., Gudermann, T., & Steinritz, D. (2018). Anti-apoptotic and moderate anti-inflammatory effects of berberine in sulfur mustard exposed keratinocytes. *Toxicology Letters*, *293*, 2–8. doi:10.1016/j.toxlet.2017.09.004 PMID:28916288

Larsson, M., & Foley, B. (2023). The king's spice cabinet–Plant remains from Gribshunden, a 15th century royal shipwreck in the Baltic Sea. *PLoS One*, *18*(1), e0281010. doi:10.1371/journal.pone.0281010 PMID:36701280

Li, J., Chen, W., Liu, H., Liu, H., Xiang, S., You, F., Jiang, Y., Lin, J., Zhang, D., & Zheng, C. (2023). Pharmacologic effects approach of essential oils and their components on respiratory diseases. *Journal of Ethnopharmacology*, *304*, 115962. doi:10.1016/j.jep.2022.115962 PMID:36529244

Li, S., Huang, X., Li, Y., Ding, R., Wu, X., Li, L., ... Gu, R. (2024). Spectrum-effect relationship in Chinese herbal medicine: Current status and future perspectives. *Critical Reviews in Analytical Chemistry*, *xx*(x), 1–22. doi:10.1080/10408347.2024.2339955 PMID:38127670

Liang, Y., Dai, X., Cao, Y., Wang, X., Lu, J., Xie, L., Liu, K., & Li, X. (2023). The neuroprotective and antidiabetic effects of trigonelline: A review of signaling pathways and molecular mechanisms. *Biochimie*, *206*, 93–104. doi:10.1016/j.biochi.2022.10.009 PMID:36257493

Liu, C., Wang, W., & Gu, J. (2023). Targeting ferroptosis: New perspectives of Chinese herbal medicine in the treatment of diabetes and its complications. *Heliyon*, *9*(12), e22250. doi:10.1016/j.heliyon.2023.e22250 PMID:38076182

Lucas-Sánchez, M., Fadhlaoui-Zid, K., & Comas, D. (2023). The genomic analysis of current-day North African populations reveals the existence of trans-Saharan migrations with different origins and dates. *Human Genetics*, *142*(2), 305–320. doi:10.1007/s00439-022-02503-3 PMID:36441222

Machado, A. M., Lopes, V., Barata, A. M., Póvoa, O., Farinha, N., & Figueiredo, A. C. (2023). Chemical variability of the essential oils from two Portuguese Apiaceae: Coriandrum sativum L. and Foeniculum vulgare Mill. *Plants*, *12*(14), 2749. doi:10.3390/plants12142749 PMID:37514362

Madhu, S. V., Rao, P. V., Chandalia, H. B., Jothydev, K., & Gupta, A. (2023). A multicentric, randomized, controlled trial of yoga and fenugreek in prevention of type 2 diabetes mellitus: Methodological details—the Indian Prevention of Diabetes Study (IPDS). *International Journal of Diabetes in Developing Countries*, *43*(6), 875–882. doi:10.1007/s13410-023-01183-7

Mahajan, M., Vaidya, V., Farande, P., Bhagde, S., & Jadhav, R. (2023). Review on essential oils and ways to use them for the treatment of arthritis. *Biosciences Biotechnology Research Asia*, *20*(4), 1181–1194. doi:10.13005/bbra/3167

Malhotra, S. K. (2006). 15 - Caraway. In K. V. Peter (Ed.), *Handbook of Herbs and Spices* (pp. 270–298). Woodhead Publishing. doi:10.1533/9781845691717.3.270

Mapeka, T., Sandasi, M., Ncube, E., Viljoen, A., & van Vuuren, S. (2024). Enhancing the antimicrobial efficacy of common herbs and spices through an optimized polyherbal approach. *South African Journal of Botany*, *164*, 91–99. doi:10.1016/j.sajb.2023.11.030

Maqbool, M., Chaudhary, K., Chalotra, R., Chauhan, S., & Dahiya, R. S. (2024). Phyto-pharmacology of most common Indian culinary spices and their potential in developing new pharmaceutical therapies. *Current Traditional Medicine*, *10*(2), 117–140. doi:10.2174/2215083810666230330161740

Marchant, R. (2021). Trading Language, New Crops, New Relationships: Digging Anthropocene Foundations. In R. Marchant (Ed.), *East Africa's Human Environment Interactions: Historical Perspectives for a Sustainable Future* (pp. 115–162). Springer International Publishing. doi:10.1007/978-3-030-88987-6_3

Mariappan, B., Kaliyamurthi, V., & Binesh, A. (2023). Chapter 8 - Medicinal plants or plant derived compounds used in aquaculture. In J. Mathew, M. S. Jose, R. E.K, & A. Kumar (Eds.), Recent Advances in Aquaculture Microbial Technology (pp. 153-207). Academic Press. doi:10.1016/B978-0-323-90261-8.00003-1

Máthé, Á., & Turgut, K. (2023). Introduction to Medicinal and Aromatic Plants in Türkiye. In Á. Máthé & K. Turgut (Eds.), *Medicinal and Aromatic Plants of Turkey* (pp. 1–30). Springer International Publishing. doi:10.1007/978-3-031-43312-2_1

Memet, I., Kirici, S., Giray, E. S., Murat, T., & Taghikhani, H. (2014). Determination of suitable coriander (*Coriandrum sativum* L.) cultivars for eastern mediterranean region. *Turkish Journal of Field Crops*, *19*(1), 1–6. doi:10.17557/tjfc.05577

Mipeshwaree Devi, A., Mutum, R. D., Lakshmipriyari Devi, M., Khomdram, K. D., Premi Devi, P., Singh, L. H., & Das, S. (2023). Traditional Practices of Ethnomedicinal Plants in the North-Eastern Region of India and Their Conservation for Sustainable Utilization. In S. Jha & M. Halder (Eds.), *Medicinal Plants: Biodiversity, Biotechnology and Conservation* (pp. 343–374). Springer Nature Singapore. doi:10.1007/978-981-19-9936-9_14

Mishra, A. K., Pandey, M., Pannu, A., Dewangan, H. K., & Sahoo, P. (2024). Review on diabetes mellitus: An insight into the current scenarios, the challenges of therapy, and application of traditional drugs. *Current Traditional Medicine*, *10*(3), 107–128. doi:10.2174/2215083810666230501212125

Moulick, S. P., Jahan, F., Mamun, M. Z. U. A., Hossain, M. I. S., Waliullah, M., & Sathee, R. A. (2023). Analysis of indigenous spices widely consumed in Bangladesh: An assessment to explore its proximate contents, minerals, phytochemical compositions, and antioxidant activities. *Journal of Agriculture and Food Research*, *14*, 100720. doi:10.1016/j.jafr.2023.100720

Mutinda, E. S., Kimutai, F., Mkala, E. M., Waswa, E. N., Odago, W. O., Nanjala, C., Ndungu, C. N., Gichua, M. K., Njire, M. M., Gituru, R. W., & Hu, G. W. (2023). Ethnobotanical uses, phytochemistry and pharmacology of pantropical genus *Zanthoxylum* L. (Rutaceae): An update. *Journal of Ethnopharmacology*, *303*, 115895. doi:10.1016/j.jep.2022.115895 PMID:36513263

Nath, M., & Debnath, P. (2023). Therapeutic role of traditionally used Indian medicinal plants and spices in combating COVID-19 pandemic situation. *Journal of Biomolecular Structure & Dynamics*, *41*(12), 5894–5913. doi:10.1080/07391102.2022.2093793 PMID:35773779

Nautiyal, O. H. (2019). Black Seed (*Nigella sativa*) Oil. In M. F. Ramadan (Ed.), *Fruit Oils: Chemistry and Functionality* (pp. 839–857). Springer International Publishing. doi:10.1007/978-3-030-12473-1_46

Nedungadi, P., Salethoor, S. N., Puthiyedath, R., Nair, V. K., Kessler, C., & Raman, R. (2023). Ayurveda research: Emerging trends and mapping to sustainable development goals. *Journal of Ayurveda and Integrative Medicine*, *14*(6), 100809. doi:10.1016/j.jaim.2023.100809 PMID:37832213

Nguyen, T., Nandasiri, R., & Eskin, N. M. (2023). Phenolics of mustard seeds: A review on composition, processing effect and their bioactvities. *Journal of the American Oil Chemists' Society*, *xx*(x), xxx–xxx. doi:10.1002/aocs.12783

Nutmakul, T., & Chewchinda, S. (2023). Synergistic effect of Trikatuk, a traditional Thai formulation, on antioxidant and alpha-glucosidase inhibitory activities. *Heliyon*, *9*(1), e13063. doi:10.1016/j.heliyon.2023.e13063 PMID:36711300

Nyirenda, J., & Chipuwa, M. (2024). An ethnobotanical study of herbs and medicinal plants used in western, copperbelt, central and northern provinces of Zambia. *Phytomedicine Plus : International Journal of Phytotherapy and Phytopharmacology*, *4*(1), 100514. doi:10.1016/j.phyplu.2023.100514

Okayasu, T., Mitani, K., & Kitahara, T. (2024). Reviewing Kampo medicine (Traditional Japanese Herbal Medicine) for otology/neurotology diseases. *Auris, Nasus, Larynx*, *51*(1), 25–30. doi:10.1016/j.anl.2023.04.009 PMID:37137794

Orouji, N., Asl, S. K., Taghipour, Z., Habtemariam, S., Nabavi, S. M., & Rahimi, R. (2023). Glucosinolates in cancer prevention and treatment: Experimental and clinical evidence. *Medical Oncology (Northwood, London, England)*, *40*(12), 344. doi:10.1007/s12032-023-02211-6 PMID:37921869

Ouasti, I., Bussmann, R. W., & Elachouri, M. (2023). Apium Graveolens L.Apiaceae. In R. W. Bussmann, M. Elachouri, & Z. Kikvidze (Eds.), *Ethnobotany of Northern Africa and Levant* (pp. 1–8). Springer International Publishing. doi:10.1007/978-3-031-13933-8_34-1

Öztürk, M., Altay, V., & Keskin, M. (2023). Folkloric Knowledge of Endemic Plants: Their Traditional Uses as Spices, Food and Herbal Teas by the Local Communities in Turkiye. In Ethnic Knowledge and Perspectives of Medicinal Plants (pp. 137-152). Apple Academic Press.

Pal, D., & Lal, P. (2024). Herbal Drugs and Medicinal Plants in Controlling and Treatment of Diseases Caused by Dengue Virus (DEN-1 & 2): Ethnopharmacology, Chemistry, and Clinical and Preclinical Studies. In D. Pal (Ed.), *Anti-Viral Metabolites from Medicinal Plants* (pp. 683–746). Springer International Publishing. doi:10.1007/978-3-031-12199-9_20

Panchuk, M. (2023). Power and protest: A christian liturgical response to religious trauma. *Philosophies of Liturgy: Explorations of Embodied Religious Practice*, *xx*(x), 77. doi:10.5040/9781350349292.ch-5

Park, Y., Jin, S., Noda, I., & Jung, Y. M. (2023). Continuing progress in the field of two-dimensional correlation spectroscopy (2D-COS), part II. Recent noteworthy developments. *Spectrochimica Acta. Part A: Molecular and Biomolecular Spectroscopy*, *284*, 121750. doi:10.1016/j.saa.2022.121750 PMID:36030669

Patel, M., Vishnoi, S., & Neelima, A. (2023). An empirical review of fundamental principles of Ayurveda for women's reproductive health and diseases. *Journal of Ayurveda and Integrated Medical Sciences*, *8*(9), 105–111. doi:10.21760/jaims.8.9.17

Pathak, K., Pathak, M. P., Saikia, R., Gogoi, U., Sahariah, J. J., Das, A., Ahmad, M. Z., Paul, T., Das, J., & Alqahtani, S. A. M. (2024). Cardioprotective activities of some Indian spices: An insight into pharmacology and phytochemical investigation. *Current Traditional Medicine*, *10*(1), 52–66. doi:10.2174/2215083809666230210105442

Poddar, K. H., Sikand, G., Kalra, D., Wong, N., & Duell, P. B. (2022). Mustard oil and cardiovascular health: Why the controversy? *Journal of Clinical Lipidology*, *16*(1), 13–22. doi:10.1016/j.jacl.2021.11.002 PMID:34924350

Qari, S. H., Alqethami, A., & Qumsani, A. T. (2024). Ethnomedicinal evaluation of medicinal plants used for therapies by men and women in rural and urban communities in Makkah district. *Saudi Pharmaceutical Journal*, *32*(1), 101881. doi:10.1016/j.jsps.2023.101881 PMID:38130903

Rafieian, F., Amani, R., Rezaei, A., Karaça, A. C., & Jafari, S. M. Exploring fennel (*Foeniculum vulgare*): Composition, functional properties, potential health benefits, and safety. *Critical Reviews in Food Science and Nutrition*, 1-18. doi:10.1080/10408398.2023.2176817

Ramalingam, P. S., Mishra, R. A. K., Mekala, J. R., Gothandam, K. M., & Arumugam, S. (2024). Nigella Sativa L.–A Note on Phytochemistry, Extraction, Pharmacological Activities, and Therapeutic Utility. In Pharmacological Aspects of Essential Oils (pp. 243-264). CRC Press.

Ramineedu, K., Sankaran, K. R., Mallepogu, V., Rendedula, D. P., Gunturu, R., Gandham, S., . . . Meriga, B. (2023). Thymoquinone mitigates obesity and diabetic parameters through regulation of major adipokines, key lipid metabolizing enzymes and AMPK/p-AMPK in diet-induced obese rats. *3 Biotech*, *14*(1), 16. doi:10.1007/s13205-023-03847-x

Ramírez-Alarcón, K., Martorell, M., Gürer, E. S., Laher, I., Lam, H. L., Mohieldin, E. A. M., ... Shafique, H. (2023). Myristicin: From its biological effects in traditional medicine in plants to preclinical studies and use as ecological remedy in plant protection. *eFood*, *4*(3), e90. doi:10.1002/efd2.90

Rani, D. M., Wongso, H., Purwoko, R. Y., Winarto, N. B., Shalas, A. F., Triatmoko, B., Pratama, A. N. W., Keller, P. A., & Nugraha, A. S. (2023). Anticancer bioprospecting on medicinal plants from Indonesia: A review. *Phytochemistry*, *216*, 113881. doi:10.1016/j.phytochem.2023.113881 PMID:37827225

Rashwan, H. K., Mahgoub, S., Abuelezz, N. Z., & Amin, H. K. (2023). Black cumin seed (*Nigella sativa*) in inflammatory disorders: Therapeutic potential and promising molecular mechanisms. *Drugs and Drug Candidates*, *2*(2), 516–537. doi:10.3390/ddc2020027

Ren, J., Yang, L., Qiu, S., Zhang, A.-H., & Wang, X.-J. (2023). Efficacy evaluation, active ingredients, and multitarget exploration of herbal medicine. *Trends in Endocrinology and Metabolism*, *34*(3), 146–157. doi:10.1016/j.tem.2023.01.005 PMID:36710216

Rietjens, I. M. C. M., Cohen, S. M., Eisenbrand, G., Fukushima, S., Gooderham, N. J., Guengerich, F. P., Hecht, S. S., Rosol, T. J., Davidsen, J. M., Harman, C. L., & Taylor, S. V. (2023). FEMA GRAS assessment of natural flavor complexes: Allspice, anise, fennel-derived and related flavoring ingredients. *Food and Chemical Toxicology*, *174*, 113643. doi:10.1016/j.fct.2023.113643 PMID:36739890

Rizwan, D., & Masoodi, F. A. (2023). Brassica-derived isothiocyanates as anticancer therapeutic agents and their nanodelivery. *Phytotherapy Research*. doi:10.1002/ptr.8042 PMID:37882581

Roa Solís, C., Delgado Orellana, A., Fuller, D. Q., & Capparelli, A. (2023). Wild foods, woodland fuels, and cultivation through the Ceramic and Early Historical periods in Araucanía, Southern Chile (400–1850 ce). *Vegetation History and Archaeobotany*. doi:10.1007/s00334-023-00969-3

Saad, B. (2024). History, Present and Prospect of Greco-Arab and Islamic Herbal Medicine. In History, Present and Prospect of World Traditional Medicine (pp. 235-300). World Scientific. doi:10.1142/9789811282171_0004

Sachan, N., & Chandra, P. (2023). A systematic review on *Sesamum indicum* L.: Bridging traditional information and pharmacological indications for future translational research. *Current Traditional Medicine*, *9*(3), 87–104. doi:10.2174/2215083808666220928142242

Sakkaravarthy, A., Bhavadharani, P. V., Pushparaj, P., & Parameswaran, G. (2023). A review of processing novelties of extraction of essential oil from cardamom. *Lampyrid: The Journal of Bioluminescent Beetle Research, 13*, 73–82.

Salah, A., Sleem, R., Abd-Elaziz, A., & Khalil, H. (2023). Regulation of NF-κB expression by thymoquinone; A role in regulating pro-inflammatory cytokines and programmed cell death in hepatic cancer cells. *Asian Pacific Journal of Cancer Prevention, 24*(11), 3739–3748. doi:10.31557/APJCP.2023.24.11.3739 PMID:38019231

Sanker, V. M., Thippeswamy, D., Chikkanna, U., & Ramakrishna, K. K. (2023). Chapter Twenty One - Ayurveda herbs and herbal preparations in the management of neurological disorders. In M. Muralidhara & P. S. Rajini (Eds.), *Ayurvedic Herbal Preparations in Neurological Disorders* (pp. 577–586). Academic Press. doi:10.1016/B978-0-443-19084-1.00028-4

Sarand, I., Traksmaa, A., Klava, D., Kunkulberga, D., Straumite, E., Galoburda, R., & Bartkiene, E. (2023). Traditional Breads from the Baltic Countries (Estonia, Latvia, Lithuania). In M. Garcia-Vaquero, K. Pastor, G. E. Orhun, A. McElhatton, & J. M. F. Rocha (Eds.), *Traditional European Breads: An Illustrative Compendium of Ancestral Knowledge and Cultural Heritage* (pp. 41–59). Springer International Publishing. doi:10.1007/978-3-031-23352-4_2

Schultz, F., & Garbe, L. A. (2023). How to approach a study in ethnopharmacology? Providing an example of the different research stages for newcomers to the field today. *Pharmacology Research & Perspectives, 11*(4), e01109. doi:10.1002/prp2.1109 PMID:37497567

Shabani, H., Karami, M. H., Kolour, J., Sayyahi, Z., Parvin, M. A., Soghala, S., Baghini, S. S., Mardasi, M., Chopani, A., Moulavi, P., Farkhondeh, T., Darroudi, M., Kabiri, M., & Samarghandian, S. (2023). Anticancer activity of thymoquinone against breast cancer cells: Mechanisms of action and delivery approaches. *Biomedicine and Pharmacotherapy, 165*, 114972. doi:10.1016/j.biopha.2023.114972 PMID:37481931

Shaik, M. I., Hamdi, I. H., & Sarbon, N. M. (2023). A comprehensive review on traditional herbal drinks: Physicochemical, phytochemicals and pharmacology properties. *Food Chemistry Advances, 3*, 100460. doi:10.1016/j.focha.2023.100460

Shayani Rad, M., Moohebati, M., & Mohajeri, S. A. (2023). Beneficial effects of celery seed extract (Apium graveolens), as a supplement, on anxiety and depression in hypertensive patients: A randomized clinical trial. *Inflammopharmacology, 31*(1), 395–410. doi:10.1007/s10787-022-01083-y PMID:36334223

Sidhu, H. S. (2023). Transforming India's Agrarian Society in a Post-COVID World: Prospects and Challenges. In A. Fazli & A. Kundu (Eds.), *Reimagining Prosperity: Social and Economic Development in Post-COVID India* (pp. 133–152). Springer Nature Singapore. doi:10.1007/978-981-19-7177-8_8

Singh, N., Yadav, S. S., Kumar, S., & Narashiman, B. (2021). A review on traditional uses, phytochemistry, pharmacology, and clinical research of dietary spice Cuminum cyminum L. *Phytotherapy Research*, *35*(9), 5007–5030. doi:10.1002/ptr.7133 PMID:33893678

Singh, S., Chaurasia, P. K., & Bharati, S. L. (2023). Hypoglycemic and hypocholesterolemic properties of Fenugreek: A comprehensive assessment. *Applied Food Research*, *3*(2), 100311. doi:10.1016/j.afres.2023.100311

Skeels, S. (2017). *Northern Africa*. John Wiley & Sons Ltd. doi:10.1002/9781118961933

Sobhy, H. A., Elissawy, A. E., Eldahshan, O. A., & Singab, A. N. B. (2024). Role of Egyptian Essential Oils in Health Care and Industry. In *Pharmacological Aspects of Essential Oils* (pp. 107–119). CRC Press.

Sorkin, B., & Gahche, J. (2020). Application of p-curve analysis to dietary supplement clinical trials: Case study of trials of curcumin products for arthritis. *Current Developments in Nutrition, 4*. doi:10.1093/cdn/nzaa056_026

Spence, C. (2024). Cinnamon: The historic spice, medicinal uses, and flavor chemistry. *International Journal of Gastronomy and Food Science*, *35*, 100858. doi:10.1016/j.ijgfs.2023.100858

Srikanth, N., & Goel, S. (2023). Systematic Documentation and Drug Development from Local Health Traditions (LHTs) and Ethnomedical Practices (EMPs): Challenges and Way Forward. In S. Reddy, N. Guite, & B. Subedi (Eds.), *Ethnomedicine and Tribal Healing Practices in India: Challenges and Possibilities of Recognition and Integration* (pp. 237–249). Springer Nature Singapore. doi:10.1007/978-981-19-4286-0_15

Sulieman, A. M. E., Abdallah, E. M., Alanazi, N. A., Ed-Dra, A., Jamal, A., Idriss, H., Alshammari, A. S., & Shommo, S. A. M. (2023). Spices as sustainable food preservatives: A comprehensive review of their antimicrobial potential. *Pharmaceuticals (Basel, Switzerland)*, *16*(10), 1451. doi:10.3390/ph16101451 PMID:37895922

Sultan, M. T., Saeed, F., Raza, H., Ilyas, A., Sadiq, F., Musarrat, A., Afzaal, M., Hussain, M., Raza, M. A., & Al Jbawi, E. (2023). Nutritional and therapeutic potential of nutmeg (Myristica fragrans): A concurrent review. *Cogent Food & Agriculture*, *9*(2), 2279701. doi:10.1080/23311932.2023.2279701

Šunić, L., Ilić, Z. S., Stanojević, L., Milenković, L., Stanojević, J., Kovač, R., Milenković, A., & Cvetković, D. (2023). Comparison of the essential oil content, constituents and antioxidant activity from different plant parts during development stages of wild fennel (*Foeniculum vulgare* Mill.). *Horticulturae*, *9*(3), 364. doi:10.3390/horticulturae9030364

Suttisansanee, U., Thiyajai, P., Inthachat, W., Pruesapan, K., Wongwathanarat, K., Charoenkiatkul, S., Sahasakul, Y., & Temviriyanukul, P. (2023). Exploration of the nutritional and carotenoids profiles of vegetables in Thai cuisine as potential nutritious ingredients. *Heliyon*, *9*(5), e15951. doi:10.1016/j.heliyon.2023.e15951 PMID:37205996

Tabaza, Y., & Aburjai, T. (2024). Ethnopharmacological survey of medicinal plants used by local herbalists and traditional healers for the treatment of cancer in Jordan. *Current Traditional Medicine*, *10*(3), 6–19. doi:10.2174/221508381066623051561319

Takooree, H., Aumeeruddy, M. Z., Rengasamy, K. R. R., Venugopala, K. N., Jeewon, R., Zengin, G., & Mahomoodally, M. F. (2019). A systematic review on black pepper (*Piper nigrum* L.): from folk uses to pharmacological applications. *Critical Reviews in Food Science and Nutrition, 59*(sup1), S210-S243. doi:10.1080/10408398.2019.1565489

Tejeda, L., Vasquez Iriarte, P. E., Ortiz, J. V., Aliaga-Rossel, E., Mollinedo, P., & Peñarrieta, J. M. (2022). Characterization of Bolivian chili peppers; antioxidant capacity, total phenolic compounds, capsaicin and β-carotene concentration. *Nutrition & Food Science*, *52*(8), 1314–1327. doi:10.1108/NFS-11-2021-0341

Thahira Banu, A., & Aswini, M. (2023). Chapter 4 - Role of Piper nigrum in functional food and nutraceuticals. In A. Amalraj, S. Kuttappan, K. Varma A.C, & A. Matharu (Eds.), Herbs, Spices and Their Roles in Nutraceuticals and Functional Foods (pp. 41-60). Academic Press. doi:10.1016/B978-0-323-90794-1.00019-3

Trifan, A., Zengin, G., Korona-Glowniak, I., Skalicka-Woźniak, K., & Luca, S. V. (2023). Essential oils and sustainability: In vitro bioactivity screening of *Myristica fragrans* Houtt. postdistillation byproducts. *Plants*, *12*(9), 1741. doi:10.3390/plants12091741 PMID:37176799

Usman, I., Saif, H., Imran, A., Afzaal, M., Saeed, F., Azam, I., Afzal, A., Ateeq, H., Islam, F., Shah, Y. A., & Shah, Y. A. (2023). Innovative applications and therapeutic potential of oilseeds and their by-products: An eco-friendly and sustainable approach. *Food Science & Nutrition, 11*(6), 2599–2609. doi:10.1002/fsn3.3322 PMID:37324916

Usmani, K., Jain, S. K., & Yadav, S. (2023). Mechanism of action of certain medicinal plants for the treatment of asthma [Review]. *Journal of Ethnopharmacology, 317*, 116828. doi:10.1016/j.jep.2023.116828 PMID:37369335

Vallet, A., Locatelli, B., Valdivia-Díaz, M., Quispe Conde, Y., Matencio García, G., Ramos Criales, A., Valverde Huamanñahui, F., Ramos Criales, S., Makowski, D., & Lavorel, S. (2023). Knowledge coproduction to improve assessments of nature's contributions to people. *Conservation Biology, 14182*(6), e14182. Advance online publication. doi:10.1111/cobi.14182 PMID:37889094

Vannabhum, M., Ziangchin, N., Thepnorarat, P., & Akarasereenont, P. (2023). Metabolomic analysis of Thai herbal analgesic formula based on ultrahigh-performance liquid chromatography-quadrupole time-of-flight mass spectrometry. *Heliyon, 9*(7), e18296. doi:10.1016/j.heliyon.2023.e18296 PMID:37539319

Varghese, R., & Ray, J. G. (2023). Sustainable agriculture of tropical spices: Arbuscular mycorrhizal fungi as an ecotechnological tool – A critical review. *Journal of Applied Research on Medicinal and Aromatic Plants, 36*, 100507. doi:10.1016/j.jarmap.2023.100507

Vieira, L. V., & Juvenato, M. E., Krause, M., Heringer, O. A., Ribeiro, J. S., Brandão, G. P., . . . Carneiro, M. T. W. D. (2022). The effects of drying methods and harvest season on piperine, essential oil composition, and multielemental composition of black pepper. *Food Chemistry, 390*, 133148. doi:10.1016/j.foodchem.2022.133148 PMID:35551027

von Maydell, D., Beleites, C., Stache, A.-M., Riewe, D., Krähmer, A., & Marthe, F. (2024). Genetic variation of annual and biennial caraway (Carum carvi) germplasm offers diverse opportunities for breeding. *Industrial Crops and Products, 208*, 117798. doi:10.1016/j.indcrop.2023.117798

Wang, W., Nguyen, K. T. K., Zhao, C., & Hung, H. (2023). Earliest curry in Southeast Asia and the global spice trade 2000 years ago. *Science Advances, 9*(29), eadh5517. Advance online publication. doi:10.1126/sciadv.adh5517 PMID:37478176

Wardani, R. S., Schellack, N., Govender, T., Dhulap, A. N., Utami, P., Malve, V., & Wong, Y. C. (2023). Treatment of the common cold with herbs used in Ayurveda and Jamu: Monograph review and the science of ginger, liquorice, turmeric and peppermint. *Drugs in Context*, *12*, 1–10. Advance online publication. doi:10.7573/dic.2023-2-12 PMID:37342461

Warinhomhoun, S., Sanpinit, S., Chonsut, P., Ngamdokmai, N., Ahmad, A. R., & Wetchakul, P. (2023). Effects of black seed oil combined with olive oil or honey on antioxidant activities, phenolic content, and identification and quantification of thymoquinone, a key bioactive compound. *Journal of Agriculture and Food Research*, *14*, 100891. doi:10.1016/j.jafr.2023.100891

Waswa, E. N., Ding, S.-X., Wambua, F. M., Mkala, E. M., Mutinda, E. S., Odago, W. O., Amenu, S. G., Muthui, S. W., Linda, E. L., Katumo, D. M., Waema, C. M., Yang, J.-X., & Hu, G.-W. (2024). The genus Actinidia Lindl. (Actinidiaceae): A comprehensive review on its ethnobotany, phytochemistry, and pharmacological properties. *Journal of Ethnopharmacology*, *319*, 117222. doi:10.1016/j.jep.2023.117222 PMID:37793579

Weluwanarak, T., Changbunjong, T., Leesombun, A., Boonmasawai, S., & Sungpradit, S. (2023). Effects of *Piper nigrum* L. fruit essential oil toxicity against stable fly (Diptera: Muscidae). *Plants*, *12*(5), 1043. doi:10.3390/plants12051043 PMID:36903904

Yadav, S., Tyagi, A., Kumari, R., Srivastava, H., Rangan, P., Wankhede, D., & Gaikwad, A. B. (2023). Comparative transcriptome profiling of fruit tissue provides novel insights into piperine biosynthesis in black pepper (*Piper nigrum* L.). *Scientia Horticulturae*, *322*, 112451. doi:10.1016/j.scienta.2023.112451

Yagoub, S. O. (2023). Chapter four - Black cumin: morphology, physiology, growth, and agricultural yield. In A. A. Mariod (Ed.), *Biochemistry, Nutrition, and Therapeutics of Black Cumin Seed* (pp. 19–25). Academic Press. doi:10.1016/B978-0-323-90788-0.00016-0

Yan, S., Liu, Y., Zhao, W., Zhao, H., & Xue, X. (2023). Chemical markers of a rare honey from the traditional spice plant Amomum tsao–ko Crevost et Lemarié, via integrated GC–MS and LC–MS approaches. *Food Research International*, *172*, 113234. doi:10.1016/j.foodres.2023.113234 PMID:37689964

Zafar, S., Khan, M. K., Perveen, S., Iqbal, M., & Al-Huqail, A. A. (2023). Fennel. In M. Zia-Ul-Haq, A. Abdulkreem Al-Huqail, M. Riaz, & U. Farooq Gohar (Eds.), *Essentials of Medicinal and Aromatic Crops* (pp. 483–514). Springer International Publishing. doi:10.1007/978-3-031-35403-8_19

Zeng, M., Zhang, M., Chen, J., He, Z., Qin, F., Hu, C., Xu, H., Tao, G., Zhang, S., & Chen, J. (2017). UPLC–MS/MS and multivariate analysis of inhibition of heterocyclic amine profiles by black pepper and piperine in roast beef patties. *Chemometrics and Intelligent Laboratory Systems*, *168*, 96–106. doi:10.1016/j.chemolab.2017.06.012

Zhang, Y. (2021). Diet According to Traditional Chinese Medicine for Health and Longevity. In S. I. S. Rattan & G. Kaur (Eds.), *Nutrition, Food and Diet in Aging and Longevity* (pp. 331–356). Springer International Publishing. doi:10.1007/978-3-030-83017-5_17

Zhang, Y., Jiao, X., Liu, J., Feng, G., Luo, X., Zhang, M., Zhang, B., Huang, L., & Long, Q. (2024). A new direction in Chinese herbal medicine ameliorates for type 2 diabetes mellitus: Focus on the potential of mitochondrial respiratory chain complexes. *Journal of Ethnopharmacology*, *321*, 117484. doi:10.1016/j.jep.2023.117484 PMID:38012971

Chapter 7
Ethnobotanical, Pharmacological, and Therapeutic Significance of Dill Plants and Their Role in the Biogenic Synthesis of Crucial Chemical Substances

Shahrukh Akhtar
The Women University Multan, Pakistan

Zukhruf Nadeem
The Women University Multan, Pakistan

Shehla Perveen
Sadiq College Women University, Bahawalpur, Pakistan

ABSTRACT

Dill, also known as Anethum graveolens, is a botanical species with an annual life cycle and a branching structure. It is a significant part of Ayurvedic medicine and is used as a spice, flavor enhancer, and essential oil supplier. Dill seeds have carminative, stomachic, and diuretic properties. The volatile components of dill seeds and plants are diverse, with carvone being the main odorant. Other substances found in seeds include steroids, phenolic acids, flavonoids, and coumarins. Dill is used in traditional herbal therapy to treat various ailments, including respiratory issues, stomach discomfort, digestive disorders, and low blood sugar and cholesterol. It also has pharmacological benefits, including hypolipidemic, anticancer, antidiabetic, antispasmodic, antibacterial, anti-inflammatory, antioxidant, and anti-infertility benefits. Numerous studies have confirmed these effects.

DOI: 10.4018/979-8-3693-6105-4.ch007

1. INTRODUCTION

Anethum graveolens is a native of southern Europe and a member of the Umbelliferae family. It is a seasonal herb that grows in central and southern Asia and the Mediterranean region (Al-Snafi, 2014). The genus name Anethum comes from the Greek words "aneeton" or "aneeson," which denote a strong smell. It is also referred to as soyaa (Unani), sadakuppai (Siddha), dill (English), Shatpushpaa (Ayurvedic), and sowa (Hindi) in Punjabi (Chahal et al., 2017). Two domesticated dill species are closely related: Indian dill (*Anethum graveolens*) and European dill (*Anethum graveolens*) (Gupta et al., 2012).

2. BOTANICAL PROFILE OF DILL PLANT

2.1 Taxonomy for *Anethum graveolens*

A. graveolens commonly known as dill belongs to Kingdom-Plantae, Division-Magnoliophyta, Class-Magnoliopsida, Order-Apiales, Family-Umbelliferae, Genus-Anethum, and Species-*Graveolens* (Chahal et al., 2017; Salman et al., 2023).

2.2 Morphological Characteristics

Dill is an annual herb that typically reaches a height of 50–150 cm. Its stems are thin and have alternating leaves that are eventually separated into three or four pinnate portions. The leaves of Anethum are considerably wider than those of fennels (*Foeniculum vulgare*). The golden blossoms develop into umbel clusters (Warrier et al., 1995). These are not actual seeds; rather, they are the separated halves of tiny,

Figure 1. Anethum graveolens (dill) plant (Goodarzi et al., 2016)

dehydrated fruits known as schizocarps. Dill fruits are ovoid-shaped, flattened, and have distinct wings. They are about a tenth of an inch wide. These fruits taste a bit like caraway, and their seeds smell nice and aromatic despite being smaller and flatter than caraway seeds (Pullaiah, 2002).

2.3 Cultivation

Dill grows best in loose-textured, well-drained soil that is rich and has a pH range of 5.3 to 7.8. It also enjoys full sun exposure. Warm to hot summers with lots of sunlight is necessary to maximize its output; even a small amount of partial shade can drastically lower the harvest. The plant prefers to yield seeds quickly in dry conditions, and in suitable spots, it may self-sow. Through the use of seeds, which remain viable for three to ten years, propagation is carried out. When seed ripening begins, the flower heads must be cut off from the stalks as part of the harvesting procedure (Jana & Shekhawat, 2010a).

3. ETHNOBOTANICAL USES OF DILL PLANT

3.1 Traditional Healing Practices

Traditional uses for the dill plant include treating indigestion and flatulence, stimulating milk secretion, preventing convulsions, preventing vomiting, healing wounds, boosting appetite, fortifying the stomach, and producing essential oils (Hamza, 2017). A. graveolens is applied topically to relieve colic pain and lessen baby flatulence (Mohammed et al., 2019). The food and pharmaceutical industries make extensive use of A. graveolens seed. Dill seeds are used to treat rheumatism and digestive issues in traditional medicine. This study aimed to assess the anti-inflammatory and antioxidant properties of A. graveolens essential oil and correlate

Figure 2. (a) Seeds, (b) plants, (c) inflorescence

them with their chemical composition, in light of the plant's use in folk medicine. (Kazemi, 2015).

3.2 Culinary Uses

Dill, or *Anethum graveolens* L. (Apiaceae), is a widely used culinary herb in the Mediterranean region, Europe, and Central and Southern Asia. *Anethum graveolens* is a fragrant herb that's frequently used to season and enhance a variety of dishes, including salads, pickles, sauces, and soups. Dill oil is extracted from the seeds, leaves, and stems. Its antioxidant action is thought to be attributed to phenolic chemicals. Dill oil is used as a flavoring in food products (Hajji et al., 2021). Dill was employed as a culinary spice and an anti-flatulent in children's Gripe Water in Iraq (Sahib et al., 2014).

4. PHYTOCHEMICAL ANALYSIS OF DILL PLANTS

Essential oils, produced by plants, are essential metabolites essential for their own needs and have various applications in cosmetics, food, medicine, and agriculture, beyond feeding (Potterat & Hamburger, 2008). The dill plant's exceptional antioxidant activity is caused by the existence of terpenes, phenolics, saponins, tannins, flavonoids, and cardiac glycosides in its essential oils (Halvorsen et al., 2002). In perfumery, Dill seed essential oil adds a unique fragrance to soaps, detergents, and cosmetics (Słupski et al., 2005).

Table 1. Anethum graveolens' phytochemicals and their therapeutic properties (Mohammed et al., 2019)

Sr #	Compounds	Medicinal Characteristics	References
1	α-Phellandrene	Autophagy activation	(Hsieh et al., 2015)
2	Limonene	Anticancer, antifungal, and antioxidant	(Fitsiou et al., 2016)
3	Carvone	Hepatoprotective	(Thoppil & Bishayee, 2011)
4	Pinene	Anticancer	(Chen et al., 2015)
5	Cineole	Cytotoxic and antimicrobial	(Fitsiou et al., 2016)
6	Caffeic acid	Antidiabetic	(Goodarzi et al., 2016)
7	Quercetin	Anticancer	(Li et al., 2013)
8	Ferulic acid	Anticancer	(Al-Asmari et al., 2014)
9	Kaempferol	Anticancer	(Al-Asmari et al., 2014)

4.2 Proximate Properties

Dill seed's composition of chemicals changes depending on the kind of crop, the growing region, and the stage of harvest. The seeds contain minerals (9.09%), moisture (6.6%), and volatile oil (2.7%) (Nijjar & Singh, 1979). Upon deeper inspection, dill seeds are composed of the following ingredients: For every 100 grams, there are 6.6 grams of moisture, 17.9 grams of fat, 13.1 grams of protein, and 35.7 grams of carbohydrate. Furthermore, 6.0 grams of minerals are present in these seeds, of which 9.09 grams are mineral matter and 20.7 grams are crude fiber (Kaur et al., 2018).

4.3 Methods of Extraction of Essential Oils From Dill Seed

4.3.1 Preparation of Extracts

By employing hexane, dichloromethane, and methanol as distinct solvents, the various extracts of dill seed were produced via Soxhlet extraction. The extract of 100 grams of powdered dill seeds was prepared using hexane, dichloromethane, and methanol in separate flasks. The crude extracts were stored at 4°C after solvent removal via distillation. A stock solution of 20 mg of each component was created using methanol. After that, Methanol was used as a solvent to generate the dilutions of each chemical that were required, which were 1.0, 0.5, 0.25, 0.1, and 0.05 mg/ml, from the stock solution (Kaur et al., 2018).

4.3.2 Phytochemical Analysis

Standard procedures were used to identify the phytochemicals present in the extracts, including tannins, reducing sugar, phenolics, flavonoids, and saponins (Anyasor et al., 2010). Table 2 lists all of the tests that were performed along with the methodology, findings, and conclusions (Jana & Shekhawat, 2010b).

4.3.3 Determination of Overall Phenolic and Flavonoid Content

Using the 'Follin Ciocalteu method', the total phenolic content of the dill extract was ascertained (Lachman et al., 2000). The phenol content of dill extracts was determined using the gallic acid standard curve, while the flavonoid content was assessed using the aluminum chloride method. Results were expressed as mg of gallic acid equivalent/gram (Eom et al., 2007).

Table 2. Phytochemicals analysis

Phytochemicals	Procedures	Observations	Inferences
Flavonoids	One milliliter of the extract + added 1 percent of $AlCl_3$	An intense yellow color appears.	Present
Reducing sugars	Two milliliters of the extract and Fehling's solutions A and B were added in equal amounts.	A red precipitate was produced.	Present
Saponins	Add 2 or 3 drops of olive oil after 50 mg of extract was heated to a boil in ten milliliters of distilled water and thoroughly shaken.	Emulsion formed.	Present
Terpenoids	2 millimeter of $CHCl_3$ and 3 milliliter of H_2SO_4 were added to 5ml of extract.	A reddish-brown color became visible at the interface.	Present
Anthraquinones	1% HCl was used to boil 100 milligrams of extract, and 3 milliliters of C_6H_6 and 2 milliliters of 10% NH_3 solution were added.	Pink, violet, or red color became visible.	Present
Cardiac glycosides	a few drops of $FeCl_3$ solution and 2 ml of glacial acetic acid were added to 1 ml of extract.	A brown ring showed up at the interface.	Present
Phenolics	0.5g of extract in distilled water was added to 5% $FeCl_3$.	A dark green color became visible.	Present
Phlobatannins	50mg of extract was boiled with 1% HCl.	A red precipitate was produced.	Present

(Jana & Shekhawat, 2010; Kaur et al., 2018)

4.4 Important Metabolites

Indian dill leaves contain seventeen different compounds (Raghavan et al., 1994). The following are some of the compounds found in dill plants:

4.4.1 Terpenoids

The natural chemicals known as the 10-carbon members of the isoprenoid family, known as monoterpenes, are broadly spread across the plant kingdom and are frequently in charge of giving plants their distinctive scents (Bouwmeester et al., 1998).

1. The terpenoid carvone is found in dill seeds. Carvone has various uses, such as flavoring and aroma, antibacterial agent, building block, and biochemical environment. It also inhibits the sprouting of potatoes (Sorce et al., 1997).

   The molecular formula is $C_{10}H_{14}O$.
   Structure:

Figure 3.
(Dhalwal et al., 2008; Singh et al., 2005)

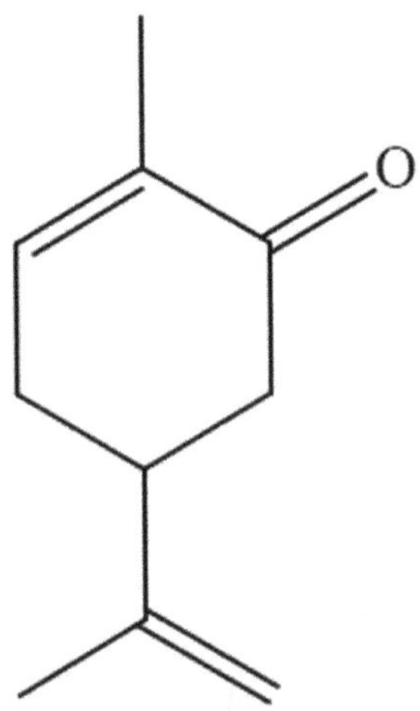

2. Dill seeds contain a cyclic monoterpene called limonene. D-limonene, a common terpene found in citrus oils like orange and lemon, is a cholesterol solvent and has been used in clinical settings to treat cholesterol-containing gallstones, making it a vital ingredient in citrus oils. Additionally, it exhibits chemopreventive and chemotherapeutic properties, and pre-clinical research indicates that its toxicity is modest (Vigushin et al., 1998).

   The molecular formula is $C_{10}H_{16}$.
   Structure:

3. α-phellandrene is a cyclic monoterpene found in dill seed.

   The molecular formula is $C_{10}H_{16}$.
   Structure:

4. β-phellandrene is also a cyclic monoterpene present in dill seed.

   The molecular formula is $C_{10}H_{16}$.
   Structure:

5. Terpenoid anethole is found in trace amounts in Anethum as well as in the essential oils of plants including fennel and anise plants (Newberne et al., 1999). It's a flavoring agent.

Figure 4.
(Blank & Grosch, 1991; Blank et al., 1992)

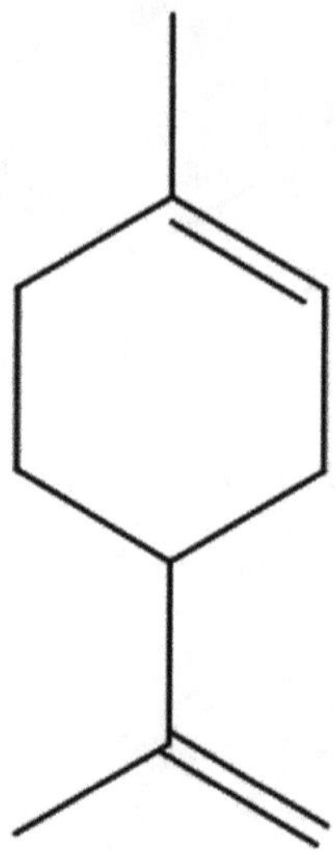

Figure 5.
(Bailer et al., 2001; Blank & Grosch, 1991)

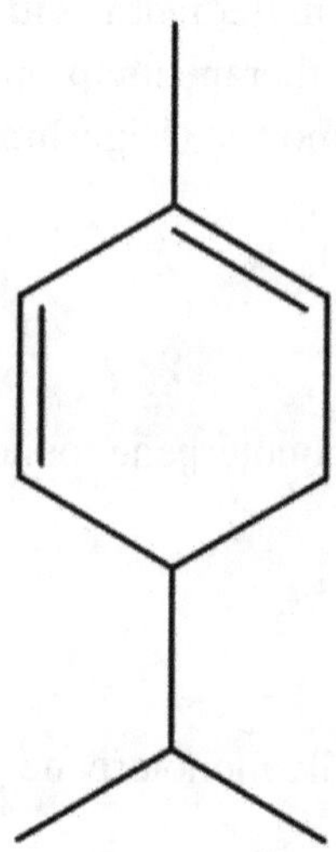

4.4.2 Hydrocarbons

1. Myristicin is a phenylpropene present in dill seeds. Myristicin, a key component of the essential oil found in dill plants, is a naturally occurring pesticide(Dhalwal

Figure 6.
(Blank & Grosch, 1991; Blank et al., 1992; Raghavan et al., 1994)

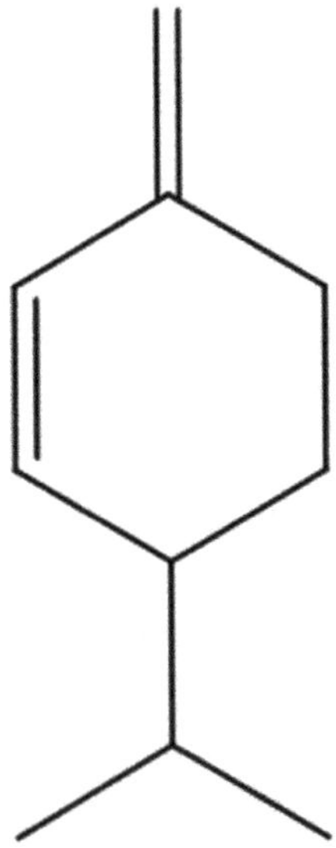

Figure 7.
(Newberne et al., 1999)

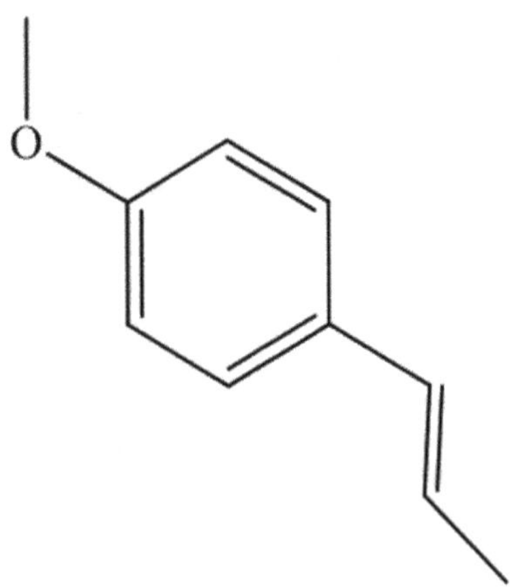

et al., 2008; Duke, 2002).

The molecular formula is $C_{11}H_{12}O_{3}$.
Structure:

Figure 8.
(Dhalwal et al., 2008; Lee et al., 2005; Raghavan et al., 1994)

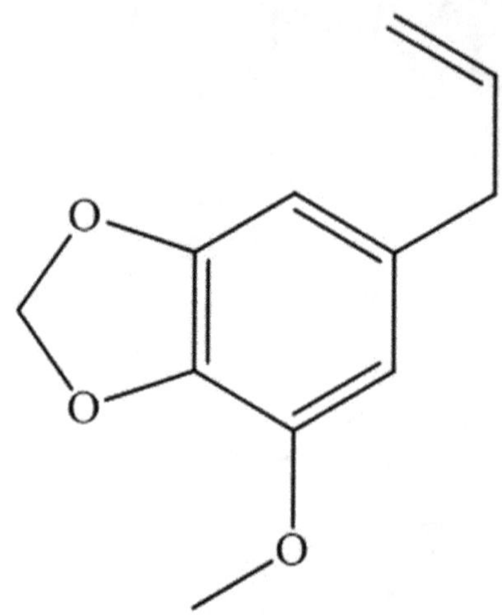

2. Apiole is a phenylpropene present in dill seed.

The molecular formula is $C_{12}H_{14}O_{4.}$
Structure:

4.4.3 Coumarin

Umbelliferone is a 7-hydroxycoumarin present in dill seed.
The molecular formula is $C_9H_6O_{3.}$
Structure:

Figure 9.
(Bailer et al., 2001; Santos et al., 2002; Shulgin & SARGENT, 1967)

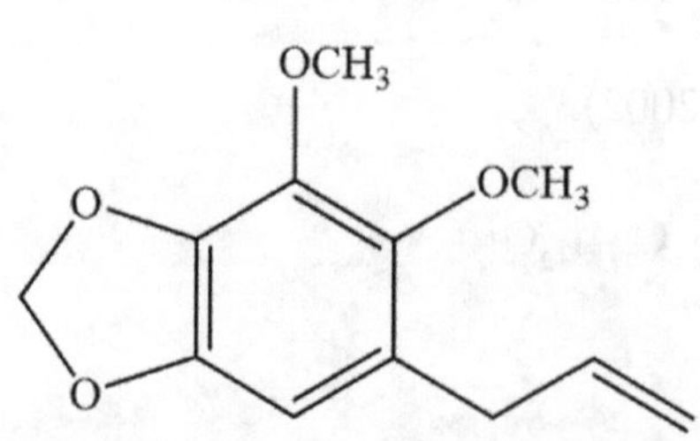

Figure 10.
(Dhalwal et al., 2008)

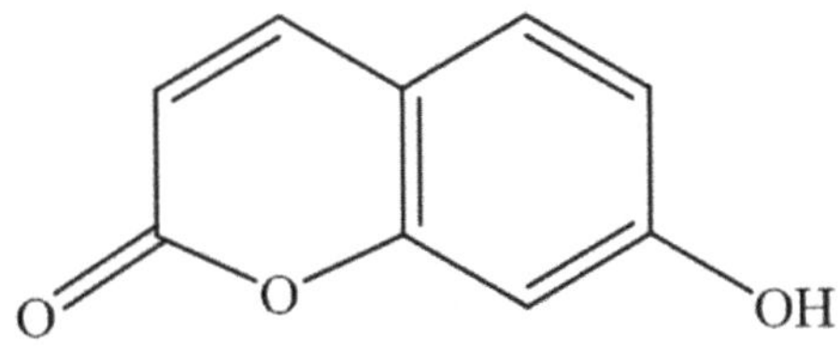

5. PHARMACOLOGICAL PROPERTIES

There are several different anti-inflammatory properties found in *Anethum graveolens* or dill plants. Dill seeds have antibacterial, antihypercholesterolemic, and antihyperlipidemic properties, according to pharmacological research (Chaurasia & Jain, 1978; Delaquis et al., 2002; Nair & CHANDA, 2007; Yazdanparast & Alavi, 2001).

5.1 Bone Health and Osteoporosis

Dill is a valuable herb in preventing the deterioration of bone and bone mineral density because of its high calcium content (Słupski et al., 2005). Osteoporosis affects millions of people annually, calcium and other necessary minerals play a critical role in the proper development and growth of bones as well as the healing of broken or damaged ones (Dixit et al., 2020).

5.2 Preventive Healthcare for Insomnia

Herbs contain essential oils that have unique and potent qualities. They have sedative, hypnotic, and simulative effects. They are both stimulating and calming (Yili et al., 2006). Nothing is more so than the essential oils derived from dill seeds. The essential oils contain flavonoids and vitamin B complex, which trigger the release of specific hormones and enzymes with sedative and hypnotic properties, assisting individuals in achieving restful sleep.

5.3 Managing Diabetes

Dill has long been linked to diabetes, as well as controlling insulin levels to prevent and treat the illness. Although there has been little research done in this field, especially on humans, studies on dill plants have suggested that they can help lessen the variations in insulin and blood lipid levels in patients with corticosteroid-induced diabetes.

A metabolic disease known as diabetes mellitus is a leading factor in fatalities and morbidity. Diabetes comes in two varieties and is just as deadly as cancer:

1. Type 1 Diabetes
2. Type 2 Diabetes

Hyperglycemia, frequent urination, intense hunger, excessive thirst, weight loss, abrupt changes in vision, fatigue, extremely dry skin, nausea, and vomiting are the symptoms of this deadly illness. It also ranks among the main causes of heart disorders (Goodarzi et al., 2016).

5.4 Carminative Benefits

The well-known carminative dill may assist avoid the awkward situation of having too much gas. In addition to being uncomfortable and embarrassing to deal with in public, excessive gas can potentially become deadly if it builds up and presses against the sensitive organs within the chest cavity (Kaur & Arora, 2009). Dill functions as a carminative, meaning it pushes gas out of the body through the digestive tract and out of the body safely, away from any health risks.

5.5 Immunity Elevation

Dill has long been known to possess antibacterial properties that prevent bodily damage from microorganisms. It has been shown to defend against a range of microbial illnesses that impact the body as well as infections that could result in small cuts or open sores on the skin due to its antibacterial activity (Svoboda & Hampson).

5.6 Diarrhea-Controlling

The two main causes of diarrhea are indigestion and microbial activity. Dill has excellent digestive qualities, therefore it may be useful for indigestion. Second, the bactericidal and germicidal properties of the flavonoids and monoterpenes found in dill's essential oils can aid in protecting against germs. These substances can aid in the treatment of diarrhea by preventing harmful microbial diseases from attacking the body (Zaidi et al., 2009).

5.7 Hiccupping Complications

A trapped gas bubble that repeatedly rises through the food pipe is the main cause of hiccups, though there are other causes as well. Second-order causes include hyperactivity, hypersensitivity, specific allergies, and nervous system malfunctions. These moments of interruption can be eased with dill. While dill acts as a sedative to ease hiccups brought on by allergies, hyperactivity, or nervous problems, it also functions as a carminative to help the body expel gasses and reduce the development of gas in the body (Meena et al., 2013).

5.8 Anticancerous

Dill contains monoterpenes, which are chemo-preventive by nature. Additionally, because of their stimulating properties, they can trigger the enzyme's secretion known as glutathione-S-transferase, which is capable of neutralizing carcinogens and preventing cancer. Glutathione is a powerful antioxidant. It shields the body against cancer by effectively neutralizing free radicals and derivatives of benzo and cyano. The additional antioxidants included in dill essential oils also somewhat aid in the prevention and treatment of cancer (Fleming, 2000).

5.9 Anti-Oxidative Properties

Because of its antioxidant qualities, the dill plant, which is recognized for its leaves and seeds, is used medicinally. The active components of the plant are extracted using extracts after the leaves are ground, dried, and powdered. More important than alkaloids, saponins, and proteins are the flavonoids, glycosides, tannins, and carbohydrates found in the leaf extract. Redox processes impact the anti-oxidant capabilities of flavonoids and other phenolic chemicals. Moreover, flavonoids contain anti-oxidative qualities that control lipid peroxidation and inhibit the development of atherogenesis, cancer, and thrombosis. While carbohydrates and glycosides affect the immune system, tannins have anti-ulcer qualities and can help prevent stomach cancer. Because of the components that donate hydrogen, *Anethum graveolens* leaf extract has anti-oxidative and therapeutic properties (Singh et al., 2005; Yadav et al., 2014).

6.BIOGENIC SYNTHESIS

6.1 Copper Nanoparticles

Development inhibition against fungi is one of the many biological actions of essential oils (EOs) that have been discovered (Karimi et al., 2016). In antimicrobial research, using different EOs to fight microorganisms is a promising strategy (Knezevic et al., 2016; Yap et al., 2014). Against a wide range of bacteria, copper demonstrated exceptional antimicrobial action (Grass et al., 2011). An inventive method with multiple uses across numerous industries is nanoencapsulation. The EOs that were isolated from the thyme and dill plant seeds and leaves in this study prevented the growth of C. nymphaea mycelium and the germination of conidia. At the highest tested concentration of EOs, the inhibitory impact peaked (Umer et al., 2014).

A procedure outlined in the literature was used to collect copper nanoparticles. After adding a solution of 2 grams of L-ascorbic acid gradually at room temperature to the reaction blend, which contained 0.4 g of cupric chloride that was soluble in 100 milliliters of deionized water, the mixture was refluxed, heated, and stirred until color changes occurred, including yellow, orange, brown, and finally dark brown-black. The entire procedure was finished in a single day. The original stock solution was diluted to create four different concentrations of copper NPs (30 ppm, 50 ppm, 150 ppm, and 200 ppm). The product did not show any signs of sedimentation over the few days (more than two months) it was stored. After the powder was gathered by using centrifugation and cleaned, it was baked using a vacuum oven for three hours so that FTIR and SEM images could be taken (Umer et al., 2014).

Copper nanoparticle encapsulation affected the composition of dill and thyme essential oils. P-cymene, carvacrol, trans-caryophyllene, γ-terpinene, thymol, camphene, thymol, and α-pinene, all increased as a result. Additionally, it has been demonstrated that encasing EOs in copper nanoparticles strengthens the phytopathogenic fungus C. nymphaea's growth suppression (Weisany et al., 2019).

6.2 Silver Nanoparticles

Chemical synthesis is the conventional method used to create nanoparticles. Nonetheless, more environmentally friendly techniques aid in mitigating the harmful consequences of conventional procedures. The production of nanoparticles by plants and microbes presents several advantages for environmentally acceptable techniques. Green synthesis provides control over crystal development and size and is economical and environmentally benign. During the green synthesis process, the active ingredients in medicinal plants cap the nanoparticles, strengthening their potential for use in clinical settings. Numerous medical uses for nanosilver exist, including device development, diagnostics, and treatment. Nanosilver is utilized in the fields of orthopedics, surgery, dentistry, and ophthalmology to coat medical instruments and supplies. It is also employed in diabetic socks and wound dressings (Menaga et al., 2017).

6.2.1 Botanical Material and Extraction Process

Anethum graveolens seeds were mechanically ground into powder, allowed to air dry in the shade, and then sealed in receptacles. Dill seed material that had been coarsely ground up was extracted using water. 200 ml of water and 25 g of coarse powder were combined, and the mixture was stirred with a magnetic stirrer for a full day. After filtering the filtrate, the water was evaporated until it was completely dry. An airtight container was used to keep the dried filtrate (Menaga et al., 2017).

6.2.2 Production of Nanoparticles

The study used an environmentally friendly method to produce silver nanoparticles (AgNPs). A solution of silver nitrate was mixed with plant extract and microwaved for five minutes at 300 W. The solution changed color and was recorded for an hour. The silver was completely reduced, and the solution was kept in the dark for further research.

6.2.3 Inhibition of Proton Pump Activity by AgNPs

Ulcers are a common lifestyle issue linked to stress, anti-inflammatory drugs, and bacteria contact. Mucosal injury is the initial stage, causing oxidative stress and reactive oxygen species generation. Defenses against ulcers include scavenging reactive oxygen species and inhibiting H+K+ATPase, a proton pump for acid release in the stomach's gastric mucosa.

Anti-secretory medications like histamine H2 receptor blockers, omeprazole, lansoprazole, ranitidine, and H+K+ATPase inhibitors negatively impact human health. Although proton pump inhibitors have improved stomach ulcer treatment, there is no known cure and long-term use can cause harmful side effects.(Chellappan & Pemiah, 2014). Therefore, finding safer and more effective anti-ulcer medicines is urgently needed. Researchers are investigating far safer anti-ulcer medicines as a result of a broad search that has been launched. Herbs, medicinal plants, and unrefined pharmaceuticals are thought to have the ability to treat several illnesses, including stomach ulcers. It has been demonstrated that dill seed extracts shield the stomach mucosal lining in rats with ethanol-induced gastric ulcers (Al & Ali, 2014).

6.2.4 Outcomes of Ag NPs

In comparison to the typical proton-pump inhibitor medication omeprazole, which showed 71 percent inhibition, the extract's 67% proton-pump inhibition was comparable. Therefore, the work provides a fresh line of inquiry into the application of A. graveolens as an anti-ulcer properties in vivo. It is still unknown whether these nanosilver particles have a gastroprotective effect in vivo, and there is still no proof that they inhibit the proton pump. Thus, our investigation into the inhibitory effect of the proton pump by the nanosilver that was biosynthesized using a plant extract is fresh research (Menaga et al., 2017).

7. CONCLUSION

Anethum graveolens L., often known as dill, is a powerful spice and therapeutic herb that contains essential oils. It is grown for its seeds and leaves, which are then used as a pickling, condiment, spice, flavoring ingredient, and fragrance in detergents and soaps. Many substances can be found in Dill's essential oil, such as quinones, tannins, iridoids, terpenes, glycosides, carotenoids, flavonoids, alkaloids, sterols, and anthracene derivatives. It has anti-oxidative, anti-ulcer, anti-diabetic, anti-carminative, and anti-cancer pharmacological qualities. Encapsulating essential oils of dill and thyme with copper nanoparticles inhibits the growth of Colletotrichum nymphaeae mycelium and acts as an antifungal.

REFERENCES

Al, S., & Ali, E. (2014). The Pharmacological Importance of*Anethum graveolens. A Review. International Journal of Pharmacy and Pharmaceutical Sciences*, *6*(4), 11–13.

Al-Asmari, A. K., Al-Elaiwi, A. M., Athar, M. T., Tariq, M., Al Eid, A., & Al-Asmary, S. M. (2014). A review of hepatoprotective plants used in Saudi traditional medicine. *Evidence-Based Complementary and Alternative Medicine*, *2014*, 2014. doi:10.1155/2014/890842 PMID:25587347

Al-Snafi, A. E. (2014). The pharmacological importance of*Anethum graveolens*–A review. *International Journal of Pharmacy and Pharmaceutical Sciences*, *6*(4), 11–13.

Anyasor, G. N., Ogunwenmo, O., Oyelana, O. A., & Akpofunure, B. E. (2010). Phytochemical constituents and antioxidant activities of aqueous and methanol stem extracts of Costus afer Ker Gawl.(Costaceae). *African Journal of Biotechnology*, *9*(31), 4880–4884.

Bailer, J., Aichinger, T., Hackl, G., de Hueber, K., & Dachler, M. (2001). Essential oil content and composition in commercially available dill cultivars in comparison to caraway. *Industrial Crops and Products*, *14*(3), 229–239. doi:10.1016/S0926-6690(01)00088-7

Blank, I., & Grosch, W. (1991). Evaluation of potent odorants in dill seed and dill herb (Anethum graveolens L.) by aroma extract dilution analysis. *Journal of Food Science*, *56*(1), 63–67. doi:10.1111/j.1365-2621.1991.tb07976.x

Blank, I., Sen, A., & Grosch, W. (1992). Sensory study on the character-impact flavour compounds of dill herb (Anethum graveolens L.). *Food Chemistry*, *43*(5), 337–343. doi:10.1016/0308-8146(92)90305-L

Bouwmeester, H. J., Gershenzon, J., Konings, M. C., & Croteau, R. (1998). Biosynthesis of the monoterpenes limonene and carvone in the fruit of caraway: I. Demonstration of enzyme activities and their changes with development. *Plant Physiology*, *117*(3), 901–912. doi:10.1104/pp.117.3.901 PMID:9662532

Chahal, K., Kumar, A., Bhardwaj, U., & Kaur, R. (2017). Chemistry and biological activities of*Anethum graveolens* L.(dill) essential oil: A review. *Journal of Pharmacognosy and Phytochemistry*, *6*(2), 295–306.

Chaurasia, S., & Jain, P. (1978). Antibacterial activity of essential oils of four medicinal plants. *The Indian Journal of Hospital Pharmacy*, *15*(6), 166–168.

Chellappan, D., & Pemiah, B. (2014). Pharmacognostical, phytochemical and in vivo gastro-protective investigation of Gmelina arborea. *International Journal of Pharmacy and Pharmaceutical Sciences*, *6*(4), 153–157.

Chen, W., Liu, Y., Li, M., Mao, J., Zhang, L., Huang, R., Jin, X., & Ye, L. (2015). Anti-tumor effect of α-pinene on human hepatoma cell lines through inducing G2/M cell cycle arrest. *Journal of Pharmacological Sciences*, *127*(3), 332–338. doi:10.1016/j.jphs.2015.01.008 PMID:25837931

Delaquis, P. J., Stanich, K., Girard, B., & Mazza, G. (2002). Antimicrobial activity of individual and mixed fractions of dill, cilantro, coriander and eucalyptus essential oils. *International Journal of Food Microbiology*, *74*(1-2), 101–109. doi:10.1016/S0168-1605(01)00734-6 PMID:11929164

Dhalwal, K., Shinde, V. M., & Mahadik, K. R. (2008). Efficient and sensitive method for quantitative determination and validation of umbelliferone, carvone and myristicin in*Anethum graveolens* and Carum carvi seed. *Chromatographia*, *67*(1-2), 163–167. doi:10.1365/s10337-007-0473-6

Duke, J. A. (2002). *Handbook of medicinal herbs*. CRC Press. doi:10.1201/9781420040463

Eom, S.-H., Jin, C.-W., Park, H.-J., Kim, E.-H., Chung, I.-M., Kim, M.-J., ... Cho, D.-H. (2007). Far infrared ray irradiation stimulates antioxidant activity in Vitis flexuosa THUNB. Berries. *Korean Journal of Medicinal Crop Science*, *15*(5), 319–323.

Fitsiou, E., Mitropoulou, G., Spyridopoulou, K., Tiptiri-Kourpeti, A., Vamvakias, M., Bardouki, H., ... Chlichlia, K. (2016). Phytochemical profile and evaluation of the biological activities of essential oils derived from the Greek aromatic plant species Ocimum basilicum, Mentha spicata, Pimpinella anisum and Fortunella margarita. *Molecules (Basel, Switzerland)*, *21*(8), 1069. doi:10.3390/molecules21081069 PMID:27537869

Fleming, T. (2000). Article. In T. Fleming (Ed.), *PDR® for Herbal Medicines* (pp. 224–225). Medical Economics Company.

Goodarzi, M. T., Khodadadi, I., Tavilani, H., & Abbasi Oshaghi, E. (2016). The role of *Anethum graveolens* L.(Dill) in the management of diabetes. *Journal of Tropical Medicine*. PMID:27829842

Grass, G., Rensing, C., & Solioz, M. (2011). Metallic copper as an antimicrobial surface. *Applied and Environmental Microbiology*, *77*(5), 1541–1547. doi:10.1128/AEM.02766-10 PMID:21193661

Gupta, R., Anwer, M., & Sharma, Y. (2012). Dill. In *Handbook of Herbs and Spices* (pp. 275–285). Elsevier. doi:10.1533/9780857095671.275

Hajji, M., Falcimaigne-Gordin, A., Ksouda, G., Merlier, F., Thomasset, B., & Nasri, M. (2021). A water-soluble polysaccharide from*Anethum graveolens* seeds: Structural characterization, antioxidant activity and potential use as meat preservative. *International Journal of Biological Macromolecules*, *167*, 516–527. doi:10.1016/j.ijbiomac.2020.12.006 PMID:33279565

Halvorsen, B. L., Holte, K., Myhrstad, M. C., Barikmo, I., Hvattum, E., Remberg, S. F., ... Andersen, L. F. (2002). A systematic screening of total antioxidants in dietary plants. *The Journal of Nutrition*, *132*(3), 461–471. doi:10.1093/jn/132.3.461 PMID:11880572

Hamza, L. F. (2017). *Anethum graveolens*: Physicochemical properties, medicinal uses, antimicrobial effects, antioxidant effect, anti-inflammatory and analgesic effects: A review. *World*, *5*(7).

Hsieh, L.-C., Hsieh, S.-L., Chen, C.-T., Chung, J.-G., Wang, J.-J., & Wu, C.-C. (2015). Induction of α-phellandrene on autophagy in human liver tumor cells. *The American Journal of Chinese Medicine*, *43*(01), 121–136. doi:10.1142/S0192415X15500081 PMID:25649747

Jana, S., & Shekhawat, G. S. (2010a). *Anethum graveolens*: An Indian traditional medicinal herb and spice. *Pharmacognosy Reviews*, *4*(8), 179. doi:10.4103/0973-7847.70915 PMID:22228959

Jana, S., & Shekhawat, G. S. (2010b). Phytochemical analysis and antibacterial screening of in vivo and in vitro extracts of Indian medicinal herb:*Anethum graveolens*. *Research Journal of Medicinal Plant*, *4*(4), 206–212. doi:10.3923/rjmp.2010.206.212

Karimi, K., Ahari, A. B., Weisany, W., Pertot, I., Vrhovsek, U., & Arzanlou, M. (2016). Funneliformis mosseae root colonization affects*Anethum graveolens* essential oil composition and its efficacy against Colletotrichum nymphaeae. *Industrial Crops and Products*, *90*, 126–134. doi:10.1016/j.indcrop.2016.06.024

Kaur, G. J., & Arora, D. S. (2009). Antibacterial and phytochemical screening of*Anethum graveolens*, Foeniculum vulgare and Trachyspermum ammi. *BMC Complementary and Alternative Medicine*, *9*(1), 30. doi:10.1186/1472-6882-9-30 PMID:19656417

Kaur, N., Chahal, K., & Singh, R. (2018). Phytochemical screening and antioxidant activity of*Anethum graveolens* L. seed extracts. *Pharma Innov. J*, *7*, 324–329.

Kazemi, M. (2015). Phenolic profile, antioxidant capacity and anti-inflammatory activity of*Anethum graveolens* L. essential oil. *Natural Product Research*, *29*(6), 551–553. doi:10.1080/14786419.2014.951934 PMID:25154406

Knezevic, P., Aleksic, V., Simin, N., Svircev, E., Petrovic, A., & Mimica-Dukic, N. (2016). Antimicrobial activity of Eucalyptus camaldulensis essential oils and their interactions with conventional antimicrobial agents against multi-drug resistant Acinetobacter baumannii. *Journal of Ethnopharmacology*, *178*, 125–136. doi:10.1016/j.jep.2015.12.008 PMID:26671210

Lachman, J., Hamouz, K., Orsak, M., & Pivec, V. (2000). Potato tubers as a significant source of antioxidants in human nutrition. *Rostlinná Výroba*, *46*(5), 231–236.

Lee, B. K., Kim, J. H., Jung, J. W., Choi, J. W., Han, E. S., Lee, S. H., Ko, K. H., & Ryu, J. H. (2005). Myristicin-induced neurotoxicity in human neuroblastoma SK-N-SH cells. *Toxicology Letters*, *157*(1), 49–56. doi:10.1016/j.toxlet.2005.01.012 PMID:15795093

Li, T., Zhu, J., Guo, L., Shi, X., Liu, Y., & Yang, X. (2013). Differential effects of polyphenols-enriched extracts from hawthorn fruit peels and fleshes on cell cycle and apoptosis in human MCF-7 breast carcinoma cells. *Food Chemistry, 141*(2), 1008–1018. doi:10.1016/j.foodchem.2013.04.050 PMID:23790880

Meena, S., Balraj, S., Dheraj, S., Ranjan, J., & Meena, R. (2013). Pre and post harvest factors effecting yield and quality of seed spices: A review. *International J. Seed Spices, 3*(1), 1–11. doi:10.56093/IJSS.v12i1.1

Menaga, S., Kripa, K., & Sangeetha, R. (2017). Evaluation of in vitro inhibition of proton-pump by nanosilver particles synthesized using seeds of*Anethum graveolens. Evaluation, 10*(6).

Mohammed, F. A., Razvi, S. S., Abdul, W. M., Mohammed, K., Hakeem, K. R., Banaganapalli, B., & Elkady, A. I. (2019). *Protective role of medicinal herbAnethum graveolens (Dill) against various human diseases and metabolic disorders. Plant and Human Health* (Vol. 3). Pharmacology and Therapeutic Uses.

Nair, R., & Chanda, S. (2007). Antibacterial activities of some medicinal plants of the western region of India. *Turkish Journal of Biology, 31*(4), 231–236.

Newberne, P., Smith, R., Doull, J., Goodman, J., Munro, I., Portoghese, P., ... Adams, T. (1999). The FEMA GRAS assessment of trans-anethole used as a flavouring substance. *Food and Chemical Toxicology, 37*(7), 789–811. doi:10.1016/S0278-6915(99)00037-X PMID:10496381

Nijjar, G., & Singh, N. (1979). Effect of Nitrogen, Phosphorus and Potassium Fertilization on the Growth, Yield and Quality of Thompson Seedless Grapes. *Indian Journal of Horticulture, 36*(1), 114–118.

Potterat, O., & Hamburger, M. (2008). Drug discovery and development with plant-derived compounds. In *Natural Compounds as Drugs* (Vol. I, pp. 45–118). Springer. doi:10.1007/978-3-7643-8117-2_2

Pullaiah, T. (2002). *Medicinal Plants in Andhra Pradesh*. Daya Books.

Raghavan, B., Abraham, K., Shankaranarayana, M., & Koller, W. (1994). Studies on flavor changes during drying of dill (Anethum sowa Roxb.) leaves. *Journal of Food Quality, 17*(6), 457–466. doi:10.1111/j.1745-4557.1994.tb00166.x

Sahib, A. S., Mohammed, I. H., & Sloo, S. A. (2014). Antigiardial effect of*Anethum graveolens* aqueous extract in children. *Journal of Intercultural Ethnopharmacology, 3*(3), 109. doi:10.5455/jice.20140523104104 PMID:26401357

Salman, T. A.-S., Ahmed, A. T., Oleiwi, G., & Al-Ibrahemi, N. (2023). Study of the Effect of Oil Extract of Dill (Anethumgraveolens L.) Plant on Oxidative Stress Parameters of Liver Enzymes in Male Rats. IOP Conference Series: Earth and Environmental Science.

Shulgin, A. T., & Sargent, T. (1967). Psychotropic phenylisopropylamines derived from apiole and dillapiole. *Nature*, *215*(5109), 1494–1495. doi:10.1038/2151494b0 PMID:4861200

Singh, G., Maurya, S., De Lampasona, M., & Catalan, C. (2005). Chemical constituents, antimicrobial investigations, and antioxidative potentials of*Anethum graveolens* L. essential oil and acetone extract: Part 52. *Journal of Food Science*, *70*(4), M208–M215. doi:10.1111/j.1365-2621.2005.tb07190.x

Słupski, J., Lisiewska, Z., & Kmiecik, W. (2005). Contents of macro and microelements in fresh and frozen dill (Anethum graveolens L.). *Food Chemistry*, *91*(4), 737–743. doi:10.1016/j.foodchem.2004.06.046

Sorce, C., Lorenzi, R., & Ranalli, P. (1997). The effects of (S)-(+)-carvone treatments on seed potato tuber dormancy and sprouting. *Potato Research*, *40*(2), 155–161. doi:10.1007/BF02358241

Svoboda, K. P., & Hampson, J. B. (n.d.). Bioactivity of essential oils of selected temperate aromatic plants: antibacterial, antioxidant, antiinflammatory and other related pharmacological activities. Academic Press.

Thoppil, R. J., & Bishayee, A. (2011). Terpenoids as potential chemopreventive and therapeutic agents in liver cancer. *World Journal of Hepatology*, *3*(9), 228. doi:10.4254/wjh.v3.i9.228 PMID:21969877

Umer, A., Naveed, S., Ramzan, N., Rafique, M. S., & Imran, M. (2014). A green method for the synthesis of copper nanoparticles using L-ascorbic acid. *Matéria (Rio de Janeiro)*, *19*(3), 197–203. doi:10.1590/S1517-70762014000300002

Vigushin, D. M., Poon, G. K., Boddy, A., English, J., Halbert, G. W., Pagonis, C., ... Committee, C. R. C. P. I. I. C. T. (1998). Phase I and pharmacokinetic study of D-limonene in patients with advanced cancer. *Cancer Chemotherapy and Pharmacology*, *42*(2), 111–117. doi:10.1007/s002800050793 PMID:9654110

Warrier, P., Nambiar, V., Ramankutty, C., & Vaidya, S. A. (1995). *Indian Medicinal Plants*. Orient Longman.

Yadav, M., Chatterji, S., Gupta, S. K., & Watal, G. (2014). Preliminary phytochemical screening of six medicinal plants used in traditional medicine. *International Journal of Pharmacy and Pharmaceutical Sciences*, *6*(5), 539–542.

Yap, P. S. X., Yiap, B. C., Ping, H. C., & Lim, S. H. E. (2014). Essential oils, a new horizon in combating bacterial antibiotic resistance. *The Open Microbiology Journal*, *8*(1), 6–14. doi:10.2174/1874285801408010006 PMID:24627729

Yazdanparast, R., & Alavi, M. (2001). Antihyperlipidaemic and antihypercholesterolaemic effects of*Anethum graveolens* leaves after the removal of furocoumarins. *Cytobios*, *105*(410), 185. PMID:11409638

Yili, A., Yimamu, H., Maksimov, V., Aisa, H., Veshkurova, O., & Salikhov, S. I. (2006). Chemical composition of essential oil from seeds of*Anethum graveolens* cultivated in China. *Chemistry of Natural Compounds*, *42*(4), 491–492. doi:10.1007/s10600-006-0190-7

Zaidi, S. F., Yamada, K., Kadowaki, M., Usmanghani, K., & Sugiyama, T. (2009). Bactericidal activity of medicinal plants, employed for the treatment of gastrointestinal ailments, against Helicobacter pylori. *Journal of Ethnopharmacology*, *121*(2), 286–291. doi:10.1016/j.jep.2008.11.001 PMID:19041711

Chapter 8
Nutritional and Therapeutic Potential of Coriandеrum Sativum

Zahra Asghar
The Women University Multan, Pakistan

ABSTRACT

Coriandеrum sativum L. commonly known as coriander is an annual herb. It belongs to the family Apiaceae (Umbelliferae). It is a well-known herb all around the world because of its food, culinary, and medicinal uses. It is a source of many phytochemicals such as polyphenols, essential oil, reducing sugars, carotinoids, terpenoids, glycosides, sterols, isocoumarins, flavonoids, tannins, alkaloids, fatty acids, and coumarins. Coriander sativum also contains high nutritional values of carbohydrates, fibers, oils, proteins, along with wide range of vitamins, minerals, and trace elements. Due to the easy collection of the plant and its widespread uses and also its remarkable biological activities, this plant has become the food and medicine in many parts of the world. The primary objective of this study is to present comprehensive analyzed information on the botanical, chemical, pharmacological aspects of C. sativum along with its applications in the biogenic synthesis.

1. INTRODUCTION

Scientific of name coriander is *Coriandrum Sativum L.* Long ago it is used as a culinary herb (Mandal, Mandal, 2015). The name of coriander originates from "korannon". Korannon is also originates from two Greece words "koris" and "annon". But the genus "Coriandrum" is first used by Pliny the Elder (Burdock,

DOI: 10.4018/979-8-3693-6105-4.ch008

Carabin, 2009). Coriander belongs to the family Apiaceae, which is previously named as Umbelliferae due to the reason the flowers are produced in the form of umbels/clusters (Laribi, 2015; Davis, 1970). Coriander is small and an annual herb. It originates from the Middle East, India, Mediterranean countries, Latin America, Africa, and Southeast Asia cuisine. It is called as dhania, cilantro and cilantrillo in some countries, while in other countries it is called as Arab parsley, Chinese parsley, Mexican parsley and Yuen sai as well (Aissaoui, 2008). However, coriander and cilantro are used for the same thing, but in fact they have the change meanings. Green freshly harvested coriander leaves are called cilantro and coriander is used for the herb when dried fruits are harvested (Abascal, Yarnell, 2012). Depending upon the climate conditions its crop can be cultivated in summer or in winter. The coriander plant grows worldwide (Randall, 2013). At first coriander originates from the Mediterranean areas, but now mostly grown in the equatorial regions such as Romania, Ukraine, Mexico, Morocco, Russia, Argentina, and India (Priyadarshi, 2016). In different areas the coriander has the different regional names. Some regional names are as; it is called as Dhaniwal in Kashmiri; Dhania in Oddiya and Punjabi; Dhane in Bengali; Kothimbir in Marathi; Viral dhania in Tamil; kisnis in Turkish; Coriander fruits in English; Dhaniya in Hindi etc (Evans, 2002; Diederichsen, 1996; Nimish, 2011). The suitable temperature range for the coriander to germinate and grow is 20-25 °C a cool and comparatively frost dry places. The germination of coriander is temperature dependent and takes about 10-15 days to germinate. Mostly coriander grows well in dried weather conditions. But coriander can grow in any type of soil such as loamy soil, light, moist, well- drained, light soil and as well as in heavy black soil (Bhat, 2014). Coriander is a soft herb. It usually grows up to the height of 20 in. Leaves of coriander are irregular in shape. Mostly, at the base of the plant the leaves are lobed in shape and slender and feathery leaves are present on the flowering stems of the plant. All parts of coriander are used for eating but most commonly seeds and leaves are used in cooking (Samuelson,). The flowers are produced in small clusters and are asymmetrical. These flowers are white or light pink in colour. The flower petals which originate away from the center of the umbel are longer (5–6 mm) than those pointing toward it (only 1–3 mm). *C. sativum* seeds are ovate and globular dry schizocarp and have two mericarps. Seeds have many lengthwise ridges on its surface. Mostly they possess a sweetened, a bit suffocating, citrus-like flavor as sage (Mandal, Mandal, 2015; Burdock, Carabin, 2009; Laribi, 2015; Davis, 1970).

The *C. sativum* leaves have different taste than fruits/seeds (McGee, 2010). The fresh leaves of *C. sativum* are used in various foods such as in chutneys, salads and salsa. They are used for garnishing to many dishes (Moulin, 2002). High temperature reduces the taste and smell of coriander leaves. So, they are added just before serving the dish. In some countries the leaves of coriander are used in huge quantity and cooked till the taste diminished (Kaur, 2019). Roots of coriander have strong taste than leaves. Roots of coriander are used in various dishes specially in Thai dishes such as in curry pastes or soups (Laribi, 2015) The essential oils (EOs) and fatty acids are present in the ripened fruits of *C. sativum*. The EOs of *C. sativum* have linalool (67.7%), α- pinene (10.5%), γ- terpine (9.0%), Geranyl acetate (Diederichsen, 1996). (4.0%), Geraniol (1.9%) as main compounds and β- pinene Camphene, Myrcene, Limonene, p-cymol, Dipentene,α- terpinene, Borenol, n-decylaldehyde as minor compounds in it (Fatty acids also have some major and minor compounds in it. The major compounds are Petroselinic acid (68.8%), Linoleic acid (16.6%), Oleic acid (7.5%), Palmitic acid (3.8%) and the minor components are Stearic acid, Vaccenic acid, Myristic acid (Diederichsen, 1996).

2. PHYTOCHEMICALS

"Phyto" is used for plants. It is a Greek word. Phytochemicals have the protective and disease preventive properties. Due to these properties, these chemicals protect the human from a host of diseases. Essential and fatty oils are the very important phytochemicals of *C. sativum* fruits (Coşkuner, Karababa, 2007) EOs content of seeds is from 0.03 to 2.6% and fatty oil contents are from 9.9 to 27.7%. Some other

Figure 1. (i) flowers, (ii) leaves, (iii) seeds, (iv) roots of C. sativum

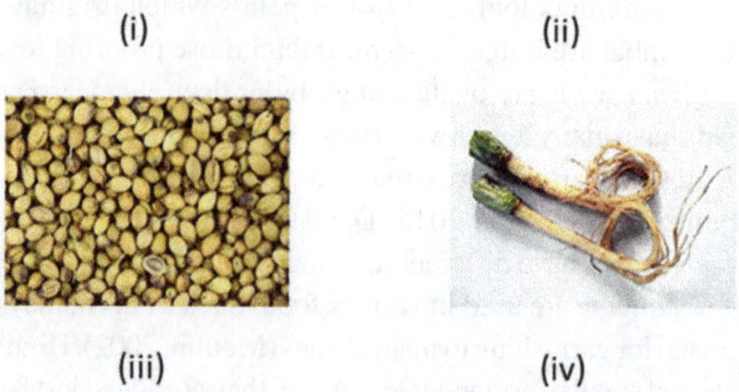

compounds such as crude protein, fat, fiber, and ash contents are also present in the coriander. Their concentration vary from 11.5 to 21.3%, 17.8 to 19.15%, 28.4 to 29.1%, and 4.9 to 6.0%, respectively (Coşkuner, Karababa, 2007). The fruits/seeds of *corianderum sativum* also have sugar, resins, tannins, alkaloids, sterols, anthraquinones, flavones and fixed oils (Barros, 2012; Uitterhaegen, 2016). The fatty acids in *C. sativum* seeds/fruits are *cis*-6-octadecenoic acid, linoleic acid, oleic acid, and palmitic acid (Kaur, 2019; Coşkuner, Karababa, 2007). Like all other green leaf containing vegetables, coriander have large amount of vitamins. It has large quantity of vitamin A and β-carotene as well as vitamin C in it. It also contain the minerals and iron (Girenko, 1982). *C. sativum* contain very low amount of cholesterol and saturated fats. Large amount of dietary fiber, thiamine and zinc are also present in *C. sativum*. About 84% water is present in unripen coriander (Bhat, 2014). The smell and taste of the mature coriander fruits is totally different from the fresh coriander herb, because open chain aldehyde such as C10 to C16 chain length aldehydes with foul smelling are prominent in fresh coriander oil (Steinmetz, Potter, 1996). Monoterpene hydrocarbons and oxygen containing monoterpenes are major compounds in the oil extracted from the coriander seed/fruit (Bhuiyan, Begum, Sultana, 2009). Coriandrone A and B which are two new isocoumarins along with two known isocoumarins, coriandrin and dihydrocoriandrin were extracted from the coriander parts which are present in air (Baba, 1991). Three other isocoumarins such as coriandrones C, D, and E are extracted from complete plant of coriander (Taniguchi, 1996). Caffeic acid, glycitin and protocatechinic acid were also present and these are main polyphenolics of *C. sativum* aerial parts (Melo, Filho, Guerra, 2005).

Here are the most important phytochemicals which are present in coriander.

2.1 Essential Oils (EOs)

Different localities performed much work on the constituents present in C. *sativum* essential oil. In fact, coriander includes in one of the very important spices containing essential oils (EOs). The EOs can be present in all parts of coriander i.e. in its leaves, fruits, roots and bark. But the composition of oil present in different parts is different. During different periods of growth of coriander chemical composition of oil undergoes changes. These changes affect the smell of the plant. That's why the smell and taste of ripened coriander (seeds) is totally different from the smell of the fresh plant. Due to trans-tridecene presence in the coriander essential oil, unripen fruits/seeds and leaves have an unpleasant smell (Mandal, Mandal, 2015)

But the aroma in ripen fruits is pleasantly reminiscent of citrus peel and sage [1, 26]. According to FCC (Food Chemical Codex), Burdock and co-workers have given

Table 1. Characteristics Metrics

Characteristic	Metric
Appearance	Pale yellow liquid or appears colorless
Fragrance	Coriander like
Heavy metals	Passes test
Alcohol's solubility test	Passes test
Angular rotation	Between +8 and +15
Specific gravity	Between 0.863 and 0.875
Identification	Infrared absorption spectrum
Refractive index	Between 1.462 and 1.472 at 20°C

specifications about coriander oil (Burdock, Carabin, 2009). These specifications are given in the following table:

C. sativum contain about 13–20% vegetable oil and 0.2–1.5% essential oil in its fruits or seeds. However, it has been reported that some cultivations may contain up to 2.6% of essential oil. Other research work has shown that about 1% of essential oil is present in coriander. The main compound in oil is monoterpenoid linalool (i) which is up to 50% and significant amounts of limonene (ii), camphor (iii) and geraniol (iv) (Fig.2) are also present (Pavlić, 2015).

In some regions coriander EOs extracted from the seeds/fruits has been reported to be around 0.5-2.5%. This means that a different ratio of essential oil concentrations has been presented by various cultivations and various areas. However, most studies have shown that linalool is the major component in essential oils which is present about

Figure 2. (i) Linalool, (ii) limonene, (iii) camphor, (iv) geraniol

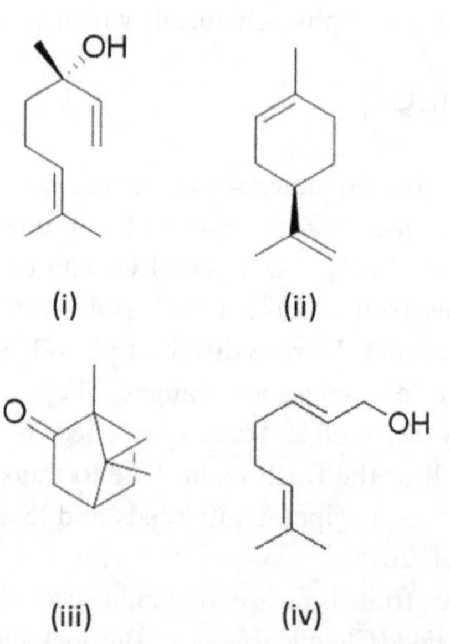

60-70% [1, 26]. Sometimes it is present at about 87.54%. Some other components are also present in *C. sativum* such as α-Pinene, camphor and geraniol. These are the main components other than linalool and also responsible for the characteristic aroma and smell of the herb (Pavlić, 2015).

2.2 Fatty Acids

Large amount of monounsaturated fatty acids about 1.8% is present in vegetable oil/fatty oil obtained from *C. sativum* fruits. Particularly 73% of petroselinic acid is present in vegetable oil [19, 28]. The coriander oil has been assumed to be a Novel Food Constituent as it is used easily by healthy individuals as a dietary ingredient at a higher doses of 600 milligram in a day (Uitterhaegen, 2016). Petroselinic acid (Fig.3) is present in coriander. Due to the petroselinic acid presence in EOs of coriander it is called triglyceride.

The *C. sativum* is considered a main source of essential oil and lipids because it is rich in linalool and petroselinic acid extracted from seeds and other aerial parts (Rajeshwari, 2011). Petroselinic acid presence in coriander makes it valuable and interesting. It is a monounsaturated fatty acid. In coriander EOs, petroselinic acid occurs as an omega-12 fatty acid. Because petroselinic acid is the positional isomer of octadecenoic acid due to presence of its double bond at position 6, but in octadecenoic acid it is present at position 9. About 68-83% petroselinic acid is present in coriander oil (Kiralan, 2009).

2.3 Carotenoids

Coriander when analyzed by HPLC/MS, it contains carotenoids such as *β*-carotene (Fig.4). *β*-carotene is the precursor of vitamin-A. When all coriander cultivars were grown under same situations, *β*-carotene is present in higher amount in mature leaves than in fruits /seeds (Divya, Puthusseri, Neelwarne, 2012). Carotinoids other than *β*-carotene (*β*-cryptoxanthin epoxide, lutein-5,6-epoxide, violaxanthin, and neoxanthin) are also present in coriander. But *β*-carotene content is high about 61.14% of all carotenes present in the ether extract of coriander (Guerra, Melo, Filho, 2005).

Figure 3. Petroselinic acid

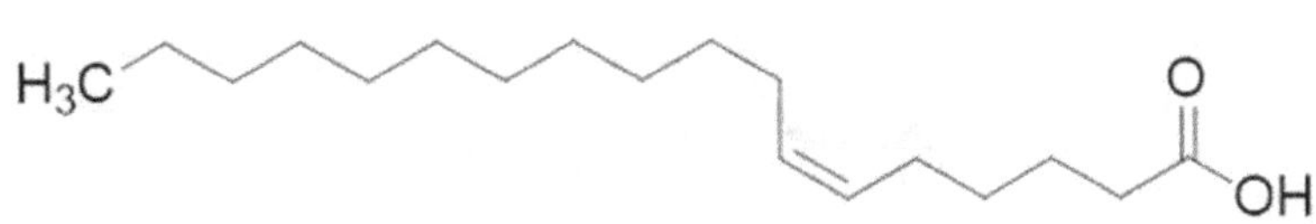

Figure 4. Beta-carotene

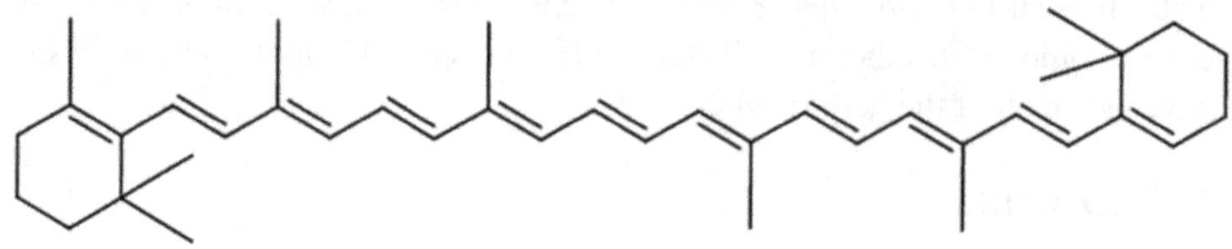

2.4 Polyphenols

Polyphenols are present in *C. sativum* fruits/seeds and leaves. These are very important secondary metabolites. Polyphenols show strong biological properties. Common polyphenols such as flavones, tannins, and anthraquinones are present in coriander fruits (Barros, 2012). LC/MS techniques can be used for the characterization of phenolic compounds in coriander leaves. Quercetin-glucuronide (Fig.5) is a main phenolic compound present in leaves, seeds/fruits and other plant parts (Kaiser, Kammerer, Carle, 2013). Moreover, chlorogenic acids, gallic, ferulic, and caffeic, were also present in high amount in coriandr leaves (Abascal, Yarnell, 2012).

2.5 Isocoumarins

The coriander parts which are present outside the ground contain isocoumarins. These isocoumarins are coriandrin (Fig.6), coriandrones A and B and dihydrocoriandrin (Baba, 1991). When the coriander's aerial parts obtained from the garden of the OU (Osaka University) were treated with methanol solution and then analyzed, the extract contain the coriandrones C-E (Taniguchi, 1996).

Figure 5. Quercetin-glucuronide

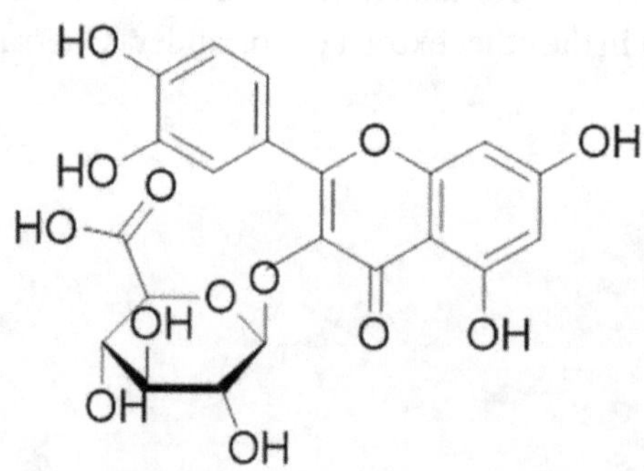

Figure 6. Coriandrin

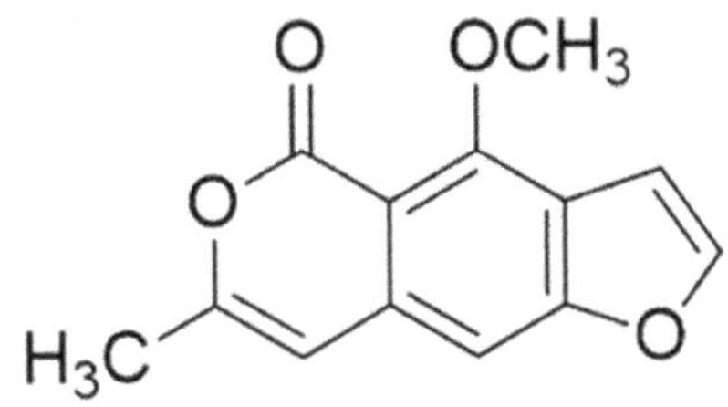

3. PHARMACOLOGICAL PROPERTIES OF C. SATIVUM

It has been reported that the coriander show many biological/pharmacological activities i.e. as anti-oxidant, anti-diabetic, anti-mutagenic, anti-lipidemic, anti-spasmodic. Coriander leaves are used as hypotonic and analgesics. These are useful in suppuration, toothache, hiccough, piles, inflammation, scabies, jaundice and glandular tuberculosis. The *C. sativum* fruits are used as to reduce fever, treat constipation, treat gastric ulcers, aphrodisiac diuretic, stimulant, chic, anathematic and for the treatment of biliousness, bronchitis and vomiting. Coriander fruits show a strong hypoglycemic effect in rats (Dash, 2011). The most important pharmacological properties of coriander will be described in the following order:

3.1 Anti-Diabetic Effects

When coriander extract (20 mg/kg) was orally administered for relatively repeated short periods in hyperlipidemic and streptozotocinthe-diabetic mice, it normalized the glycemia. It reduces the increased levels of insulin and insulin resistance. It also reduces the cholesterol and LDL cholesterol level, and triglycerides. Some components of metabolic syndrome can be reduced by coriander extract. The extract also have the cardiovascular protective effect (Aissoui, 2011). Therefore, coriander is accepted as anti-hyperglycemia food supplement and considered an orally active ingredient for diabetes (Gray, Flatt, 1999). The fasting blood glucose level can be decreased significantly by the uptake of coriander fruit and concentration of glycogen increased in liver in mice. Some other activities in the liver of mice such as glycogen synthase activity, hexokinase, glucose-6-phosphate dehydrogenase activity and phosphoglucomutase activity increased significantly but the glycogen phosphorylase and glucose-6-phosphatase activities are decreased (Chithra, Leelamma, 1999). The coriander's fruits can be accepted for the treatment of chronic hepatic diseases and hepatic fibrosis (Wijayagunawardanea, Wijerathnea, Herathb, 2015). In general, the hyperglycemia decreases and glucose uptake, metabolism and insulin secretion increases by the use of *C. sativum* extract (Gray, Flatt, 1999; Dash, 2011; Aissaoui, 2011; Chithra, 1999; Wijayagunawardanea, Wijerathnea, Herathb, 2015, Swanston-Flatt, 1990).

3.2 Hepatoprotective Activity

Extracts of coriander save the liver from oxidative damage. Tetrachloromethane (CCl4) induce the oxidative stress. Level of thiobarbituric acid responsive compounds, serum glutamate oxaloacetate and pyruvate transaminase reduces in mice which are pretreated with different concentrations of coriander extract against tetrachloromethane treated mice. Hepatic enzymes such as superoxide dismutase, catalase and glutathione peroxidase were significantly increased by plant extract treatment against tetrachloromethane treated mice. Oral administration of coriander leaf extract at a dosage of 200 mg/kg significantly reduces the tetrachloromethane's induced harmful effects. The leaf extract activity at this concentration level was similar to the standard drug silymarin (Sreelatha, Padma, Umadevi, 2009).

3.3 Anti-Bacterial and Anti-Fungal Properties

Coriander's anti-microbial activity is due to its essential oil. Coriander essential oil exhibits a strong anti-microbial properties against oral pathogens and the dental gel formulations (Alves-Silva, 2013). Coriander essential oil (EOs) is used to stop microorganism's broad spectrum. Effective anti-bacterial activity of coriander EOs against *Staphylococcus aureus* and Gram-negative bacterial strains including *Escherichia coli, Klebsiella pneumoniae, Salmonella typhimurium* and *Pseudomonas aeruginosa* and two clinical multidrug-resistant isolates of *Acinetobacter baumannii* was demonstrated. Coriander oil's main mechanism is to damage the cell membrane until the cell die (Silva, 2011). Coriander aqueous extract has been found to be excellent for acne-causing bacteria such as Propionibacterium acne and Staphylococcus epidermidis. Commercially developed acne treatments have the same activity (Vats, Sharma, 2012). Coriander's essential oil is considered the best one for the development of advanced antiacne products due to its anti-bacterial activity (Vats, Sharma, 2012). *C. sativum* is also found a good herbal remedy for diaper dermatitis, a common skin disorder, due to its anti-oxidant, anti-inflammatory, analgesic and anti-microbial properties (Dastgheib, 2017). The pathogenic bacteria associated to foodborne diseases and hospital infections can be killed effectively by the use of coriander oil. Due to this the use of coriander oil is promoted in many anti-bacterial products (Silva, 2011). Coriander essential oil has been investigated for its anti-bacterial and anti-fungal activity in many studies and good efficacy of anti-microbial activity has been found (Mandal, Mandal, 2015; Alves-Silva, 2013, Soares, 2012). Coriander's essential oil obtained by hydro distillation was investigated for its anti-fungal activity against various Candida species. Oil showed strong anti-fungal activity against all tested Candida species but the essential oil showed no activity against C. tropicalis CBS 94. So it is concluded that the oil can be used as a strong anti-microbial agent to cure and prevent yeast infections caused Candida species (Begnami, 2010). The essential oil extracted from *C. sativum* leaves shows anti-microbial activity against clinical isolates of bacteria i.e. Gram (+) and Gram (−) bacteria, but do not shows any activity against the pathogenic fungus i.e. *C. albicans* (Matasyoh, 2009). It was investigated that the coriander fruit oil shows excellent activity against *S. aureus, Listeria monocytogenes, S. haemolyticus, E. coli* and *P. aeruginosa* Abascal, Yarnell, 2012). When *C. sativum* essential oils (EOs) obtained by two different distillation techniques i.e. distillation and microwave-assisted hydrodistillation, were compared, they show the same anti-microbial activity. Apart from saving energy and time, no significant differences in activity were found (Sourmaghi, 2015).

3.4 Anti-Oxidant Activity

An anti-oxidant is defined as "any substance that is used to stop or reduce the oxidation of oxidizable substance". It can reduce the oxidation even when it is present in small concentrations (Rhee, Park, Cho, 2009; Halliwell, 1995; Wiseman, Balentine, Frei, 1997; Matés, Pérez-Gómez, De Castro, 1999). The products of natural substances and their derivatives show effective anti-oxidant properties, which are subsequently associated with hypolipidemic, anti-cancer, anti-inflammatory and anti-aging properties [65-71]. Polyphenolic compounds are well known anti-oxidants present in *C. sativum* (Hashim, 2005). Phenolic compounds are present in leaf and seed extracts and also in the coriander oil. Coriander leaf extract showed the strongest anti-oxidant activity as compared to seeds. Ethyl acetate present in both parts of coriander is the main cause of anti-oxidant activity. Therefore, it is believed that anti-oxidant content of food would be increased by the addition of coriander to food and thereby inhibiting the unwanted oxidation process as it act a natural anti-oxidant (Wangensteen, Samuelsen, Malterud, 2004). The biochemical effects of coriander seeds were studied in rats. These rats were suffering colon cancer which is induced by 1,2-dimethylhydrazine (DMH). A spice diet was administered during the first 15 weeks. When compared the 1,2-dimethylhydrazine (DMH) control group and the spice-treated group, the study shows that cholesterol concentration and the phospholipid: cholesterol ratio decreased, while phospholipid levels were increased mainly in 1,2-dimethylhydrazine control group. Stool dry weight, faecal bile acids and neutral sterols increased in spice treated group as compared to the DMH-treated group. This study showed, the *C. sativum* give protection against the harmful impacts on lipid metabolism in colon cancer (Chithra, Leelamma, 2000). According to Kaur and Kapoor (Kaur, Kapoor, 2002) the anti-oxidant activity is due to the total phenolics content in *C. sativum.*

3.5 Anti-Aging Effects

The long-chain fatty acids present in coriander extracts are very important in topical anti-aging products. These fatty acids help in preventing moisture loss from the epidermis and also restore the barrier properties of the epidermis. Therefore, fatty acids which have long-chain can be considered the main anti-aging agents in *C. sativum* extract. These fatty acids are present in coriander oil. So the coriander fruit oil is used to treat various skin diseases and is used as part of herbal sunscreens (Abascal, Yarnell, 2012). The ceramides of petroselinic acid are also present in essential oils (EOs) of coriander. The extract of coriander is also used as an anti-irritant. This coriander extract maintain the skin tone and skin texture. A specifically formulated extract of coriander fruit i.e.Umbelliferin® is used to support the skin's barrier functions. It contains triglycerides of petroselinic acid (Majeed, Prakash, 2005).

3.6 Metal-Detoxification

C. sativum has the ability to detoxify metals. This detoxification property was caused by the binding of carboxyl group on mercury. Due to this ability, it is useful as a natural cleanser as it has the ability to remove harmful metals from the body. Many chemical components present in *C. sativum* binds to harmful metals and then gets removed from the cells. Bhat et al. investigated that *C. sativum* herb is very useful for removal of mercury ion (Hg^{2+}) and methyl mercury ($HgCH_3$) from aqueous solutions (Bhat, 2014). These investigations clearly indicated that the coriander can be used as a sorbent to remove inorganic ions and methylmercury ($HgCH_3$) from polluted water (Cho, 2006).

3.7 Anthelmintic Effects

In vitro and *in vivo* anthelmintic properties of coriander seeds/fruits extract were studied. *In vitro* studies of raw aqueous extract and hydro-alcoholic extract were tested on the egg and adult *Haemonchus contortus* parasite and also *in vivo* anthelmintic properties of the aqueous extract of coriander were tested in sheep which is also infected by *Haemonchus contortus* parasite. Both extracts of coriander completely inhibit the egg from hatching (Pathak, Kasture, Bhatt, 2011; Eguale, 2007). Moreover, all doses of essential oil exhibit a significant level of toxicity to an insect "Sitophilus granarius" after 5 days in chickpea grains (Zoubiri, Baaliouamer, 2010).

3.8 Antiulcer Effects

C. sativum is used to stop the peptic ulcers and *H. pylori* in stomach. Heavy metals absorption and other toxins absorption can be prevented in the body by the use of coriander. In many research studies, it can be concluded that the anti-gastric ulcer and anti-secretory activity of coriander is due to the anti-oxidant properties of various components which are present in *C. sativum*. These constituents are also helpful in scavenging reactive oxygen species on the surface of the gastric mucosa or may also form a protective layer through hydrophobic interactions. Therefore, it save the cells in stomach from damage (Al-Mofleh, 2006). Latest studies have shown that coriander fruits protect animals from the ulcer causing effects of ethanol (CH_3OH), salt, indomethacin, sodium hydroxide (NaOH) and pyloric ligation in a dosage-dependent manner (Abascal, Yarnell, 2012).

4. SOME OTHER BENEFICIAL HEALTH EFFECTS

C. sativum also has many beneficial health effects. It is used as controlling swelling, diarrhea, menstrual disorders, mouth ulcers, smallpox, anemia, eye care, skin disorders, conjunctivitis, etc (Shahzad, Shakil-ur-Rehman, Fafique, 2019).

4.1 Swellings

Cineole, a coriander essential oil constituent and linoleic acid are present in *C. sativum*. These compounds have anti-arthritic and anti-rheumatic activities. They are much effective in treating swelling, which is caused by kidney failure or anemia. Because some coriander's ingredients are helpful in eliminating excess water from the body.

4.2 Control of Diarrhea and Digestive Health

C. sativum, because of its strong fragrance which is due to its EOs, is helpful in proper secretion of enzymes and gastric juices in stomach. These enzymes stimulate the digestive process and the peristaltic movement. It helps in treating some other problems like anorexia. Borneol and linalool are also the constituents of EOs help in the digestion process, proper hepatic function and intestinal binding. They are also helpful in treating diarrhea which is due to fungal and microbial action. Other constituents such as cineol, borneol, limonene, alpha-pinene and beta-phelandrene have anti-bacterial properties. The fresh coriander leaves are also very good appetizers.

4.3 Mouth Ulcers

Citronelol is an excellent antiseptic. It is a component of coriander EOs. Other components of coriander essential oil have healing and disinfectant effects. These components not allow the cut and sore in mouth to go bad. They are helpful in the healing of ulcers and refresh the breath.

4.4 Anemia

Coriander contains iron, which directly helps in the treatment of anemia.

4.5 Small Pox

The coriander EOs contains a large amount of acids, anti-microbial, anti-infective, free-radical scavengers and cleansing agents. The iron and vitamin C presence in coriander also give strength to the immune system. These coriander properties are helpful in the prevention and curing of smallpox. They are also used to give relief from pain and have a calming effect on smallpox patients.

4.6 Menstruation Disorders

Some components of coriander help in the maintenance of menstrual cycle. They are helpful in proper production of hormones from endocrine glands. They help in inducing of proper menstrual cycles. They also help in reducing pain during menstruation.

4.7 Eye Care

Coriander has a lot of anti-oxidants in its essential oils. It also contains vitamins (A and C) and minerals (phosphorus). These vitamins and minerals prevent eye aging. These components reduce macular degeneration and helps in soothing of eyes from stress. *C. sativum* is an excellent germicide which has strong anti-microbic activities that saves the eyes from dangerous diseases e.g. conjunctivitis.

4.8 Skin Disorders

Coriander essential oils are good for treating skin disorders such as eczema, fungal infections and dryness due to its antiseptic, detoxifying, germicide, anti-fungal and anti-oxidant activities. Some components of *C. sativum* helps in the treatment of ulcers, spasms, and inflammation. These components act as an expectorant, protecting and soothing of the liver. Some components of coriander are anticarcinogenic, antihistamine anticonvulsant, and anesthetics. *C. sativum* is considered a natural stimulant and was formerly it widely used in certain preparations in combination with other herbs. Afifi et al conducted a study on *C. sativum* essential oil and showed that oil has significant analgesic properties in mice.

5. SILVER NANOPARTICLES OF *C. SATIVUM*

Nanoparticles have very important applications in various fields. Due to greater surface area and smaller particle size nanoparticle show better photo catalytic activities. These particles are synthesized by using nanotechnology. All nanoscale materials and devices can be synthesized and designed by the use of nanotechnology. Silver nanoparticles of coriander can be synthesized by using dried seed extract of coriander. These particles are performing a main part in area of biomedical nanotechnology. Nanoparticles are used for storage and data transmission (Zhang, Liu, Sun, 2010) delivery of drug (Li, 2009) catalyst (Shiraishi, Toshima, 2000) surface-enhanced Raman scattering (Matejka, 1992) electronics (Gittins, 2000) chemistry (Manisha, 2014) biotechnology (Salata, 2004) water treatment (Tiwari, Behari, Sen, 2008). and conversion of solar energy Oelhafen, Schüler, 2005). Different methods can be used for the preparation of nanoparticles like polysaccharide method, irradiation method, tollens method, etc. But for the preparation of silver nanoparticles a domestic microwave method can be used. The reaction mixture of dried coriander fruits and silver nitrate solution was prepared in 1: 9 ratios respectively. This mixture was placed in domestic microwave for incubation. The reaction mixture was kept for 2min. The mixture changes its colour from light yellow to red-brown. Colour changes of reaction mixture indicate the reduction of silver ions. FTIR and UV spectral analysis was performed for the characterization of silver nanoparticles Skoog, Holler, Crouch, 2007; Vogel, 1989; Pavia, 2005). SEM studies were used for the morphology of nanoparticles.

Figure 7. SEM structure of AgNPs

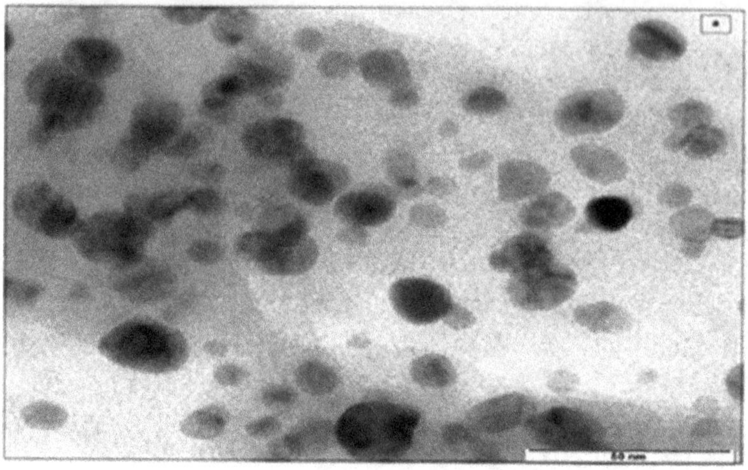

6. TRADITIONAL USES OF CORIANDER

Coriander is a common flavoring substance. It is one of the first spices to be used as flavoring agent (culinary purposes) almost in all dishes. The small plant is used in making chutneys. Leaves of *C. sativum* are useful to flavor curries, soups and sauces. Coriander leaves, stalks and seeds/fruits have good smell. Coriander essential oil (EOs) and oleoresin are mostly used in seasoning meat products as well as other sausages. They are also used in curry mixes, pastries, chewable gum and in spices (Moawad, 2015). Coriander has been traditionally used as anti-diabetic plant in some parts of Europe (Rajeshwari, Andallu, 2011; Melnyk, Marcone, 2011). Coriander fruits have good effect on health. In healing spices, coriander is considered as a main compound. Due to its anti-inflammation activities, it has been traditionally used in India. In the US, *C. sativum* has currently been studied for its cholesterol level lowering properties (Msaada, 2007). The EOs of coriander is used in many foods. It is also used in perfumes, flavors and carminatives. Due to its powerful anti-septic, aromatic, stimulant, carminative, stomachic, expectorant, anti-spasmodic and diuretic properties, it is used in various medicines (Ahmed, Abadi, Mohammed, 2018). The fruits are the main components which give the medical importance to coriander. These fruits are useful as a drug such as for joints pain, indigestion, rheumatism and against worms (Wangensteen, Samuelsen, Malterud, 2004). Essential oil of coriander fruit extract is used as anti-bacterial agent in many lotions and shampoos (Bhuiyan, Begum, Sultana, 2009). Coriander is used to give relief from spasm, insomnia and anxiety (Emam, Heydari, 2008). It is useful for other diseases such as sub-acid gastritis, diarrhea, and dyspepsia Platel,

Srinivasan, 2004). It is used in many folk medicine such as used against intestinal parasites (Wichtl, 1994). Leaves of coriander are mostly useful as local medicine. Leaves are used as carminative, spasmolytic, digestive, and galactagogue. Coriander shows strong lipolytic properties (Leung, Foster, 1996). As coriander belongs to apiaceae family, it contain furanocoumarins which cause allergic reactions, so it has been used with cautions (Burdock, Carabin, 2009). In Pakistan, all parts of plant are useful in many treatments. These are used to treat flatulence, cough, dysentery, diarrhea, jaundice, and vomiting. However, in history, it is mentioned that *C. sativum* like many others spices has an aphrodisiac effect [103]. In Turkey, the fruit infusions of *C. sativum* are used as an appetizer and in indigestion (Laribi, 2015). In some regions of India, the seeds of coriander are used for the treatment of spermatorrhea, leucorrhea and rheumatic fever (Rajeshwari, Andallu, 2011; Melnyk, Marcone, 2011). However, the usual dosage of coriander fruit powder is about 1 to 5g. It is used three times in a day in traditional medicines (Abascal, Yarnell, 2012). In short, *C. sativum* is very important plant as it has been integrated between the traditional and scientific usages.

7. CONCLUSION

Pharmaceutical and phytochemical studies have been conducted on various parts of *C. sativum*. These studies support the medicinal potential of *C. sativum* plant. Coriander essential oil have a great amount of useful phyto-nutrients which have the beneficial health effects. Coriander is included in healing spices. The healing properties is due to the essential oils. The essential oil have applications in the field of food, in cosmetic and pharmacological industries. It is used as flatus-relieving, cytostatics, lipolytic, antifungal, stimulant and stomachic. It have the antibacterial, antifungal, anticarcinogenic and antidiabetic properties. Ultimately it is suggested that, the *C. sativum* is amazingly a safe herb and it is very good to use *C. sativum* in diet.

REFERENCES

Abascal, K., & Yarnell, E. (2012). Cilantro—Culinary Herb or Miracle Medicinal Plant? *Alternative and Complementary Therapies*, *18*(5), 259–264. doi:10.1089/act.2012.18507

Afifi, N. (1994). Some pharmacological activities of essential oils of certain umbelliferous fruits. Veterinary Medical Journal Giza, Cairo Univ.

Ahmed, E. H., Abadi, R. S., & Mohammed, A. M. (2018). Phytochemical screening, chemical composition and antioxidant activity of seeds essential oil of Coriandrum sativum L. from the Sudan. *International Journal of Herbal Medicine*, *6*(1), 1–4.

Aissaoui, A., El-Hilaly, J., Israili, Z. H., & Lyoussi, B. (2008). Acute diuretic effect of continuous intravenous infusion of an aqueous extract of Coriandrum sativum L. in anesthetized rats. *Journal of Ethnopharmacology*, *115*(1), 89–95. doi:10.1016/j.jep.2007.09.007 PMID:17961943

Aissaoui, A., Zizi, S., Israili, Z. H., & Lyoussi, B. (2011). Hypoglycemic and hypolipidemic effects of Coriandrum sativum L. in Meriones shawi rats. *Journal of Ethnopharmacology*, *137*(1), 652–661. doi:10.1016/j.jep.2011.06.019 PMID:21718774

Al-Mofleh, I., Alhaider, A. A., Mossa, J. S., Al-Sohaibani, M. O., Rafatullah, S., & Qureshi, S. (2006). Protection of gastric mucosal damage by Coriandrum sativum L. pretreatment in Wistar albino rats. *Environmental Toxicology and Pharmacology*, *22*(1), 64–69. doi:10.1016/j.etap.2005.12.002 PMID:21783688

Ali, B., Al-Wabel, N. A., Shams, S., Ahamad, A., Khan, S. A., & Anwar, F. (2015). Essential oils used in aromatherapy: A systemic review. *Asian Pacific Journal of Tropical Biomedicine*, *5*(8), 601–611. doi:10.1016/j.apjtb.2015.05.007

Alves-Silva, J. M., Dias dos Santos, S. M., Pintado, M. E., Pérez-Álvarez, J. A., Fernández-López, J., & Viuda-Martos, M. (2013). Chemical composition and in vitro antimicrobial, antifungal and antioxidant properties of essential oils obtained from some herbs widely used in Portugal. *Food Control*, *32*(2), 371–378. doi:10.1016/j.foodcont.2012.12.022

Aruoma, O. I. (2003). Methodological considerations for characterizing potential antioxidant actions of bioactive components in plant foods. *Mutation Research. Fundamental and Molecular Mechanisms of Mutagenesis*, *523*, 9–20. doi:10.1016/S0027-5107(02)00317-2 PMID:12628499

Baba, K., Xiao, Y.-Q., Taniguchi, M., Ohishi, H., & Kozawa, M. (1991). Isocoumarins from Coriandrum sativum. *Phytochemistry*, *30*(12), 4143–4146. doi:10.1016/0031-9422(91)83482-Z

Barros, L., Dueñas, M., Dias, M. I., Sousa, M. J., Santos-Buelga, C., & Ferreira, I. C. F. R. (2012). Phenolic profiles of in vivo and in vitro grown Coriandrum sativum L. *Food Chemistry*, *132*(2), 841–848. doi:10.1016/j.foodchem.2011.11.048 PMID:23107699

Begnami, A., Duarte, M. C. T., Furletti, V., & Rehder, V. L. G. (2010). Antimicrobial potential of Coriandrum sativum L. against different Candida species in vitro. *Food Chemistry*, *118*(1), 74–77. doi:10.1016/j.foodchem.2009.04.089

Bhat, S. (2014). Coriander (Coriandrum sativum L.): Processing, nutritional and functional aspects. *African Journal of Plant Science*, *8*(1), 25–33. doi:10.5897/AJPS2013.1118

Bhuiyan, M. N. I., Begum, J., & Sultana, M. (2009). *Chemical composition of leaf and seed essential oil of Coriandrum sativum L. from Bangladesh.* III. *Bangladesh Journal of Pharmacology*, *4*(2), 150–153. doi:10.3329/bjp.v4i2.2800

Burdock, G. A., & Carabin, I. G. (2009). Safety assessment of coriander (Coriandrum sativum L.) essential oil as a food ingredient. *Food and Chemical Toxicology*, *47*(1), 22–34. doi:10.1016/j.fct.2008.11.006 PMID:19032971

Ceska, O., Chaudhary, S. K., Warrington, P., Ashwood-Smith, M. J., Bushnell, G. W., & Poultont, G. A. (1988). Coriandrin, a novel highly photoactive compound isolated from Coriandrum sativum. *Phytochemistry*, *27*(7), 2083–2087. doi:10.1016/0031-9422(88)80101-8

Chithra, V., & Leelamma, S. (1999). Coriandrum sativum—Mechanism of hypoglycemic action. *Food Chemistry*, *67*(3), 229–231. doi:10.1016/S0308-8146(99)00113-2

Chithra, V., & Leelamma, S. (2000). Coriandrum sativum—Effect on lipid metabolism in 1, 2-dimethyl hydrazine induced colon cancer. *Journal of Ethnopharmacology*, *71*(3), 457–463. doi:10.1016/S0378-8741(00)00182-3 PMID:10940583

Cho, J. Y., Park, S.-C., Kim, T.-W., Kim, K.-S., Song, J.-C., Lee, H.-M., Sung, H.-J., Rhee, M.-H., Kim, S.-K., Park, H.-J., Song, Y.-B., Yoo, E.-S., & Lee, C.-H. (2006). Radical scavenging and anti-inflammatory activity of extracts from Opuntia humifusa Raf. *The Journal of Pharmacy and Pharmacology*, *58*(1), 113–119. doi:10.1211/jpp.58.1.0014 PMID:16393471

Cortés-Eslava, J., Gómez-Arroyo, S., Villalobos-Pietrini, R., & Espinosa-Aguirre, J. J. (2004). Antimutagenicity of coriander (Coriandrum sativum) juice on the mutagenesis produced by plant metabolites of aromatic amines. *Toxicology Letters*, *153*(2), 283–292. doi:10.1016/j.toxlet.2004.05.011 PMID:15451560

Coşkuner, Y., & Karababa, E. (2007). Physical properties of coriander seeds (Coriandrum sativum L.). *Journal of Food Engineering*, *80*(2), 408–416. doi:10.1016/j.jfoodeng.2006.02.042

Darughe, F., Barzegar, M., & Sahari, M. (2012). Antioxidant and antifungal activity of Coriander (Coriandrum sativum L.) essential oil in cake. *International Food Research Journal*, *19*(3), 1253–1260.

Dash, B. (2011). *Antibacterial activities of methanol and acetone extracts of fenugreek (Trigonella foenum) and coriander (Coriandrum sativum)*. Life Sciences and Medicine Research.

Dastgheib, L., Pishva, N., Saki, N., Khabnadideh, S., Kardeh, B., Torabi, F., Arabnia, S., & Heiran, A. (2017). Efficacy of topical coriandrum sativum extract on treatment of infants with diaper dermatitis: A single blinded non-randomised controlled trial. *The Malaysian Journal of Medical Sciences : MJMS*, *24*(4), 97–101. doi:10.21315/mjms2017.24.4.11 PMID:28951694

Davis, P. H. (1970). Flora of Turkey and the East Aegean Islands. Vol. 3. Flora of Turkey and the East Aegean Islands.

de Almeida Melo, E., Mancini Filho, J., & Guerra, N. B. (2005). Characterization of antioxidant compounds in aqueous coriander extract (Coriandrum sativum L.). *Lebensmittel-Wissenschaft + Technologie*, *38*(1), 15–19. doi:10.1016/j.lwt.2004.03.011

Diederichsen, A. (1996). *Coriander: Coriandrum Sativum L* (Vol. 3). Bioversity International.

Divya, P., Puthusseri, B., & Neelwarne, B. (2012). Carotenoid content, its stability during drying and the antioxidant activity of commercial coriander (Coriandrum sativum L.) varieties. *Food Research International*, *45*(1), 342–350. doi:10.1016/j.foodres.2011.09.021

Eguale, T., Tilahun, G., Debella, A., Feleke, A., & Makonnen, E. (2007). In vitro and in vivo anthelmintic activity of crude extracts of Coriandrum sativum against Haemonchus contortus. *Journal of Ethnopharmacology*, *110*(3), 428–433. doi:10.1016/j.jep.2006.10.003 PMID:17113738

Eidi, M., Eidi, A., Saeidi, A., Molanaei, S., Sadeghipour, A., Bahar, M., & Bahar, K. (2009). Effect of coriander seed (Coriandrum sativum L.) ethanol extract on insulin release from pancreatic beta cells in streptozotocin-induced diabetic rats. *Phytotherapy Research*, *23*(3), 404–406. doi:10.1002/ptr.2642 PMID:19003941

Eikani, M. H., Golmohammad, F., & Rowshanzamir, S. (2007). Subcritical water extraction of essential oils from coriander seeds (Coriandrum sativum L.). *Journal of Food Engineering*, *80*(2), 735–740. doi:10.1016/j.jfoodeng.2006.05.015

Emam & Heydari. (2008). *Effect of extract and essential oil of Coriandrum sativum seed against pentylenetetrazole-induced seizure.* Academic Press.

Evans, W. (2002). *Trease and Evans: Pharmacognocy* (Fifteenth International Edition). WB Saunders.

Girenko, M. (1982). *Initial material and basic trends in breeding of some uncommon species of vegetables.* J. Bull. VIR im. *Vavilova, 120*, 33–37.

Gittins, D. I., Bethell, D., Nichols, R. J., & Schiffrin, D. J. (2000). Diode-like electron transfer across nanostructured films containing a redox ligand. *Journal of Materials Chemistry, 10*(1), 79–83. doi:10.1039/a902960e

Gray, A. M., & Flatt, P. R. (1999). Insulin-releasing and insulin-like activity of the traditional anti-diabetic plant Coriandrum sativum (coriander). *British Journal of Nutrition, 81*(3), 203–209. doi:10.1017/S0007114599000392 PMID:10434846

Grosso, C., Ferraro, V., Figueiredo, A. C., Barroso, J. G., Coelho, J. A., & Palavra, A. M. (2008). Supercritical carbon dioxide extraction of volatile oil from Italian coriander seeds. *Food Chemistry, 111*(1), 197–203. doi:10.1016/j.foodchem.2008.03.031

Guerra, N. B., de Almeida Melo, E., & Mancini Filho, J. (2005). Antioxidant compounds from coriander (Coriandrum sativum L.) etheric extract. *Journal of Food Composition and Analysis, 18*(2-3), 193–199. doi:10.1016/j.jfca.2003.12.006

Halliwell, B. (1995). Antioxidant characterization: Methodology and mechanism. *Biochemical Pharmacology, 49*(10), 1341–1348. doi:10.1016/0006-2952(95)00088-H PMID:7763275

Hashim, M. (2005). Effect of polyphenolic compounds from Coriandrum sativum on H2O2-induced oxidative stress in human lymphocytes. *Food Chemistry, 92*(4), 653–660. doi:10.1016/j.foodchem.2004.08.027

Hogg, N. (!998). Free radicals in disease. In Seminars in reproductive endocrinology. Thieme Medical Publishers, Inc. doi:10.1055/s-2007-1016284

Kaiser, A., Kammerer, D. R., & Carle, R. (2013). Impact of blanching on polyphenol stability and antioxidant capacity of innovative coriander (Coriandrum sativum L.) pastes. *Food Chemistry, 140*(1-2), 332–339. doi:10.1016/j.foodchem.2013.02.077 PMID:23578650

Kaur, C., & Kapoor, H. C. (2002). Anti-oxidant activity and total phenolic content of some Asian vegetables. *International Journal of Food Science & Technology, 37*(2), 153–161. doi:10.1046/j.1365-2621.2002.00552.x

Kaur, R., & Pujari, M. (2019). Comparative study of control of seed mycoflora of Coriandrum sativum l. by herbal and chemical methods. *Plant Archives*, *19*(2), 1082–1084.

Kiralan, M., Calikoglu, E., Ipek, A., Bayrak, A., & Gurbuz, B. (2009). Fatty acid and volatile oil composition of different coriander (Coriandrum sativum) registered varieties cultivated in Turkey. *Chemistry of Natural Compounds*, *45*(1), 100–102. doi:10.1007/s10600-009-9240-2

Koyyati, R. (2013). *Biological synthesis of silver nanoparticles using Raphanus sativus var. longipinnatus leaf extract and evaluation of their antioxidant and antibacterial activity.* International Journal of Medicine and Pharmaceutical Sciences, 3(4), 89-100.

Koyyati, R. (2014). Antibacterial activity of silver nanoparticles synthesized using Amaranthus viridis twig extract. *International Journal of Research in Pharmaceutical Sciences*, *5*(1), 32–39.

Kudle, K. R. (2013). Green synthesis of silver nanoparticles using water soluble gum of Sterculia foetida and evaluation of its antimicrobial activity. *International Journal of Research in Pharmaceutical Sciences*, *4*(4), 563–568.

Kudle, K. R. (2013). Fruit (epicarp and endocarp) extract mediated synthesis of silver nanoparticles from Sterculia foetida plant and evaluation of their antimicrobial activity. *Nanoscience and Nanotechnology: An International Journal*, *3*(3), 45–48.

Kudle, K. R. (2013). Microwave assisted green synthesis of silver nanoparticles using Stigmaphyllon littorale leaves their characterization and anti-microbial activity. *International Journal of Nanomaterials and Biostructures*, *3*(1), 13–16.

Kumar, C., Sarwar, M., & Dimri, B. (1977). Bulgarian coriander in India and its future prospects in export trade. *Indian Perfumer*, *21*(3), 146–150.

Laribi, B., Kouki, K., M'Hamdi, M., & Bettaieb, T. (2015). Coriander (Coriandrum sativum L.) and its bioactive constituents. *Fitoterapia*, *103*, 9–26. doi:10.1016/j.fitote.2015.03.012 PMID:25776008

Leung, A. Y., & Foster, S. (1996). Encyclopedia of common natural ingredients used in food, drugs, and cosmetics. *Journal of the American Chemical Society*, *118*(37), 8988.

Li, J.-L., Wang, L., Liu, X.-Y., Zhang, Z.-P., Guo, H.-C., Liu, W.-M., & Tang, S.-H. (2009). In vitro cancer cell imaging and therapy using transferrin-conjugated gold nanoparticles. *Cancer Letters*, *274*(2), 319–326. doi:10.1016/j.canlet.2008.09.024 PMID:18977071

Majeed, M., & Prakash, L. (2005). Novel natural approaches to anti-aging skin care. *Cosmetics and Toiletries Manufacture Worldwide*, *1*, 11–15.

Mandal, S., & Mandal, M. (2015). Coriander (Coriandrum sativum L.) essential oil: Chemistry and biological activity. *Asian Pacific Journal of Tropical Biomedicine*, *5*(6), 421–428. doi:10.1016/j.apjtb.2015.04.001

Manisha, D. (2014). Microwave assisted biogenic synthesis of silver nanoparticles using dried seed extract of Coriandrum sativum, characterization and antimicrobial activity. *International Journal of Chemtech Research*, *6*, 3957–3961.

Matasyoh, J., Maiyo, Z. C., Ngure, R. M., & Chepkorir, R. (2009). Chemical composition and antimicrobial activity of the essential oil of Coriandrum sativum. *Food Chemistry*, *113*(2), 526–529. doi:10.1016/j.foodchem.2008.07.097

Matejka, P., Vlckova, B., Vohlidal, J., Pancoska, P., & Baumruk, V. (1992). The role of triton X-100 as an adsorbate and a molecular spacer on the surface of silver colloid: A surface-enhanced Raman scattering study. *Journal of Physical Chemistry*, *96*(3), 1361–1366. doi:10.1021/j100182a063

Matés, J. M., Pérez-Gómez, C., & De Castro, I. N. (1999). Antioxidant enzymes and human diseases. *Clinical Biochemistry*, *32*(8), 595–603. doi:10.1016/S0009-9120(99)00075-2 PMID:10638941

McGee, H. (2010). *Cilantro haters, it's not your fault.* The New York Times.

Melnyk, J. P., & Marcone, M. F. (2011). Aphrodisiacs from plant and animal sources—A review of current scientific literature. *Food Research International*, *44*(4), 840–850. doi:10.1016/j.foodres.2011.02.043

Moawad, S. A., El-Ghorab, A. H., Hassan, M., Nour-Eldin, H., & El-Gharabli, M. M. (2015). Chemical and microbiological characterization of Egyptian cultivars for some spices and herbs commonly exported abroad. *Food and Nutrition Sciences*, *6*(07), 643–659. doi:10.4236/fns.2015.67068

Moulin, L. (2002). *Eating and drinking in Europe: a cultural history*. Mercatorfonds.

Msaada, K. (2009). *Changes in fatty acid composition of coriander (Coriandrum sativum L.) fruit during maturation.* Industrial Crops and Products, 29(2-3), 269-274.

Msaada, K., Hosni, K., Taarit, M. B., Chahed, T., Kchouk, M. E., & Marzouk, B. (2007). Changes on essential oil composition of coriander (Coriandrum sativum L.) fruits during three stages of maturity. *Food Chemistry*, *102*(4), 1131–1134. doi:10.1016/j.foodchem.2006.06.046

Msaada, K., Taarit, M. B., Hosni, K., Hammami, M., & Marzouk, B. (2009). Regional and maturational effects on essential oils yields and composition of coriander (Coriandrum sativum L.) fruits. *Scientia Horticulturae*, *122*(1), 116–124. doi:10.1016/j.scienta.2009.04.008

Nadeem, M., Muhammad Anjum, F., Issa Khan, M., Tehseen, S., El-Ghorab, A., & Iqbal Sultan, J. (2013). Nutritional and medicinal aspects of coriander (Coriandrum sativum L.). *British Food Journal*, *115*(5), 743–755. doi:10.1108/00070701311331526

Nagella, P., & … . (2012). Chemical constituents, larvicidal effects and antioxidant activity of petroleum ether extract from seeds of Coriandrum sativum L. *Journal of Medicinal Plants Research*, *6*(15), 2948–2954. doi:10.5897/JMPR11.992

Nalvothula, R., & … . (2014). Biogenic synthesis of silver nanoparticles using Tectona grandis leaf extract and evaluation of their antibacterial potential. *International Journal of Chemtech Research*, *6*(1), 293–298.

Nejad Ebrahimi, S., Hadian, J., & Ranjbar, H. (2010). Essential oil compositions of different accessions of Coriandrum sativum L. from Iran. *Natural Product Research*, *24*(14), 1287–1294. doi:10.1080/14786410903132316 PMID:20803372

Oelhafen, P., & Schüler, A. (2005). Nanostructured materials for solar energy conversion. *Solar Energy*, *79*(2), 110–121. doi:10.1016/j.solener.2004.11.004

Pathak, N., Kasture, S., & Bhatt, N. (2011). Phytochemical screening of Coriander sativum Linn. *International Journal of Pharmaceutical Sciences Review and Research*, *9*, 159–163.

Pathak Nimish, L. (2011). Phytopharmacological properties of Coriander sativum as a potential medicinal tree: An overview. *Journal of Applied Pharmaceutical Science*, *1*(4), 20–25.

Pavia, D. L. (2005). *Introduction to organic laboratory techniques: a small scale approach*. Cengage Learning.

Pavlić, B., Vidović, S., Vladić, J., Radosavljević, R., & Zeković, Z. (2015). Isolation of coriander (Coriandrum sativum L.) essential oil by green extractions versus traditional techniques. *The Journal of Supercritical Fluids*, *99*, 23–28. doi:10.1016/j.supflu.2015.01.029

Platel, K., & Srinivasan, K. (2004). Digestive stimulant action of spices: A myth or reality? *The Indian Journal of Medical Research*, *119*(5), 167. PMID:15218978

Priyadarshi, S., Khanum, H., Ravi, R., Borse, B. B., & Naidu, M. M. (2016). Flavour characterisation and free radical scavenging activity of coriander (Coriandrum sativum L.) foliage. *Journal of Food Science and Technology*, *53*(3), 1670–1678. doi:10.1007/s13197-015-2071-1 PMID:27570292

Rajeshwari, U., & Andallu, B. (2011). Medicinal benefits of coriander (Coriandrum sativum L). *Spatula DD*, *1*(1), 51–58. doi:10.5455/spatula.20110106123153

Randall, K., Drew, M. D., Øverland, M., Østbye, T.-K., Bjerke, M., Vogt, G., & Ruyter, B. (2013). Effects of dietary supplementation of coriander oil, in canola oil diets, on the metabolism of [1-14C] 18: 3n-3 and [1-14C] 18: 2n-6 in rainbow trout hepatocytes. *Comparative Biochemistry and Physiology. Part B, Biochemistry & Molecular Biology*, *166*(1), 65–72. doi:10.1016/j.cbpb.2013.07.004 PMID:23867781

Rhee, M. H., Park, H.-J., & Cho, J. Y. (2009). Salicornia herbacea: Botanical, chemical and pharmacological review of halophyte marsh plant. *Journal of Medicinal Plants Research*, *3*(8), 548–555.

Salata, O. V. (2004). Applications of nanoparticles in biology and medicine. *Journal of Nanobiotechnology*, *2*(1), 3. doi:10.1186/1477-3155-2-3 PMID:15119954

Samuelsson, M. (2003). *Aquavit: And the New Scandinavian Cuisine*. Houghton Mifflin Harcourt.

Shahzad, A. N. (2019). S. Shakil-ur-Rehman, and N. Rafique, *Effectiveness of eccentric and concentric muscle energy techniques on hamstring length in healthy population. Rawal Medical Journal*, *44*(2), 350–352.

Shiraishi, Y., & Toshima, N. (2000). Oxidation of ethylene catalyzed by colloidal dispersions of poly (sodium acrylate)-protected silver nanoclusters. *Colloids and Surfaces. A, Physicochemical and Engineering Aspects*, *169*(1-3), 59–66. doi:10.1016/S0927-7757(00)00417-9

Silva, F., Ferreira, S., Duarte, A., Mendonça, D. I., & Domingues, F. C. (2011). Antifungal activity of Coriandrum sativum essential oil, its mode of action against Candida species and potential synergism with amphotericin B. *Phytomedicine*, *19*(1), 42–47. doi:10.1016/j.phymed.2011.06.033 PMID:21788125

Silva, F., Ferreira, S., Queiroz, J. A., & Domingues, F. C. (2011). Coriander (Coriandrum sativum L.) essential oil: Its antibacterial activity and mode of action evaluated by flow cytometry. *Journal of Medical Microbiology*, *60*(10), 1479–1486. doi:10.1099/jmm.0.034157-0 PMID:21862758

Skoog, D. A., Holler, F. J., & Crouch, S. R. (2007). Instrumental analysis. Brooks/Cole, Cengage Learning.

Soares, B. V., Morais, S. M., dos Santos Fontenelle, R. O., Queiroz, V. A., Vila-Nova, N. S., Pereira, C. M. C., Brito, E. S., Neto, M. A. S., Brito, E. H. S., Cavalcante, C. S. P., Castelo-Branco, D. S. C. M., & Rocha, M. F. G. (2012). Antifungal activity, toxicity and chemical composition of the essential oil of Coriandrum sativum L. fruits. *Molecules (Basel, Switzerland)*, *17*(7), 8439–8448. doi:10.3390/molecules17078439 PMID:22785271

Sourmaghi, M. H. S., Kiaee, G., Golfakhrabadi, F., Jamalifar, H., & Khanavi, M. (2015). Comparison of essential oil composition and antimicrobial activity of Coriandrum sativum L. extracted by hydrodistillation and microwave-assisted hydrodistillation. *Journal of Food Science and Technology*, *52*(4), 2452–2457. doi:10.1007/s13197-014-1286-x PMID:25829632

Sreelatha, S., Padma, P., & Umadevi, M. (2009). Protective effects of Coriandrum sativum extracts on carbon tetrachloride-induced hepatotoxicity in rats. *Food and Chemical Toxicology*, *47*(4), 702–708. doi:10.1016/j.fct.2008.12.022 PMID:19146910

Sriti, J., Msaada, K., Talou, T., Faye, M., Vilarem, G., & Marzouk, B. (2012). Coupled extruder-headspace, a new method for analysis of the essential oil components of Coriandrum sativum fruits. *Food Chemistry*, *134*(4), 2419–2423. doi:10.1016/j.foodchem.2012.04.029 PMID:23442704

Sriti, J., Talou, T., Faye, M., Vilarem, G., & Marzouk, B. (2011). Oil extraction from coriander fruits by extrusion and comparison with solvent extraction processes. *Industrial Crops and Products*, *33*(3), 659–664. doi:10.1016/j.indcrop.2011.01.005

Steinmetz, K. A., & Potter, J. D. (1996). Vegetables, fruit, and cancer prevention: A review. *Journal of the American Dietetic Association*, *96*(10), 1027–1039. doi:10.1016/S0002-8223(96)00273-8 PMID:8841165

Sunil, C., Agastian, P., Kumarappan, C., & Ignacimuthu, S. (2012). In vitro antioxidant, antidiabetic and antilipidemic activities of Symplocos cochinchinensis (Lour.) S. Moore bark. *Food and Chemical Toxicology*, *50*(5), 1547–1553. doi:10.1016/j.fct.2012.01.029 PMID:22306921

Swanston-Flatt, S., Day, C., Bailey, C. J., & Flatt, P. R. (1990). Traditional plant treatments for diabetes. Studies in normal and streptozotocin diabetic mice. *Diabetologia*, *33*(8), 462–464. doi:10.1007/BF00405106 PMID:2210118

Taniguchi, M., Yanai, M., Xiao, Y. Q., Kido, T., & Baba, K. (1996). Three isocoumarins from Coriandrum sativum. *Phytochemistry*, *42*(3), 843–846. doi:10.1016/0031-9422(95)00930-2

Uitterhaegen, E., Sampaio, K., Delbeke, E., De Greyt, W., Cerny, M., Evon, P., Merah, O., Talou, T., & Stevens, C. (2016). Characterization of French coriander oil as source of petroselinic acid. *Molecules (Basel, Switzerland)*, *21*(9), 1202. doi:10.3390/molecules21091202 PMID:27617992

Vats, A., & Sharma, P. (2012). Formulation and evaluation of topical anti acne formulation of coriander oil. *Int J Pharm Pharm Sci Res*, *2*, 61–66.

Vats, A., & Sharma, P. (2013). Comparative study, and analysis of release kinetics of coriander formulations. *Indo American Journal of Pharmaceutical Research*, *3*(1), 1334–1348.

Vogel, A. I. (1989). *Vogel's textbook of practical organic chemistry* (Vol. 5). Longman Scientific & Technical New York.

Wangensteen, H., Samuelsen, A. B., & Malterud, K. E. (2004). Antioxidant activity in extracts from coriander. *Food Chemistry*, *88*(2), 293–297. doi:10.1016/j.foodchem.2004.01.047

Wichtl, M. (1994). *Coriandri fructus. Herbal Drugs and Phytopharmaceuticals*. CRC Press.

Wijayagunawardanea, M. P. B., Wijerathnea, C. U. B., & Herathb, C. B. (2015). Indigenous herbal recipes for treatment of liver cirrhosis. *Procedia Chemistry*, *14*, 270–276. doi:10.1016/j.proche.2015.03.038

Wiseman, S. A., Balentine, D. A., & Frei, B. (1997). Antioxidants in tea. *Critical Reviews in Food Science and Nutrition*, *37*(8), 705–718. doi:10.1080/10408399709527798 PMID:9447271

Zhang, H., Liu, Y., & Sun, S. (2010). Synthesis and assembly of magnetic nanoparticles for information and energy storage applications. *Frontiers of Physics in China*, *5*(4), 347–356. doi:10.1007/s11467-010-0104-9

Zoubiri, S., & Baaliouamer, A. (2010). Essential oil composition of Coriandrum sativum seed cultivated in Algeria as food grains protectant. *Food Chemistry*, *122*(4), 1226–1228. doi:10.1016/j.foodchem.2010.03.119

Chapter 9
Phytochemical, Ethnobotanical, and Pharmacological Activities of Aframomum melegueta

Shamsa Khan
The Women University Multan, Pakistan

Aleena Munir
The Women University Multan, Pakistan

Okasha Fayyaz
Muhammad Nawaz Sharif University of Engineering and Technology, Pakistan

Ali Hassan
Bahauddin Zakariya University, Multan, Pakistan

ABSTRACT

Aframomum melegueta (grains of paradise) is a herbaceous plant of West Africa, locally used as a spice and to treat common ailments like diarrhea, painful arthritis, snake bite, and scorpion sting. Various parts of the plant like stem, leaves, and roots have various phytochemicals like flavonoids, terpenoids, saponins, phenolic compounds, alkaloids cardiac glycosides, and saponins that give alligator pepper a great medicinal value. These phytochemicals exert anti-inflammatory, anti-cancer, anti-diabetic, antimicrobial, antioxidants, and hepatoprotective effects. However, more comprehensive studies need to be performed to reveal the medicinal purposes of alligator pepper. This chapter encapsulates findings from previous studies on the phytochemicals, ethnobotanical, pharmacological activities, and biogenic synthesis of chemicals of Aframomum melegueta.

DOI: 10.4018/979-8-3693-6105-4.ch009

1. INTRODUCTION

Herbs play a crucial part in medicine. A herbal medication is a product that contains herbs and is utilised for its medicinal or pharmacological effects. Nearly 80% of the global population relies on herbal remedies, according to the World Health Organisation (WHO). Herbal remedies are prescribed by around 70% of German doctors. Rumour has it that Germans have easy access to 600–700 different types of natural remedies. A growing number of Americans have taken an interest in herbal remedies during the past two decades. In addition, there are 450–500 medicinal plants in Bangladesh (Motaleb et al., 2011). Among the many traditional uses for herbaceous plants, *Aframomum melegueta* (AM) stands out as a member of the Zingiberaceae family that alleviates aches and pains, catarrh, diarrhoea, congestion, and rheumatism, among other conditions. Guinea, Ghana, the Ivory Coast, Sierra Leone, and Nigeria are among the countries that raise grains of paradise, a native plant of West Africa (Ono et al., 2017). It is also being cultured by Indian house gardens (Toh et al., 2019). *A. melegueta* is a fragrant plant that produces spicy, delicious fruit. This perennial herb can grow up to 4 metres long. The small stem has scars from falling aerial leaves all around it. It has a lot of branches. The lower surface, which is thin and coincident, is where the roots are located. It yields a horizontal, tuberous, fragrant rhizome with scaly leaves and sporadic buds in the axils. The huge leaves have a width of about 12 cm and a length of 30 cm. There are nearly half of the bracts overlapping. The flowers have an orange lip and a beautiful pinkish-orange section. The fruits are juicy, indehiscent, and packed with innumerable tiny seeds that are encased non a delicious coating. The seeds are extremely fragrant, angular, and granular when they are young; they are reddish-brown or golden yet darken when they dry. In West and Central Africa, traditional medicine makes substantial use of this genus (Iwu, 2014). 85.47% dry matter, 7.18% crude protein, 2.55% ether extract, 2.49% ash, 5.50% crude fibre, and 73.63% total carbohydrates have all been reported for alligator pepper seeds. In addition, the seeds are a rich source of minerals (mg/kg) including calcium, magnesium, phosphorus, iron, zinc, manganese, potassium, sodium, and copper. Additionally, it has been discovered that alligator pepper seeds contain a number of phytochemicals, including phenol and flavonoids, terpenoids, alkaloids, tannins, and saponins (Eko et al., 2022). The alligator pepper is taxonomicallyl classified as kingdom (plantae), phylum (Tracheophyta), class (Liliopsida), order (Zingiberales), family (Zingiberaceae), genus (Aframomum), and species (*Aframomum melegueta k. schum*).

2. ETHNOBOTANICAL SIGNIFICANCE

Megbe-dogboe, which translates as "never lacking for the sick," is the colloquial name for the plant in Ewe, Ghana (Quattrocchi, 2012).

- Traditionally it has been used as a spice for food flavouring and as a re-medy for digestive and intestinal health, migraine, dysentery, and fever (Hattori et al., 2017).
- It is still used to flavor vinegar and to produce wine, beer, and spirits. Originally, it was employed as a flavoring agent for the ancient wine known as "Hippocras" (Grains of paradise, 2014).
- Moreover, the plant has anthelmintic, galactogogue, purgative, and homeostatic (hemostatic) properties.
- Together with the leaves of Urera oblongifolia, the seeds are applied externally to cure malignancies.
- The seeds are combined with salt and applied to the inside of the mouth to treat sleeping sickness in Senegal. In order to relieve excruciating arthritis, *Aframomum melegueta* extract has been utilized as a massage oil (B.A.S., 2007; Toh et al., 2019).
- The seeds have traditional uses in treating stomachaches, diarrhea, flatulence, and severe inflammatory diseases.
- The roots were utilized by the native Kenyans to treat diarrhea and snake bites.
- The Gabonese use the juice produced from the leaves and stems to heal intestinal illnesses and wounds (Umukoro & Ashorobi, 2005).
- The Igbo people of Eastern Nigeria grind kola nuts and alligator pepper together, with the former's hot, spicy flavor reducing the latter's harsh flavor (Ogbunugafor et al., 2017).
- Strong antiseptic or bactericidal qualities found in *Aframomum melegueta* seed extracts, which are used to treat wounds and prevent infections (Doherty & Olaniran, 2010).
- The seeds were combined with other spices by Nigerians to make a locally renowned dish called pepper soup.
- The powdered leaves are used to make soup in the country's southwest as a treatment for fever and gastrointestinal issues.
- The Yoruba people of Nigeria utilize this plant, which has multiple grains in a single capsule and is considered a sign of favor, to invoke prayers during traditional ceremonies. This is in addition to its therapeutic significance.
- There have been reports that freshly picked fruit functions as an aphrodisiac, stimulating sexual behavior. The fruit is utilized as a tonic to stimulate sexual

desire, and the rhizome is utilized to make an infertility treatment that encourages insemination.

- It had also been shown that the seed powder possessed abortifacient properties. applied to safe abortions, then.
- There have also been reports of the cytoprotective and antiulcer properties (Owokotomo, 2014).
- The grains of paradise also found use as fat reducing seeds simply by chewing (B.A.S., 2007; Osuntokun, 2020a).
- *A. melegueta* used for Tuberculosis and a remedy for scorpion stings (Odetunde & Avungbeto, 2015).
- Additionally, it is discovered that the natives employ an extract from the stem bark of *Aframomum melegueta* as a herbal treatment to cure a variety of contagious skin conditions, including measles, smallpox, and chickenpox (Osuntokun, 2020a).
- The root extract is used to prevent over-breastfeeding in nursing mothers and to manage postpartum hemorrhage (PPH) following childbirth.
- The leaf is externally applied to treat Hansen's illness (leprosy) (Iwu, 2014).

3. PHYTOCHEMICALS

Many phytochemicals are found in *Aframomum melegueta;* the presence of saponins is minimal, while that of tannins, flavonoids, phenolic compounds, alkaloids, and terpenoids is moderate, and the presence of steroids and cardiac glycosides is considerable. *Aframomum melegueta* contains flavonoids (2.03+0.07%), alkaloids (2.17+0.29%), phenolic compounds (35.40+0.76%), and cardiac glycosides (58.67+1.1%) in terms of the percentage composition of phytochemicals (Echo et al., 2013).

Depending on the genetic makeup and geographical origin of the samples, the essential oil of *A. melegueta* exhibited considerable variety. For instance, the essential oil of seeds collected in Ibadan, Nigeria, included the primary elements humulene (60.9%) and caryophylene (21.7%) (Ajaiyeoba & Ekundayo, 1999). Oxygenated monoterpenes made up the majority of the Cameroonian sample (83.3%), with the main volatile components being α-terpineol (19.4%) and 1,8-cineole (58.5%) (Kamte et al., 2017). The Central African Republic's seeds showed large concentrations of β-pinene (>30%) and sesquiterpene hydrocarbons (50%) in them (Lamaty et al., 1993). Comparably, the abundant volatile elements in the Calabar, Nigeria sample were (E)-ocimene (23.22%), Humulene (26.23%), (E)-caryophyllene (19.17%), and (S)-2-heptyl acetate (16.22%) (Ukeh et al., 2009).

Table 1. Chemical components of essential oil of Aframomum melegueta and their %age composition (Oladunmoye, 2019)

Sr.No	Components	Molecular formula	Structural formula	%age composit-ion	Extraction solvent	Identification / Analysis
1	Humulene	$C_{15}H_{24}$	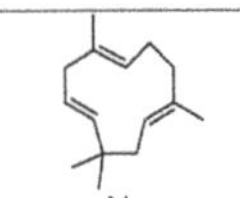	0.42%	10% NaOH	GC-MS
2	Isonicotinic acid	$C_6H_5NO_2$	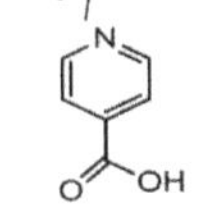	0.29%	10% NaOH	GC-MS
3	Dodecanoic acid	$C_{12}H_{24}O_2$	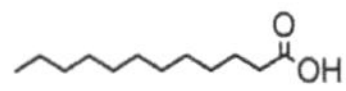	6.95%	10% NaOH	GC-MS
4	Butan-2-one-4-(3-hydroxy-2-methoxyphenyl)	$C_{11}H_{14}O_3$	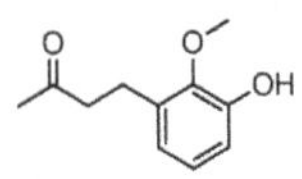	41.77%	10% NaOH	GC-MS
5	Hexahydro-3a-2(3H)-benzofuranone	$C_8H_{12}O_2$	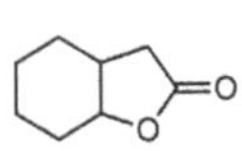	2.10%	10% NaOH	GC-MS
6	n-Hexadecanoic acid	$C_{16}H_{32}O_2$	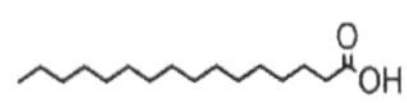	6.95%	10% NaOH	GC-MS
7	Caryophyllene	$C_{15}H_{24}$	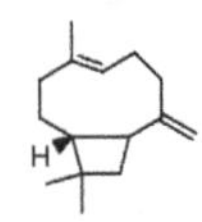	6.95%	10% NaOH	GC-MS
8	Benzaldehyde-3-hydroxy-4-methoxy	$C_8H_8O_3$	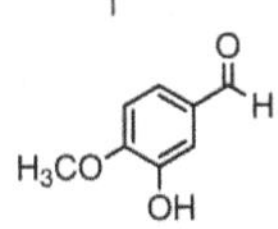	31.23%	10% NaOH	GC-MS
9	Spiro[5.6]dodecane-1,7-dione	$C_{12}H_{18}O_2$	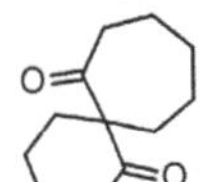	0.89%	10% NaOH	GC-MS
10	cis-Z-α-Bisabolene epoxide	$C_{15}H_{24}O$	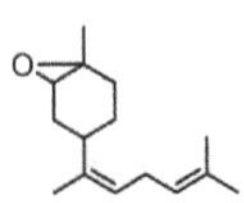	0.50%	10% NaOH	GC-MS
11	1-Oxa-2-sila-5-boracyclopent-3-ene, 4,5-diethyl-2,2,3-trimethyl	$C_9H_{19}BoSi$	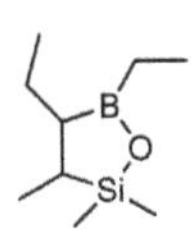	0.80%	10% NaOH	GC-MS
12	3,5-Heptadienal, 2-ethylidene-6-methyl	$C_{10}H_{14}O$	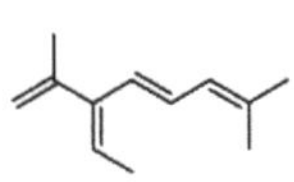	4.88%	Methanol	GC-MS

continued on following page

Table 1. Continued

13	7-Oxo-2-oxa-7-thiotricyclo [4.4.0.0(3, 8)] decan-4-ol	$C_8H_{12}O_3S$	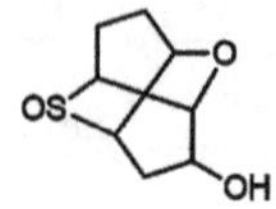	16.12%	Methanol	GC-MS
14	Aromadendrene oxide-1	$C_{15}H_{24}O$	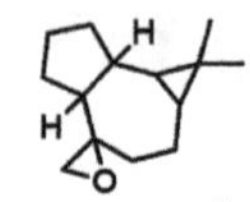	1.34%	Methanol	GC-MS
15	Ethyl homovanillate	$C_{11}H_{14}O_4$	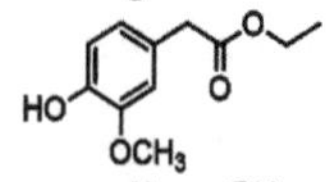	0.28%	Methanol	GC-MS
16	p-Fluorobenzaldehyde oxime	C_7H_6FNO	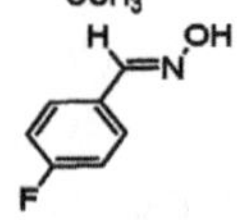	2.76%	Methanol	GC-MS
17	5, 8,11,14-Eicosatetraenoic acid, methyl ester	$C_{21}H_{34}O_2$	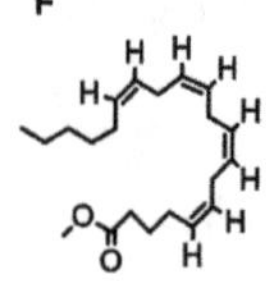	10.68%	Methanol	GC-MS
18	Myrtenyl acetate	$C_{12}H_{18}O_2$	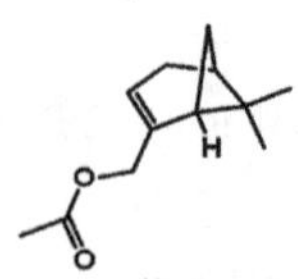	21.22%	Methanol	GC-MS
19	Caryophyllene oxide	$C_{15}H_{24}O$	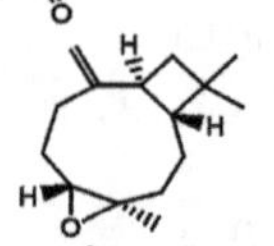	6.97%	Methanol	GC-MS
20	1, 4-Benzene diol-2-methoxy	$C_7H_8O_3$	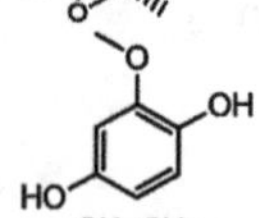	27.13%	Methanol	GC-MS
21	Sorbitol	$C_6H_{14}O_6$	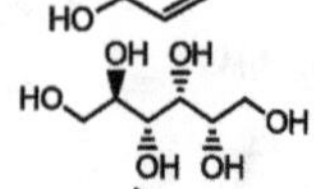	1.34%	Methanol	GC-MS
22	3,4-Dimethoxyphenylacetone	$C_{11}H_{14}O_3$	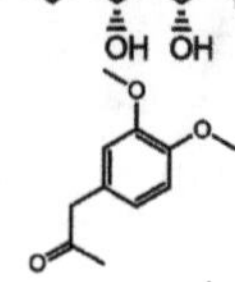	0.44%	Methanol	GC-MS
23	Trans-β-ionone	$C_{13}H_{20}O$	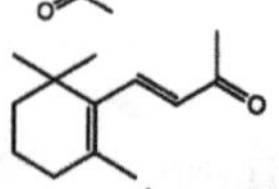	1.58%	Ethanol	GC-MS
24	2(3H)-Furanone, 5-butyldihydro-4-methyl	$C_9H_{16}O_2$	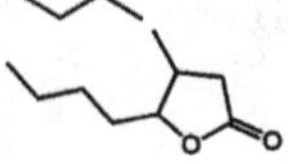	1.89%	Ethanol	GC-MS

continued on following page

Table 1. Continued

25	Corymbolone	$C_{15}H_{24}O_2$	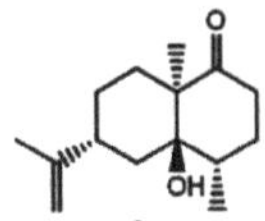	28.9%	Ethanol	GC-MS
26	2,6-dimethyl Pyridine carboxylate	$C_9H_9NO_4$	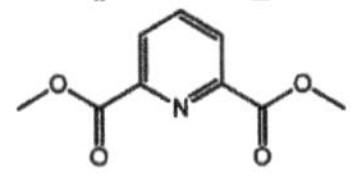	1.70%	Ethanol	GC-MS
27	Ethyl oleate	$C_{20}H_{38}O_2$	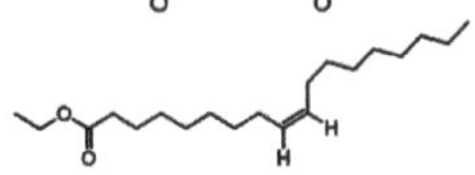	9.90%	Ethanol	GC-MS
28	n-Hexadecanoic acid ethyl ester	$C_{18}H_{36}O_2$	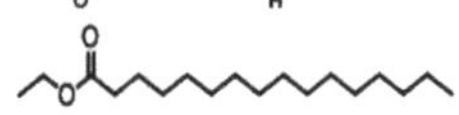	15.89%	Ethanol	GC-MS
29	Linoleic acid ethyl ester	$C_{20}H_{36}O_2$	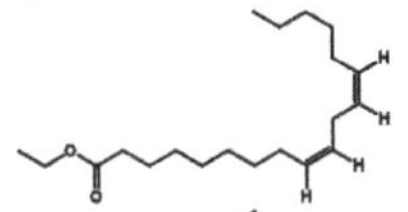	3.61%	Ethanol	GC-MS
30	γ-Sitosterol	$C_{29}H_{50}O$	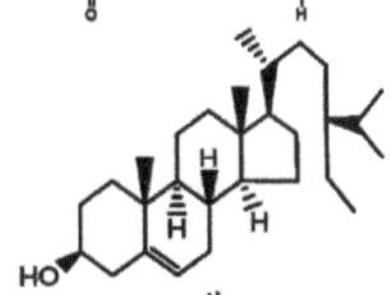	23.10%	Ethanol	GC-MS
31	Alloaromadendrene	$C_{15}H_{24}$	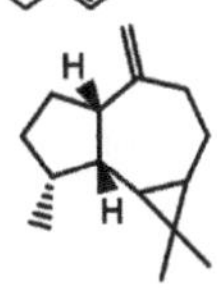	1.09%	Ethanol	GC-MS
32	1-Nortrachelogenin	$C_{20}H_{22}O_7$	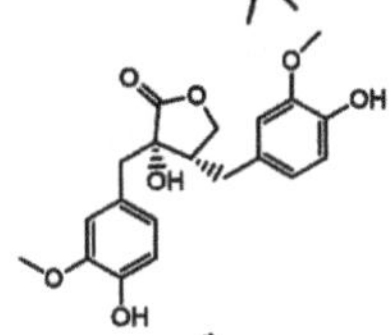	0.90%	Ethanol	GC-MS
33	Aromandendrene	$C_{15}H_{24}$	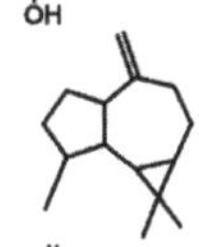	1.13%	petroleum ether	GC-MS
34	γ-Muurolene	$C_{15}H_{24}$	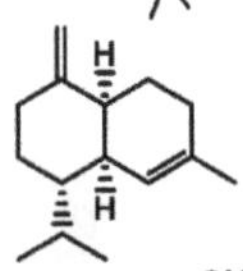	15.85%	petroleum ether	GC-MS
35	Adamantan-2-ol	$C_{10}H_{16}O$	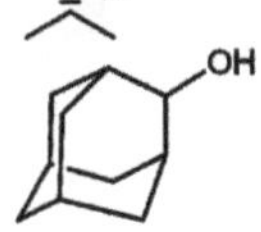	1.57%	petroleum ether	GC-MS
36	Naphthalene-2-methyl	$C_{11}H_{10}$	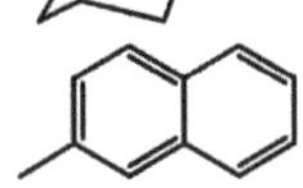	24.97%	petroleum ether	GC-MS

continued on following page

Table 1. Continued

37	9-Cyclohexylbicyclo (3.3.1) nonan-9-ol	$C_{15}H_{26}O$	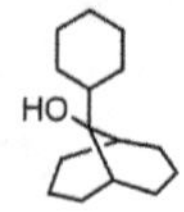	2.18%	petroleum ether	GC-MS
38	Benzene (1-ethyl-1-propenyl)	$C_{11}H_{14}$	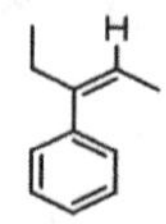	13.95%	petroleum ether	GC-MS
39	5-Hepten-3-yne-2-ol, 6-methyl-5-(1-methylethyl)-	$C_{11}H_{18}O$	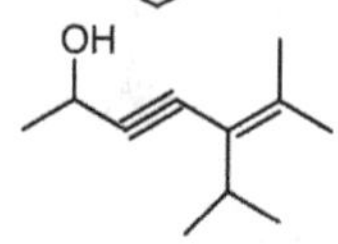	1.63%	petroleum ether	GC-MS
40	1-H-indene, 2,3, dihydro 4,7,methyl	$C_{11}H_{14}$	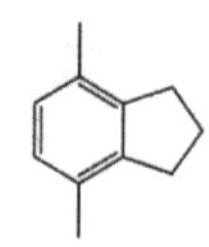	11.77%	petroleum ether	GC-MS
41	Tetramethyl (z,z,z)-humulene	$C_{15}H_{24}$	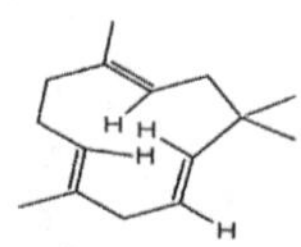	1.82%	petroleum ether	GC-MS
42	3-Azabicyclo (3.2.2) nonane	$C_8H_{15}N$	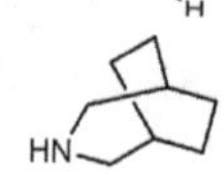	2.21%	petroleum ether	GC-MS
43	n-Decanoic acid	$C_{10}H_{20}O_2$	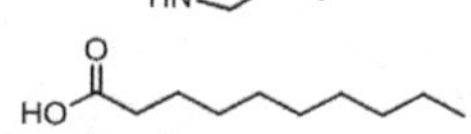	6.10%	petroleum ether	GC-MS
44	Benzene thiol-2-methoxy	C_7H_8SO	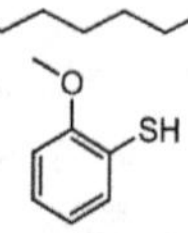	9.03%	petroleum ether	GC-MS
45	Capsaicin	$C_{18}H_{27}NO_3$	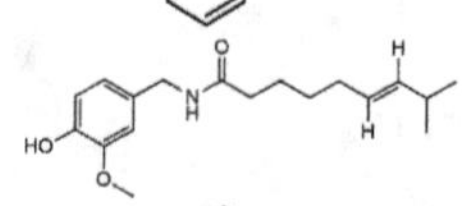	2.16%	n-hexane	GC-MS
46	Isolongifolene 9-hydroxy	$C_{15}H_{24}O$	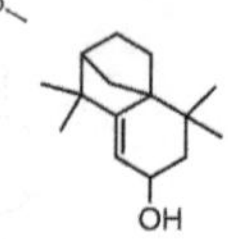	4.80%	n-hexane	GC-MS
47	Oxirane, tetradecyl	$C_{16}H_{32}O$	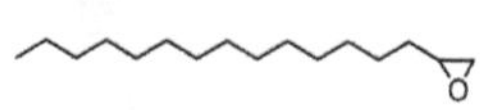	4.69%	n-hexane	GC-MS
48	Santolina triene	$C_{10}H_{16}$	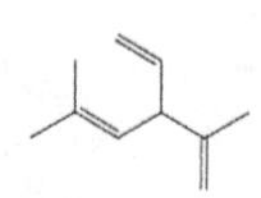	4.90%	n-hexane	GC-MS
49	3,7-Nonadien-2-Ol,4,8-dimethyl-	$C_{11}H_{20}O$	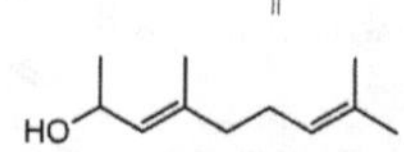	4.20%	n-hexane	GC-MS

continued on following page

Table 1. Continued

50	2-Acetyl-5-methyl furane	$C_7H_8O_2$	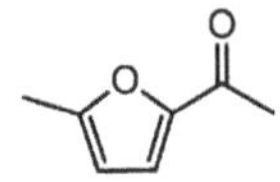	2.81%	n-hexane	GC-MS
51	Nerolidol-2	$C_{15}H_{26}O$	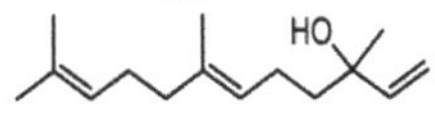	4.40%	n-hexane	GC-MS
52	2,6-Dimethyl-5-hepten-1-ol	$C_9H_{18}O$	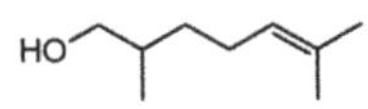	3.08%	n-hexane	GC-MS
53	Tetradecanoic acid ethyl ester	$C_{16}H_{32}O_2$	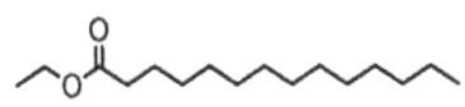	9.04%	n-hexane	GC-MS
54	Dodecanoic acid ethyl ester	$C_{14}H_{28}O_2$	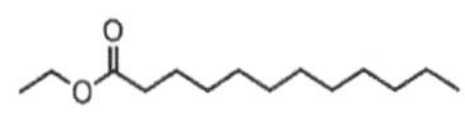	8.07%	n-hexane	GC-MS
55	4, 8-Methanoazulen-9-ol	$C_{11}H_8O$	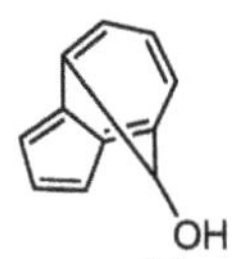	3.51%	n-hexane	GC-MS
56	4-Amino-2,3-xylenol	$C_8H_{11}NO$	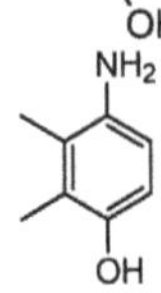	1.59%	n-hexane	GC-MS
57	Oleic acid	$C_{18}H_{34}O_2$	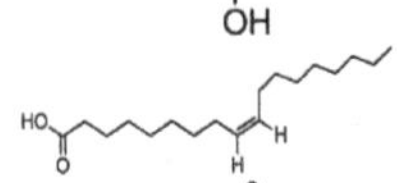	2.41%	n-hexane	GC-MS
58	Ethanone, 1-(cyclohexen-1-yl)	$C_8H_{12}O$	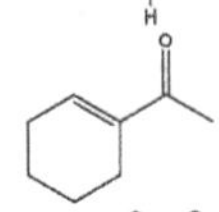	3.19%	n-hexane	GC-MS
59	Cyclotetracosane	$C_{24}H_{48}$	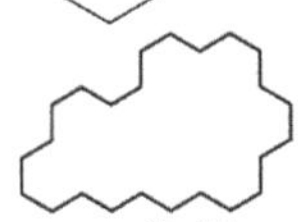	1.40%	n-hexane	GC-MS
60	Gingerol	$C_{17}H_{26}O_4$	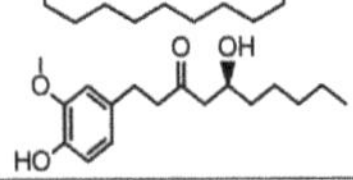	2.09%	n-hexane	GC-MS

(PAHs): Fresh alligator pepper does not contain any polyaromatic hydrocarbons (PAHs). However, smoked dried alligator pepper contains six PAHs (benzo [b] fluorene, benzo [a] anthracene, chrysene, benzo [k] fluorene, dibenzo [a,h] anthracene, and benzo [g,h,i] perylene). While these concentrations were found to be lower than quantification levels, it is possible to use smoke-dried alligator pepper for a variety of purposes without risking one's health (Young & Tarawou, 2016).

Vitamins: *Aframomum melegueta* was shown to have significant amounts of vital vitamins, including vitamin B, vitamin B3, vitamin D, vitamin E, and vitamin K. Major components of the plant include vitamins A, C, and B12 (Olumekun et al., 2020).

Figure 1.

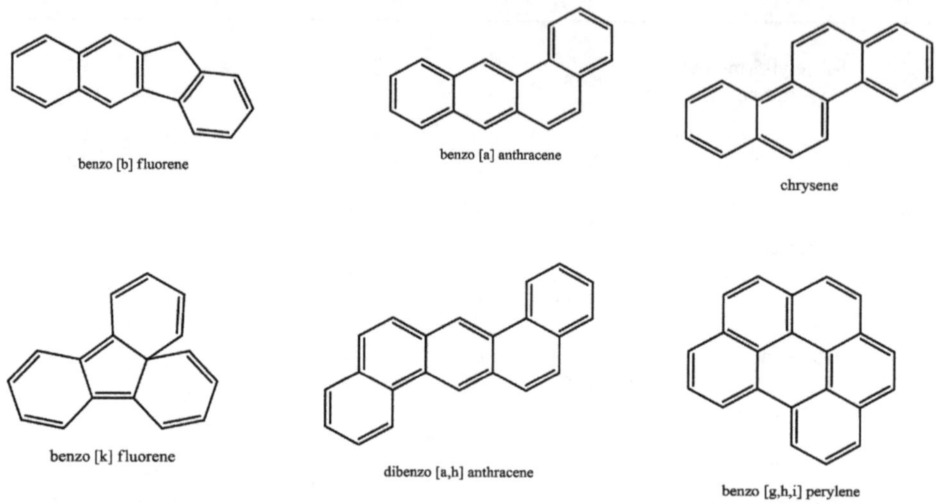

Figure 2.

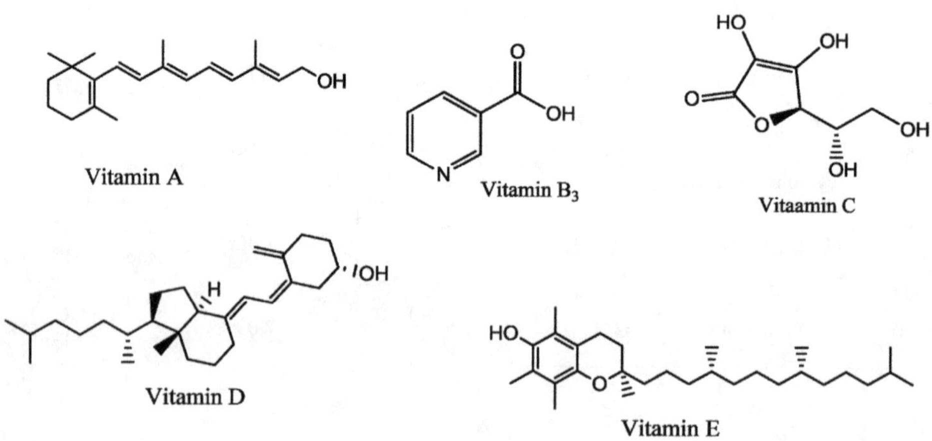

The quality and amount of essential oils derived from distinct plant sections differed in their composition (Owokotomo, 2014). Phenol and flavonoids are abundant in *Aframomum melegueta* seed extract, whereas Superoxide Dismutase (SOD) and Catalase (CAT) are well-represented in *A. meleguata* leaf extract, and Reduced Glutathione (GSH) as well as Glutathione Peroxidise (GPX) are more prevalent in the stem bark (Osuntokun et al., 2020).

The active ingredients of *Aframomum melegueta* include oleanolic acid, gingerol, shogaol, paradol, rac-6-dihydroparadol, and gingeredione (Mohammed, Awolola, Koorbanally et al, 2017; Toh et al., 2019).

Minerals Analysis

Aframomum melegueta seeds were powdered after being dried. Atomic absorption spectrometer analysis of dried ash samples that were dissolved in 10 ml of 3% ferric chloride was used to determine the mineral contents. The resulting mixture was heated on a hot plate, and then distilled water was added to the filtered solution up to the 100 mL mark in a volumetric flask. The results showed that iron and phosphorus were present in higher quantities, whereas potassium, calcium, magnesium, zinc, and manganese were also detected in reasonable amounts. However, chromium, cobalt, and cadmium were found to be absent (Alaje et al., 2014). While potassium and phosphorus were not traceable in another investigation, the highest concentrations of calcium, magnesium, zinc, manganese, and iron were found (Odebunmi et al., 2009).

Nutritional Analysis

The bigger percentage is made up of dry matter plus total carbohydrates, but there was also a significant amount of moisture, ash, crude fat, crude protein, and crude fiber. Nutritional evaluation of moisture was done by heating it in an oven at 105°C until the weight remained constant. For crude protein, the Kjeldahl method was used, and for crude fat, a soxhlet apparatus was used for a thorough extraction. Ash content through burning and crude fiber through sample digestion using H_2SO_4 and NaOH before burning it (Alaje et al., 2014; Odebunmi et al., 2009).

Figure 3.

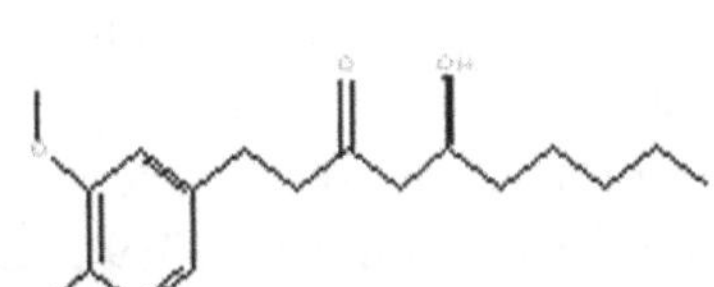

Gingerol

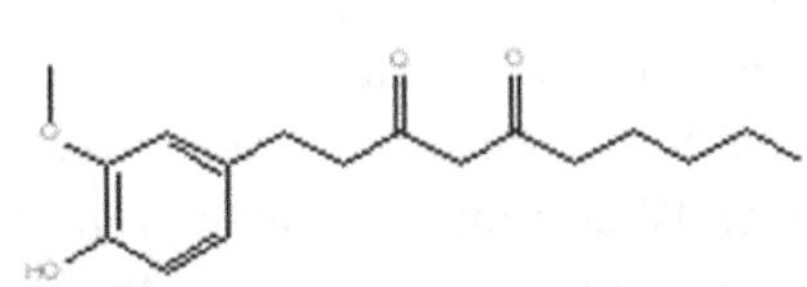

Gingeredione

4. PHARMACOLOGICAL IMPORTANCE

In the West, *Aframomum Melegueta* is traditionally used with other herbs to cure common disorders such rheumatism, diarrhea, sore throats, catarrh, and congestion. Cultivated for its rich medical and pharmacological properties, including antibacterial, hepatoprotective, anti-cancer, and anti-diabetic actions, this perennial herb is (Osuntokun et al., 2021). With components primarily investigated via single treatment mechanisms, medicinal herb *Aframomum melegueta* has been investigated and reported to have intriguing pharmacological potentials contrary to inflammation, nociception, Alzheimer's disease, Parkinson's disease, and some other complex disorders that threaten global healthcare (NaAllah et al., 2021). Phytochemicals include flavonoids, phenolic compounds, tannins, steroids, saponin, terpenoids, cardiac glycosides, and alkaloids are abundant in the seeds. In folk medicine, the seeds were known to enhance sexual function. Additionally, it is frequently employed in West Africa to treat hyperprolactinemia and is utilized as a medicine for a number of illnesses, including stomachaches, snake bites, and diarrhea. Gismondi claims that the plant's extracts have differentiative and anti-proliferative properties. The extracts were shown to have therapeutic action against specific tested bacterial strains in a study by, indicating that they may be useful in treating gastrointestinal tract infections, nausea, respiratory issues, colds, fevers, allergies, urinary tract infections, and fungal infections (Ajayi & Omodele, 2021).

4.1. Anti-Oxidant and Anti-Inflammatory Activity

Antioxidants are the source of anti-inflammatory activity. These antioxidants prevent prostaglandins from acting. It has been established that triterpenoids, phenolics, and alkaloids have anti-inflammatory effect due to their antioxidative qualities in reducing oxides, as well as through plasma eruption and cell movement, particularly of leukocytes (Amadi et al., 2016). Melanodialdehyde is produced by the antioxidants in AM, which also have anti-inflammatory properties. (MDA) (Ogbunugafor et al., 2017). Because 6-paradol inhibits the cyclooxigenase-2 (COX-2) enzyme and 6-shogoal is implicated in the development of pro-inflammatory IL1β genes, the natural extract from grains of paradise possesses anti-inflammatory properties. Thus, it is possible to speculate that the extract acts both at the genetic and enzyme levels. Tannins aid in the healing process of wounds. 6. The anti-inflammatory properties of gingerol (Oyagbemi et al., 2010). When combined with 6-shogoal, gingerols' anti-inflammatory properties are similar to those of aspirin (Ilic, Dey, Poulev, Logendra, Kuhn, & Raskin, 2014; Osuntokun, 2020a; Toh et al., 2019), i.e., it was discovered that [6]-Shogaol's effectiveness was comparable to aspirin (Ilic, Dey, Poulev, Logendra, Kuhn, & Raskin, 2014).

4.2. Antidiabetic Activity

The root cause of Type-2 diabetes is elevated blood glucose levels brought on by the body's resistance to insulin, which stays in the blood and causes the pancreas to stop producing insulin. Meanwhile, the constant breakdown of carbohydrates as well as polysaccharides into simple sugar by the enzymes α-amylase plus α-glucosidase raises blood glucose levels (Toh et al., 2019). *A. melegueta* leaf extract has anti-diabetic properties (Mojekwu et al., 2011). *A. melegueta* is reported to include paradol, 6-shogoal 6-gingerol, oleanolic acid, and acarbose (Osuntokun, 2020a). The two compounds that inhibit α-amylase and α-glucosidase more effectively are oleanolic acid and 6-gingerol (Mohammed, Awolola, Koorbanally et al, 2017). According to a different study, *A. melegueta's* aqueous leaf extract increases the release of insulin from either regenerated or residual β-cells (Amadi et al., 2016). According to Nosiri et al., *A. melegueta* ethanolic extract of seeds decreased blood glucose levels and stimulated the regeneration of pancreatic ß-cells in diabetic rats triggered with alloxan; nevertheless, a greater amount of the extract shown a more restorative effect than a lower one. Compared to the control group, the 200/400 mg/kg *A. melegueta* seed extract in water dramatically lowered glycemia in rats with alloxan-induced diabetes. Using alligator pepper guarantees that the body's blood sugar is regulated (Mustapha et al., 2023).

Anti-glycation activity: Glycation is an essential therapeutic target for the treatment of diabetes. The process of glycation is the non-enzymatic union of reducing sugars and protein amino groups, which typically rearranges to form stable ketoamines and produces advanced glycation end products (AGEs) (Ahmed, 2005). This natural response is dependent on the degree and duration of hyperglycemia, protein half-lives, and tissue permeability to free glucose (Berrou et al., 2009). Because high levels of advanced glycation end products (AGEs) and enhanced glycation can affect immunogenicity, change enzymatic function, diminish ligand binding, and modify protein half-lives, they can lead to diabetes problems (Kostolanská et al., 2009). It is discovered that *A. melegueta's* polyphenolic extract has strong anticoagulation activity, with an Ic50 of 0.125 mg/mL that rises with concentration (Kazeem et al., 2012). Glycation and AGE-induced toxicity is caused by an increase in the generation of free radicals. Therefore, effective antioxidants that remove free radicals are also effective anti-glycation agents (Ahmad & Ahmed, 2006; Nakagawa et al., 2002). We may now conclude that *A. melegueta* is a beneficial spice option for those with diabetes.

4.3. Anti-Cancer Activity

Aframomum melegueta extract of root flavonoids are helpful for pancreatic cancer (-)-buplerol from AM extract causes pancreatic cancer cells to undergo apoptosis. Paradols help to suppress skin cancer. In addition to flavonoids, common plant phenolic chemicals called paradols cause humans promyelocytic leukemia cells to undergo apoptosis, which has anticancer properties (Ogbunugafor et al., 2017; Osuntokun, 2020a). [6]-gingerol has unfathomable potential advantages for the management and control of cancer as a chemo-protective drug (Oyagbemi et al., 2010), and it is among *A.melegueta's* main constituents.

4.4. Anti-Bacterial Activity

Soft tissue infections and urinary tract infections are treated with *Aframomum melegueta*. Numerous bacteria, including Escherichia coli, Klebsiella pneumoniae, Enterococcus faecalis, Staphylococcus saprophyticus, Proteus mirabilis, and Staphylococcus aureus, are commonly responsible for urinary tract infections. It has been discovered that *Aframomum melegueta* works well against a few clinically significant bacterial isolates, including Staphylococcus aureus (Okwu et al., 2019), pseudomonas aeruginosa, resistant to methicillin, Salmonella spp., Bacillus spp., Salmonella spp., Escherichia coli, Proteus spp., and Klebsiella spp. Shigella species. Candida, Aspergillus, and Saccharomyces sp. The cause of this activity is terpenoids like aframodial (Toh et al., 2019). Additionally, it has been discovered that *A. melegueta's* cold water extract (CWE), hot water extract (HWE), and methanol extract (ME) all suppress the development of Salmonella typhi but not Escherichia coli or Klebsiella pneumoni. The growth of Escherichia coli is inhibited by ethanol extract (EE), however Salmonella typhi and Klebsiella pneumonia are not inhibited (Alo et al., 2012). According to Odo, Christian, *A. melegueta* seed extracts in water, methanol, and ethanol inhibited the development of bacteria in the following order: *E. coli* > *S. typhi* > *Shigella* ssp > *K. pneumonia* (Odo, 2015).

Furthermore, *Aframomum melegueta's* ethanol seed extract stops all asymptomatic bacteria from growing. The *Aframomum melegueta* extracts showed greater antibacterial activity against *Enterobacter cloacae, Proteus mirabilis, Proteus vulgaris, E. coli, and Klebsiella spp.* (Osuntokun et al., 2021). The seeds and leaves of AM contain phytochemicals called tannins, saponins, alkaloids, steroids, and cardiac glycoside that have been shown to have remarkable therapeutic effects in the treatment of respiratory issues, fungal infections, gastrointestinal tract infections, allergies, nausea, fever, and colds. The extract's terpenoids have been shown to be efficient against bacteria (Kirimuhuzya & Anywar, 2015). Zerumin A (A) and (E)-labda-

Figure 4.

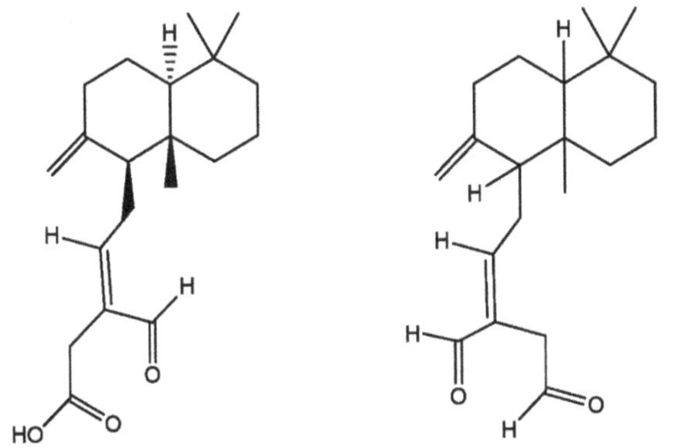

Zerumin A (E)-labda-8(17),12-diene-15,16-dial

(A) (B)

8(17),12-diene-15,16-dial (B) found in the rhizome extract of AM may efficiently destroy Gram-positive bacteria like Escherichia coli and Listeria monocytogenes.

The age of the plant utilized, environmental variables (such as temperature, light, and water), the extraction solvent, and the extraction technique can all have an impact on a plant extract's varying antibacterial qualities. It is discovered that acetone is a better solvent to utilize when trying to extract diterpenoids from the plant's rhizome (Amadi et al., 2016).

4.5. Activity Against Castor Oil Induced Diarrhea

The colonic cells' production of prostaglandin is linked to the diarrhea brought on by castor oil. It lends credence to the theory that the AM seed extract's antidiarrheal properties could be linked to its prevention of prostaglandin production.The phytochemicals 6-paradol, 7-paradol, 6-shogoal, gingerol, and zingerone are responsible for this activity (Ogbunugafor et al., 2017; Umukoro & Ashorobi, 2005).

4.6. Antiobesity Activity

The seeds of *A. melegueta* are extracted ethanolically to provide 6-gin-gerol, 6-paradol, 6-shogaol, and 6-gingeredione.By stimulating brown adipose tissue, a location of sympathetically mediated metabolic thermogenesis, these substances raise energy expenditure. Individuals with negligible amounts of brown fat adipose tissue did not exhibit any reaction when exposed to cold. In addition to its potential benefits in managing obesity in people and animals, grains of paradise extract may also facilitate relaxation (Amadi et al., 2016; Hattori et al., 2017). *A. melegueta's* anti-obesity action is based on its 90% lipase inhibitory activity, which inhibits the pancreatic lipase enzyme and, as a result, lowers fat absorption. This mechanism may be utilized for the pharmacologic and dietary control of obesity (Ekanem et al., 2007).

In rats on a high-lipid diet (HLD), alligator pepper aqueous extract was also found to be highly beneficial in reducing body weight, blood cholesterol level (CHOL), low density lipoproteins (LDL), and triacylglycerides (TG). It also prevented the buildup of fat in blood vessels, a condition known as atherosclerosis. Nevertheless, T-protein and albumin show little change (Tijjani & Luka, 2013).

The herb's ability to reduce obesity is mostly attributed to the presence of vanilloids, specifically 6-gingerol and 6-paradol. Six-paradol and six-gingerol specifically regulated multiple expressions of genes and AMPK phosphorylation in the liver as well as adipose tissue. It seems that 6-gingerol, in particular, may improve hepatic metabolism by enhancing fatty acid β-oxidation, and 6-paradol may impact adipose differentiation by blocking free fatty acid translocation. By controlling genes involved in biosynthesis, these substances prevent lipid buildup and fatty acid intake during adipogenic differentiation. Adipocyte size decreases due to increased lipolysis, fatty acid β-oxidation, and decreased TG accumulation, which prevents WAT from growing (Kumar et al., 2022).

4.7. Hematopoietic Effect

The effect of methanolic extracts of seeds of AM on 2,4-dinitrophenylhydrazine-induced anaemic rats is a less well-known finding. The treatment increased platelet count and hemoglobin levels, suggesting its erythropoietic potential in treating anemia. The effects of ethanol seed extract on the counts of various blood cells varied depending on the dose; at a high dose of 400 mg/kg, the counts of hemoglobin (Hb), red blood cells (RBCs), platelets, erythrocyte sedimentation rate, and neutrophils increased; at a low dose of 200 mg/kg, the counts of white blood cells (WBCs), leukocytes, and packed cell volume (PCV) increased (Omoboyowa et al., 2017).

In contrast, past observations have indicated that higher amounts of the extract cause haematotoxicity (Akpanabiatu et al., 2013). *Aframomum melegueta* significantly decreased the amount of white blood cells (WBCs) in the fluid in a dose-associated way, according to another study. Additionally, the extract significantly reduced the breakdown of rats red blood cells when subjected to phenyl hydrazine, suggesting that it had a membrane-stabilizing action (Solomon & Ashorobi, 2008).

4.8. Treatment of Infertility in Males

Innate plus acquired urogenital malformations, genital tract abnormalities, varicocele (high scrotal temperature), endocrine disorders, immunological abnormalities, and genetic abnormalities are among the reasons that lower male fertility (Rowe et al., 2000). The penile erection index, mounting frequency, and frequency of genital grooming and sniffing all increased after taking an aqueous extract of *A. melegueta* seeds every day for eight days. The ejaculation latency increased (delayed) but the intromission latency decreased (Amadi et al., 2016). The IFA Nkari People of Ini Local Government Area, Akwa Ibom State, Nigeria, have traditionally treated low libido and weak erection in men by simply soaking crushed AM seeds in fresh palm wine (at least a bottle) and drinking three glasses of it throughout the day (Erhabor et al., 2013).

4.9. Anti-Hypertensive Activity

Systolic blood pressure (SBP), diastolic blood pressure (DBP), pulse pressure, and mean arterial pressure are all lowered in normotensive and hypertensive individuals by *A. melegueta* seeds. While percentage reductions across normotensives and hypertensives were found to be similar, SBP averages (15–16%) showed larger percentage reductions than DBP (9–10%), and SBP took longer to decrease than DBP. This is due to the fact that sympathetic activation will counteract the effects of AM on BP. The inadequate adrenergic receptors in hypertensives may be the cause of this discrepancy. A downregulation of the α-receptor may result from elevated norepinephrine concentrations. There is also a research indicating that a high norepinephrine concentration can lower the α1-receptor mRNA level in vascular smooth muscles, which can cause a very sluggish reflex reversal of lowered blood pressure. Since the amount of blood pressure drop was shown to be between safe bounds, both young and old hypertensive individuals may find it helpful in managing their hypertension (Amadi et al., 2016; B.A.S., 2007).

4.10. Abortifacient Property

In the first trimester of pregnancy, pregnant women may be at risk for health problems due to *Afromomum melegueta* (Inegbenebor et al., 2009; Osuntokun, 2020a). However, there haven't been any recorded negative side effects for guys (Osuntokun, 2020a). Since *Aframomum melegueta* has anti-progesterone properties, greater concentrations of saline extract have been proven to work better in this regard.Progesterone inhibition may be caused by the β-caryophyllene element of the sesquiterpene substance found in alligator pepper (Inegbenebor & Ebomoyi, 2014). Owing to the low concentration of hydrophilic components in *Aframomum melegueta,* the aqueous extract of the plant did not negatively impact rat fertility. Granulated alligator pepper was fed to the rats, but they did not litter. Therefore, AM may be to blame for rats' infertility. Not that I support abortion, but if the dosage was effective in people, AM might be used in fertility clinics as a harmless abortifacient drug in cases where the mother's life is at grave risk due to the pregnancy (Inegbenebor et al., 2011).

4.11. Hepatoprotective Effect vs. Toxicological Evaluation

Drinking alcohol leads to a number of liver conditions. Alcohol is broken down by the liver into simple final molecules such as acetaldehyde. Extremely reactive substances produced when alcohol is metabolized, like reactive oxygen species (ROS), which can lead to liver conditions such acute alcoholic hepatitis, simple steatosis, and cirrhosis. Overexposure to oxidants damages the mitochondria, which lowers the amount of energy produced. It has been demonstrated that the common grains of paradise can counteract such harmful effects. The antioxidant qualities of phytochemicals like flavonoids and phenols allow them to exhibit hepatoprotective effects. Moreover, these substances reduce the amounts of caspases 3 and 9, ALT (alanine aminotransferase), TNF-α (tumor necrosis factor alpha), and IL-1β (tumor necrosis factor alpha). Hepatocyte apoptosis brought on by CCl4 is likewise prevented (El-Halawany, Dine, El Sayed, & Hattori, 2014; Osuntokun et al., 2018). Another study refuted the claim that rats subjected to a sublethal dose of methanol-extracted AM seeds had significantly higher levels of ALT (alanine aminotransferase) (Nwaehujor, 2014; Ogbunugafor et al., 2017). Thus, extended overuse of alligator pepper need to be avoided.

4.12. Spasmolytic Activity

ACMA (*Aframomum melegueta* and *Citrus aurantifoliamix.*) have been shown to have relaxing effects on rat tracheal smooth muscles. The extract did not include muscarinic receptors, but it most likely inhibited cellular calcium (Ahounou et al., 2012). Intragastrically infused grains of paradise extract has also been shown to reduce BAT SNA (brown adipose tissues sympathetic nerve activity). Humans would experience a calming feeling as a result of this declining BAT SNA increasing parasympathetic nerve activity (Hattori et al., 2017).

4.13. Insecticidal Activity

A significant pest of wheat is the smaller grain borer *Rhyzopertha dominica* (Fabricius). Storing wheat can be protected against *R. dominica* in a number of ways. Its populations shown resilience to malathion, as demonstrated by Ramesh Babu (Babu et al., 2017). In this sense, a botanical extract of *A. melegueta* was helpful since it had a repellent effect on *R. dominica*. The strong, pungent smell of the *A.melegueta* plant's steam-distilled oil may be the cause of its repellent properties, as beetles showed avoidance behavior when they encountered it. Even after adding five grams of the host plant's winter wheat kernel to the wheat grains, the oil continued to resist *R. dominica* (Ukeh, 2008).

The adult *Sitophilus zeamais* (greater grain weevil) is repelled by vacuum-distilled extracts of *A. melegueta* and ginger, *Zingiber officinale* (Zingiberaceae), both in the presence and absence of maize (Zea mays) grains. The distillates' liquid chromatographic separation revealed that the fractions containing oxygenated chemicals are responsible for the repelling activity. The scientific basis for the repellent qualities of Z. officinale and *A. melegueta* is that they contain 1,8-cineole, geranial, and neral in a ratio of 5.48:2.13:1, and (S)-2-heptanol, (S)-2-heptyl acetate, and (R)-linalool in a ratio of 1:63. It suggests its possible application in protecting stored goods at the level of small-scale African farmers (Ukeh et al., 2009).

Larvicidal and Pupicidal Potential

The Anopheles stephensi Liston mosquito is one of the main carriers of malaria. *Aframomum melegueta* can be used to treat malaria since methanol extracts from its leaves and seeds shown larvicidal and pupicidal ability against the larvae as well as pupae of Anopheles species. The percentage of Anopheles species larval and pupal mortality rose along with the concentration of leaves and seed extract. When compared to leaf extract, seed extract exhibits a higher proportion of deaths on both larvae and pupa at a given dose (Ileke et al., 2017).

4.14. Anti-Fungal Activity

Basal stem rot in cowpeas (Vigna unguiculata) and numerous other crops, including tobacco and tomatoes, is caused by the fungus Sclerotium rolfsii. Using *A.melegueta* extract, its growth in these plants can be suppressed. It is possible to conclude that the plant seed extract has a greater inhibitory impact (86.67%) than the leaf extract (43.47%) when comparing their respective inhibitory activities against the Sclerotium rolfsii fungus. *Aframomum melegueta* seed extract's chemical makeup inhibits the most potent phytoconstituents, including amines, ketones, aldehydes, and phenolic compounds, which helps to inhibit Sclerotium rolfsii (Okwu & Mgbeahuruike, 2018). Bamidele discovered that *Aframomum melegueta* aqueous extract was useful in preventing Aspergillus niger, Aspergillus flavus, Rhizopus stolonifer, and Rhizoctonia species from growing their colonies when he investigated the antifungal potency of *Aframomum melegueta* extracts of seeds on postharvest rot fungi of two citrus species. Against the mycelial development of Fusarium species along with other fungi, the aqueous extract of *Aframomum melegueta* exhibited a more potent antifungal effect (Gyasi et al., 2022).

4.15. Neuroprotective Effect

It is commonly known that in humans who have experienced traumatic brain injury (TBI), oxidative stress and inflammation play a major role in the detrimental biochemical cascades of secondary injuries that lead to an increase in long-term neurodegeneration and lifetime impairment (Ansari et al., 2008). *Aframomum melegueta* have strong anti-oxidant (Amadi et al., 2016), anti-inflammatory (Oyagbemi et al., 2010), anti-microbial (Toh et al., 2019), anti-apoptotic (El-Halawany, Dine, El Sayed, & Hattori, 2014), anti-diabetic (Mohammed, Awolola, Koorbanally et al, 2017) and anti-nociceptive (Solomon & Ashorobi, 2007) properties. All of these characteristics point to the possibility of analgesic and neuroprotective benefits following traumatic brain injury (TBI). Because of its pleiotropic qualities—that is, its anti-oxidative and anti-inflammatory effects—a natural new therapeutic drug candidate called PMI-006 derived from *Aframomum melegueta* seeds has been shown to have neuroprotective effect, significantly improving functional outcomes following traumatic brain injury (TBI) (Kumar et al., 2015). A unique chemical with antidepressant qualities against TBI-induced depression, PMI-006 reverses the impact of TBI on a group of tiny, non-coding microRNAs (miRNAs), which is altered by TBI (Sell et al., 2019) and utilized to treat alzheiemer disease (G.K., 2017).

4.16. Anti-Stress Potential

There is no evidence to suggest that A. melegueta seeds are helpful under stress (Solomon, 2005).

4.17. Anticorrosion Activity

Activity against chemical induced corrosion: The active components of *Aframomum melegueta* seeds are paradol (1-(4-Hydroxy-3-methoxyphenyl) decan-3-one) and shogaol ((E)-1-(4-Hydroxy-3-methoxyphenyl) dec-4-en-3-one). It's interesting to note that paradol and shogoal have several chemical structures with common organic corrosion inhibitors. *A. melegueta* (AM) extracts successfully prevented mild steel corrosion between 30 and 60 degrees Celsius in oxygenated mixtures of 1MHCl and 0.5MH2SO4. In acidic solutions, certain of the organic ingredients of AM extract undergo protonation, while the remaining constituents persist in their molecular form. Consequently, the involvement of both protonated and molecule species to the adsorption process can be attributed to AM's observed inhibitory efficacy.The corrosion inhibition effectiveness of 1MHCl solution differs slightly from that of 0.5MH2SO4 solution. This difference can be explained by the extract's protonated components, which prevent external protonated species from adhering to the mild steel surface. In contrast, the sulfate ions in the HCl solution do not as strongly impede the Adsorption of protonated species on the mild steel surface. On the mild steel surface sunk in 1M HCl, it guarantees an impressive adsorption of protonated inhibitor species; in contrast, the metal specimen sunk in 0.5M H2SO4 is more activated to adsorb non-protonated species.

Factors affecting: Corrosion inhibition efficacy is dependent on temperature and concentration. 1. AM extract is more successful at inhibiting corrosion at high temperatures in 1M HCl and at lower temperatures in 0.5M H2SO4.

2. AM extract's ability to inhibit corrosion on mild steel surfaces increases with concentration. The concentration of AM extract enhanced the inhibitory efficacy, reaching maximum values of 96.1% as well as 55% in 0.5M H2SO4 and 1M HCl, respectively (Oguzie et al., 2013).

Activity against fungal induced corrosion: Through their metabolic activity, a wide variety of microbes can accelerate the pace at which metals corrode. Fungal physiological activity and the wide range of metabolites they produce allow them to attach themselves to metals such as copper, iron, aluminum, zinc, and nickel modified by polyaniline (Juzeliūnas et al., 2006). Even at inadequate moisture levels, fungus adhere to the metal surface due to adhesion forces and begin to grow. They consequently establish a chemical connection with metal (Lugauskas et al., 2008). Fungi use acids during their metabolic activities to make formic, citric, and acetic acids, which erode metals (Little & Ray, 2001). Fungi that grow on metal surfaces have the ability to consume gaseous products such as oxygen and hydrogen, as well as sulfur and nitrate that have accumulated on degraded poles like iron. This depolarizes the metals and intensifies corrosion (Obuekwe, 1981). The investigation into the antifungal activity of *A. melegueta* cold water extract on corroded metal caused by *Aspergillius fumigates* and *Acremonium kiliense* showed that the two types of fungi corroded the metal in different ways based on their capacity to multiply on the metal and produce metabolites that alter the polarization resistance. It has been demonstrated that A. fumigatus is more caustic, and this rate of corrosion rises with time. For the two distinct fungi, the inhibitory impact of the cold water extract varied. With an inhibitory efficiency of 85% at the 25 mg/ml concentration, it was discovered that the anti-corrosion activity was more successful against *A. kiliense*. Though it was not as successful, an inhibitory effect against *Aspergillius fumigates* was also seen. *A. melegueta* contains tannins, saponins, and alkaloids as phytochemicals. Therefore, it is possible that the phytochemicals found in the seed extract have fungicidal and anti-corrosion properties. The authors further asserted that saponins constitute fungicidal effects against A. fumigatus species. Nevertheless, a thorough examination of the mechanism behind *A. melegueta's* fungicidal effects has not been conducted (Chinedu, n.d.). The ability of saponin moieties to dissolve lipids may also play a role in inhibiting fungal development (Imo et al., 2018; Oguzie et al., 2012). It results in a decrease in the amount of cells and a decrease in tannins' ability to bind proteins, which interferes with the cells' basic metabolic processes (Chinedu, n.d.).

4.18. Anti-Viral Activity

Poxvirus is not as virulence and is not as deadly in tissues of chick embryos (ECEs) (Singh & Tripathy, 2003). Less than a "atypical" strain of the Kenyan virus known as the Fowlpox virus (FPV). In mature birds, the viral strain caused mild paralysis, increased epithelial proliferation, endotheliosis, and lymphoproliferative disease (Nyaga et al., 1979). Because of its high virulence, atypical FPV may potentially cause minor infections in humans (Singh & Tripathy, 2003).

The extract of *Aframomium meleguta* functions as a preventive measure against the Fowl Pox Virus (FPV). Two common phenols, butan-2-one-4-(3-hydroxy-2-methoxyphenyl (B) and benzaldehyde-3-hydroxy-4-methoxy (A), are the source of the prophylactic activity of AM extract. It has been shown that they have antiviral effect against the chicken pox virus in embryonic chick eggs (ECEs). It decreased the virus concentration in addition to the embryo death rate; the percentage of virus reduction was reported to be 91.7% (Mk et al., 2019). *A. melegueta* extract's phenols directly interact with the FPV lipid envelope, changing its identifying components to some extent (Liu et al., 2015; Reeves et al., 2011). Nevertheless, it has recently been shown that phenols ionize host tissue surface receptors, reducing the capacity of viruses to adhere directly to host tissues. Due to this, viral membranes were suspended in a non-fusogenic manner and were able to penetrate the FPV lipid bilayer envelope (Liu et al., 2015).

There are several explanations for why the extracts killed the chick embryos (Mk et al., 2019). Hepatic steatosis in the hepatocytes of Wister albino rats may be caused by a hazardous dose greater than 1000 μm of a certain nucleoside analogue, which would cause the rats to lose weight (Jena et al., 2009). When plant phenol concentrations surpass 700 μm, specific tissue functions see a decline in strength (Tapiero et al., 2002). Additionally, phenols have the potential to generate inhibitory complexes that block protease-dependent activators, so impeding the chemical components involved in the early prophase phase of mitosis and making them harmful (El Gharras, 2009). However, reducing the amount of phenols in the extract can be a very useful way to prevent these fatal effects on chick embryos (Mk et al., 2019).

4.19. Activity Against COVID-19

SARS-CoV-2 (COV-19) is a global pandemic that has impacted everyone. Lungs are the organ most affected by COVID-19 because they have a high concentration of type I and type II alveolar cells, which contain the enzyme angiotensin–converting enzymes 2 (ACE2), which aids in the virus's entry into host cells. The virus has glycoprotein spikes that it uses to bind to the ACE2 enzyme and penetrate the host cells (Letko et al., 2020). The amount of ACE2 present in a tissue determines how infected it is (Xu et al., 2020). ACE2 has an impact on the digestive system as well. It is highly expressed in the small intestine's enterocytes, duodenal, and rectal epithelium, as well as in the stomach glandular cells (Hamming et al., 2004). Increased levels of IL-2, IL-6, IL-7, and granulocyte macrophage stimulating factors (GM-CSF), which can induce cytokine releasing syndrome, have been observed in laboratory trials. These factors include interferons gamma, inducible protein 10 (IP-10), macrophage inflammatory protein Alpha 1 (MIP-1Alpha), and tumor necrosis factor alpha (Huang et al., 2020). It will be able to treat COVID-19 if all of these factors are prevented. Because of its ingredients, a natural source such as *Aframomum melegueta* can be particularly significant in this regard. The combination of all these compounds can be used to treat COVID-19. These compounds include 8-gingerol, 6-Shogaol, 6-Gingerol, Methyl-6-Gingerol, 2-(5 butylfuran-2-yl) ethyl-2-methoxyphenol, 6-Rac-6-Dihydro paradol, 6-Gingeredione, and 6-Paradol. Secondary metabolites include (-)-buplerol, 16-oxo-8(17), 12(E)-labdadien-15-oic acid, (-)-arctigenin, labda-8(17), 12-dien-15, 16-dial, (E)-14-hydroxy-15-norlabda-8(17), 12-dien-16-al, 5-hydroxy-7-methoxy flavone, and apigenin (Osuntokun, 2020b). As 1-(4′-hydroxy-3′, 5′-dimethoxyphenyl)-3-(S)-acetyl *Aframomum melegueta's* 7-(3,4,5,6-trihydroxyphenyl) heptane and dihydrogingerenone chemicals bind free radicals generated by liver hepatocytes, preventing GSH from being reduced. Furthermore, these substances also lower the amounts of caspases 3 and 9, ALT, TNF-α, and IL-1β. Hepatocyte apoptosis produced by CCl4 is also suppressed (El-Halawany, Dine, El Sayed, & Hattori, 2014; Osuntokun et al., 2018). Additionally, the new secondary metabolites have the ability to cause pancreatic cancer cells to undergo apoptosis (Balaji et al., 2022; Osuntokun, 2020a).

Because *Aframomum melegueta* contains all of the important vitamins in significant amounts—Vitamins A, C, and B12 being the main components—as well as Vitamins B, B3, D, E, and K, it can be used as an immunity booster (Olumekun et al., 2020). Vitamin D decreases inflammation and boosts innate immunity (Osuntokun, 2020a). In human monocytes and neutrophils, the active ingredient of vitamin D (1,25-dihydroxyvitamin D; 1,25-OH2D) induces the synthesis of antimicrobial peptides (AMPs) as "endogenous antibiotics." AMPs shield epithelial surfaces from harm by breaking down the lipoprotein membranes of microorganisms like influenza viruses (Martinez et al., 2013). Thus, vitamin D increases the epithelium's ability to create AMPs upon microbial exposure. In light of the current epidemic, *Aframomum melegueta's* therapeutic value for treating SARS-CoV-2 (COVID-19) has increased.

5. BIOGENIC SYNTHESIS

Around the world, a lot of work is done to lessen the amount of hazardous waste produced. In this sense, contemporary advancements in science and business are constantly merging with "green" chemistry and chemical processes (Sharma et al., 2008).

5.1. Silvernano Particles (AgNPs) Synthesis

The synthesis of nanoparticles utilizing green materials, such as plants and algae, etc., shows that these materials enclose the particles and prevent the particles from reacting instantly with their surroundings. The size of these nanoparticles falls between 1 and 100 nm.Biogenic synthesis produces nanoparticles with high stability, size, and polydispersity (Chen et al., 2008).

The process of creating silver nanoparticles (SNPs) using an aqueous extract of the seeds of *Aframomum melegueta* requires treating the extract with a 1 mM AgNO3 solution and heating the resultant solution at 70 0C for 30 minutes. This is possible because the phytochemicals' active functional groups, such as aldehyde, hydroxyl, amine, and carboxyl units, can convert Ag+ into Ag0. There was a direct correlation found between the size of the particles and the reaction time, with an average particle size of 27 nm and an onset span of 10 minutes for 7.6 nm particles (Dare et al., 2015).

These silver nanoparticles have found remarkable applications in the fields of highly sensitive biomolecular detection, diagnostics (Schultz et al., 2000), photocatalysis, lithography (Xia et al., 1999) and antibacterial activities (Thirumurugan et al., 2009). Smaller the particle size greater will be the antibacterial activity (Dong et al., 2019). The fastest antibacterial activity is observed with AgNPs size of 5 nm (Loo et al., 2018). In general, AgNPs with an average size below 19 nm has better bactericidal effect (Korshed et al., 2019).

5.2. Synthesis of Phenolic Compounds

The most active ingredients of *Aframomum melegueta* are phenolic compounds, which have a wide range of uses including anti-oxidants, anti-inflammatory, antitumor, anti-cancer, anti-proliferative, anti-metastatic, and pro-apoptotic properties (Uemori, 2018).

5.2.1. Soxhlet Extraction Technique

The antioxidants were extracted using the soxhlet extraction process (Adegoke et al., 2003). Using the following technique, nine different types of phenolic compounds were extracted from the *Aframomum melegueta* seed extract and analyzed using NMR spectroscopy (Gröblacher et al., 2012).

Extraction and Fractionation

Aframomum melegueta seeds that had been crushed were extracted using the Soxhlet extraction procedure (1 kg). Using silica gel as the solid phase and solvents with increasing polarity (first 100% n-hexane, then 2% to 50% EtOAc, then infusions of 2% to 75% MeOH, and finally 100% MeOH), the extract was then subjected with gradient elution column chromatography, yielding 15 fractions. MeOH/H2O or MeCN/H2O gradients were used in preparatory and semipreparative HPLC for the following four fractions: 4 (110 mg), 6 (330 mg), 10 (355 mg), and 14 (210 mg).

Isolation and purification

4-[2-(5-butylfuran-2-yl) ethyl]-2-methoxyphenol

4-[2-(5-butylfuran-2-yl) ethyl]-2-methoxyphenol (11 mg) was isolated from fraction 4, using prepative HPLC with gradient MeCN/H2O in ratios 70:30 to 90:10 to finally 100% MeCN.

- 6-gingerdione and 6-Paradol

After fraction 6 was run through a preparatory HPLC elution gradient of MeCN/H_2O, 75:25, and ultimately 100% MeCN, 6-gingerdione and 6-Paradol were extracted. Following that, isocratic solutions of MeCN/H_2O 70:30 and 75:25 were used to purify 6-gingerdione (33.4 mg) and 6-Paradol (27.1 mg) using semi-preparative HPLC.

- 6-gingerol (187.8 mg) and methyl-6-gingerol, 8-gingerol, rac-6-dihydroparadol (57.4 mg)

All four of these compounds were extracted from fraction 10 using preparative HPLC, eluted with a gradient of MeCN/H_2O, 65:35, and then 100% MeCN. Methyl-6-gingerol (18.9 mg) and 8-gingerol (30.6 mg) were further purified using semi-preparative HPLC using an isocratic solvents system of MeOH/H2O (75:25) or MeCN/H_2O (75:25).

- 3S, 5S-1-(4-hydroxy-3-methoxyphenyl) decane-3, 5-diol (29.1 mg) And 3R, 5S-1-(4-hydroxy-3-methoxyphenyl) decane-3,5-diol (24.9 mg)

These compounds were obtained via fraction 16 utilizing preparative HPLC with MeCN/H_2O, 50:50, and then 100% MeCN.

NMR spectroscopy was used to describe four phenolic compounds—6-paradol, 6-shogaol, 6-gingerol, and a pentacyclic triterpene called oleanolic acid—that were produced using the same isolation and purification technique (Mohammed, Gbonjubola, Koorbanally et al, 2017). Nonetheless, there were differences in the gradient solvents utilized in terms of both quality and quantity.

5.2.2. Supercritical Carbon Dioxide (SCCO2) Extraction

Comparatively speaking, the $SCCO_2$ extraction method is more beneficial and environmentally benign than other traditional extraction methods, including the soxhlet extraction method. High diffusivity and high solvent strength that are easily adjustable by changing temperature and pressure make $SCCO_2$ an appealing extraction medium (Goto et al., 2015). *Aframomum melegueta* has the ability to biosynthesize 6-gingerol, 6-shogaol, and 6-paradol by the use of an extraction technique that is briefly explained as follows (Ono et al., 2017; Uemori, 2018).

Extraction

Using a semi-continuous flow-type system with a CO_2 flow rate of 3 mL min–1, 6-gingerol, 6-shogaol, and 6-paradol were obtained from *Aframomum melegueta* seeds by $SCCO_2$ at temperatures between 40 and 80 0C and pressures between 20 and 40 MPa. After being collected for 180 minutes and weighed, the extraction products contained in 10 milliliters of ethanol were immediately refrigerated at 5 degrees Celsius.

Analytical Method

The extracted polyphenolic compounds were analysed qualitatively by GC-MS analysis and quantitative analysis was carried out by HPLC using gradient elution technique with solvents water and acetonitrile, 90% acetonitrile for 0-25 min to 100% acetonitrile for 25-30 mint and finally 30% acetonitrile for 30-40 min. These analysis were performed under isothermal condition at 40 ^{0}C.

Characterization

FT-IR was used to describe the residues that had been oven dried at 60°C. The SEM micrographs revealed a physical alteration in the surface of the paradise seed grains upon extraction with SCCO2.

CONCLUSION

The research that is now available indicates that grains of paradise, because of their phytochemicals, have significant medical value. There is currently very little research on the pharmacological effects of AM and its function in biogenic synthesis, despite its enormous potential. Consequently, in order to maximize the extraction of bioactive chemicals from the plant, a thorough investigation is needed. It may also be investigated how it affects other biological processes including immunomodulation.

REFERENCES

Adegoke, G. O., Makinde, O., Falade, K. O., & Uzo-Peters, P. I. (2003). Extraction and characterization of antioxidants from Aframomum melegueta and Xylopia aethiopica. *European Food Research and Technology*, *216*(6), 526–528. doi:10.1007/s00217-003-0683-6

Ahmad, M. S., & Ahmed, N. (2006). Antiglycation Properties of Aged Garlic Extract: Possible Role in Prevention of Diabetic Complications. *The Journal of Nutrition*, *136*(3), 796S–799S. doi:10.1093/jn/136.3.796S PMID:16484566

Ahmed, N. (2005). Advanced glycation endproducts—Role in pathology of diabetic complications. *Diabetes Research and Clinical Practice*, *67*(1), 3–21. doi:10.1016/j.diabres.2004.09.004 PMID:15620429

Ahounou, J. F., Ouedraogo, G. G., Gbenou, J. D., Ouedraogo, S., Agbodjogbe, W. K., Dansou, P. H., & Moudachirou, M. (2012). Spasmolytic effects of aqueous extract of mixture from Aframomumum melegueta (K Schum) - Citrus aurantifolia (Christm and Panzer) on isolated trachea from rat. *African Journal of Traditional, Complementary, and Alternative Medicines*, *9*(2), 228–233. doi:10.4314/ajtcam.v9i2.7 PMID:23983339

Ajaiyeoba, E. O., & Ekundayo, O. (1999). Essential oil constituents of Aframomum melegueta (Roscoe) K. Schum. seeds (alligator pepper) from Nigeria. *Flavour and Fragrance Journal*, *14*(2), 109–111. doi:10.1002/(SICI)1099-1026(199903/04)14:2<109::AID-FFJ775>3.0.CO;2-M

Ajayi, J., & Omodele, I. (2021). Concentration and Differential fractions of Saponin Rich Aframomum melegueta Extract Influences In-Vitro Antioxidant Activities, and Silver Nitrate can Stimulate such Activities in the Plant. *Biochemical Pharmacology*, *10*, 186.

Akpanabiatu, M., Ekpo, N., Ufot, U., Malachy, N., Akpan, E., & Etuk, E. (2013). Acute toxicity, biochemical and haematological study of Aframomum melegueta seed oil in male Wistar albino rats. *Journal of Ethnopharmacology*, *150*(2), 150. doi:10.1016/j.jep.2013.09.006 PMID:24045175

Alaje, D. O., Owolabi, K., Olakunle, T., Oluoti, O. J., & Adetuberu, I. A. (2014). Nutritional, Minerals and Phytochemicals composition of Garcinia cola [Bitter cola] and Aframomum melegueta [Alligator pepper]. *IOSR Journal of Environmental Science, Toxicology and Food Technology*, *8*, 86–91. doi:10.9790/2402-08158691

Alo, M. N., Anyim, C., Igwe, J. C., Elom, M., & Uchenna, D. S. (2012). Antibacterial activity of water, ethanol and methanol extracts of Ocimum gratissimum, Vernonia amygdalina and Aframomum melegueta. *Advances in Applied Science Research*, *3*, 844–848.

Amadi, S., Zhang, Y., & Wu, G. (2016). Research progress in phytochemistry and biology of Aframomum species. *Pharmaceutical Biology*, *54*(11), 1–10. doi:10.3109/13880209.2016.1173068 PMID:27159353

Ansari, M. A., Roberts, K. N., & Scheff, S. W. (2008). Oxidative stress and modification of synaptic proteins in hippocampus after traumatic brain injury. *Free Radical Biology & Medicine*, *45*(4), 443–452. doi:10.1016/j.freeradbiomed.2008.04.038 PMID:18501200

Babu, S., Kumar, M., Madhumathi, T., & Ram, S. (2017). Status of insecticide resistance in lesser grain borer, Rhyzopertha dominica to Malathion and Deltamethrin in Andhra Pradesh. *Journal of Entomology and Zoology Studies*, *5*, 741–746.

Balaji, A., Bhuvaneswari, S., Raj, L. S., Bupesh, G., Meenakshisundaram, K. K., & Saravanan, K. M. (2022). A review on the potential species of the zingiberaceae family with anti-viral efficacy towards enveloped viruses. *Journal of Pure & Applied Microbiology*, *16*(2), 796–813. doi:10.22207/JPAM.16.2.35

B.A.S., L. (2007). Hypotensive and Antihypertensive Effects of Aframomum melegueta Seeds in Humans. *International Journal of Pharmacology*, 3.

Berrou, J., Tostivint, I., Verrecchia, F., Berthier, C., Boulanger, E., & Mauviel, A. (2009). Advanced glycation end products regulate extracellular matrix protein and protease expression by human glomerular mesangial cells. *International Journal of Molecular Medicine*, *23*(4), 513–520. PMID:19288028

Chen, H., Roco, M. C., Li, X., & Lin, Y. (2008). Trends in nanotechnology patents. *Nature Nanotechnology*, *3*(3), 123–125. doi:10.1038/nnano.2008.51 PMID:18654475

Chinedu, I. (2018). *Fungal Corrosion of Mild Steel and the Application of Aframomum melegueta Biomass Extract as Natural Source of Inhibition*. Academic Press.

Dare, E. O., Oseghale, C. O., Labulo, A. H., Adesuji, E. T., Elemike, E. E., Onwuka, J. C., & Bamgbose, J. T. (2015). Green synthesis and growth kinetics of nanosilver under bio-diversified plant extracts influence. *Journal of Nanostructure in Chemistry*, *5*(1), 85–94. doi:10.1007/s40097-014-0139-5

Doherty, F., & Olaniran, O. (2010). Antimicrobial Activities of Aframomum Melegueta (Alligator Pepper). *International Journal of Biology*, 2. doi:10.5539/ijb.v2n2p126

Dong, Y., Zhu, H., Shen, Y., Zhang, W., & Zhang, L. (2019). Antibacterial activity of silver nanoparticles of different particle size against Vibrio Natriegens. *PLoS One*, *14*(9), e0222322. doi:10.1371/journal.pone.0222322 PMID:31518380

Echo, A., Osuagwu, A., Agbor, R., Okpako, E. C., & Ekanem, B. E. (2013). Phytochemical composition of Aframomum melegueta and Piper guineense seeds. *World Journal of Applied Environmental Chemistry.*, *2*, 17–21.

Ekanem, A. P., Wang, M., Simon, J. E., & Moreno, D. A. (2007). Antiobesity properties of two African plants (Afromomum meleguetta and Spilanthes acmella) by pancreatic lipase inhibition. *Phytotherapy Research*, *21*(12), 1253–1255. doi:10.1002/ptr.2239 PMID:17705140

Eko, P. M., Afolabi, K. D., Olajide, R., Enyenihi, G. E., Mbaba, E. N., & Efffiong, M. E. (2022). *Haematology and Serum Metabolites of Pre-laying Isa Brown Hens fed Varying Dietary Levels of Alligator Pepper (Aframomum melegueta, k. Schum).* Seed Meal.

El Gharras, H. (2009). Polyphenols: Food sources, properties and applications – a review. *International Journal of Food Science & Technology*, *44*(12), 2512–2518. doi:10.1111/j.1365-2621.2009.02077.x

El-Halawany, A. M., Dine, R. S. E., El Sayed, N. S., & Hattori, M. (2014). Protective Effect of Aframomum melegueta phenolics Against CCl4-Induced Rat Hepatocytes Damage; Role of Apoptosis and Pro-inflammatory Cytokines inhibition. *Scientific Reports*, *4*(1), 5880. doi:10.1038/srep05880 PMID:25077538

El-Halawany, A. M., Dine, R. S. E., El Sayed, N. S., & Hattori, M. (2014). Protective Effect of Aframomum melegueta phenolics Against CCl4-Induced Rat Hepatocytes Damage; Role of Apoptosis and Pro-inflammatory Cytokines inhibition. *Scientific Reports*, *4*(1), 5880. doi:10.1038/srep05880 PMID:25077538

Erhabor, J., Idu, M., & Udo, F. O. (2013). Ethnomedicinal Survey of Medicinal Plants Used in the Treatment of Male Infertilty among the IFA Nkari People of Ini Local Government Area of AkwaIbom State, Nigeria. *Research Journal of Recent Sciences*, *2*, 5–11.

G.K., P. (2017). Alzheimer's disease: A challenge in managing with certain medicinal plants - A review. *International Journal of Pharmaceutical Sciences and Research*, *4*, 4960–4972.

Goto, M., Kanda, H., Wahyudiono, & Machmudah, S. (2015). Extraction of carotenoids and lipids from algae by supercritical CO2 and subcritical dimethyl ether. *The Journal of Supercritical Fluids*, *96*, 245–251. doi:10.1016/j.supflu.2014.10.003

Grains of paradise. (2014). Encyclopædia Britannica, Inc.

Gröblacher, B., Maier, V., Kunert, O., & Bucar, F. (2012). Putative Mycobacterial Efflux Inhibitors from the Seeds of Aframomum melegueta. *Journal of Natural Products*, *75*(7), 1393–1399. doi:10.1021/np300375t PMID:22789014

Gyasi, E., Kotey, D.A., Adongo, B.A., Adams, F.K., Owusu, E.O., & Mohammed, A. (2022). Management of major seed-borne fungi of cowpea (Vigna unguiculata (L.) Walp) with four selected botanical extracts. *Advances in Agriculture*.

Hamming, I., Timens, W., Bulthuis, M. L., Lely, A. T., Navis, G., & van Goor, H. (2004). Tissue distribution of ACE2 protein, the functional receptor for SARS coronavirus. A first step in understanding SARS pathogenesis. *The Journal of Pathology*, *203*(2), 631–637. doi:10.1002/path.1570 PMID:15141377

Hattori, H., Yamauchi, K., Onwona-Agyeman, S., & Mitsunaga, T. (2017). Effect of Grains of Paradise (GP) Extract Intake on Obesity and Sympathetic Nerve Activity. *American Journal of Plant Sciences*, *08*(2), 85–95. doi:10.4236/ajps.2017.82007

Huang, C., Wang, Y., Li, X., Ren, L., Zhao, J., Hu, Y., Zhang, L., Fan, G., Xu, J., Gu, X., Cheng, Z., Yu, T., Xia, J., Wei, Y., Wu, W., Xie, X., Yin, W., Li, H., Liu, M., ... Cao, B. (2020). Clinical features of patients infected with 2019 novel coronavirus in Wuhan, China. *Lancet*, *395*(10223), 497–506. doi:10.1016/S0140-6736(20)30183-5 PMID:31986264

Ileke, K., Adesina, J., & Okunola, O. (2017). Larvicidal and Pupicidal Potential of Aframomum melegueta K. Schum Extracts Against Mosquito, Anopheles Species. *Journal of Entomological Research Society*, *19*, 121–127.

Ilic, N. M., Dey, M., Poulev, A. A., Logendra, S., Kuhn, P. E., & Raskin, I. (2014). Anti-inflammatory activity of grains of paradise (Aframomum melegueta Schum) extract. *Journal of Agricultural and Food Chemistry*, *62*(43), 10452–10457. doi:10.1021/jf5026086 PMID:25293633

Ilic, N. M., Dey, M., Poulev, A. A., Logendra, S., Kuhn, P. E., & Raskin, I. (2014). Anti-inflammatory activity of grains of paradise (Aframomum melegueta Schum) extract. *Journal of Agricultural and Food Chemistry*, *62*(43), 10452–10457. doi:10.1021/jf5026086 PMID:25293633

Imo, E., Orji, J., & Nweke, C. (2018). *Corrosion and Fungal Growth Inhibiting Effects of Piper guineense Extracts*. Academic Press.

Inegbenebor, U., Amadi, K., Ebomoyi, M., & Onyia, K. (2011). Abortifacient properties of alligator pepper (Aframomum melegueta) seeds. *Journal of Pharmacy & Bioresources*, *6*(2), 6. doi:10.4314/jpb.v6i2.63327

Inegbenebor, U., & Ebomoyi, M. (2014). Effect of saline extract of alligator pepper (zingiberaceae aframomum melegueta) on serum progesterone in pregnant sprague dawley rats. *International Journal of Herbs and Pharmacological Research.*, *3*, 46–51.

Inegbenebor, U., Ebomoyi, M. I., Onyia, K. A., Amadi, K., & Aigbiremolen, A. E. (2009). Effect of alligator pepper (Zingiberaceae aframomum melegueta) on first trimester pregnancy in Sprague Dawley rats. *Nigerian Journal of Physiological Sciences; Official Publication of the Physiological Society of Nigeria*, *24*(2), 161–164. PMID:20234758

Iwu, M. M. (2014). *Handbook of African medicinal plants*. doi:10.1201/b16292

Jena, A., Sachdeva, R. K., Sharma, A., & Wanchu, A. (2009). Adverse drug reactions to nonnucleoside reverse transcriptase inhibitor-based antiretroviral regimen: A 24-week prospective study. *Journal of the International Association of Physicians in AIDS Care (Chicago, Ill.)*, *8*(5), 318–322. doi:10.1177/1545109709343967 PMID:19721097

Juzeliūnas, E., Ramanauskas, R., Lugauskas, A., Leinartas, K., Samulevičienė, M., & Sudavičius, A. (2006). Influence of wild strain Bacillus mycoides on metals: From corrosion acceleration to environmentally friendly protection. Electrochimica Acta -. *Electrochimica Acta*, *51*(27), 6085–6090. doi:10.1016/j.electacta.2006.01.067

Kamte, S. L. N., Ranjbarian, F., Campagnaro, G. D., Nya, P. C. B., Mbuntcha, H., Woguem, V., Womeni, H. M., Ta, L. A., Giordani, C., Barboni, L., Benelli, G., Cappellacci, L., Hofer, A., Petrelli, R., & Maggi, F. (2017). Trypanosoma brucei Inhibition by Essential Oils from Medicinal and Aromatic Plants Traditionally Used in Cameroon (Azadirachta indica, Aframomum melegueta, Aframomum daniellii, Clausena anisata, Dichrostachys cinerea and Echinops giganteus). *International Journal of Environmental Research and Public Health*, *14*(7), 737. doi:10.3390/ijerph14070737 PMID:28684709

Kazeem, M. I., Akanji, M. A., Hafizur, R. M., & Choudhary, M. I. (2012). Antiglycation, antioxidant and toxicological potential of polyphenol extracts of alligator pepper, ginger and nutmeg from Nigeria. *Asian Pacific Journal of Tropical Biomedicine*, *2*(9), 727–732. doi:10.1016/S2221-1691(12)60218-4 PMID:23570003

Kirimuhuzya, C., & Anywar, G. (2015). Phytochemical profile and antibacterial activity of crude extracts of the pod of Aframomum angustifolium (Sonn.) K. Schum. *European Journal of Biological Research*, *5*, 36–41.

Korshed, P., Li, L., Liu, Z., Mironov, A., & Wang, T. (2019). Size-dependent antibacterial activity for laser-generated silver nanoparticles. *Journal of Interdisciplinary Nanomedicine*, *4*(1), 24–33. doi:10.1002/jin2.54

Kostolanská, J., Jakus, V., & Barák, L. (2009). Monitoring of Early and Advanced Glycation in Relation to the Occurrence of Microvascular Complications in Children and Adolescents with Type 1 Diabetes Mellitus. *Physiological research / Academia Scientiarum Bohemoslovaca, 58*, 553-61.

Kumar, A., Kennedy-Boone, D., Weisz, H., Capra, B., Uchida, T., Jennings, K., & ... . (2015). Neuroprotective Effects of Aframomum melegueta Extract after Experimental Traumatic Brain Injury. *Natural Products Chemistry & Research*, *3*(1), 3. doi:10.4172/2329-6836.1000167

Kumar, M., Kaushik, D., Kaur, J., Proestos, C., Oz, F., Oz, E., Gupta, P., Kundu, P., Kaur, A., Anisha, A., & Ritika, R. (2022). A critical review on obesity: Herbal approach, bioactive compounds, and their mechanism. *Applied Sciences (Basel, Switzerland)*, *12*(16), 8342. doi:10.3390/app12168342

Lamaty, G., Menut, C., Koudou, J., Régnier, P., & Bessière, J.-M. (1993). Aromatic Plants of Tropical Central Africa. XI. Essential Oils of Leaf and Seed of Aframomum melegueta (Roscoe) K. Schum. from Central African Republic. *The Journal of Essential Oil Research*, *5*(1), 81–83. doi:10.1080/10412905.1993.9698174

Letko, M., Marzi, A., & Munster, V. (2020). Functional assessment of cell entry and receptor usage for SARS-CoV-2 and other lineage B betacoronaviruses. *Nature Microbiology*, *5*(4), 562–569. doi:10.1038/s41564-020-0688-y PMID:32094589

Little, B., & Ray, R. (2001). A review of fungal influenced corrosion. *Corrosion Reviews*, *19*(5-6), 401–417. doi:10.1515/CORRREV.2001.19.5-6.401

Liu, S.-W., Katsafanas George, C., Liu, R., Wyatt Linda, S., & Moss, B. (2015). Poxvirus Decapping Enzymes Enhance Virulence by Preventing the Accumulation of dsRNA and the Induction of Innate Antiviral Responses. *Cell Host & Microbe*, *17*(3), 320–331. doi:10.1016/j.chom.2015.02.002 PMID:25766293

Loo, Y. Y., Rukayadi, Y., Nor-Khaizura, M.-A.-R., Kuan, C. H., Chieng, B. W., Nishibuchi, M., & Radu, S. (2018). In Vitro Antimicrobial Activity of Green Synthesized Silver Nanoparticles Against Selected Gram-negative Foodborne Pathogens. *Frontiers in Microbiology*, *9*(1555), 1555. doi:10.3389/fmicb.2018.01555 PMID:30061871

Lugauskas, A., Leinartas, K., & Binkauskienė, E. (2008). Possibility of micromycetes detected in dust to grow on metals (Al, Fe, Cu, Zn) and on polyaniline-modified Ni. *Ekologija (Lietuvos Mokslu Akademija)*, *54*(3), 149–157. doi:10.2478/V10055-008-0023-z

Martinez, S. E., Davies, N. M., & Reynolds, J. K. (2013). *Toxicology and safety of flavonoids. Flavonoid Pharmacokinetics: Methods of Analysis*. Preclinical and Clinical Pharmacokinetics, Safety, and Toxicology.

Mk, O., Oi, A., Bo, O., Ia, A., & Bi, O. (2019). Derivatized Extracts from Aframomium melegueta K. Schum. and Vernonia amygdalina Delile Contain Organic Compounds that Showed Antiviral Effects against Atypical Fowl Pox Virus (FPV Kabete). *Journal of Antivirals & Antiretrovirals*, *11*, 1–10.

Mohammed, A., Awolola, V., Koorbanally, N., & Islam, M. (2017). Inhibition of key enzymes linked to type 2 diabetes by compounds isolated from Aframomum melegueta fruit. *Pharmaceutical Biology*, *55*(1), 1010–1016. doi:10.1080/13880209.2017.1286358 PMID:28176546

Mohammed, A., Gbonjubola, V. A., Koorbanally, N. A., & Islam, M. S. (2017). Inhibition of key enzymes linked to type 2 diabetes by compounds isolated from Aframomum melegueta fruit. *Pharmaceutical Biology*, *55*(1), 1010–1016. doi:10.1080/13880209.2017.1286358 PMID:28176546

Mojekwu, T., Oshiozokhai, Y., Ojokuku, S., Oyebadejo, S., & Eboetse, Y. (2011). Hypoglyceamic effects of aqueous extract of aframomum melegueta leaf on alloxan-induced diabetic male albino rats hypoglyceamic effects of aqueous extract of aframomum melegueta leaf on alloxan-induced diabetic male albino rats. *Pacific Journal of Medical Sciences, 8*(1).

Motaleb, M.A., Hossain, M.K., & Office, I.B.C. (2011). *Selected medicinal plants of Chittagong Hill Tracts*. Academic Press.

Mustapha, B. O., Ademoyegun, O. T., & Ahmed, R. S. (2023). Horticultural crops as natural therapeutic plants for the therapy of diabetes mellitus. *Egyptian Journal of Basic and Applied Sciences*, *10*(1), 380–395. doi:10.1080/2314808X.2023.2217387

NaAllah, AAyipo, Y.OKomolafe, D.ISolihu, SBamidele, BAlabi, M.A. (2021). Phytochemical screening and in silico pharmacological profiling of ethanolic extract of Aframomum melegueta for prostate carcinoma. *Journal of Applied Pharmaceutical Science*, *11*(7), 132–145.

Nakagawa, T., Yokozawa, T., Terasawa, K., Shu, S., & Juneja, L. R. (2002). Protective Activity of Green Tea against Free Radical- and Glucose-Mediated Protein Damage. *Journal of Agricultural and Food Chemistry*, *50*(8), 2418–2422. doi:10.1021/jf011339n PMID:11929306

Nwaehujor, C. (2014). Hepatotoxicity of Methanol Seed Extract of Aframomum melegueta [Roscoe] K. Schum. (Grains of paradise) in Sprague-Dawley Rats. *African Journal of Biomedical Research*, *2*, 61–66.

Nyaga, P. N., Kaminjolo, J. S., Mutiga, E. R., & Bebora, L. C. (1979). Occurrence of Atypical Fowlpox in Poultry Farms in Kenya. *Avian Diseases*, *23*(3), 745–752. doi:10.2307/1589751 PMID:230810

Obuekwe, C. O. (1981). *Microbial corrosion of pipelines: University of Alberta*. Edmondton.

Odebunmi, E., Oluwaniyi, O., Awolola, V., & Adediji, O. D. (2009). Proximate and nutritional composition of kola nut (Cola nitida), bitter cola (Garcinia cola) and alligator pepper (Afromomum melegueta). *African Journal of Biotechnology*, 8.

Odetunde, A., & Avungbeto, L. (2015). Antimicrobial effect and phytochemical analysis of Aframomum melegueta on some selected bacteria and fungi. *European Journal of Biotechnology and Bioscience.*, *3*, 15–19.

Odo, C. (2015). Anti-bacterial Effects of Aframomum melegueta Seed Extracts on Some Enterobacteriaceae Organisms. *World Journal of Pharmaceutical Research*, *4*, 304–309.

Ogbunugafor, H., Ugochukwu, C., & Kyrian-Ogbonna, A. (2017). The role of spices in nutrition and health: A review of three popular spices used in Southern Nigeria. *Food Quality and Safety*, *1*(3), 1. doi:10.1093/fqsafe/fyx020

Oguzie, E., Ogukwe, C., Jude-anthony, O., Nwanebu, F., Adindu, B., & Udeze, I. (2012). Broad spectrum corrosion inhibition: Corrosion and microbial (SRB) growth inhibiting effects of Piper guineense extract. *Journal of Materials Science*, *47*(8), 3592–3601. doi:10.1007/s10853-011-6205-1

Oguzie, E. E., Iheabunike, Z. O., Oguzie, K. L., Ogukwe, C. E., Chidiebere, M. A., Enenebeaku, C. K., & Akalezi, C. O. (2013). Corrosion Inhibiting Effect of Aframomum melegueta Extracts and Adsorption Characteristics of the Active Constituents on Mild Steel in Acidic Media. *Journal of Dispersion Science and Technology*, *34*(4), 516–527. doi:10.1080/01932691.2012.682008

Okwu, D., & Mgbeahuruike, E. (2018). *Chemical Composition and in Vitro Antifungal Activity Screening of Seed and Leaf Extracts from Aframomum meleguata and Monodora myristica against Sclertium rolfsii of Cowpea Plant (Vigna unguiculata L. Walp).* Academic Press.

Okwu, M. U., Olley, M., Akpoka, A. O., & Izevbuwa, O. E. (2019). Methicillin-resistant Staphylococcus aureus (MRSA) and anti-MRSA activities of extracts of some medicinal plants: A brief review. *AIMS Microbiology*, *5*(2), 117–137. doi:10.3934/microbiol.2019.2.117 PMID:31384707

Oladunmoye, M.K. (2019). Characterization of Organic Compounds in Aframomum melegueta K. Schum Using GC-MS. *Medicinal & Aromatic Plants, 8*.

Olumekun, V.O., Osuntokun, O. T., Ajayi, A. O., Omotuyi, I. O., & Olonisakin, A. (2020). Estimation of Vitamins in Aframomum melegueta [Roscoe] K. Schum and Its Relevance as Natural Immune Booster. *Journal of Applied Life Sciences International, 23*(7), 25-39.

Omoboyowa, D., Agha Aja, O., Eluu, F., & Ngobidi, K. (2017). Effects of Methanol Seed Extract of Aframomum melegueta (Alligator Pepper) on Wistar Rats with 2,4-Dinitrophenylhydrazine Induced Hemolytic Anemia. *Recent Advances in Biology and Medicine.*, *03*, 11. doi:10.18639/RABM.2017.03.443648

Ono, M., Kawamoto, Y., Uemori, C., Diono, W., Kanda, H., & Goto, M. (2017). Extraction of Phytochemicals from Grains of Paradise Using Supercritical Carbon Dioxide. *Engineering Journal (New York)*, *21*, 53–64.

Osuntokun, O. (2020a). *Aframomum Melegueta (Grains of Paradise)*. Academic Press.

Osuntokun, O. (2020b). Synergistic efficacy of Aframomum melegueta [Roscoe] K. Schum and Spondias mombin(Linn), A predictive treatment of SARS-CoV-2 (COVID-19) Infection. *Journal of Bioscience & Biomedical Engineering*.

Osuntokun, O., Balogun, J., Akele, E., Adedayo, S., & Bello, O. (2021). Aframomum melegueta [Roscoe] K. Schum) ethanolic seed extract against asymptomatic bacteriuria isolates from pregnant women attending antenatal clinic in Ondo State, Nigeria. *International Journal of Pathogen Research.*, *8*(2), 12–43. doi:10.9734/ijpr/2021/v8i230200

Osuntokun, Ot., Baraldi, C., & Gamberini, M. C. (Eds.). (2018). Evaluation of quantitative elemental compositions and antioxidant potentials of spondias mombin extracts (linn), a precursor against infectious diseases. Academic Press.

Osuntokun, O., Olumekun, V., Ajayi, O., Omotuyi, O., & Adebisi, O. (2020). Assessment of in-vitro Antioxidant/Enzymes Inhibitory Potentials of Aframomum melegueta [Roscoe] K. Schum (Grains of Paradise) Leaf, Stem Bark, Seed Bark and Seed Extracts. *International Journal of Current Research*, *20*, 40–57.

Owokotomo, A. (2014). Chemical Constituents of the Leaf, Stem, Root and Seed Essential Oils of Aframomum melegueta (K. Schum) from South West Nigeria. *International Research Journal of Pure and Applied Chemistry*, *4*(4), 395–401. doi:10.9734/IRJPAC/2014/7397

Oyagbemi, A. A., Saba, A. B., & Azeez, O. I. (2010). Molecular targets of [6]-gingerol: Its potential roles in cancer chemoprevention. *BioFactors (Oxford, England)*, *36*(3), 169–178. doi:10.1002/biof.78 PMID:20232343

Quattrocchi, U. (2012). *CRC world dictionary of medicinal and poisonous plants: common names, scientific names, eponyms, synonyms, and etymology*. CRC, Taylor & Francis Group.

Reeves, P. M., Smith, S. K., Olson, V. A., Thorne, S. H., Bornmann, W., Damon, I. K., & Kalman, D. (2011). Variola and Monkeypox Viruses Utilize Conserved Mechanisms of Virion Motility and Release That Depend on Abl and Src Family Tyrosine Kinases. *Journal of Virology, 85*(1), 21–31. doi:10.1128/JVI.01814-10 PMID:20962097

Rowe, P. J., Comhaire, F. H., Hargreave, T. B., & Mahmoud, A. M. A. (2000). *WHO manual for the standardized investigation, diagnosis and management of the infertile male*. World Health Organization.

Schultz, S., Smith, D. R., Mock, J. J., & Schultz, D. A. (2000). Single-target molecule detection with nonbleaching multicolor optical immunolabels. *Proceedings of the National Academy of Sciences of the United States of America, 97*(3), 996–1001. doi:10.1073/pnas.97.3.996 PMID:10655473

Sell, S. L., Boone, D. R., Weisz, H. A., Cardenas, C., Willey, H. E., Bolding, I. J., Micci, M.-A., Falduto, M. T., Torres, K. E. O., DeWitt, D. S., Prough, D. S., & Hellmich, H. L. (2019). MicroRNA profiling identifies a novel compound with antidepressant properties. *PLoS One, 14*(8), e0221163. doi:10.1371/journal.pone.0221163 PMID:31442236

Sharma, V., Yngard, R., & Lin, Y. (2008). Silver nanoparticles: Green synthesis and their antimicrobial activities. Adv. Colloid Interface Sci. 145, 83-96. *Advances in Colloid and Interface Science, 145*, 83–96. doi:10.1016/j.cis.2008.09.002 PMID:18945421

Singh, P., & Tripathy, D. N. (2003). Fowlpox Virus Infection Causes a Lymphoproliferative Response in Chickens. *Viral Immunology, 16*(2), 223–227. doi:10.1089/088282403322017956 PMID:12828873

Solomon, U. (2005). Anti-stress potential of aqueous seed extract of Aframomum melegueta. *African Journal of Biomedical Research*, 8.

Solomon, U., & Ashorobi, B. (2008). Further pharmacological studies on aqueous seed extract of Aframomum melegueta in rats. *Journal of Ethnopharmacology, 115*(3), 489–493. doi:10.1016/j.jep.2007.10.019 PMID:18060710

Solomon, U., & Ashorobi, R. (2007). Further studies on the antinociceptive action of aqueous seed extract of Aframomum melegueta. *Journal of Ethnopharmacology, 109*(3), 501–504. doi:10.1016/j.jep.2006.08.025 PMID:17023130

Tapiero, H., Tew, K. D., Ba, G. N., & Mathé, G. (2002). Polyphenols: Do they play a role in the prevention of human pathologies? *Biomedicine and Pharmacotherapy, 56*(4), 200–207. doi:10.1016/S0753-3322(02)00178-6 PMID:12109813

Thirumurugan, G., Shaheedha, S. M., & Dhanaraju, M. D. (2009). In vitro evaluation of anti-bacterial activity of silver nanoparticles synthesised by using Phytophthora infestans. *International Journal of Chemtech Research*, *1*, 714–716.

Tijjani, H., & Luka, C. (2013). Effects of Afromomum melegueta, Zingiber officinale and Piper nigrum on Some Biochemical and Haematological Parameters in Rats Fed with High Lipid Diet. *International Journal of Pure & Applied Bioscience.*, *1*, 61–67.

Toh, E. Y. S., Chooi Ling, L., Ling, A., Soi Moi, C., & Koh, R. Y. (2019). Overview of the pharmacological activities of aframomum melegueta. *Pertanika. Journal of Tropical Agricultural Science*, *42*, 1–13.

Uemori, C. (2018). *Extraction and nanoparticle preparation of natural materials by using supercritical carbon dioxide*. Graduate School of Engineering, Nagoya University.

Ukeh, D. (2008). Bioactivities of Essential Oils of Aframomum melegueta and Zingiber officinale both (Zingiberaceae) Against Rhyzopertha dominica (Fabricius). *Journal of Entomology*, *5*(3), 193–199. doi:10.3923/je.2008.193.199

Ukeh, D., Birkett, M., Pickett, J., Bowman, A., & Luntz, A. (2009). Repellent activity of alligator pepper, Aframomum melegueta, and ginger, Zingiber officinale, against the maize weevil, Sitophilus zeamais. *Phytochemistry*, *70*(6), 751–758. doi:10.1016/j.phytochem.2009.03.012 PMID:19394981

Umukoro, S., & Ashorobi, R. B. (2005). Effect of Aframomum melegueta. Seed Extract on Castor Oil-Induced Diarrhea. *Pharmaceutical Biology*, *43*(4), 330–333. doi:10.1080/13880200590951748 PMID:28925843

Xia, Y., Rogers, J. A., Paul, K. E., & Whitesides, G. M. (1999). Unconventional Methods for Fabricating and Patterning Nanostructures. *Chemical Reviews*, *99*(7), 1823–1848. doi:10.1021/cr980002q PMID:11849012

Xu, H., Zhong, L., Deng, J., Peng, J., Dan, H., Zeng, X., Li, T., & Chen, Q. (2020). High expression of ACE2 receptor of 2019-nCoV on the epithelial cells of oral mucosa. *International Journal of Oral Science*, *12*(1), 8. doi:10.1038/s41368-020-0074-x PMID:32094336

Young, E., & Tarawou, T. (2016). Determination of PAHs in Fresh and Smoke-dried Alligator Pepper. *International Journal of Scientific Research in Science and Technology*, *2*, 204–207.

Chapter 10
The Nutritional and Therapeutic Aspects of Australian Macadamia Nuts

Fakhr Un Nisa
The Women University Multan, Pakistan

Tahira Batool
The Women University Multan, Pakistan

ABSTRACT

Macadamia nuts, native to Australia, are a nutritional powerhouse with numerous health benefits. They are rich in heart-healthy monounsaturated fats, antioxidants, essential vitamins and minerals, and antioxidants. They help regulate blood sugar levels, provide a sensation of fullness, and have anti-inflammatory properties. Macadamia nuts also hydrate skin and may prevent signs of ageing. They are used in skin care products due to their hydrating and restorative properties. However, some people have severe allergic reactions, potentially leading to respiratory issues. Overall, Australian macadamia nuts offer a fun and health-conscious option for those seeking a holistic approach to well-being.

1. INTRODUCTION

In order to improve health and well-being, the food industry is now very interested in launching new products. In this regard, there's a rising need for functional food items that have been fortified with long-chain fatty acids.

DOI: 10.4018/979-8-3693-6105-4.ch010

Tree seeds with a hard, frequently lignified shell are called nuts. Hazelnut, Brazil nut, and chestnut are examples of true nuts. They are often classified alongside other nuts like almond, cashew, and peanut. Nuts and seeds come from diverse plants, with a varied nutritional composition. They have a food supply that is intended to satisfy the requirements of the growing plant embryo, often fat or starch or polysaccharides. These foods are concentrated sources of dietary energy, protein, unsaturated fatty acids, micronutrients, and fiber (K C, 2021).

Macadamia nut trees can reach heights of 2 to 12 meters. They are indigenous to tropical rain forests in Sulawesi, Indonesia, New Caledonia, and eastern Australia. Their preferred conditions include temperatures above 10°C, 1000–2000 mm of rainfall, and rich, well-drained soils. The only naturally occurring food plants in Australia with economic significance are the two commercially significant species, Macadamia integrifolia and Macadamia tetraphylla. In 2008, the International Nut and Dried Fruit Foundation reported an annual production of around 26,000 tonnes worldwide (Ballhorn, 2011).

Around the world, people eat raw (natural) or roasted tree nuts as healthy snacks. Their high concentration of polyunsaturated fatty acids and necessary amino acids makes them very attractive (Bailey & Stein, 2020). Macadamia (Macadamia tetraphylla) nuts are creamy in texture and have a mild taste similar to butter (Moreno-Pérez, Sánchez-García et al., 2011). China's macadamia nut business has grown quickly in recent years, and 15,000 hectares of planting land are anticipated; the trees are primarily planted in the southern provinces of Yunnan and Guangxi (Hong, Groven et al., 2018; Zhong, Xiang et al., 2020). Cream-coloured macadamia kernels have an oil content of 72–78% and a moisture content of 1.5%. Browning of the kernels following roasting and the development of an unpleasant flavor as a result of oxidation and rancidity during storage are the two main flaws. It is essential to gradually dry items down to 1.5% moisture content in order to avoid browning and prolong shelf life (M. M. Wall, 2013). A nut's edible kernel is its seed, which is mainly made up of embryonic cotyledons and is protected from the environment by a fibrous husk and a hard, brown shell. Nuts are harvested following fruit falls or tree removal, then dried, roasted, and salted to a 1.5% moisture level. They are then often dry-roasted or fried in coconut oil (Gray, 2013). The subtropical Australian rainforests are home to the macadamia tree, which prefers temperatures between 25 and 35 degrees Celsius. For brief periods, it can tolerate frosts as low as -6°C, but prolonged durations or colder temperatures can harm or kill mature trees. Frost damage is especially prone to inflorescence. A minimum temperature of 20°C is needed for flowering, while prolonged exposure to temperatures higher than 35°C might result in deformed and chlorotic growth. The deep, proteoid roots and sclerophyllous leaves of the macadamia tree improve the uptake of nutrients from deficient soils. Survival conditions are not the same as ideal production conditions, though. Deep, well-drained soils with a high organic matter content, a medium cation exchange capacity, and a pH of 5.0 to 6.0 are ideal for macadamia growth (R. A. Stephenson, 2003).

2. HISTORY, ORIGIN, DISTRIBUTION, AND PRODUCTION

The macadamia nut originated in Australia, specifically in the rain forests of Queensland and New South Wales. It developed in Hawaii and spread to East Africa before being isolated in several Southeast Asian countries. In the 1850s, German-Australian physician and botanist Ferdinand von Mueller discovered the macadamia nut and named it after his colleague, John Macadam. The genus consists of eight species, with two producing edible nuts. Most of these species produce poisonous or inedible nuts. Macadamia nuts were imported to Hawaii in the 1880s for windbreak cultivation and later developed commercially for dessert nuts in 1930. Today, most macadamia plantations thrive in Hawaii, and the nut is distributed mainly in tropical climates of Australia, Brazil, Indonesia, Peru, Fiji, Kenya, Israel, Colombia, Venezuela, New Zealand, South Africa, and Tanzania (Verma, Yadav et al.). The evergreen tree species known as macadamia (Macadamia integrifolia Maiden and Betche) is native to Australia's coastal rain forests, but it was initially grown for commercial purposes in Hawaii (Alam, Hardner et al., 2019). The production of macadamia nuts has been steadily rising worldwide in recent years. Still, less than 2 percent of the world's tree nut production comes from macadamia nuts. Australia, Brazil, Colombia, China, Costa Rica, Guatemala, Kenya, Malawi, Hawaii, and South Africa are the countries where the crop is mostly cultivated (INC, 2020). There have been reports of smaller-scale farming in Tanzania, Venezuela, Vietnam, Argentina, Jamaica, Fiji, Mexico, Mozambique, Swaziland, New Zealand and Zimbabwe. Two years after transplantation, macadamia trees can begin to yield fruit with proper care. On the other hand, as macadamia trees age, their output generally increases. Accordingly, trees that are 1 to 5 years old can give around 5 kg of nuts annually per tree, while older trees that are 5 years old or older have been known to generate up to 20 kg of nuts annually, depending on the growing environment (White & Hanan, 2014). Macadamia trees grow under different soil conditions, requiring varying irrigation and fertilizer recommendations. In Hawaii, volcanic lava-rock soils require more water than sedimentary soils, which is reducing yield and growth in places like south Kona. Macadamia trees need 0.5 cm of water per day from rainfall and/or irrigation, hence water sources should be constructed to allow for irrigation. Analysis of leaf tissue is used to calculate fertilizer requirements. illnesses include Phytopthora, Botrytis blossom illnesses, and husk spot disease caused by *Pseudocercospera macadamiae* can have an impact on macadamia productivity. Disease rates are lower in drier production areas. Macadamia nuts can sustain significant damage from insect pests such as the banana spotted bug, fruit spotting bug, and southern green stink bug. Integrated pest management is frequently used to minimize the need for pesticides (Wells, 2013).

3. BOTANICAL DESCRIPTION

The genus Macadamia, which comprises about ten species in the family Proteaceae, yields edible nuts in two varieties: Macadamia integrifolia and Macadamia tetraphylla, or their hybrids. M. integrifolia is the main species used in trading. Australia is the natural home of macadamia nuts, which are also grown in Hawaii (Hokmabadi & Sedaghati, 2014). The latest revised botanical classification by Smith classifies the smooth-shell type of macadamia, which has three leaves at a node, as Macadamia integrifolia, while the rough-shell type with long, spiny leaves is considered M. tetraphylla, a different species. Some authorities initially considered both types as the same species, M. ternifolia, but designated the smooth-shell type as M. ternifolia var. integrifolia. The smooth-shell type was designated as a distinct botanical variety due to the similarity in leaf characteristics of young, smooth-shell seedlings to mature leaf characteristics of rough-shell trees. In Australia, the same types are recognized, and Wills described four different forms of *M. ternifolia*, one of which is likely synonymous with *M. tetraphylla* (Hamilton & Fukunaga, 1959).

Kingdom: Plantae
Phylum: Magnoliophyta
Class: Magnoliopsida
Order: Proteales
Family: Proteaceae
Genus: Macadamia
Species: (i) integrifolia,(ii) ternifolia,(iii) tetraphylla (Verma, Yadav et al.)

3.1 Species

Costello et al. (2008) state that there are nine species in this genus(Costello, Plan 20082012. Report to Department of the Environment et al.). The species native to Australia are *Macadamia integrifolia, M. tetraphylla, M. ternifolia, M. jansenii, M. whelani, M. claudiensis and M. grandis, whereas M neurophylla* is native to New Caledonia and *M. hildebrandii* is native to Sulawesi in Indonesia. However, only

Figure 1. Macadamia nuts

Table 1. List of macadamia species

S.No.	Species	Origin
1	M. hildebrandii Steenis	Celebes
2	M. Sleumer	New Caledonia
3	M. veillardii (Brongn. and Gris.) Sleumer	
4	M. francii (Guill.) Sleumer	
5	M. whelani F.M. Bail.	Eastern Australia
6	M. ternifolia F. Muell.	
7	M. integrifolia Maiden and Betche	
8	M. tetraphylla LAS. Johnson	
9	M. prealta (F. Muell.) F. M. Bail	
10	M. heyana (F. M. Bail.) Sleumer	

the smooth shelled *(Macadamia integrifolia)* and the rough shelled (*Macadamia tetraphylla*) are cultivated for their edible nuts (Peace, Vithanage et al., 2003).

4. ECONOMIC SIGNIFICANCE AND GLOBAL IMPORTANCE

Macadamia, an Australian plant, has been domesticated on a commercial scale as a food crop, primarily in Australia, the US, and South Africa. The plant has expanded into Brazil, Guatemala, Kenya, New Zealand, Malawi, and Paraguay. World production in 2009 was over 26,000 tonnes of kernel, with Australia producing 10,000 tons, South Africa 5,000 tons, and Hawaii 3,000 tons. Macadamia nut production in Australia increased from 4400 tons in 1987 to over 37,000 tonnes in 2009. The annual value of macadamia exports from Australia is estimated at $20.1 million for nut-in shell and $64.5 million for kernel in 2008. Macadamia cultivars are propagated through grafting, with Hawaiian cultivars still forming the basis of the industry worldwide. In Australia, two HAES cultivars, 344 and 741, are the most commonly grown. Breeding programs have focused on cultivars selected for local conditions, such as 'Hidden Valley A16' (H. M. Wallace & D. Walton, 2011).

5. PRODUCTS RELATED TO MACADAMIA NUTS

The market for plant-based beverages has grown significantly in the last several years. This is due to the fact that they are marketed as the primary alternative to cow's milk for customers who are on a restricted diet, either due to health concerns or a choice for non-animal products (Penha, Santos et al., 2021). Macadamia integrifolia's excellent nutritional content and advantageous health characteristics make it a suitable option for the manufacturing of plant-based drinks. The nut is high in fats, particularly monounsaturated fatty acids (MUFAs),

Figure 2. World macadamia kernel production from 2012 to 2022
(Source: Nuts & Dried Fruits Statistical Yearbook 2022/2023)

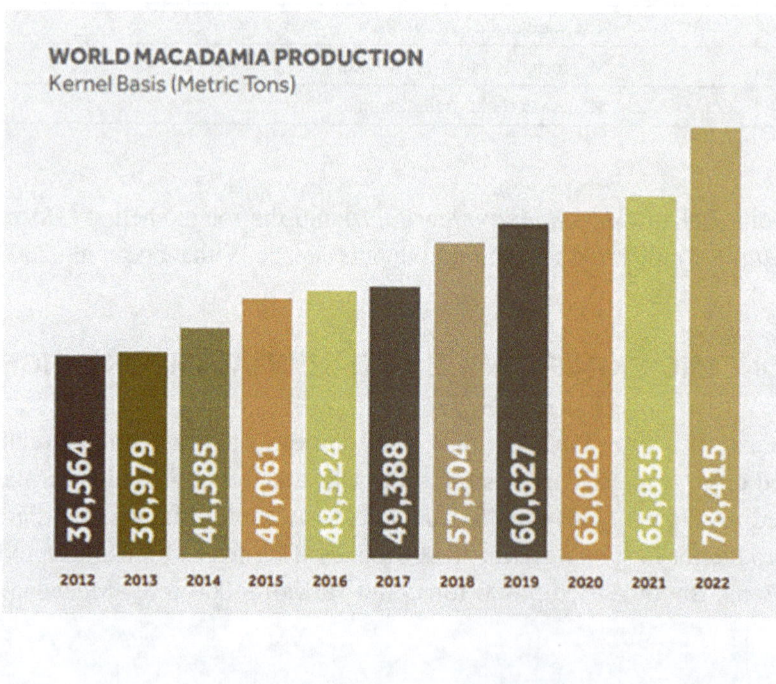

protein, fiber, vitamins, carbs, and minerals (Wall, 2010; Zavala, Hernández-Martínez et al., 2023).

5.1 Nut Snacks, Oils, and Flours

Healthy and innovative snacks enhanced with long-chain omega-3 fatty acids derived from fish oil (LCω3PUFAs) has been synthesized. Macadamia integrifolia nut beverage industry waste was used to synthesize these snacks. In terms of proteins, lipids, and carbs, the product's macro component concentrations were more than anticipated (Lamas & Álvarez, 2023).

The kernels of certain macadamia species are poisonous due to cyanogen chemicals, making them unfit for human consumption. Macadamia integrifolia and Macadamia tetraphylla, together with their hybrids, are the commercial macadamia species used for edible oil production and human consumption. The high oil content in the kernels of both species and their hybrids is a defining characteristic. Macadamia integrifolia has a smooth, spherical form with a smooth shell, while Macadamia tetraphylla has a rough, spindle-shaped shell (D. Walton, 2005). When comparing mature macadamia kernels (>60%) to immature ones (1%–48%), the former has a substantially higher oil content (KOAZE, KARANJA et al., 2002).

Macadamia kernels are a rich source of fat, with 69%-76% of their content being fat. Macadamia oil is extracted from these kernels using cold-pressing to produce extra-virgin oil, while the remaining oil is extracted using solvents. The major fatty acids are oleic acid, palmitoleic acid, and palmitic acid. Macadamia oil has the highest monounsaturated oil content (80%) among common edible oils, with low saturated fatty acid content (13%). A monounsaturated fatty acid-rich diet can lower the risk of coronary heart disease and prevent atherosclerosis. Palmitoleic acid improves hyperglycemia and hypertriglyceridemia by increasing insulin sensitivity and suppressing proinflammatory gene expressions. Macadamia oil's unique fatty acid composition makes it suitable for cosmetic applications, particularly skin-care products (Stolp & Kodali, 2022). Macadamia oil cake, used alone or mixed with soybean meal, is an effective protein source for cattle and fish feed without adverse effects. It may also provide nutrient benefits for humans with minimal antinutrients. The flour production process is simple, with less than 10% moisture and 13% lipid content. The remaining lipid content is comparable to full-fat soy flour, with about 20% lipid content. Removing lipids from the flour may improve its functional properties and alleviate rancid flavor development during storage (Tananuwong & Jitngarmkusol, 2019). Functional properties of flours, such as water and oil absorption capacities, emulsification/stabilization, and foam formation, are influenced by interactions between their chemical composition and other food components. Proteins and carbohydrates are key substances due to their hydrophilic and hydrophobic portions and charged functional groups. Variations in these macromolecules can contribute to different properties. Understanding these aspects is crucial for effectively using flours in food products (Jitngarmkusol, Hongsuwankul et al., 2008).

6. NUT HARVESTING AND STORAGE

Macadamia nuts, which fall from the tree when they are mature, are harvested either manually or by machines. To prevent mold and deterioration, the husk must be removed as soon as possible. The nuts are then dried to 10% moisture or less before being processed. They are further dried in silos to 10-15% moisture for longer storage. Drying is performed in stages to avoid adverse effects on kernel quality. After cracking and separating the shell from the kernel, the product can be lightly roasted and salted or packaged raw in bulk in vacuum-filled, foil-laminate bags. Traditionally, kernels have been roasted in coconut oil, but dry-roasting techniques are now more popular. The packaged product is then kept in cold storage to prolong shelf-life, allowing kernels to be safely stored for at least a year (R. A. Stephenson, 2003).

6.1 Factors Affecting Safety of Macadamia Nuts

The quality of macadamia nuts and their ability to maintain their original composition are determined by their original composition upon harvest, as well as the techniques used in their processing, packing, and storing. Nuts supplied to the processor vary greatly in terms of ripeness, moisture content, disease, and insect damage due to production and harvesting methods (Laura Graciela, Mario et al., 2022).

Additional environmental and horticultural elements that may impact oil accumulation during nut maturity include temperature, water stress, and nitrogen fertility (R. Stephenson, Gallagher et al., 2003). Macadamia nuts are dried to a moisture content of 3.5% before cracking, resulting in a brittle shell and a kernel moisture content of 1.5%. Rapid drying can cause brittleness, physical damage, and loss of flavor. Slow drying can cause mould, germination, and browning of the kernel's center. Brown centres are a major quality defect and cost the Australian industry around $2 million per annum. Incorrect drying regimes and poorly ventilated silos can contribute to brown centres. Temperature during drying is crucial for the dried product's quality. An incremental drying process of 2 days at 38°C, 45°C, and 60°C is recommended, a variation of a method developed by Prichavudhi and Yamamoto (1965) (H. M. Wallace & D. A. Walton, 2011).

(a) Pre- harvest factors

(i) Environmental factors

Climate factors that deviate from the present and historical growth circumstances found in the crop's natural environment, such as abrupt temperature fluctuations and variations in precipitation, might harm the crop. Therefore, only specific geographic and climatic ranges allow for the profitable cultivation of macadamias. The ideal daily and seasonal temperatures for macadamia are between 14 to 30 degrees Celsius; extended stays outside of these ranges have a negative impact on the plant's ability to develop, produce, and maintain quality (Barrueto, Merz et al., 2018; R. Stephenson & Gallagher, 1986b). In terms of precipitation, macadamia thrives and produces well in regions where rainfall is evenly distributed and averages 1500 mm annually (Du Preez, 2015). Macadamia nut output and quality are negatively impacted by water stress during nut maturity. Macadamias need up to four months of severe temperature contrasts and mild water stress to promote blooming and nut set (R. Stephenson & Gallagher, 1986a; R. Stephenson, Gallagher et al., 2003). This illustrates the extent to which climate affects macadamia production, with geographical factors like height, aspect, and slope only being thought to be significant when it comes to influencing temperature and water needs (Powell, 2009).

(ii) Harvest intervals

Macadamia nuts are harvested from the ground by hand or mechanically, with shake-harvesting being an alternative method. In Hawaii, 90% of the crop is harvested between mid-September and mid-December, with the peak in October. In Australia, harvesting usually extends from March to June. Early harvesting can improve kernel quality and yield, but early harvests may result in lower quality in size and roasted flavor. Harvest intervals for macadamia nuts vary from once a week in wet weather to once a month in dry weather. Factors such as labor availability, rats, insect and wild pig damage, mold development, and germination also affect harvest intervals. An interval of four weeks is a good starting point, with the main cause of quality loss being exposure to 15 sunlight. However, loss of eating quality due to time on the ground (0 to 5 months) is not clearly demonstrated (Tsumura, 1988).

(b) Post- harvest factors

(i) Dehusking

Postharvest handling of macadamias can cause physical damage to the kernels, including shoulder damage, breaking into pieces, bruising, dust formation, and excessive browning during roasting (D. A. Walton & Wallace, 2010). Fruit is squeezed by mechanical dehuskers to remove the fibrous husk; for example, a novel dehusker was assessed by (Luan & Liang, 1983). After-roasting darkening of kernels, a postharvest condition in which normal kernels turn abnormally black after roasting, is linked to cellular damage in macadamia (Albertson, De Giovanni et al., 2005).

(ii) Sorting

Macadamia nuts undergo visual sorting after dehusking to remove small, defective, or moldy nuts. However, there are no distinct differences in nut appearance between immature and mature nuts at harvest, making grading wet nuts difficult. Immature nuts may exhibit increased browning after roasting due to higher sugar contents. Some growers use water flotation to separate immature nuts from the post-harvest stream, but marginal nuts may be missed, resulting in brown or off-flavor kernels. Post-harvest handling of macadamia nuts can increase the potential for kernel damage, with shoulder damage, kernel pieces, and unsound kernels increasing when dropped more than twice. Dropping low moisture content nuts led to more shoulder damage, dusty kernels, bruised, oily kernels, and darkened or mottled kernels. Impact damage to macadamia nuts lowers the quality of raw and roasted kernels, causing commercial losses (M. Wall, 2013).

7. QUALITY COMPONENTS AND INDICES

Macadamia nuts are gaining recognition as a valuable source of nutrients and health benefits due to their high content of monounsaturated fatty acids. With a fat content of approximately 72%, macadamia nuts are rich in oleic acid and palmitoleic acid, which have been found to prevent beta-cell apoptosis and improve circulating lipid profiles in both animal and human subjects. Palmitoleic acid functions as an adipose tissue-derived lipid hormone that stimulates muscle insulin action and suppresses hepatosteatosis in mice deficient in fatty acid binding protein. Other bioactive constituents of macadamia nuts include phytosterols, primarily β-sitosterol, with small amounts of campstrol, stigmasterol, and squalene (Hernandez, 2016).

7.1. Macronutrient Content

The moisture percentage of most nuts when eaten is low because green nuts when picked, can contain 50% or more water. However, these nuts must be cured or semi dried for storage. From Table 1 macronutrients of macadamia are clear.

Table 2. Structure of chemical compounds from macadamia genus

Compounds	Name
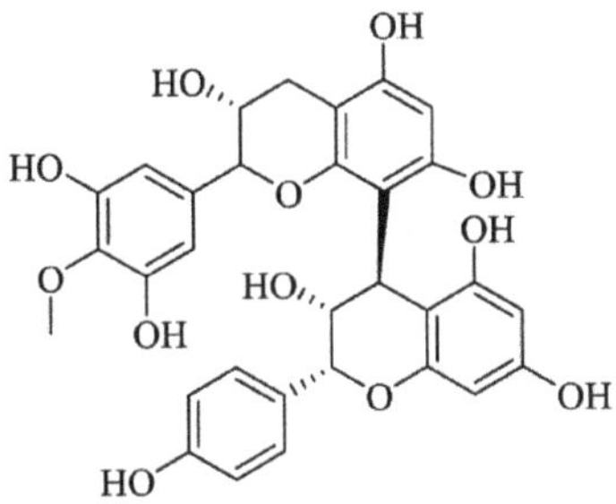	**Proanthocyanidin**
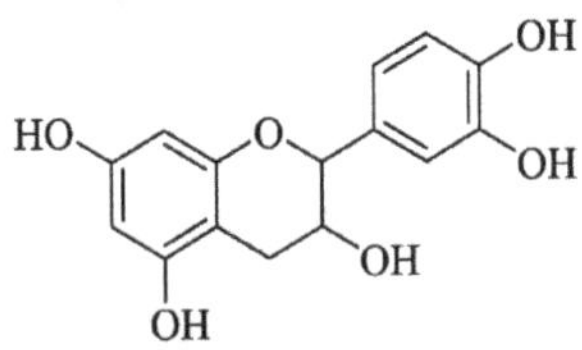	Catechin
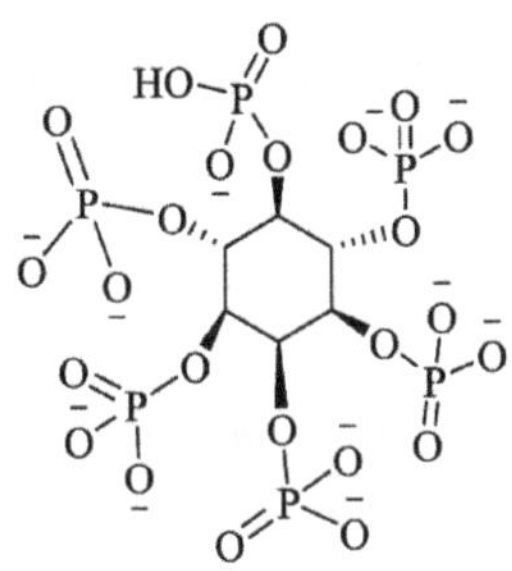	**Phytate**
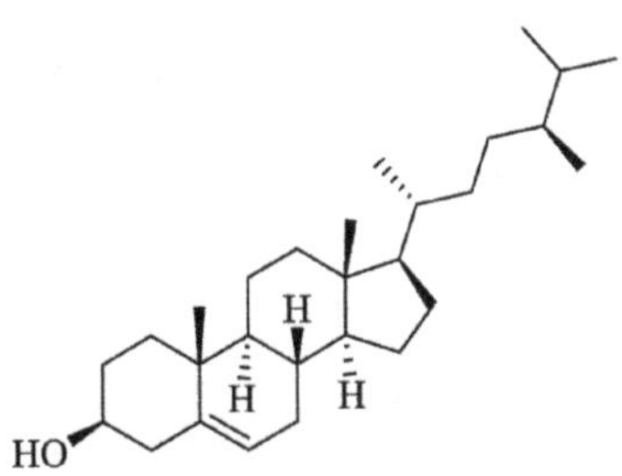	**Campesterol**
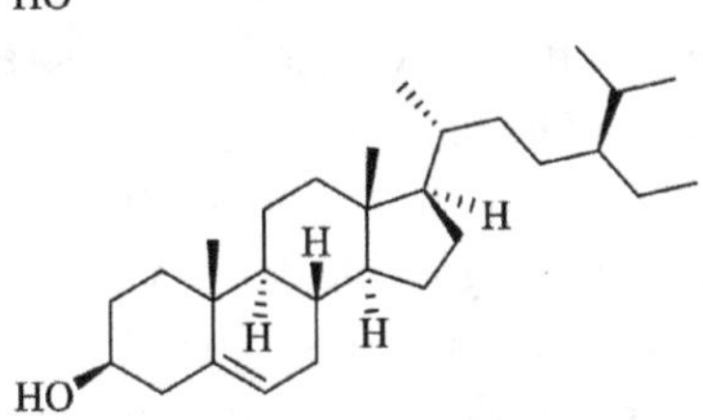	**β-sitosterol**
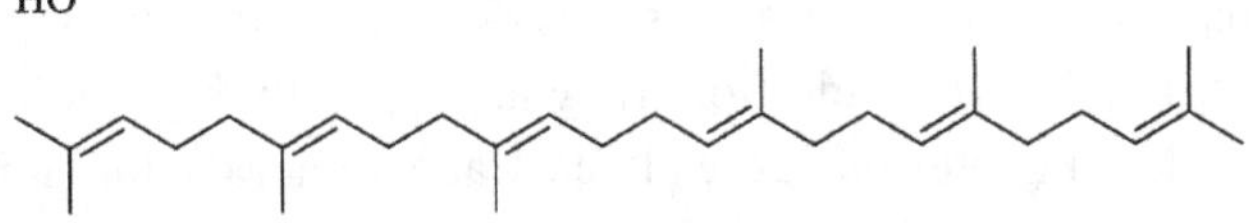	**Squalene**

continued on following page

Table 2. Continued

Structure	Compound
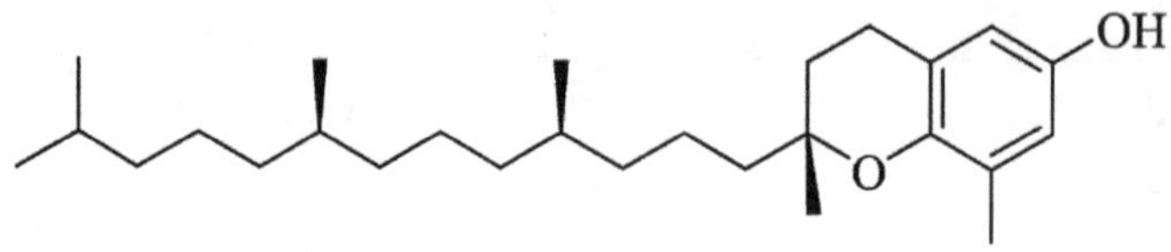	**Γ-tocopherol**
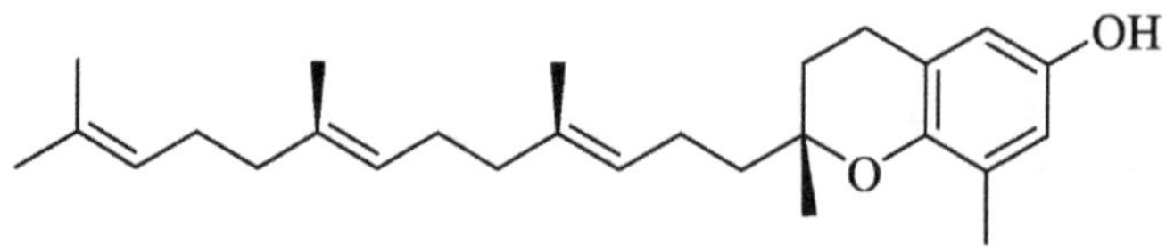	**Γ-tocotrienol**
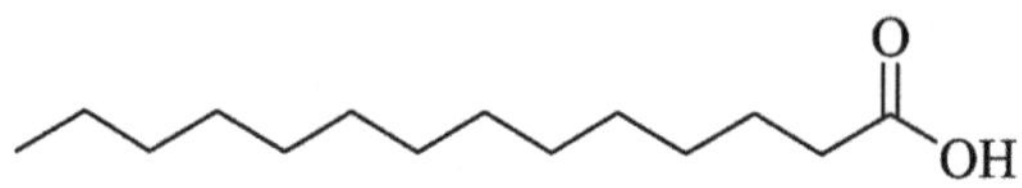	**Myristic acid**
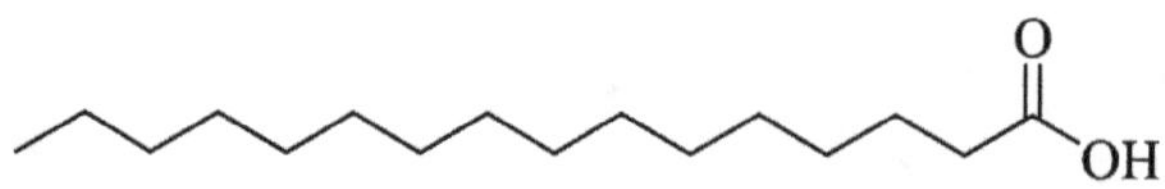	Palmitic acid
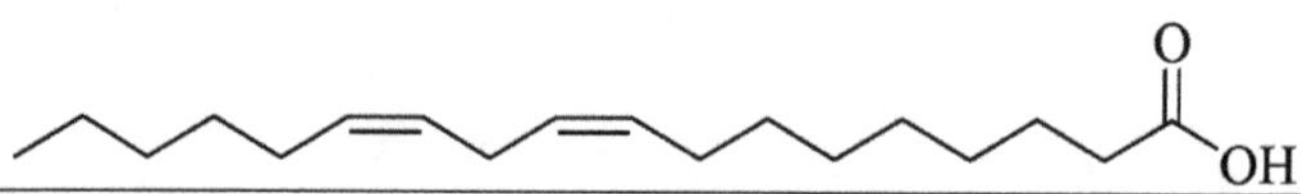	**Linoleic acid**

Table 3. Water, macronutrient, and energy content of macadamia nuts (per 100 g, kernel only) (Caballero, 2005)

Water (g)	Protein (g)	Fat (g)	Carbohydrate (g)	Energy (kcal)
1.3	7.9	77.6	4.8	748

7.2. Micro Nurients

Tocopherols, or vitamin E, and B vitamins, including folic acid, are generally found in nuts and seeds; however, some, like sunflower seeds, almonds, and hazelnuts, have significantly higher levels of vitamin E than others. Vitamin C is absent from nuts and seeds, while many nuts have negligible or no vitamin A activity. Nuts and seeds are rich sources of numerous minerals. Specifically, a variety of nuts and seeds—particularly sesame seeds—are excellent providers of calcium (Rizvi, Raza et al., 2014).In addition, they are typically high in iron, potassium, magnesium, and phosphorus, as well as trace minerals including chromium, manganese, copper, and zinc. Brazil nuts have an espeecially high selenium content.In macadamia nuts Vitamin E is present 1.49mg, Sodium(280 mg),Potassium (300 mg),Calcium (47 mg), Magnesium (100 mg), Phosphorus (200 mg), Iron (1.6 mg), Copper (0.43 mg), Zinc (1.1 mg) and Manganese (5.5 mg) per 100 g kernel only (Caballero, 2005).

7.3. Oil Content and Immaturity

Oil content in macadamia nuts is crucial for quality, with mature nuts having 72% or more oil content floating in water, while immature ones have less oil, leading to rejection due to early abscission caused by preharvest factors like pests and diseases (H. M. Wallace & D. Walton, 2011).

7.4. Quality Defects

Processors pay macadamia growers based on kernel quality, including defects like mould, germination, insect damage, and discoloration, which affect the percentage of nut-in-shell (Guthrie, Greensill et al., 2004).

Mould, a discolouration of fruiting bodies, is common in orchards, often found after prolonged storage or wet weather, and germination is identified by a green or discoloured kernel crest. Insect damage can cause a depression, brown or translucent lesion in the kernel, or a chewed appearance(Evans) (Sardella, Muscat et al., 2016).

7.5. Appearance

The return to the processor is also determined by the presence of macadamia kernels. Although processor types differ, premium models typically feature a full kernel devoid of defects, dust, discoloration, and shoulder damage (O'Hare, Bell et al., 2000). Shoulder damage refers to torn areas on the kernel's micropylar hemisphere caused by tissue adhering to the inner surface. Drying causes kernel shrinkage, leaving an air gap, which is desirable for minimal kernel damage during cracking (Chan Jr, 1983).

7.6. Rancidity and Shelf Life

Rancidity is a crucial quality parameter for nuts, referring to 'off' odours and flavors resulting from lipid oxidation or lipolysis, which breaks down oils into fatty acids. Oxidation, a process where oxygen attacks glycerides, can be initiated by heat, pro-oxidants, certain enzymes, or light (Robards, Kerr et al., 1988).

7.7. The Importance of Antioxidants

Antioxidants are substances that protect living tissues from damage by reactive oxygen species (ROS). However, their action is temporary as they delay oxidation without preventing it entirely. Antioxidants that delay oxidative rancidity will not affect FFA formation due to chemical hydrolysis. Free radicals and ROS are key chemical species that contribute to food quality changes and disease development. Oxidation of unsaturated lipids is a major cause of food quality deterioration (Wanasundara, Shahidi et al., 1997). The factors determining the oxidative stability of macadamia kernels are largely unknown, and several antioxidant-property phenolic compounds have been identified at low concentrations in macadamia oil, but their possible synergistic effects or the effects of other unidentified compounds are not known (Quinn & Tang, 1996).

8. NUTRIENT COMPOSITION OF MACADAMIA NUTS

The composition and amount of oil in macadamia nuts determine their quality. For best eating and processing quality, mature nuts have a specific gravity (SG < 1.0) of at least 72% oil. Although they are categorized as second-grade, kernels with an SG of 1.0–1.025 can be used to lower-grade goods. Commercially, third-grade kernels (SG > 1.025) are unsatisfactory. Oil does not start to accumulate until the nuts reach full size and the shell solidifies. When the reducing sugar concentration lowers in the late summer, it builds up quickly in the kernel(R. A. Stephenson, 2003). Nutrient-rich macadamia nuts may be found in abundance. This nut has a high energy content (737±18 kcal/100g) due to its high lipid content (73.5±1.4 g/100g). Monounsaturated fatty acids (MUFA) were found in high concentration (49.4-58.7 mg/100g), mostly oleic (34.5-47.0 mg/100g) and palmitoleic (7.1-12.8 mg/100g). When compared to nutritional criteria, the content of the minerals Mg, Zn, Cu, and Mn is good. The levels of total phenolic compounds (TPC) (77.9-96.3 mg galic acid equivalent/100g), α-tocopherol (0.2-18.4 mg/100g), and total antioxidant capacity (TAC) (21.0-36.8 μM Trolox equivalent/g) are associated with its antioxidant activities. The cultivar HAES 344 had the highest TAC level and TPC content (Mereles Ceuppens, Ferro et al., 2017).

8.1 Phytochemical Compounds of Macadamia Genus

Macadamia oil, which is generated from Macadamia integrifolia and Macadamia tetraphylla, has a 75% monounsaturated fatty acid content, with oleic acid having the greatest concentration and followed by palmitoleic and eicosenoic acids.

Macadamia oil also contains 10% and 15% of polyunsaturated and saturated fatty acids, respectively. Macadamia nuts have been shown to have high quantities of palmitoleic acid (15–22%) (Aquino-Bolaños, Mapel-Velazco et al., 2017).

According to Hu et al. (2019), palmitoleic acid has an impact on health and is linked to several disorders, including cancer, diabetes, cardiovascular disease, insulin resistance, non-alcoholic fatty liver disease (NAFLD), obesity, non-alcoholic steatohepatitis (NASH), and diabetes. According to earlier research, palmitoleic acid may reduce inflammation, diabetes, and cardiovascular disorders (Hu, Fitzgerald et al., 2019).

Roasting enhanced the amount of chemical components in macadamia nuts, including squalene, sterol, tocopherol, and total phenolic content. Meanwhile, roasting reduced the levels of thiamine and tocopherol. The oxidative stability index of the kernels increased by 21.6%, while the total polyphenol content increased by 25.6% (Tu, Wu et al., 2021).

9. FOOD APPLICATIONS

Macadamia nuts are a gourmet nut, valued for their use in snack nuts, candies, confections, nut butters, and culinary oils. They can be salted, dry roasted, chocolate covered, honey-roasted, toffee coated, or coffee glazed. Whole kernels are enjoyed in various ways, while kernel pieces are added to ice cream, nut brittles, cookies, cakes, and pies. Macadamia oil may be useful in skincare products (M. M. Wall, 2013). Macadamia nuts are used in roasted snacks, blanched nuts, and confectionery products like cakes and cookies. They come in whole, chopped, and sliced forms. Roasting increases nut flavor, reduces moisture, and hardens texture. Thermal treatment increases oil migration, and if used in chocolate bars, softening may

Table 4. Chemical components found in the macadamia genus nuts

Macadamia Species	Compounds	References
M. integrifolia	Proanthocyanidins, minerals (Na, B,Al, Ca, Cu, K, Mg, Fe,Mn, Na,S, Zn P,), phenolic acids, lipids (myristic, palmitic, oleic, linoleic, palmitoleic,stearic, elaidic, eicosenoic, arachidic),phytates,	(Alasalvar & Bolling, 2015; Richards, Kämper et al., 2020)
Macadamia sp.	Proanthocyanidins, phytosterols, flavonoids, amino acids (glycine, valine, leucine, isoleucine, cysteine, methionine, proline, phenylalanine, serine, tyrosine, aspartic acid, threonine, glutamic acid, lysine, histidine, arginine), lipids (myristic, palmitic, lauric, stearic, arachidic, palmitoleic, oleic, behenic, α-linolenic, gadoleic, linoleic,), proteins (, legumins, vicilins, serotonin and 2S albumins), polyphenols, γ-tocopherol, α-tocotrienol, epicatechin, glucose, fructose	(Lara, Vilcacundo et al., 2017; Yılmaz, Taş et al., 2019)

occur. Whole, chopped, and sliced nuts are used to meet various requirements (Bhattacharya, 2023).

10. PHARMACOLOGICAL ACTIVITIES OF MACADAMIA NUTS

Owing to their high nutrient content, macadamia nuts can play a significant role in a balanced diet that helps to avoid and cure disease. Despite having a high fat and energy content, macadamia nuts are largely high in monounsaturated fat, which has been linked to several health advantages. Other significant nutritional components found in macadamia nuts include proteins, dietary fibre, vitamins, minerals, and phytochemicals. There needs to be more information available on macadamia nut health advantages.

10.1. Antioxidant Activity

Apart from their nutritional attributes, macadamia nuts have been claimed by certain writers to possess antioxidant qualities due to the presence of tocopherols, tocotrienols, and squalene. Antioxidants are a class of molecules known for their capacity to scavenge free radicals; as a result, they are regarded as a substitute for fighting deficits brought on by oxidative stress, including aging and rheumatic and cardiovascular disorders (Rengel, Pérez et al., 2015). Using the DPPH technique, macadamia nut peptides of varying molecular weights were assessed for their antioxidant activity. The results indicated 80.97% DPPH free radical scavenging activity. Its antioxidant activity increased together with the quantity of the protein-peptide component (Insanu, Hartati et al., 2021).

10.2. Antimicrobial Activity

The antibacterial action of M. integrifolia's seeds, leaves, and flowers has been studied in certain studies. MiAMP1 (β-barrelins), an antimicrobial peptide, is present in M. integrifolia seeds. With a dosage range of 0.2 - 2 μM, the MiAMP1 could provide protection against a variety of phytopathogenic fungi and Gram-positive bacteria (dos Santos-Silva, Zupin et al., 2020).

Nuts and leaves of M. integrifolia that were extracted using methanol and water had antibacterial qualities against P. mirabilis. For nuts and leaves, the methanol extracts' minimal inhibitory concentrations (MICs) were 15 μg/ml and 2790 μg/ml, respectively. Additionally, the MIC values of water extracts for nuts and leaves were 558 μg/ml and 4378.2 μg/ml, respectively (Cock, Winnett et al., 2015).

10.3. Anti-Dyslipidemia Activity

It has been demonstrated that consuming a lot of macadamia nuts lowers LDL and total cholesterol in hypercholesterolemic men and women. Macadamia nuts have been found to lower total and LDL cholesterol in individuals with hypercholesterolemia. Studies have shown that a diet high in macadamia nuts can affect lipid profiles, with plasma monounsaturated fatty acid levels being elevated. This leads to a decrease in plasma cholesterol and LDL cholesterol levels, while HDL cholesterol levels increase. These results suggest that the consumption of macadamia nuts, which result in a higher intake of monounsaturated fats, may reduce the proportion of saturated fats ingested from the diet. This suggests that diets high in monounsaturated fatty acids may be more strongly associated with lowering cholesterol and reducing CVD risk than low-fat diets (Yamagata, 2022).

Numerous studies of the literature have shown how macadamia nuts can lower LDL cholesterol, total cholesterol, and triglycerides. In 5–10 weeks, ingestion of macadamia nuts by overweight and hypercholesterolemic individuals helped lower triglycerides (Mejia, Kendall et al., 2014). For five weeks, a macadamia nut-enriched diet decreased both total and LDL cholesterol in 14 hypercholesterolemic patients (Kim, Keogh et al., 2017).The effectiveness of macadamia nuts in decreasing cholesterol has been shown by recent well-controlled intervention trials, which included LDL-cholesterol reductions ranging from 4% to 11% when compared to comparator diets (Ros, 2010). Macadamia nuts have been shown to have beneficial effects on circulating lipid levels in animal feeding studies and six human clinical trials. A study in hamsters found that macadamia oil-fed hamsters had lower non-HDLC and triglyceride concentrations compared to palm and coconut oil-fed hamsters. HDLC levels were higher in macadamia-oil-fed hamsters compared to coconut, canola, and safflower oil-fed hamsters. Another study in hyperlipidemic rats found that macadamia nuts led to significantly lower levels of total cholesterol and triglyceride, and higher levels of HDLC levels compared to the control group (Wood & Garg, 2011).

10.4. Anti-Inflammatory Activity

Chromatographic analysis was used in cell culture experiments with murine macrophages (RAW264.7) to investigate the ability of macadamia oily nut extract to modify inflammatory processes. The findings demonstrated that the lipopolysaccharide (LPS)-induced expression of iNos, Tnfα, Il1β, and Il6 mRNAs of RAW264.7 macrophages was considerably reduced by macadamia oily nut extract up to 71%, 29%, 57%, and 27%, respectively (Schlörmann, Birringer et al., 2016).

Furthermore, consuming macadamia nuts for four weeks (corresponding to 15% calorie intake) reduced the concentration of leukotriene B4 LTB4, a plasma marker of inflammation, in plasma by 22.5% (Garg, Blake et al., 2007).

10.5. Dietary Controlled

According to earlier studies, rats who consumed both macadamia oil and high fat had a lower body mass index (5.8 kg/m2) than rats that just consumed high fat (6.2 kg/m2) (Poudyal, Kumar et al., 2013). According to Kim et al. (2017), giving 101 overweight subjects a nut-enriched diet for 18 months could reduce their body weight, BMI, and waist circumference by 1.6 kg/m2, 6.9 cm, and 4.1 kg for the moderate-fat group and 1.4 kg/m2, 2.6 cm, and 2.9 kg for the low-fat group (Kim, Keogh et al., 2017).

Macadamia oil dietary supplement capsules included significant levels of monounsaturated fatty acids (80%) and polyunsaturated fatty acids (2.8–4.7%), among other health-maintaining and nutritious qualities (Monika & Anna, 2019).

10.6. Anti-Tyrosinase Activity

The anti-tyrosinase activity, fractions, and isolated chemical from macadamia were assessed by El Hawary et al. (2020). The tyrosinase inhibitory activity of the extract was found to be IC50 of 85 mg/ml, whereas the n-butanol fraction was 75 mg/ml and the ethyl acetate fraction was 60 mg/ml. In the meantime, gallic acid (IC50 = 56 mg/ml) shown potent anti-tyrosinase activity (Monika & Anna, 2019).

10.7. Chemopreventive Activity

One of the primary short-chain fatty acids, butyrate acid, was created by macadamia nut fermentation in amounts of 19.59 mmol/l. On the other hand, colon cancer cells exposed to this histone deacetylase inhibitor showed inhibition of proliferation, apoptosis, and differentiation (Schlörmann, Birringer et al., 2016). With EC50 values of 86.9 mg/ml, the macadamia nut extract demonstrated mild antiproliferative potential to prevent the proliferation of Caco-2 colon cancer cells (Yang, Liu et al., 2009).

10.8. Non-Alcoholic Fatty Liver Disease (NAFLD) Prevention Activity

Hepatic steatosis is associated with fructose/sucrose-induced hepatic lipogenesis, which is linked to non-alcoholic fatty liver disease (NAFLD). Macadamia nut oils' high amount of unsaturated fatty acids has been shown by Siddiqui et al. (2015) to be able to prevent the development of sucrose/fructose-induced hepatic steatosis in C57BL/6 mice (Siddiqui, Xu et al., 2015).

10.9. Anti Cardiovascular Disease

Monounsaturated fatty acids (MUFA) are abundant in macadamia nuts, with oleic (about 60%) and palmitoleic (20%) acids making up the majority. There is no other dietary source that compares to macadamia nuts in terms of MUFA content. Diets heavy in MUFA-rich foods lower plasma LDL cholesterol levels and lower the risk of cardiovascular disease. Together with other bioactive ingredients like tocopherols, phytosterols, and squalene, the high proportion of unsaturated fatty acids may be responsible for this cholesterol-lowering and coronary-protective impact (Wall, 2010).

11. ADVERSE EFFECTS AND REACTIONS (ALLERGIES AND TOXICITY)

There have been reports of macadamia nut allergies. Skin signs, respiratory symptoms, and gastrointestinal hypersensitivity are examples of such adverse food reactions; nonetheless, none of these symptoms were severe. Nonetheless, there have been a few reported occurrences of anaphylaxis, which can be lethal (Knott, Gürer et al., 2008).

Figure 3. Summary of physiological changes that can be induced with the consumption of macadamia nut

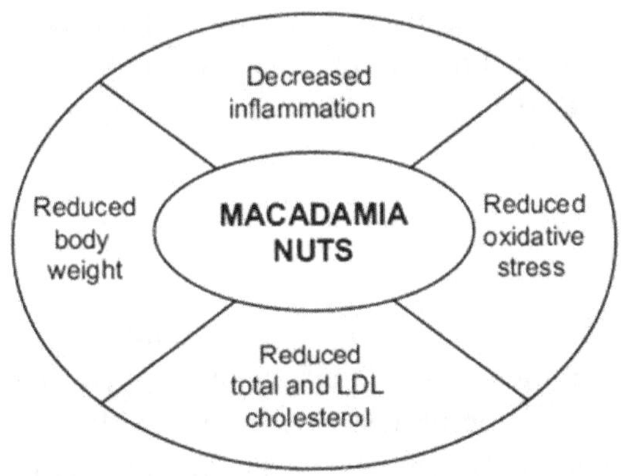

12. TOXINS AND CONTAMINANTS

Macadamia is becoming more and more important commercially in the food, cosmetics, and pharmaceutical sectors. It is thought that macadamia's bioactive components improve blood lipid profiles, reduce inflammation and oxidative stress, and lower the risk factors for cardiovascular disease (Aquino-Bolaños, Mapel-Velazco et al., 2017; Preedy & Watson, 2020). On the other hand, certain species in the Macadamia genus have chemicals called cyanogenic glycosides, which are secondary metabolites that cause cyanogenesis, the production of hydrogen cyanide (Bolarinwa, Orfila et al., 2014). There have been reports of the acute and long-term toxicity of hydrogen cyanide in food obtained from plants (Onojah & Odin, 2015). Macadamia nuts, originating from Australia, are a popular snack food with 75% fat by weight, with 80% being monounsaturated. Consuming a high-monounsaturated-fat diet can lower cholesterol and low-density lipoprotein levels. The majority of oil-free components are proteins (27%-30%) and sugars (6%-8%). Macadamia nuts are also used in candies, cakes, and cookies. However, potential toxicants like chocolate, xylitol, and grapes/raisins may need to be considered. Macadamia nut toxicosis has only been reported in dogs (Gwaltney-Brant, 2013). The amounts of cyanogenic glycosides, such proteacin and dhurrin, differ amongst Macadamia plant species (Dahler, Mcconchie et al., 1995; Miller & Tuck, 2013). Moreover, the amounts of these substances vary at different stages of plant growth, from seed germination to plant maturity, and they are unevenly distributed inside the various plant sections (such as nuts, seeds, and roots) (Jones, 1972; Kriedeman, 1964). Nuts and seeds can be contaminated by mould growth during storage, producing toxic secondary metabolites called mycotoxins, including aflatoxin B1. These mycotoxins, produced by three species of mould, Aspergillus flavus, A. parasiticus, and A. nomius, can contaminate various food commodities, including tree nuts and peanuts. Aflatoxins are toxic to humans and animals, and may be involved in human liver cancer. Some mould species can proliferate within growing crops, forming an endophytic relationship with the plant. Regulations limit aflatoxin levels in foods, but their significance in public health is small (Caballero, 2005).

13. CONCLUSION

The Macadamia genus had various phytochemical compounds and pharmacological activities. Macadamia genus's phytochemical compounds had few pharmacological activities such as antioxidant, anti-inflammatory, anti-dyslipidemia, and antimicrobial. Further research is required to explore the other parts of Macadamia genus plants' potential chemical constituents and pharmacological activities. Macadamia nuts are a valuable source of nutrients, in particular monounsaturated fatty acids. Modern cultivation practices have resulted in a reliable and plentiful supply of macadamianuts, from countries such as the United States of America and Australia. Macadamia nuts have been shown to improve various risk factors for cardiovascular disease, including hypercholesterolemia, body weight, oxidative stress, and inflammation. Macadamia nuts may be useful when consumed alone, or in combination with traditiona pharmacotherapy to reduce risk of cardiovascular disease. Long-term health benefits of macadamia nut consumption warrant further investigation.

REFERENCES

Alam, M., Hardner, C., Nock, C., O'Connor, K., & Topp, B. (2019). Historical and molecular evidence of genetic identity of macadamia cultivars HAES741 and HAES660. *HortScience*, *54*(4), 616–620. doi:10.21273/HORTSCI13318-18

Alasalvar, C., & Bolling, B. W. (2015). Review of nut phytochemicals, fat-soluble bioactives, antioxidant components and health effects. *British Journal of Nutrition*, *113*(S2), S68–S78. doi:10.1017/S0007114514003729 PMID:26148924

Albertson, P., De Giovanni, C., Cocksedge, R., Forrester, R., Rae, A., Mason, R., & McConchie, C. (2005). Inducing biochemical changes to simulate after-roast darkening in macadamia kernel. *Australian Journal of Experimental Agriculture*, *45*(10), 1315–1323. doi:10.1071/EA04176

Aquino-Bolaños, E. N., Mapel-Velazco, L., Martín-del-Campo, S. T., Chávez-Servia, J. L., Martínez, A. J., & Verdalet-Guzmán, I. (2017). Fatty acids profile of oil from nine varieties of Macadamia nut. *International Journal of Food Properties*, *20*(6), 1262–1269. doi:10.1080/10942912.2016.1206125

Bailey, H. M., & Stein, H. H. (2020). Raw and roasted pistachio nuts (Pistacia vera L.) are 'good'sources of protein based on their digestible indispensable amino acid score as determined in pigs. *Journal of the Science of Food and Agriculture*, *100*(10), 3878–3885. doi:10.1002/jsfa.10429 PMID:32323331

Ballhorn, D. J. (2011). Cyanogenic Glycosides in Nuts and Seeds. In V. R. Preedy, R. R. Watson, & V. B. Patel (Eds.), *Nuts and Seeds in Health and Disease Prevention* (pp. 129–136). Academic Press. doi:10.1016/B978-0-12-375688-6.10014-3

Barrueto, A. K., Merz, J., Hodel, E., & Eckert, S. (2018). The suitability of Macadamia and Juglans for cultivation in Nepal: An assessment based on spatial probability modelling using climate scenarios and in situ data. *Regional Environmental Change*, *18*(3), 859–871. doi:10.1007/s10113-017-1225-2

Bhattacharya, S. (2023). Edible nuts. In S. Bhattacharya (Ed.), *Snack Foods* (pp. 173–192). Academic Press. doi:10.1016/B978-0-12-819759-2.00006-9

Bolarinwa, I. F., Orfila, C., & Morgan, M. R. (2014). Amygdalin content of seeds, kernels and food products commercially-available in the UK. *Food Chemistry*, *152*, 133–139. doi:10.1016/j.foodchem.2013.11.002 PMID:24444917

Caballero, B. (2005). *Encyclopedia of human nutrition*. Elsevier.

Chan, H. T. Jr. (1983). *Handbook of tropical foods*.

Cock, I., Winnett, V., Sirdaarta, J., & Matthews, B. (2015). The potential of selected Australian medicinal plants with anti-Proteus activity for the treatment and prevention of rheumatoid arthritis. *Pharmacognosy Magazine*, *11*(42, Suppl 1), S190. doi:10.4103/0973-1296.157734 PMID:26109767

Costello, G., Gregory, M., & Donaitu, P. (2008). *Southern Macadamia Species Recover, Plan 20082012. Report to Department of the Environment, W., Heritage and the, & Arts, C. b.* H. A. L.

Dahler, J., Mcconchie, C., & Turnbull, C. (1995). Quantification of cyanogenic glycosides in seedlings of three Macadamia (Proteaceae) species. *Australian Journal of Botany*, *43*(6), 619–628. doi:10.1071/BT9950619

dos Santos-Silva, C. A., Zupin, L., Oliveira-Lima, M., Vilela, L. M. B., Bezerra-Neto, J. P., Ferreira-Neto, J. R., ... Aburjaile, F. F. (2020). Plant antimicrobial peptides: State of the art, in silico prediction and perspectives in the omics era. *Bioinformatics and Biology Insights*, *14*, 14. doi:10.1177/1177932220952739 PMID:32952397

Du Preez, A. B. (2015). *Studies on macadamia nut quality*. Stellenbosch University.

Evans, R. H. P. J. (2005). Product description manual: Australian macadamia kernel. Australian Macadamia Society

Garg, M. L., Blake, R. J., Wills, R. B., & Clayton, E. H. (2007). Macadamia nut consumption modulates favourably risk factors for coronary artery disease in hypercholesterolemic subjects. *Lipids*, *42*(6), 583–587. doi:10.1007/s11745-007-3042-8 PMID:17437143

Gray, J. (2013). Nuts and seeds. In B. Caballero (Ed.), *Encyclopedia of Human Nutrition* (3rd ed., pp. 329–335). Academic Press. doi:10.1016/B978-0-12-375083-9.00208-7

Guthrie, J., Greensill, C., Bowden, R., & Walsh, K. (2004). Assessment of quality defects in macadamia kernels using NIR spectroscopy. *Australian Journal of Agricultural Research*, *55*(4), 471–476. doi:10.1071/AR03179

Gwaltney-Brant, S. M. (2013). Macadamia Nuts. In M. E. Peterson & P. A. Talcott (Eds.), *Small Animal Toxicology* (3rd ed., pp. 625–627). W.B. Saunders. doi:10.1016/B978-1-4557-0717-1.00056-9

Hamilton, R. A., & Fukunaga, E. T. (1959). *Growing macadamia nuts in Hawaii.* Academic Press.

Hernandez, E. M. (2016). 4 - Specialty Oils: Functional and Nutraceutical Properties. In T. A. B. Sanders (Ed.), *Functional Dietary Lipids* (pp. 69–101). Woodhead Publishing. doi:10.1016/B978-1-78242-247-1.00004-1

Hokmabadi, H., & Sedaghati, E. (2014). Safety of Food and Beverages: Nuts. In Y. Motarjemi (Ed.), *Encyclopedia of Food Safety* (pp. 340–348). Academic Press. doi:10.1016/B978-0-12-378612-8.00292-4

Hong, M. Y., Groven, S., Marx, A., Rasmussen, C., & Beidler, J. (2018). Anti-inflammatory, antioxidant, and hypolipidemic effects of mixed nuts in atherogenic diet-fed rats. *Molecules (Basel, Switzerland)*, *23*(12), 3126. doi:10.3390/molecules23123126 PMID:30501043

Hu, W., Fitzgerald, M., Topp, B., Alam, M., & O'Hare, T. J. (2019). A review of biological functions, health benefits, and possible de novo biosynthetic pathway of palmitoleic acid in macadamia nuts. *Journal of Functional Foods*, *62*, 103520. doi:10.1016/j.jff.2019.103520

INC. (2020). *Nuts and dried fruits statistical yearbook 2019/2020.* International Nut and Dried Fruit Council Spain.

Insanu, M., Hartati, R., Bajri, F., & Fidrianny, I. (2021). Macadamia genus: An updated review of phytochemical compounds and pharmacological activities. *Biointerface Research in Applied Chemistry*, *11*, 14898–14905. doi:10.33263/BRIAC116.1489814905

Jitngarmkusol, S., Hongsuwankul, J., & Tananuwong, K. (2008). Chemical compositions, functional properties, and microstructure of defatted macadamia flours. *Food Chemistry*, *110*(1), 23–30. doi:10.1016/j.foodchem.2008.01.050 PMID:26050161

Jones, D. (1972). Phytochemical ecology. *Cyanogenic glycosides and their function*, 103-124.

K C., M. (2021). *Material uses of nut and other fruit shells.* Retrieved from https://nbn-resolving.org/urn:nbn:de:hbz:1383-opus4-8755

Kim, Y., Keogh, J. B., & Clifton, P. M. (2017). Benefits of nut consumption on insulin resistance and cardiovascular risk factors: Multiple potential mechanisms of actions. *Nutrients*, *9*(11), 1271. doi:10.3390/nu9111271 PMID:29165404

Knott, E., Gürer, C. K., Ellwanger, J., Ring, J., & Darsow, U. (2008). Macadamia nut allergy. *Journal of the European Academy of Dermatology and Venereology*, *22*(11), 1394–1395. doi:10.1111/j.1468-3083.2008.02657.x PMID:18331311

Koaze, H., Karanja, P. N., Kojima, M., Baba, N., & Ishibashi, K.KOAZE. (2002). Lipid accumulation of macadamia nuts during kernel development. *Nihon Shokuhin Hozo Kagakkaishi*, *28*(2), 67–73. doi:10.5891/jafps.28.67

Kriedeman, P. (1964). Cyanide formation in Sorghum almum in relation to nitrogen and phosphorus nutrition. *Australian Journal of Experimental Agriculture*, *4*(12), 15–16. doi:10.1071/EA9640015

Lamas, D. L., & Álvarez, S. T. (2023). Development of snacks based on Macadamia integrifolia nuts enriched with omega-3. *Food Chemistry Advances*, *3*, 100564. doi:10.1016/j.focha.2023.100564

Lara, D., Vilcacundo, E., Carrillo, C., Carpio, C., Silva, M., Alvarez, M., & Carrillo, W. (2017). Obtention of protein concentrate and polyphenols from macadamia (Macadamia integrifolia) with aqueous extraction method. *Asian Journal of Pharmaceutical and Clinical Research*, *10*(2), 138–142. doi:10.22159/ajpcr.2017.v10i2.14808

Laura Graciela, M., Mario, S., Karen Patricia, M., Eva Eugenia Soledad, C., Edelira, V., & Laura, C. (2022). Challenges and Advances in the Production of Export-Quality Macadamia and Its Integral Use with Green Technologies. In K. Muhammad Sarwar (Ed.), Tropical Plant Species and Technological Interventions for Improvement. Rijeka: IntechOpen.

Luan, J.-M., & Liang, T. (1983). An improved macadamia nut husker. *Journal of Agricultural Engineering Research, 28*(4), 337–347. doi:10.1016/0021-8634(83)90065-3

Mejia, S. B., Kendall, C. W., Viguiliouk, E., Augustin, L. S., Ha, V., Cozma, A. I., ... Leiter, L. A. (2014). Effect of tree nuts on metabolic syndrome criteria: A systematic review and meta-analysis of randomised controlled trials. *BMJ Open, 4*(7), e004660. doi:10.1136/bmjopen-2013-004660 PMID:25074070

Mereles Ceuppens, L. G., Ferro, E. A., Alvarenga, N. L., Caballero, S. B., Wiszovaty, L., Piris Jara, P. A., & Michajluk, B. J. (2017). Chemical composition of Macadamia integrifolia (Maiden and Betche) nuts from Paraguay. *International Food Research Journal, 24*(6).

Miller, R. E., & Tuck, K. L. (2013). The rare cyanogen proteacin, and dhurrin, from foliage of Polyscias australiana, a tropical Araliaceae. *Phytochemistry, 93*, 210–215. doi:10.1016/j.phytochem.2013.03.004 PMID:23566716

Monika, M., & Anna, K.-D. (2019). Nut oils and their dietetic and cosmetic significance: A review. *Journal of Oleo Science, 68*(2), 111–120. doi:10.5650/jos.ess18216 PMID:30651417

Moreno-Pérez, A. J., Sánchez-García, A., Salas, J. J., Garcés, R., & Martínez-Force, E. (2011). Acyl-ACP thioesterases from macadamia (Macadamia tetraphylla) nuts: Cloning, characterization and their impact on oil composition. *Plant Physiology and Biochemistry, 49*(1), 82–87. doi:10.1016/j.plaphy.2010.10.002 PMID:21071236

O'Hare, P., Bell, D., & Salmon, T. (2000). *Australian macadamia industry code of sound orchard practices*. Australian Macadamia Society.

Onojah, P., & Odin, E. (2015). Cyanogenic glycoside in food plants. *Int. J. Innov. Sci. Math, 3*, 2347–9051.

Peace, C. P., Vithanage, V., Turnbull, C. G., & Carroll, B. J. (2003). A genetic map of macadamia based on randomly amplified DNA fingerprinting (RAF) markers. *Euphytica, 134*(1), 17–26. doi:10.1023/A:1026190529568

Penha, C. B., Santos, V. D. P., Speranza, P., & Kurozawa, L. E. (2021). Plant-based beverages: Ecofriendly technologies in the production process. *Innovative Food Science & Emerging Technologies, 72*, 102760. doi:10.1016/j.ifset.2021.102760

Poudyal, H., Kumar, S. A., Iyer, A., Waanders, J., Ward, L. C., & Brown, L. (2013). Responses to oleic, linoleic and α-linolenic acids in high-carbohydrate, high-fat diet-induced metabolic syndrome in rats. *The Journal of Nutritional Biochemistry, 24*(7), 1381–1392. doi:10.1016/j.jnutbio.2012.11.006 PMID:23333092

Powell, M. A. (2009). *Predicting the geographic distributions of Macadamia integrifolia and Macadamia ternifolia*. University of the Sunshine Coast.

Preedy, V. R., & Watson, R. R. (2020). *Nuts and seeds in health and disease prevention*. Academic press.

Quinn, L., & Tang, H. (1996). Antioxidant properties of phenolic compounds in macadamia nuts. *Journal of the American Oil Chemists' Society, 73*(11), 1585–1588. doi:10.1007/BF02523529

Rengel, A., Pérez, E., Piombo, G., Ricci, J., Servent, A., Tapia, M. S., . . . Montet, D. (2015). *Lipid profile and antioxidant activity of macadamia nuts (Macadamia integrifolia) cultivated in Venezuela*. Academic Press.

Richards, T. E., Kämper, W., Trueman, S. J., Wallace, H. M., Ogbourne, S. M., Brooks, P. R., Nichols, J., & Hosseini Bai, S. (2020). Relationships between nut size, kernel quality, nutritional composition and levels of outcrossing in three macadamia cultivars. *Plants, 9*(2), 228. doi:10.3390/plants9020228 PMID:32053871

Rizvi, S., Raza, S. T., Ahmed, F., Ahmad, A., Abbas, S., & Mahdi, F. (2014). The role of vitamin e in human health and some diseases. *Sultan Qaboos University Medical Journal, 14*(2), e157–e165. PMID:24790736

Robards, K., Kerr, A. F., & Patsalides, E. (1988). Rancidity and its measurement in edible oils and snack foods. A review. *Analyst, 113*(2), 213–224. doi:10.1039/an9881300213 PMID:3288002

Ros, E. (2010). Health benefits of nut consumption. *Nutrients, 2*(7), 652–682. doi:10.3390/nu2070652 PMID:22254047

Sardella, D., Muscat, A., Brincat, J.-P., Gatt, R., Decelis, S., & Valdramidis, V. (2016). A Comprehensive Review of the Pear Fungal Diseases. *International Journal of Fruit Science, 16*(4), 351–377. doi:10.1080/15538362.2016.1178621

Schlörmann, W., Birringer, M., Lochner, A., Lorkowski, S., Richter, I., Rohrer, C., & Glei, M. (2016). In vitro fermentation of nuts results in the formation of butyrate and c 9, t 11 conjugated linoleic acid as chemopreventive metabolites. *European Journal of Nutrition, 55*(6), 2063–2073. doi:10.1007/s00394-015-1020-0 PMID:26286349

Siddiqui, R. A., Xu, Z., Harvey, K. A., Pavlina, T. M., Becker, M. J., & Zaloga, G. P. (2015). Comparative study of the modulation of fructose/sucrose-induced hepatic steatosis by mixed lipid formulations varying in unsaturated fatty acid content. *Nutrition & Metabolism*, *12*(1), 1–14. doi:10.1186/s12986-015-0038-x PMID:26583036

Stephenson, R., & Gallagher, E. (1986a). Effects of night temperature on floral initiation and raceme development in macadamia. *Scientia Horticulturae*, *30*(3), 213–218. doi:10.1016/0304-4238(86)90099-3

Stephenson, R., & Gallagher, E. (1986b). Effects of temperature during latter stages of nut development on growth and quality of macadamia nuts. *Scientia Horticulturae*, *30*(3), 219–225. doi:10.1016/0304-4238(86)90100-7

Stephenson, R., Gallagher, E., & Doogan, V. (2003). Macadamia responses to mild water stress at different phenological stages. *Australian Journal of Agricultural Research*, *54*(1), 67–75. doi:10.1071/AR02108

Stephenson, R. A. (2003). NUTS. In B. Caballero (Ed.), *Encyclopedia of Food Sciences and Nutrition* (2nd ed., pp. 4205–4212). Academic Press. doi:10.1016/B0-12-227055-X/00848-8

Stolp, L. J., & Kodali, D. R. (2022). Naturally occurring high-oleic oils: Avocado, macadamia, and olive oils. In F. J. Flider (Ed.), *High Oleic Oils* (pp. 7–52). AOCS Press. doi:10.1016/B978-0-12-822912-5.00003-4

Tananuwong, K., & Jitngarmkusol, S. (2019). *Macadamia flours: Nutritious ingredients for baked goods. In Flour and Breads and their Fortification in Health and Disease Prevention*. Elsevier.

Tsumura, T. (1988). *Factors Affecting Macadamia Nut Stability*. University of Hawai'i at Manoa.

Tu, X. H., Wu, B., Xie, Y., Xu, S. L., Wu, Z. Y., Lv, X., Wei, F., Du, L.-Q., & Chen, H. (2021). A comprehensive study of raw and roasted macadamia nuts: Lipid profile, physicochemical, nutritional, and sensory properties. *Food Science & Nutrition*, *9*(3), 1688–1697. doi:10.1002/fsn3.2143 PMID:33747479

Wall, M. (2013). *Improving the quality and safety of macadamia nuts. In Improving the safety and quality of nuts*. Elsevier.

Wall, M. M. (2010). Functional lipid characteristics, oxidative stability, and antioxidant activity of macadamia nut (Macadamia integrifolia) cultivars. *Food Chemistry*, *121*(4), 1103–1108. doi:10.1016/j.foodchem.2010.01.057

Wall, M. M. (2013). 12 - Improving the quality and safety of macadamia nuts. In L. J. Harris (Ed.), *Improving the Safety and Quality of Nuts* (pp. 274–296). Woodhead Publishing. doi:10.1533/9780857097484.2.274

Wallace, H. M., & Walton, D. (2011). *Macadamia (Macadamia integrifolia, Macadamia tetraphylla and hybrids). In Postharvest biology and technology of tropical and subtropical fruits*. Elsevier.

Wallace, H. M., & Walton, D. A. (2011). 19 - Macadamia (Macadamia integrifolia, Macadamia tetraphylla and hybrids). In E. M. Yahia (Ed.), *Postharvest Biology and Technology of Tropical and Subtropical Fruits* (pp. 450–474e). Woodhead Publishing. doi:10.1533/9780857092885.450

Walton, D. (2005). *Anatomy and handling: Implications for macadamia nut quality*. University of the Sunshine Coast.

Walton, D. A., & Wallace, H. M. (2010). Dropping macadamia nuts-in-shell reduces kernel roasting quality. *Journal of the Science of Food and Agriculture*, *90*(13), 2163–2167. doi:10.1002/jsfa.4063 PMID:20632387

Wanasundara, P., Shahidi, F., & Shukla, V. (1997). Endogenous antioxidants from oilseeds and edible oils. *Food Reviews International*, *13*(2), 225–292. doi:10.1080/87559129709541106

Wells, M. L. (2013). 1 - Agricultural practices to reduce microbial contamination of nuts. In L. J. Harris (Ed.), *Improving the Safety and Quality of Nuts* (pp. 3–21). Woodhead Publishing. doi:10.1533/9780857097484.1.3

White, N., & Hanan, J. (2014). *A model of macadamia with application to pruning in orchards.* Paper presented at the XXIX International Horticultural Congress on Horticulture: Sustaining Lives, Livelihoods and Landscapes (IHC2014).

Wood, L. G., & Garg, M. L. (2011). *Macadamia Nuts (Macadamia integrifolia and tetraphylla) and their Use in Hypercholesterolemic Subjects. In Nuts and Seeds in Health and Disease Prevention*. Elsevier.

Yamagata, K. (2022). Chapter 1 - Effects of nut consumption on the inhibition of obesity-related endothelial cell dysfunction and prevention of cardiovascular diseases. In Studies in Natural Products Chemistry (Vol. 75, pp. 1-39). Elsevier.

Yang, J., Liu, R. H., & Halim, L. (2009). Antioxidant and antiproliferative activities of common edible nut seeds. *Lebensmittel-Wissenschaft + Technologie*, *42*(1), 1–8. doi:10.1016/j.lwt.2008.07.007

Yılmaz, C., Taş, N. G., Kocadağlı, T., & Gökmen, V. (2019). Determination of serotonin in nuts and nut containing products by liquid chromatography tandem mass spectrometry. *Food Chemistry, 272*, 347–353. doi:10.1016/j.foodchem.2018.08.064 PMID:30309554

Zavala, J. S., Hernández-Martínez, D., & Osorio-Revilla, G. (2023). Elaboración de una bebida de nuez de macadamia. *Investigación y Desarrollo en Ciencia y Tecnología de Alimentos, 8*(1), 471–476. doi:10.29105/idcyta.v8i1.63

Zhong, Y., Xiang, X., Wang, X., Zhang, Y., Hu, M., Chen, T., & Liu, C. (2020). Fabrication and characterization of oil-in-water emulsions stabilized by macadamia protein isolate/chitosan hydrochloride composite polymers. *Food Hydrocolloids, 103*, 105655. doi:10.1016/j.foodhyd.2020.105655

Compilation of References

Abascal, K., & Yarnell, E. (2012). Cilantro—Culinary Herb or Miracle Medicinal Plant? *Alternative and Complementary Therapies*, *18*(5), 259–264. doi:10.1089/act.2012.18507

Abdallah, E. G., Amer, A. M., Omar, H. S., & Hussein, M. H. (2023). In vitro propagation and Agrobacterium-mediated genetic transformation of caraway (*Carum carvi* L.). *Egyptian Pharmaceutical Journal*, *22*(1), 94–104. doi:10.4103/epj.epj_136_22

Abdel-Salam, O. M. E., & Mózsik, G. (2023). Capsaicin, The Vanilloid Receptor TRPV1 Agonist in Neuroprotection: Mechanisms Involved and Significance. *Neurochemical Research*, *48*(11), 3296–3315. doi:10.1007/s11064-023-03983-z PMID:37493882

Aberoumand, A., & Deokule, S. S. (2010). Total Phenolic Contents of Some Plant Foods as a Antioxidant Compound. *Journal of Food Technology*, *8*(3), 131–133. doi:10.3923/jftech.2010.131.133

Abou El-Soud, N., El-Laithy, N., El-Saeed, G., Wahby, M. S., Khalil, M., Morsy, F., & Shaffie, N. (2011). Antidiabetic activities of Foeniculum vulgare Mill. essential oil in streptozotocin-induced diabetic rats. *Macedonian Journal of Medical Sciences*, *4*(2), 139–146.

Abouri, M., El Mousadik, A., Msanda, F., Boubaker, H., Saadi, B., & Cherifi, K. (2012). An ethnobotanical survey of medicinal plants used in the Tata Province, Morocco. *Int J Med Plants Res*, *1*(7), 99–123.

Abubakar, A. R., & Haque, M. (2020). Preparation of medicinal plants: Basic extraction and fractionation procedures for experimental purposes. *Journal of Pharmacy & Bioallied Sciences*, *12*(1), 1. doi:10.4103/jpbs.JPBS_175_19 PMID:32801594

Aćimović, M. (2019). Nutraceutical potential of Apiaceae. *Reference Series in Phytochemistry*, 1311-1341.

Aćimović, M., Jeremić, K., Salaj, N., Gavarić, N., Kiprovski, B., Sikora, V., & Zeremski, T. (2020). Marrubium vulgare L.: A phytochemical and pharmacological overview. *Molecules (Basel, Switzerland)*, *25*(12), 2898. doi:10.3390/molecules25122898 PMID:32599693

Adamu, M., Naidoo, V., & Eloff, J. N. (2014). The Antibacterial Activity, Antioxidant Activity and Selectivity Index of Leaf Extracts of Thirteen South African Tree Species Used in ethnoveterinary Medicine to Treat helminth Infections. *BMC Veterinary Research*, *10*(1), 52. doi:10.1186/1746-6148-10-52 PMID:24589020

Adegoke, G. O., Makinde, O., Falade, K. O., & Uzo-Peters, P. I. (2003). Extraction and characterization of antioxidants from Aframomum melegueta and Xylopia aethiopica. *European Food Research and Technology*, *216*(6), 526–528. doi:10.1007/s00217-003-0683-6

Adetutu, A., Olorunnisola, O. S., & Owoade, O. A. (2015). Nutritive Values and Antioxidant Activity of *Citrulluslanatus* Fruit Extract. *Food and Nutrition Sciences*, *6*(11), 1056–1064. doi:10.4236/fns.2015.611109

Afifi, N. (1994). Some pharmacological activities of essential oils of certain umbelliferous fruits. Veterinary Medical Journal Giza, Cairo Univ.

Afzali, M., Ghaeli, P., Khanavi, M., Parsa, M., Montazeri, H., Ghahremani, M. H., & Ostad, S. N. (2015). Non-addictive opium alkaloids selectively induce apoptosis in cancer cells compared to normal cells. *Daru : Journal of Faculty of Pharmacy, Tehran University of Medical Sciences*, *23*(1), 16. doi:10.1186/s40199-015-0101-1 PMID:25890335

Agarwal, V., Sharma, A. K., Upadhyay, A., Singh, G., & Gupta, R. (2012). Hypoglycemic effects of Citrullus colocynthis roots. *Acta Poloniae Pharmaceutica*, *69*(1), 75–79. PMID:22574509

Aggarwal, G., Sharma, M., Singh, R., & Sharma, U. (2024). Ethnopharmacologically important highly subsidized Indian medicinal plants: Systematic review on their traditional uses, phytochemistry, pharmacology, quality control, conservation status and future prospective. *Journal of Ethnopharmacology*, *320*, 117385. doi:10.1016/j.jep.2023.117385 PMID:37951375

Agrawal, M., Singhal, M., Jasoria, Y., Chaudhary, H., & Prajapati, B. G. (2024). Pharmacological Aspects of Coriander Essential Oils. In *Pharmacological Aspects of Essential Oils* (pp. 227–242). CRC Press.

Ahmad, M. S., & Ahmed, N. (2006). Antiglycation Properties of Aged Garlic Extract: Possible Role in Prevention of Diabetic Complications. *The Journal of Nutrition*, *136*(3), 796S–799S. doi:10.1093/jn/136.3.796S PMID:16484566

Ahmad, N., Riaz, S., Ali, A., & Riaz, T. (2023). Fenugreek (*Trigonella foenum-graecum*): An Overview of Food Uses and Health Benefits. In T. Ismail, S. Akhtar, & C. E. Lazarte (Eds.), *Neglected Plant Foods Of South Asia: Exploring and valorizing nature to feed hunger* (pp. 453–471). Springer International Publishing. doi:10.1007/978-3-031-37077-9_18

Ahmed, S. S. T., Fahim, J., & Abdelmohsen, U. R. (2021). Chemical and biological potential of Ammi visnaga [L.] Lam. and Apium graveolens L.: A review. *Journal of advanced Biomedical and Pharmaceutical Sciences, 4*(3), 160-176.

Ahmed, A. G., Hussein, U. K., Ahmed, A. E., Kim, K. M., Mahmoud, H. M., Hammouda, O., Jang, K. Y., & Bishayee, A. (2020). Mustard seed (*Brassica nigra*) extract exhibits antiproliferative effect against human lung cancer cells through differential regulation of apoptosis, cell cycle, migration, and invasion. *Molecules (Basel, Switzerland), 25*(9), 2069. doi:10.3390/molecules25092069 PMID:32365503

Ahmed, E. H., Abadi, R. S., & Mohammed, A. M. (2018). Phytochemical screening, chemical composition and antioxidant activity of seeds essential oil of Coriandrum sativum L. from the Sudan. *International Journal of Herbal Medicine*, *6*(1), 1–4.

Ahmed, N. (2005). Advanced glycation endproducts—Role in pathology of diabetic complications. *Diabetes Research and Clinical Practice*, *67*(1), 3–21. doi:10.1016/j.diabres.2004.09.004 PMID:15620429

Ahounou, J. F., Ouedraogo, G. G., Gbenou, J. D., Ouedraogo, S., Agbodjogbe, W. K., Dansou, P. H., & Moudachirou, M. (2012). Spasmolytic effects of aqueous extract of mixture from Aframomumum melegueta (K Schum) - Citrus aurantifolia (Christm and Panzer) on isolated trachea from rat. *African Journal of Traditional, Complementary, and Alternative Medicines, 9*(2), 228–233. doi:10.4314/ajtcam.v9i2.7 PMID:23983339

Ai, C., Zou, Y., Liu, H., Yang, Z., & Xi, J. (2023). Traditional Chinese herbal medicine for allergic diseases: A review. *The American Journal of Chinese Medicine*, *51*(4), 779–806. doi:10.1142/S0192415X23500374 PMID:37060193

Aissaoui, A., El-Hilaly, J., Israili, Z. H., & Lyoussi, B. (2008). Acute diuretic effect of continuous intravenous infusion of an aqueous extract of Coriandrum sativum L. in anesthetized rats. *Journal of Ethnopharmacology*, *115*(1), 89–95. doi:10.1016/j.jep.2007.09.007 PMID:17961943

Aissaoui, A., Zizi, S., Israili, Z. H., & Lyoussi, B. (2011). Hypoglycemic and hypolipidemic effects of Coriandrum sativum L. in Meriones shawi rats. *Journal of Ethnopharmacology*, *137*(1), 652–661. doi:10.1016/j.jep.2011.06.019 PMID:21718774

Ajaiyeoba, E. O., & Ekundayo, O. (1999). Essential oil constituents of Aframomum melegueta (Roscoe) K. Schum. seeds (alligator pepper) from Nigeria. *Flavour and Fragrance Journal*, *14*(2), 109–111. doi:10.1002/(SICI)1099-1026(199903/04)14:2<109::AID-FFJ775>3.0.CO;2-M

Ajayi, J., & Omodele, I. (2021). Concentration and Differential fractions of Saponin Rich Aframomum melegueta Extract Influences In-Vitro Antioxidant Activities, and Silver Nitrate can Stimulate such Activities in the Plant. *Biochemical Pharmacology*, *10*, 186.

Akhtari, E., Ram, M., Zaidi, S. M. A., Marques, A. M., Rahimi, R., & Bahramsoltani, R. (2024). Fenugreek (*Trigonella foenum-graecum* L.) in women's health: A review of clinical evidence and traditional use. *Journal of Herbal Medicine*, *43*, 100816. doi:10.1016/j.hermed.2023.100816

Akhtar, M. J., & Khan, S. A. (2024). Chemistry and biological activity of mustard oil: Therapeutic benefits and risk to healthcare. *Revista Brasileira de Farmacognosia*, *34*(1), 65–79. doi:10.1007/s43450-023-00450-2

Akhzari, M., Mohammadi, S., Jalili, S., Barazesh, M., & Noorizadeh, K. (2023). Nigella sativa, a jack of all trades plant in medicine: Pharmacological aspects in diseases treatment and prevention. *The Natural Products Journal*, *13*(6), 2–24. doi:10.2174/2210315513666221005093047

Akpanabiatu, M., Ekpo, N., Ufot, U., Malachy, N., Akpan, E., & Etuk, E. (2013). Acute toxicity, biochemical and haematological study of Aframomum melegueta seed oil in male Wistar albino rats. *Journal of Ethnopharmacology*, *150*(2), 150. doi:10.1016/j.jep.2013.09.006 PMID:24045175

Alaje, D. O., Owolabi, K., Olakunle, T., Oluoti, O. J., & Adetuberu, I. A. (2014). Nutritional, Minerals and Phytochemicals composition of Garcinia cola [Bitter cola] and Aframomum melegueta [Alligator pepper]. *IOSR Journal of Environmental Science, Toxicology and Food Technology*, *8*, 86–91. doi:10.9790/2402-08158691

Alam, M., Hardner, C., Nock, C., O'Connor, K., & Topp, B. (2019). Historical and molecular evidence of genetic identity of macadamia cultivars HAES741 and HAES660. *HortScience*, *54*(4), 616–620. doi:10.21273/HORTSCI13318-18

Alaoui, A., & Laaribya, S. (2017). Etude ethnobotanique et floristique dans les communes rurales Sehoul et Sidi-Abderrazak. *Nature & Technology*.

Alasalvar, C., & Bolling, B. W. (2015). Review of nut phytochemicals, fat-soluble bioactives, antioxidant components and health effects. *British Journal of Nutrition*, *113*(S2), S68–S78. doi:10.1017/S0007114514003729 PMID:26148924

Al-Asmari, A. K., Al-Elaiwi, A. M., Athar, M. T., Tariq, M., Al Eid, A., & Al-Asmary, S. M. (2014). A review of hepatoprotective plants used in Saudi traditional medicine. *Evidence-Based Complementary and Alternative Medicine*, *2014*, 2014. doi:10.1155/2014/890842 PMID:25587347

Al-Asmari, A. K., Athar, M. T., & Kadasah, S. G. (2017). An updated phytopharmacological review on medicinal plant of Arab region: Apium graveolens linn. *Pharmacognosy Reviews*, *11*(21), 13. doi:10.4103/phrev.phrev_35_16

Al-Azzawie, H. F., & Alhamdani, M. S. S. (2006). Hypoglycemic and antioxidant effect of oleuropein in alloxan-diabetic rabbits. *Life Sciences*, *78*(12), 1371–1377. doi:10.1016/j.lfs.2005.07.029 PMID:16236331

Albertson, P., De Giovanni, C., Cocksedge, R., Forrester, R., Rae, A., Mason, R., & McConchie, C. (2005). Inducing biochemical changes to simulate after-roast darkening in macadamia kernel. *Australian Journal of Experimental Agriculture*, *45*(10), 1315–1323. doi:10.1071/EA04176

Ali, B., Al-Wabel, N. A., Shams, S., Ahamad, A., Khan, S. A., & Anwar, F. (2015). Essential oils used in aromatherapy: A systemic review. *Asian Pacific Journal of Tropical Biomedicine*, *5*(8), 601–611. doi:10.1016/j.apjtb.2015.05.007

Ali, T., & Sultan, H. (2023). An Islamic perspective on infection treatment and wound healing. *Religions*, *14*(8), 1044. doi:10.3390/rel14081044

Al-Mofleh, I., Alhaider, A. A., Mossa, J. S., Al-Sohaibani, M. O., Rafatullah, S., & Qureshi, S. (2006). Protection of gastric mucosal damage by Coriandrum sativum L. pretreatment in Wistar albino rats. *Environmental Toxicology and Pharmacology*, *22*(1), 64–69. doi:10.1016/j.etap.2005.12.002 PMID:21783688

Al-Ogaili, F. H., & Almahdawi, F. H. (2023). Green Fabrication and Characterization of Zinc Oxide Nanoparticles Using Celery Leaf Extract. *Journal of Survey in Fisheries Sciences*, *10*(4S), 2938–2945.

Alo, M. N., Anyim, C., Igwe, J. C., Elom, M., & Uchenna, D. S. (2012). Antibacterial activity of water, ethanol and methanol extracts of Ocimum gratissimum, Vernonia amygdalina and Aframomum melegueta. *Advances in Applied Science Research*, *3*, 844–848.

Al, S., & Ali, E. (2014). The Pharmacological Importance of*Anethum graveolens. A Review. International Journal of Pharmacy and Pharmaceutical Sciences*, *6*(4), 11–13.

Al-Snafi, A. E. (2014). The pharmacological importance of*Anethum graveolens*–A review. *International Journal of Pharmacy and Pharmaceutical Sciences*, *6*(4), 11–13.

Altaş, S., Kızıl, G., Kızıl, M., Ketani, A., & Haris, P. I. (2011). Protective effect of Diyarbakır watermelon juice on carbon tetrachloride-induced toxicity in rats. *Food and Chemical Toxicology*, *49*(9), 2433–2438. doi:10.1016/j.fct.2011.06.064 PMID:21740951

Alves-Silva, J. M., Dias dos Santos, S. M., Pintado, M. E., Pérez-Álvarez, J. A., Fernández-López, J., & Viuda-Martos, M. (2013). Chemical composition and in vitro antimicrobial, antifungal and antioxidant properties of essential oils obtained from some herbs widely used in Portugal. *Food Control*, *32*(2), 371–378. doi:10.1016/j.foodcont.2012.12.022

Aly, A., Maraei, R., Rezk, A., & Diab, A. (2023). Phytochemical constitutes and biological activities of essential oil extracted from irradiated caraway seeds (*Carum carvi* L.). *International Journal of Radiation Biology*, *99*(2), 318–328. doi:10.1080/09553002.2022.2078004 PMID:35549975

Amadi, S., Zhang, Y., & Wu, G. (2016). Research progress in phytochemistry and biology of Aframomum species. *Pharmaceutical Biology*, *54*(11), 1–10. doi:10.3109/13880209.2016.1173068 PMID:27159353

Ambrose, D. C., Manickavasagan, A., & Naik, R. (2016). *Leafy medicinal herbs: botany, chemistry, postharvest technology and uses*. CABI. doi:10.1079/9781780645599.0000

Amirghofran, Z. (2010). Medicinal Plants as Immunosuppressive Agents in Traditional Iranian Medicine. *Iranian Journal of Immunology*, *7*, 65–73.

Anand, A. V., Sreedevi, M. J., & Swapna, T. S. (2023). Plant Conservation Associated with Traditional Knowledge: Past and Future. In S. T. Sukumaran & K. T R (Eds.), Conservation and Sustainable Utilization of Bioresources (pp. 261-290). Springer Nature Singapore. doi:10.1007/978-981-19-5841-0_11

Anitha, T., Balakumar, C., Ilango, K. B., Jose, C. B., & Vetrivel, D. (2014). Antidiabetic activity of the aqueous extracts of Foeniculum vulgare on streptozotocin-induced diabetic rats. *Int J Adv Pharm Biol Chem*, *3*(2), 487–494.

Anlar, H. G., & Bacanli, M. (2020). Thymoquinone: the active compound of black seed (Nigella sativa). In V. R. Preedy (Ed.), *Pathology* (pp. 369–378). Academic Press. doi:10.1016/B978-0-12-815972-9.00035-4

Anonymous. (2002). Quality Control Methods for Medicinal plants materials. (An authorized publication of W.H.O. Geneva). New Delhi: A.I.T.B.S. Publishers & Distributors.

Ansari, M. A., Roberts, K. N., & Scheff, S. W. (2008). Oxidative stress and modification of synaptic proteins in hippocampus after traumatic brain injury. *Free Radical Biology & Medicine*, *45*(4), 443–452. doi:10.1016/j.freeradbiomed.2008.04.038 PMID:18501200

Anwar, T., Qureshi, H., Naeem, H., Siddiqi, E. H., Hanif, A., Anwaar, S., Noreen, Z., Iqbal, J., Ali, B., Iqbal, R., & Paray, B. A. (2024). Prospects of the sustainability of traditional and ethnic foods and their applications for basic health care in human society. *Journal of King Saud University. Science*, *36*(2), 103016. doi:10.1016/j.jksus.2023.103016

Anyasor, G. N., Ogunwenmo, O., Oyelana, O. A., & Akpofunure, B. E. (2010). Phytochemical constituents and antioxidant activities of aqueous and methanol stem extracts of Costus afer Ker Gawl.(Costaceae). *African Journal of Biotechnology*, *9*(31), 4880–4884.

Aquino-Bolaños, E. N., Mapel-Velazco, L., Martín-del-Campo, S. T., Chávez-Servia, J. L., Martínez, A. J., & Verdalet-Guzmán, I. (2017). Fatty acids profile of oil from nine varieties of Macadamia nut. *International Journal of Food Properties*, *20*(6), 1262–1269. doi:10.1080/10942912.2016.1206125

Archita, K., Ved, P., Vikesh Kumar, S., & Rohit, S. (2023). Exploring the therapeutic potential of herbal plants in managing blood sugar levels: A comprehensive evaluation. *Pharmacognosy Research*, *16*(1), 10–18. doi:10.5530/pres.16.1.2

Arunachalam, K., Sreeja, P. S., & Yang, X. (2023). Mechanisms and therapeutic actions of edible fruits in inflammatory bowel disease: A review of preclinical studies. *Food Chemistry Advances*, *3*, 100498. doi:10.1016/j.focha.2023.100498

Aruna, K., & Sivaramakrishnan, V. M. (1990). Plant products as protective agents against cancer. *Indian Journal of Experimental Biology*, *28*(11), 1008–1011. PMID:2283166

Aruoma, O. I. (2003). Methodological considerations for characterizing potential antioxidant actions of bioactive components in plant foods. *Mutation Research. Fundamental and Molecular Mechanisms of Mutagenesis*, *523*, 9–20. doi:10.1016/S0027-5107(02)00317-2 PMID:12628499

Asadikaram, G., Akbari, H., Vakili, S., Asiabanha, M., Shahrokhi, N., & Savardashtaki, A. (2018). The effects of opium addiction on thyroid and sex hormones in diabetic and non-diabetic male and female rats. *Acta endocrinologica (Bucharest, Romania: 2005)*, *14*(4), 466-472. doi:10.4183/aeb.2018.466

Asha, S., Pawan, S., & Annu, D. (2015). Determination of total phenolic content and total protein in guava (*Psidiumguajava*) and lemon (*Citrus limon*). *Int J Biol Pharm Res*, *4*(11), 827–834.

Ashokkumar, K., Vellaikumar, S., Muthusamy, M., Dhanya, M. K., & Aiswarya, S. (2021). Compositional variation in the leaf, mace, kernel, and seed essential oil of nutmeg (Myristica fragrans Houtt.) from the Western Ghats, India. *Natural Product Research*, *36*(1), 432–435. do i:10.1080/14786419.2020.1771713 PMID:32515616

Asif, H., Akram, M., Usmanghani, K., Akhtar, N., Shah, P. A., Uzair, M., ... Rehman, R. (2011). Monograph of apium graveolens Linn. *Journal of Medicinal Plants Research*, *5*(8), 1494–1496.

Atta, A., & Alkofahi, A. (1998). Anti-nociceptive and anti-inflammatory effects of some Jordanian medicinal plant extracts. *Journal of Ethnopharmacology*, *60*(2), 117–124. doi:10.1016/S0378-8741(97)00137-2

Attia, E. S., Amer, A. H., & Hasanein, M. A. (2019). The hypoglycemic and antioxidant activities of garden cress [Lepidium sativum L.] seed on alloxan-induced diabetic male rats. *Natural Product Research*, *33*(6), 901–905. doi:10.1080/14786419.2017.1413564 PMID:29237302

Augusti, K. T. (1973). Studies on the effects of a hypoglycemic principle from Allium Cepa Linn. *The Indian Journal of Medical Research*. PMID:4757998

Augustine, I. (2020). Hypolipidaemic and Antidiabetic Potency of Allium cepa [Onions] Bulb in Alloxan-Induced Diabetic Rats. Acta Scientific Nutritional Health, 4, 1-8.

Aworh, O. C. (2023). African traditional foods and sustainable food security. *Food Control*, *145*, 109393. doi:10.1016/j.foodcont.2022.109393

Aynehchi, Y., & Jaffarian, S. (1973). Determination of thebaine in various parts of Papaver bracteatum Lindl. During the growing season. *Lloydia*, *36*(4), 427–429. PMID:4773272

B.A.S., L. (2007). Hypotensive and Antihypertensive Effects of Aframomum melegueta Seeds in Humans. *International Journal of Pharmacology*, 3.

Baba, K., Xiao, Y.-Q., Taniguchi, M., Ohishi, H., & Kozawa, M. (1991). Isocoumarins from Coriandrum sativum. *Phytochemistry*, *30*(12), 4143–4146. doi:10.1016/0031-9422(91)83482-Z

Babu, S., Kumar, M., Madhumathi, T., & Ram, S. (2017). Status of insecticide resistance in lesser grain borer, Rhyzopertha dominica to Malathion and Deltamethrin in Andhra Pradesh. *Journal of Entomology and Zoology Studies*, *5*, 741–746.

Baharvand-Ahmadi, B., Bahmani, M., Tajeddini, P., Naghdi, N., & Rafieian-Kopaei, M. (2016). An ethno-medicinal study of medicinal plants used for the treatment of diabetes. *Journal of Nephropathology*, *5*(1), 44–50. doi:10.15171/jnp.2016.08 PMID:27047810

Bailer, J., Aichinger, T., Hackl, G., de Hueber, K., & Dachler, M. (2001). Essential oil content and composition in commercially available dill cultivars in comparison to caraway. *Industrial Crops and Products*, *14*(3), 229–239. doi:10.1016/S0926-6690(01)00088-7

Bailey, C. J., & Day, C. (1989). Traditional plant medicines as treatments for diabetes. *Diabetes Care*, *12*(8), 553–564. doi:10.2337/diacare.12.8.553 PMID:2673695

Bailey, H. M., & Stein, H. H. (2020). Raw and roasted pistachio nuts (Pistacia vera L.) are 'good'sources of protein based on their digestible indispensable amino acid score as determined in pigs. *Journal of the Science of Food and Agriculture*, *100*(10), 3878–3885. doi:10.1002/jsfa.10429 PMID:32323331

Bakhtiar, Z., Hassandokht, M., Naghavi, M. R., & Mirjalili, M. H. (2024). Variability in proximate composition, phytochemical traits and antioxidant properties of Iranian agro-ecotypic populations of fenugreek (*Trigonella foenum-graecum* L.). *Scientific Reports*, *14*(1), 87. doi:10.1038/s41598-023-50699-9 PMID:38167975

Balaji, A., Bhuvaneswari, S., Raj, L. S., Bupesh, G., Meenakshisundaram, K. K., & Saravanan, K. M. (2022). A review on the potential species of the zingiberaceae family with anti-viral efficacy towards enveloped viruses. *Journal of Pure & Applied Microbiology*, *16*(2), 796–813. doi:10.22207/JPAM.16.2.35

Ballhorn, D. J. (2011). Cyanogenic Glycosides in Nuts and Seeds. In V. R. Preedy, R. R. Watson, & V. B. Patel (Eds.), *Nuts and Seeds in Health and Disease Prevention* (pp. 129–136). Academic Press. doi:10.1016/B978-0-12-375688-6.10014-3

Bansal, K., Sharma, S., & Bajpai, M. (2024). Herbal medicines-a fruitful approach to periodic illness dysmenorrhoea: Evidence-based review. *Current Pharmaceutical Biotechnology*, *25*(2), 179–195. doi:10.2174/1389201024666230623161113 PMID:37357517

Banu, K. S., & Cathrine, L. (2015). General Techniques Involved in Phytochemical Analysis. *International Journal of Advanced Research in Chemical Science*, *2*(4), 25–32.

Barkaoui, M., Katiri, A., Boubaker, H., & Msanda, F. (2017). Ethnobotanical survey of medicinal plants used in the traditional treatment of diabetes in Chtouka Ait Baha and Tiznit [Western Anti-Atlas], Morocco. *Journal of Ethnopharmacology*, *198*, 338–350. doi:10.1016/j.jep.2017.01.023 PMID:28109915

Barros, L., Dueñas, M., Dias, M. I., Sousa, M. J., Santos-Buelga, C., & Ferreira, I. C. F. R. (2012). Phenolic profiles of in vivo and in vitro grown Coriandrum sativum L. *Food Chemistry*, *132*(2), 841–848. doi:10.1016/j.foodchem.2011.11.048 PMID:23107699

Barrueto, A. K., Merz, J., Hodel, E., & Eckert, S. (2018). The suitability of Macadamia and Juglans for cultivation in Nepal: An assessment based on spatial probability modelling using climate scenarios and in situ data. *Regional Environmental Change*, *18*(3), 859–871. doi:10.1007/s10113-017-1225-2

Bas, A. L., Demirci, S., Yazihan, N., Uney, K., & Ermis Kaya, E. (2012). Nerium oleander distillate improves fat and glucose metabolism in high-fat diet-fed streptozotocin-induced diabetic rats. *International Journal of Endocrinology*. PMID:23251156

Bassler, T. J. (1992). Niacin benefits. *The Western Journal of Medicine*, *156*(2), 209–209. PMID:1536086

Beaudoin, G. A., & Facchini, P. J. (2014). Benzylisoquinoline alkaloid biosynthesis in opium poppy. *Planta*, *240*(1), 19–32. doi:10.1007/s00425-014-2056-8 PMID:24671624

Begnami, A., Duarte, M. C. T., Furletti, V., & Rehder, V. L. G. (2010). Antimicrobial potential of Coriandrum sativum L. against different Candida species in vitro. *Food Chemistry*, *118*(1), 74–77. doi:10.1016/j.foodchem.2009.04.089

Behradmanesh, S., Derees, F., & Rafieian-Kopaei, M. (2013). Effect of Salvia officinalis on diabetic patients. *Journal of Renal Injury Prevention*, *2*(2), 51. PMID:25340127

Belhaj, S., Chaachouay, N., & Zidane, L. (2021). Ethnobotanical and toxicology study of medicinal plants used for the treatment of diabetes in the High Atlas Central of Morocco. *Journal of Pharmacy & Pharmacognosy Research*, *9*(5), 619–662. doi:10.56499/jppres21.1014_9.5.619

Belhaj, S., Dahmani, J., Belahbib, N., & Zidane, L. (2020). Ethnopharmacological and Ethnobotanical study of Medicinal plants in the Central High Atlas, Morocco ةساردلا طسوتملا سلطألا يف ةيبطلا تاتابنلل ةىجولوبونث إلاو ةيجولوكمرفونثيإلا ريبكلا للمغرب. *Ethnobotany Research and Applications*, *20*, 1–40. doi:10.32859/era.20.18.1-40

Belmaker, R. H., & Agam, G. (2008). Major depressive disorder. *The New England Journal of Medicine*, *358*(1), 55–68. doi:10.1056/NEJMra073096 PMID:18172175

Benali, T., Khabbach, A., Ennabili, A., & Hammani, K. (2017). Ethnopharmacological prospecting of medicinal plants from the Province of Guercif. *Moroccan Journal of Biology*, *14*(14), 1–14.

Benkhnigue, O., Ben Akka, F., Salhi, S., Fadli, M., Douira, A., & Zidane, L. (2014). Catalogue des plantes médicinales utilisées dans le traitement du diabète dans la région d'Al Haouz-Rhamna. *J Anim Plant Sci*, *23*(1), 3539–3568.

Benlamdini, N., Elhafian, M., Rochdi, A., & Zidane, L. (2014). Étude floristique et ethnobotanique de la flore médicinale du Haut Atlas oriental. *Journal of Applied Biosciences*, *78*(1), 6771–6787. doi:10.4314/jab.v78i1.17

Bennett, M. D. (2023). Caribbean plantation economies as colonial models: The case of the English East India Company and St. Helena in the late seventeenth century. *Atlantic Studies*, *20*(4), 508–539. doi:10.1080/14788810.2022.2034569

Berrou, J., Tostivint, I., Verrecchia, F., Berthier, C., Boulanger, E., & Mauviel, A. (2009). Advanced glycation end products regulate extracellular matrix protein and protease expression by human glomerular mesangial cells. *International Journal of Molecular Medicine*, *23*(4), 513–520. PMID:19288028

Bhat, S. (2014). Coriander (Coriandrum sativum L.): Processing, nutritional and functional aspects. *African Journal of Plant Science*, *8*(1), 25–33. doi:10.5897/AJPS2013.1118

Bhattacharya, S. (2023). Edible nuts. In S. Bhattacharya (Ed.), *Snack Foods* (pp. 173–192). Academic Press. doi:10.1016/B978-0-12-819759-2.00006-9

Bhuiyan, M. N. I., Begum, J., & Sultana, M. (2009). *Chemical composition of leaf and seed essential oil of Coriandrum sativum L. from Bangladesh.* |||. *Bangladesh Journal of Pharmacology*, *4*(2), 150–153. doi:10.3329/bjp.v4i2.2800

Bird, D. A., Franceschi, V. R., & Facchini, P. J. (2003). A tale of three cell types: Alkaloid biosynthesis is localized to sieve elements in opium poppy. *The Plant Cell*, *15*(11), 2626–2635. doi:10.1105/tpc.015396 PMID:14508000

Blahovec, J. (2002). Mechanics of poppy seeds. *International Journal of Food Properties*, *5*(2), 277–287. doi:10.1081/JFP-120005785

Blank, I., & Grosch, W. (1991). Evaluation of potent odorants in dill seed and dill herb (Anethum graveolens L.) by aroma extract dilution analysis. *Journal of Food Science*, *56*(1), 63–67. doi:10.1111/j.1365-2621.1991.tb07976.x

Blank, I., Sen, A., & Grosch, W. (1992). Sensory study on the character-impact flavour compounds of dill herb (Anethum graveolens L.). *Food Chemistry*, *43*(5), 337–343. doi:10.1016/0308-8146(92)90305-L

Bnouham, M., Mekhfi, H., Abdelkhaleq, L., & Ziyyat, A. (2002). Ethnopharmacology Forum Medicinal Plants used in the treatment of diabetes in Morocco. *International Journal of Diabetes and Metabolism*, *10*, 33–50.

Boeing, H., Bechthold, A., Bub, A., Ellinger, S., Haller, D., Kroke, A., Leschik-Bonnet, E., Muller, M. J., Oberritter, H., Schulze, M., Stehle, P., & Watzl, B. (2012). Critical review: Vegetables and fruit in the prevention of chronic diseases. *European Journal of Nutrition*, *51*(6), 637–663. doi:10.1007/s00394-012-0380-y PMID:22684631

Bolarinwa, I. F., Orfila, C., & Morgan, M. R. (2014). Amygdalin content of seeds, kernels and food products commercially-available in the UK. *Food Chemistry*, *152*, 133–139. doi:10.1016/j.foodchem.2013.11.002 PMID:24444917

Bontempo, P., Doto, A., Miceli, M., Mita, L., Benedetti, R., Nebbioso, A., & Altucci, L. (2011). Psidiumguajava L. anti-neoplastic effects: Induction of apoptosis and cell differentiation. *Cell Proliferation*, *45*(1), 22–31. doi:10.1111/j.1365-2184.2011.00797.x PMID:22172154

Boudjelal, A., Henchiri, C., Siracusa, L., Sari, M., & Ruberto, G. (2012). Compositional analysis and in vivo anti-diabetic activity of wild Algerian Marrubium vulgare L. infusion. *Fitoterapia*, *83*(2), 286–292. doi:10.1016/j.fitote.2011.11.005 PMID:22100836

Boufous, H., Marhoume, F., Chait, A., & Bagri, A. (2017). Ethnopharmacological survey of medicinal plants with hallucinogenic effect and used against pain, inflammatory diseases, diabetes and urinary lithiasis in Zagora "Morocco". *Journal of Intercultural Ethnopharmacology*, *6*(4), 342–350. doi:10.5455/jice.20170721062527

Bousta, D., Boukhira, S., Aafi, A., Ghanmi, M., & el Mansouri, L. (2014). Ethnopharmacological Study of anti-diabetic medicinal plants used in the Middle-Atlas region of Morocco. *International Journal of Pharma Research and Health Sciences*, *2*(1), 75–79.

Bouwmeester, H. J., Gershenzon, J., Konings, M. C., & Croteau, R. (1998). Biosynthesis of the monoterpenes limonene and carvone in the fruit of caraway: I. Demonstration of enzyme activities and their changes with development. *Plant Physiology*, *117*(3), 901–912. doi:10.1104/pp.117.3.901 PMID:9662532

Bown, D., & Herb Society of, A. (1995). *Encyclopedia of herbs & their uses*. Retrieved from https://archive.org/details/encyclopediaofhe0000bown

Brahimi, A., El Ouardi, M., Kaouachi, A., Boudboud, A., Hajji, L., Hajjaj, H., & Mazouz, H. (2022). Characterization of the Biochemical Potential of Moroccan Onions. *International Journal of Food Sciences*, *2022*, 2022. doi:10.1155/2022/2103151 PMID:35308133

Brand-Williams, W., Cuvelier, M. E., & Berset, C. (1995). Use of a free radical method to evaluate antioxidant activity. *Lebensmittel-Wissenschaft + Technologie*, *28*(1), 25–30. doi:10.1016/S0023-6438(95)80008-5

Brindha, P., Sasikala, B., & Purushothaman, K. K. (1981). Pharmacognostic studies on Merugan kizhangu. *Bulletin of Medio-Ethno Botanical Research*, *3*, 84–86.

Brindisi, L. J., Lyu, W., Juliani, H. R., Wu, Q., Tepper, B. J., & Simon, J. E. (2023). Determination of glucosinolates and breakdown products in Brassicaceae baby leafy greens using UHPLC-QTOF/MS and GC/MS. *Food Chemistry Advances*, *3*, 100389. doi:10.1016/j.focha.2023.100389

Bulduk, I., & Taktak, F. (2013). Isolation and Characterization of Antitumor Alkaloid from Poppy Capsules (Papaver somniferum). *Journal of Chemistry*, *493870*, 1–4. Advance online publication. doi:10.1155/2013/493870

Burdock, G. A. (2022). Assessment of black cumin (*Nigella sativa* L.) as a food ingredient and putative therapeutic agent. *Regulatory Toxicology and Pharmacology*, *128*, 105088. doi:10.1016/j.yrtph.2021.105088 PMID:34838871

Burdock, G. A., & Carabin, I. G. (2009). Safety assessment of coriander (Coriandrum sativum L.) essential oil as a food ingredient. *Food and Chemical Toxicology*, *47*(1), 22–34. doi:10.1016/j.fct.2008.11.006 PMID:19032971

Burin, V. M., Falcão, L. D., Gonzaga, L. V., Fett, R., Rosier, J. P., & Bordignon-Luiz, M. T. (2010). Colour, Phenolic Content and Antioxidant Activity of Grape Juice. *Food Science and Technology (Campinas)*, *30*(4), 1027–1032. doi:10.1590/S0101-20612010000400030

Busnel, R., & Manrique López, H. (2023). The political economy of a failed drug reform: Insights from Peru's main legal coca valley. *The International Journal on Drug Policy*, *117*, 104050. doi:10.1016/j.drugpo.2023.104050 PMID:37267739

Caballero, B. (2005). *Encyclopedia of human nutrition*. Elsevier.

Caceres Guido, P., Ribas, A., Gaioli, M., Quattrone, F., & Macchi, A. (2015). The state of the integrative medicine in Latin America: The long road to include complementary, natural, and traditional practices in formal health systems. *European Journal of Integrative Medicine*, *7*(1), 5–12. doi:10.1016/j.eujim.2014.06.010

Canton-Alvarez, J. A. (2019). A Gift from the Buddhist Monastery: The Role of Buddhist Medical Practices in the Assimilation of the Opium Poppy in Chinese Medicine during the Song Dynasty (960-1279). *Medical History*, *63*(4), 475–493. doi:10.1017/mdh.2019.45 PMID:31571697

Cao, M., Zhan, M., Jing, H., Wang, Z., Wang, Y., Li, X., & Miao, M. (2024). Network pharmacology and experimental evidence: MAPK signaling pathway is involved in the anti-asthma roles of Perilla frutescens leaf. *Heliyon*, *10*(1), e22971. doi:10.1016/j.heliyon.2023.e22971 PMID:38163225

Carrillo, C., Cavia Mdel, M., & Alonso-Torre, S. (2012). Role of oleic acid in immune system; mechanism of action; a review. *Nutrición Hospitalaria*, *27*(4), 978–990. doi:10.3305/nh.2012.27.4.5783 PMID:23165533

Carta, G., Murru, E., Banni, S., & Manca, C. (2017). Palmitic Acid: Physiological Role, Metabolism and Nutritional Implications. *Frontiers in Physiology*, *8*, 902. doi:10.3389/fphys.2017.00902 PMID:29167646

Cătunescu, G. M., Bodea, I. M., David, A. P., Pop, C. R., & Rotar, A. M. (2023). Essential oils from Apiaceae family (parsley, lovage, and dill). In G. A. Nayik & M. J. Ansari (Eds.), *Essential Oils* (pp. 241–308). Academic Press. doi:10.1016/B978-0-323-91740-7.00015-3

Cerón, D. A. C., de Alencar, E. R., Faroni, L. R. D. A., Silva, M. V. D. A., & Salvador, D. V. (2023). Toxicity of allyl isothiocyanate applied in systems with or without recirculation for controlling *Sitophilus zeamais, Rhyzopertha dominica, and Tribolium castaneum* in corn grains. *Journal of the Science of Food and Agriculture*, *xx*(x), xxx–xxx. doi:10.1002/jsfa.12713 PMID:37205787

Ceska, O., Chaudhary, S. K., Warrington, P., Ashwood-Smith, M. J., Bushnell, G. W., & Poultont, G. A. (1988). Coriandrin, a novel highly photoactive compound isolated from Coriandrum sativum. *Phytochemistry*, *27*(7), 2083–2087. doi:10.1016/0031-9422(88)80101-8

Chaachouay, N., Benkhnigue, O., Fadli, M., El Ibaoui, H., & Zidane, L. (2019). Ethnobotanical and ethnopharmacological studies of medicinal and aromatic plants used in the treatment of metabolic diseases in the Moroccan Rif. *Heliyon*, *5*(10), e02191. doi:10.1016/j.heliyon.2019.e02191 PMID:31720440

Chahal, K., Kumar, A., Bhardwaj, U., & Kaur, R. (2017). Chemistry and biological activities of*Anethum graveolens* L.(dill) essential oil: A review. *Journal of Pharmacognosy and Phytochemistry*, *6*(2), 295–306.

Chain, E. (2018). Update of the Scientific Opinion on opium alkaloids in poppy seeds. *EFSA journal. EFSA Journal*, *16*(5), e05243–e05243. doi:10.2903/j.efsa.2018.5243

Chakir, A. R. S., Elbadaoui, K., & Alaoui, T. I. (2015). Antidiabetic activities of methanolic extracts of Marrubium vulgare leaves in rats. *International Journal of Pharmaceutical and Phytopharmacological Research*, *4*(5), 258–263.

Chan, H. T. Jr. (1983). *Handbook of tropical foods.*

Chansiw, N., Chusri, P., Praman, S., Hawiset, T., Sukhorum, W., Champakam, S., & Charoensup, R. (2024). Anti-inflammatory potential of a Thai traditional remedy called Prabchompoothaweep in an animal model of acute and subacute inflammation. *Journal of Ethnopharmacology*, *319*, 117380. doi:10.1016/j.jep.2023.117380 PMID:37925003

Chaudhary, N., Ghimire, S. K., & Chaudhary, R. P. (2023). Traditional Amchi Medicinal Practice in Trans-Himalaya of Nepal: Conservation and Bioprospecting. In K. Arunachalam, X. Yang, & S. Puthanpura Sasidharan (Eds.), *Bioprospecting of Tropical Medicinal Plants* (pp. 307–327). Springer Nature Switzerland. doi:10.1007/978-3-031-28780-0_11

Chaudhary, S. S., Tariq, M., Zaman, R., & Imtiyaz, S. (2013). The In vitro anti-acne activity of two unani drugs. *Ancient Science of Life*, *33*(1), 35–38. doi:10.4103/0257-7941.134594 PMID:25161328

Chaudhary, S., Aslam, M., & Siddiqui, A. (2019). Tukhm Khashkhaash (Poppy Seeds): A Unani drug of multitudinous potential. *Journal of Drug Delivery and Therapeutics*, *9*(2), 528–533. doi:10.22270/jddt.v9i2.2560

Chaurasia, S., & Jain, P. (1978). Antibacterial activity of essential oils of four medicinal plants. *The Indian Journal of Hospital Pharmacy*, *15*(6), 166–168.

Cheikhyoussef, A., Cheikhyoussef, N., Sharma, R., & Hussein, A. A. (2023). Health-Promoting Effects of Cardamom (*Elettaria cardamomum*). In M. F. Ramadan (Ed.), *Cardamom (Elettaria cardamomum): Production, Processing and Properties* (pp. 133–156). Springer International Publishing. doi:10.1007/978-3-031-35426-7_11

Chellappan, D., & Pemiah, B. (2014). Pharmacognostical, phytochemical and in vivo gastroprotective investigation of Gmelina arborea. *International Journal of Pharmacy and Pharmaceutical Sciences*, *6*(4), 153–157.

Chen, H., Roco, M. C., Li, X., & Lin, Y. (2008). Trends in nanotechnology patents. *Nature Nanotechnology*, *3*(3), 123–125. doi:10.1038/nnano.2008.51 PMID:18654475

Chen, L. (2007). Clément Vigneault, GS Vijaya Raghavan, Stan Kubow, 2007. Importance of the phytochemical content of fruits and vegetables to human health. *Stewart Postharvest Review*, *3*, 2.

Chen, W., Liu, Y., Li, M., Mao, J., Zhang, L., Huang, R., Jin, X., & Ye, L. (2015). Anti-tumor effect of α-pinene on human hepatoma cell lines through inducing G2/M cell cycle arrest. *Journal of Pharmacological Sciences*, *127*(3), 332–338. doi:10.1016/j.jphs.2015.01.008 PMID:25837931

Chevallier, A. (1996). The encyclopedia of medicinal plants. Academic Press.

Chinedu, I. (2018). *Fungal Corrosion of Mild Steel and the Application of Aframomum melegueta Biomass Extract as Natural Source of Inhibition*. Academic Press.

Chithra, V., & Leelamma, S. (1999). Coriandrum sativum—Mechanism of hypoglycemic action. *Food Chemistry*, *67*(3), 229–231. doi:10.1016/S0308-8146(99)00113-2

Chithra, V., & Leelamma, S. (2000). Coriandrum sativum—Effect on lipid metabolism in 1, 2-dimethyl hydrazine induced colon cancer. *Journal of Ethnopharmacology*, *71*(3), 457–463. doi:10.1016/S0378-8741(00)00182-3 PMID:10940583

Cho, J. Y., Park, S.-C., Kim, T.-W., Kim, K.-S., Song, J.-C., Lee, H.-M., Sung, H.-J., Rhee, M.-H., Kim, S.-K., Park, H.-J., Song, Y.-B., Yoo, E.-S., & Lee, C.-H. (2006). Radical scavenging and anti-inflammatory activity of extracts from Opuntia humifusa Raf. *The Journal of Pharmacy and Pharmacology*, *58*(1), 113–119. doi:10.1211/jpp.58.1.0014 PMID:16393471

Chopra, R. N., & Chopra, R. (1969). Supplement to glossary of Indian medicinal plants. Academic Press.

Chopra, R. N., & De, P. (1929). Saussurea lappa in pharmacology and therapeutics. *The Indian Journal of Medical Research*, *17*(2), 351–359.

Choudhury, B. (2006). *Vegetables*. National Book Trust.

Chougule, M., Patel, A. R., Sachdeva, P., Jackson, T., & Singh, M. (2011). Anticancer activity of Noscapine, an opioid alkaloid in combination with Cisplatin in human non-small cell lung cancer. *Lung Cancer (Amsterdam, Netherlands)*, *71*(3), 271–282. doi:10.1016/j.lungcan.2010.06.002 PMID:20674069

Cilak, G. O., Mujdeci, G. N., & Kabak, B. (2021). Legislation on aromatic herbs in food. In C. M. Galanakis (Ed.), *Aromatic Herbs in Food* (pp. 405–438). Academic Press. doi:10.1016/B978-0-12-822716-9.00011-1

Cock, I., Winnett, V., Sirdaarta, J., & Matthews, B. (2015). The potential of selected Australian medicinal plants with anti-Proteus activity for the treatment and prevention of rheumatoid arthritis. *Pharmacognosy Magazine*, *11*(42, Suppl 1), S190. doi:10.4103/0973-1296.157734 PMID:26109767

Concheiro, M., Newmeyer, M. N., da Costa, J. L., Flegel, R., Gorelick, D. A., & Huestis, M. A. (2015). Morphine and codeine in oral fluid after controlled poppy seed administration. *Drug Testing and Analysis*, *7*(7), 586–591. doi:10.1002/dta.1742 PMID:25345619

Correia, A. G. S., Alencar, M. B., dos Santos, A. N., da Paixão, D. C. B., Sandes, F. L. F., Andrade, B., Castro, Y., & de Andrade, J. S. (2023). Effect of saffron and fenugreek on lowering blood glucose: A systematic review with meta-analysis. *Phytotherapy Research*, *37*(5), 2092–2101. Advance online publication. doi:10.1002/ptr.7817 PMID:36992660

Cortés-Eslava, J., Gómez-Arroyo, S., Villalobos-Pietrini, R., & Espinosa-Aguirre, J. J. (2004). Antimutagenicity of coriander (Coriandrum sativum) juice on the mutagenesis produced by plant metabolites of aromatic amines. *Toxicology Letters*, *153*(2), 283–292. doi:10.1016/j.toxlet.2004.05.011 PMID:15451560

Coşkuner, Y., & Karababa, E. (2007). Physical properties of coriander seeds (Coriandrum sativum L.). *Journal of Food Engineering*, *80*(2), 408–416. doi:10.1016/j.jfoodeng.2006.02.042

Costello, G., Gregory, M., & Donaitu, P. (2008). *Southern Macadamia Species Recover, Plan 20082012. Report to Department of the Environment, W., Heritage and the, & Arts, C. b.* H. A. L.

Cowan, M. M. (1999). Plant products as antimicrobial agents. *Clinical Microbiology Reviews*, *12*(4), 564–582. doi:10.1128/CMR.12.4.564 PMID:10515903

Crichton, M., Marshall, S., Isenring, E., Lohning, A., McCarthy, A. L., Molassiotis, A., ... Marx, W. (2023). Effect of a standardized ginger root powder regimen on chemotherapy-induced nausea and vomiting: A multicenter, double-blind, placebo-controlled randomized trial. *Journal of the Academy of Nutrition and Dietetics*, *xx*(x), xxx–xxx. doi:10.1016/j.jand.2023.09.003 PMID:37699474

Cseke, L. J., Kirakosyan, A., Kaufman, P. B., Warber, S., Duke, J. A., & Brielmann, H. L. (2016). *Natural products from plants*. CRC Press. doi:10.1201/9781420004472

D'Ambrosio, U., Pozo, C., Vallès, J., & Gras, A. (2023). East meets west: Using ethnobotany in ethnic urban markets of Barcelona metropolitan area (Catalonia) as a tool for biocultural exchange. *Journal of Ethnobiology and Ethnomedicine*, *19*(1), 63. doi:10.1186/s13002-023-00636-x PMID:38105250

da Silva, A. K. A., dos Santos, C. R., Carneiro, M. L. B., Joanitti, G. A., Luz, G. V. S., Rosa, S. S. R. F., . . . Tatmatsu-Rocha, J. C. (2023). Chapter 1 - Bixa orellana L. and its implications in human health: Perspectives and new trends. In R. Atta ur (Ed.), Studies in Natural Products Chemistry (Vol. 77, pp. 1-36). Elsevier. doi:10.1016/B978-0-323-91294-5.00001-4

Da Silva, J. A. T., & Hussain, A. I. (2017). Citrullus colocynthis [L.] Schrad.[colocynth]: Biotechnological perspectives. *Emirates Journal of Food and Agriculture*, *29*(2), 83–90. doi:10.9755/ejfa.2016-11-1764

Dąbrowska, J. A., Kunicka-Styczyńska, A., & Śmigielski, K. B. (2020). Biological, chemical, and aroma profiles of essential oil from waste celery seeds (Apium graveolens L.). *The Journal of Essential Oil Research*, *32*(4), 308–315. doi:10.1080/10412905.2020.1754937

Dahler, J., Mcconchie, C., & Turnbull, C. (1995). Quantification of cyanogenic glycosides in seedlings of three Macadamia (Proteaceae) species. *Australian Journal of Botany*, *43*(6), 619–628. doi:10.1071/BT9950619

Daley, D. K., & Badal, S. (2024). Plant crude drugs. In S. B. McCreath & Y. N. Clement (Eds.), *Pharmacognosy* (2nd ed., pp. 75–89). Academic Press. doi:10.1016/B978-0-443-18657-8.00002-5

Dama, G. Y., Tare, H. L., Gore, M. S., Shende, V. S., Deore, S. R., & … . (2010). Comparative cardiotonic activity of Aegle marmelos juice with digoxin on isolated frog heart. *Int J Drug Dev Res*, *2*(4), 1–8.

Damayanti, A., & Setyawan, E. (2012). Essential oil extraction of fennel seed (*Foeniculum vulgare*) using steam distillation. *International Journal of Science and Engineering*, *3*(2), 12–14. doi:10.12777/ijse.3.2.12-14

Dani, S. (2023). A study on physical and chemical characteristics of guava (Psidium guajava L.) Taiwan pink variety. *The Pharma Innovation*, *12*(5), 2023–2025. doi:10.22271/tpi.2023.v12.i5y.20175

Daradka, H. M., Abas, M. M., Mohammad, M. A., & Jaffar, M. M. (2014). Antidiabetic effect of Artemisia absinthium extracts on alloxan-induced diabetic rats. *Comparative Clinical Pathology*, *23*(6), 1733–1742. doi:10.1007/s00580-014-1963-1

Dare, E. O., Oseghale, C. O., Labulo, A. H., Adesuji, E. T., Elemike, E. E., Onwuka, J. C., & Bamgbose, J. T. (2015). Green synthesis and growth kinetics of nanosilver under bio-diversified plant extracts influence. *Journal of Nanostructure in Chemistry*, *5*(1), 85–94. doi:10.1007/s40097-014-0139-5

Darughe, F., Barzegar, M., & Sahari, M. (2012). Antioxidant and antifungal activity of Coriander (Coriandrum sativum L.) essential oil in cake. *International Food Research Journal*, *19*(3), 1253–1260.

Das, G., Das, S., Talukdar, A., Venil, C., Bose, S., Banerjee, S., ... Patra, J. (2023). Pharmacology and ethnomedicinal potential of selected plants species from Apiaceae (Umbelliferae). *Combinatorial Chemistry & High Throughput Screening*, *26*(2), 256–288. doi:10.2174/1386207325666220406110404 PMID:35388749

Dash, B. (2011). *Antibacterial activities of methanol and acetone extracts of fenugreek (Trigonella foenum) and coriander (Coriandrum sativum)*. Life Sciences and Medicine Research.

Das, P., Chandra, T., Negi, A., Jaiswal, S., Iquebal, M. A., Rai, A., & Kumar, D. (2023). A comprehensive review on genomic resources in medicinally and industrially important major spices for future breeding programs: Status, utility and challenges. *Current Research in Food Science*, *7*, 100579. doi:10.1016/j.crfs.2023.100579 PMID:37701635

Dastgheib, L., Pishva, N., Saki, N., Khabnadideh, S., Kardeh, B., Torabi, F., Arabnia, S., & Heiran, A. (2017). Efficacy of topical coriandrum sativum extract on treatment of infants with diaper dermatitis: A single blinded non-randomised controlled trial. *The Malaysian Journal of Medical Sciences : MJMS*, *24*(4), 97–101. doi:10.21315/mjms2017.24.4.11 PMID:28951694

Davis, P. H. (1970). Flora of Turkey and the East Aegean Islands. Vol. 3. Flora of Turkey and the East Aegean Islands.

de Almeida Melo, E., Mancini Filho, J., & Guerra, N. B. (2005). Characterization of antioxidant compounds in aqueous coriander extract (Coriandrum sativum L.). *Lebensmittel-Wissenschaft + Technologie*, *38*(1), 15–19. doi:10.1016/j.lwt.2004.03.011

de Vilmorin, R. L. (1950). Pascal celery and its origin. *JNY Bot Gard*, *51*, 39–41.

Debnath, S., Kumar, H., & Sharma, A. (2023). *Foeniculum vulgare* Mill. from spice to pharma: Recent advances in its medicinal value, bioactivities and perspectives. *Traditional and Integrative Medicine*, *8*(2), 217–229. doi:10.18502/tim.v8i2.13087

Delaquis, P. J., Stanich, K., Girard, B., & Mazza, G. (2002). Antimicrobial activity of individual and mixed fractions of dill, cilantro, coriander and eucalyptus essential oils. *International Journal of Food Microbiology*, *74*(1-2), 101–109. doi:10.1016/S0168-1605(01)00734-6 PMID:11929164

Delgadillo-Puga, C., Torre-Villalvazo, I., Cariño-Cervantes, Y. Y., García-Luna, C., Soberanes-Chávez, P., de Gortari, P., Noriega, L. G., Bautista, C. J., & Cisneros-Zevallos, L. (2023). Cardamom (*Elettaria cardamomum* (L.) Maton) seeds intake increases energy expenditure and reduces fat mass in mice by modulating neural circuits that regulate adipose tissue lipolysis and mitochondrial oxidative metabolism in liver and skeletal muscle. *International Journal of Molecular Sciences*, *24*(4), 3909. doi:10.3390/ijms24043909 PMID:36835337

Deshmukh, P. (2024). Utilization of herbs and spices as a conventional medicine in covid-19: A study. *Multidisciplinary Science Journal*, *6*(1), 2024008–2024008. doi:10.31893/multiscience.2024008

Destaillats, F., & Angers, P. (2002). Base-catalyzed derivatization methodology for FA analysis. Application to milk fat and celery seed lipid TAG. *Lipids*, *37*(5), 527–532. doi:10.1007/s11745-002-0928-9

Devasagayam, T. P. A., Tilak, J. C., Boloor, K. K., Sane, K. S., Ghaskadbi, S. S., & Lele, R. D. (2004). Free radicals and antioxidants in human health: Current status and future prospects. *The Journal of the Association of Physicians of India*, *52*(4), 794–804. PMID:15909857

Dey, P., Kundu, A., Kumar, A., Gupta, M., Lee, B. M., Bhakta, T., . . . Kim, H. S. (2020). Analysis of alkaloids (indole alkaloids, isoquinoline alkaloids, tropane alkaloids). *Recent Advances in Natural Products Analysis*, 505-567. doi:10.1016/B978-0-12-816455-6.00015-9

Dhalwal, K., Shinde, V. M., & Mahadik, K. R. (2008). Efficient and sensitive method for quantitative determination and validation of umbelliferone, carvone and myristicin in*Anethum graveolens* and Carum carvi seed. *Chromatographia*, *67*(1-2), 163–167. doi:10.1365/s10337-007-0473-6

Dhananjay. (2021). A retrospection of Indian grape varieties. *The Pharma Innovation Journal*, *SP-10*(11), 408–414.

Dhull, S. B., Bamal, P., Chandak, A., Nain, K. B., & Malik, A. (2024). Fenugreek (*Trigonella foenum-graecum*): An herb with impressive nutritional and antidiabetic properties. In M. Naeem & T. Aftab (Eds.), *Antidiabetic Medicinal Plants* (pp. 305–325). Academic Press. doi:10.1016/B978-0-323-95719-9.00012-4

Diederichsen, A. (1996). *Coriander: Coriandrum Sativum L* (Vol. 3). Bioversity International.

Dilrukshi, E. A. C., Nishiyama, Y., Ito, K., & Nomura, S. (2024). Alleviation of acute stress response by black pepper aroma administration. *Journal of Physiological Anthropology*, *43*(1), 3. doi:10.1186/s40101-023-00352-1 PMID:38167228

Divakaran, M., Jayasree, E., Nirmal Babu, K., & Peter, K. V. (2018). Legacy of Indian Spices: Its Production and Processing. In A. B. Sharangi (Ed.), *Indian Spices: The Legacy, Production and Processing of India's Treasured Export* (pp. 13–30). Springer International Publishing. doi:10.1007/978-3-319-75016-3_2

Divya, P., Puthusseri, B., & Neelwarne, B. (2012). Carotenoid content, its stability during drying and the antioxidant activity of commercial coriander (Coriandrum sativum L.) varieties. *Food Research International*, *45*(1), 342–350. doi:10.1016/j.foodres.2011.09.021

Djiazet, S., Kavitake, D., Ravi, R., Mezajoug Kenfack, L. B., Pullakhandam, R., Prabhavathi Devi, B. L. A., Tchiégang, C., Reddy, G. B., & Shetty, P. H. (2023). Investigation of physicochemical properties, nutritional status, and volatile compounds profiling of four underutilized spices of Cameroon origin. *Food Chemistry Advances*, *3*, 100465. doi:10.1016/j.focha.2023.100465

Djuric, Z., & Powell, L. (2001). Antioxidant capacity of lycopene-containing foods. *International Journal of Food Sciences and Nutrition*, *52*(2), 143–149. doi:10.1080/713671775 PMID:11303462

Doherty, F., & Olaniran, O. (2010). Antimicrobial Activities of Aframomum Melegueta (Alligator Pepper). *International Journal of Biology*, 2. doi:10.5539/ijb.v2n2p126

Domingo-Fernández, D., Gadiya, Y., Mubeen, S., Bollerman, T. J., Healy, M. D., Chanana, S., Sadovsky, R. G., Healey, D., & Colluru, V. (2023). Modern drug discovery using ethnobotany: A large-scale cross-cultural analysis of traditional medicine reveals common therapeutic uses. *iScience*, *26*(9), 107729. doi:10.1016/j.isci.2023.107729 PMID:37701812

Dong, Y., Zhu, H., Shen, Y., Zhang, W., & Zhang, L. (2019). Antibacterial activity of silver nanoparticles of different particle size against Vibrio Natriegens. *PLoS One*, *14*(9), e0222322. doi:10.1371/journal.pone.0222322 PMID:31518380

dos Santos, M. L., Chandran, D., Lejaniya, A. S., & da Silva, L. E. (2023). Conservation and Sustainable Use of Medicinal Plants. In S. Jha & M. Halder (Eds.), *Medicinal Plants: Biodiversity, Biotechnology and Conservation* (pp. 327–341). Springer Nature Singapore. doi:10.1007/978-981-19-9936-9_13

dos Santos-Silva, C. A., Zupin, L., Oliveira-Lima, M., Vilela, L. M. B., Bezerra-Neto, J. P., Ferreira-Neto, J. R., ... Aburjaile, F. F. (2020). Plant antimicrobial peptides: State of the art, in silico prediction and perspectives in the omics era. *Bioinformatics and Biology Insights*, *14*, 14. doi:10.1177/1177932220952739 PMID:32952397

Dreyer, K., van Rijswijk, J., Mijatovic, V., Goddijn, M., Verhoeve, H. R., van Rooij, I. A. J., Hoek, A., Bourdrez, P., Nap, A. W., Rijnsaardt-Lukassen, H. G. M., Timmerman, C. C. M., Kaplan, M., Hooker, A. B., Gijsen, A. P., van Golde, R., van Heteren, C. F., Sluijmer, A. V., de Bruin, J.-P., Smeenk, J. M. J., ... Mol, B. W. J. (2017). Oil-Based or Water-Based Contrast for Hysterosalpingography in Infertile Women. *The New England Journal of Medicine*, *376*(21), 2043–2052. doi:10.1056/NEJMoa1612337 PMID:28520519

Du Preez, A. B. (2015). *Studies on macadamia nut quality*. Stellenbosch University.

Duke, J. A. (2002). *Handbook of medicinal herbs*. CRC Press. doi:10.1201/9781420040463

Duraipandiyan, V., Al-Harbi, N. A., Ignacimuthu, S., & Muthukumar, C. (2012). Antimicrobial activity of sesquiterpene lactones isolated from traditional medicinal plant, Costus speciosus (Koen ex.Retz.) Sm. *BMC Complementary and Alternative Medicine*, *12*(1), 13. doi:10.1186/1472-6882-12-13

Dyer, A. G., Boyd-Gerny, S., McLoughlin, S., Rosa, M. G., Simonov, V., & Wong, B. B. (2012). Parallel evolution of angiosperm colour signals: common evolutionary pressures linked to hymenopteran vision. *Proc Biol Sci, 279*(1742), 3606-3615. doi:10.1098/rspb.2012.0827

Echo, A., Osuagwu, A., Agbor, R., Okpako, E. C., & Ekanem, B. E. (2013). Phytochemical composition of Aframomum melegueta and Piper guineense seeds. *World Journal of Applied Environmental Chemistry.*, *2*, 17–21.

Eddouks, M. (2017). Contribution to the study of medicinal plants used in the treatment of diabetes, obesity and hypertension in Tafilalet region. *Arabian Journal of Medicinal and Aromatic Plants*, *3*(2), 124–161.

Eddouks, M., Lemhadri, A., Hebi, M., Hidani, A. E., Zeggwagh, N. A., Bouhali, B. E., ... Burcelin, R. (2017). Capparis spinosa L. aqueous extract evokes antidiabetic effect in streptozotocin-induced diabetic mice. *Avicenna Journal of Phytomedicine*, *7*(2), 191. PMID:28348974

Eddouks, M., Maghrani, M., Lemhadri, A., Ouahidi, M. L., & Jouad, H. (2002). Ethnopharmacological survey of medicinal plants used for the treatment of diabetes mellitus, hypertension and cardiac diseases in the south-east region of Morocco. *Journal of Ethnopharmacology*, *82*(2-3), 97–103. doi:10.1016/S0378-8741(02)00164-2 PMID:12241983

Eddouks, M., Maghrani, M., Zeggwagh, N. A., & Michel, J. B. (2005). Study of the hypoglycaemic activity of Lepidium sativum L. aqueous extract in normal and diabetic rats. *Journal of Ethnopharmacology*, *97*(2), 391–395. doi:10.1016/j.jep.2004.11.030 PMID:15707780

Edeoga, H. O., Okwu, D. E., & Mbaebie, B. O. (2005). Phytochemical Constiuents ofsome Nigerian medicinal plants. *African Journal of Biotechnology*, *4*(7), 685–688. doi:10.5897/AJB2005.000-3127

Edwards, A. J., Vimyard, B. T., & Willy, E. R. (2015). Consumption of Watermelon Juice increases plasma concentration of lycopene and beta carotene in humans. *The Journal of Nutrition*, *133*(4), 1043–1050. doi:10.1093/jn/133.4.1043 PMID:12672916

Eguale, T., Tilahun, G., Debella, A., Feleke, A., & Makonnen, E. (2007). In vitro and in vivo anthelmintic activity of crude extracts of Coriandrum sativum against Haemonchus contortus. *Journal of Ethnopharmacology*, *110*(3), 428–433. doi:10.1016/j.jep.2006.10.003 PMID:17113738

Eidi, A., & Eidi, M. (2009). Antidiabetic effects of sage [Salvia officinalis L.] leaves in normal and streptozotocin-induced diabetic rats. *Diabetes & Metabolic Syndrome*, *3*(1), 40–44. doi:10.1016/j.dsx.2008.10.007

Eidi, A., Eidi, M., & Darzi, R. (2009). Antidiabetic effect of Olea europaea L. in normal and diabetic rats. *Phytotherapy Research*, *23*(3), 347–350. doi:10.1002/ptr.2629 PMID:18844257

Eidi, M., Eidi, A., Saeidi, A., Molanaei, S., Sadeghipour, A., Bahar, M., & Bahar, K. (2009). Effect of coriander seed [Coriandrum sativum L.] ethanol extract on insulin release from pancreatic beta cells in streptozotocin-induced diabetic rats. *Phytotherapy Research*, *23*(3), 404–406. doi:10.1002/ptr.2642 PMID:19003941

Eidi, M., Eidi, A., & Zamanizadeh, H. (2005). Effect of Salvia officinalis L. leaves on serum glucose and insulin in healthy and streptozotocin-induced diabetic rats. *Journal of Ethnopharmacology*, *100*(3), 310–313. doi:10.1016/j.jep.2005.03.008 PMID:16125023

Eikani, M. H., Golmohammad, F., & Rowshanzamir, S. (2007). Subcritical water extraction of essential oils from coriander seeds (Coriandrum sativum L.). *Journal of Food Engineering*, *80*(2), 735–740. doi:10.1016/j.jfoodeng.2006.05.015

Ekanem, A. P., Wang, M., Simon, J. E., & Moreno, D. A. (2007). Antiobesity properties of two African plants (Afromomum meleguetta and Spilanthes acmella) by pancreatic lipase inhibition. *Phytotherapy Research*, *21*(12), 1253–1255. doi:10.1002/ptr.2239 PMID:17705140

Eko, P. M., Afolabi, K. D., Olajide, R., Enyenihi, G. E., Mbaba, E. N., & Efffiong, M. E. (2022). *Haematology and Serum Metabolites of Pre-laying Isa Brown Hens fed Varying Dietary Levels of Alligator Pepper (Aframomum melegueta, k. Schum)*. Seed Meal.

El Azzouzi, F., & Zidane, L. (2015). La flore médicinale traditionnelle de la région de Béni-Mellal [Maroc]. *Journal of Applied Biosciences*, *91*(1), 8493–8502. doi:10.4314/jab.v91i1.8

El Fouhil, A. F., Ahmed, A. M., Atteya, M., Mohamed, R. A., Moustafa, A. S., & Darwish, H. H. (2013). An extract from date seeds stimulates endogenous insulin secretion in streptozotocin-induced type I diabetic rats. *Functional Foods in Health and Disease*, *3*(11), 441–446. doi:10.31989/ffhd.v3i11.33

El Fouhil, A. F., Ahmed, A. M., Darwish, H. H., Atteya, M., & Al-Roalle, A. H. (2011). An extract from date seeds having a hypoglycemic effect. Is it safe to use? *Saudi Medical Journal*, *32*(8), 791–796. PMID:21858387

El Gharras, H. (2009). Polyphenols: Food sources, properties and applications – a review. *International Journal of Food Science & Technology*, *44*(12), 2512–2518. doi:10.1111/j.1365-2621.2009.02077.x

El Hachlafi, N., Chebat, A., Bencheikh, R. S., & Fikri-Benbrahim, K. (2020). Ethnopharmacological study of medicinal plants used for chronic diseases treatment in Rabat-Sale-Kenitra region [Morocco]. *Ethnobotany Research and Applications*, *20*, 1–23. doi:10.32859/era.20.2.1-23

El-Assri, E.-M., Hajib, A., Choukri, H., Gharby, S., Lahkimi, A., Eloutassi, N., & Bouia, A. (2023). Nutritional quality, lipid, and mineral profiling of seven Moroccan Apiaceae seeds. *South African Journal of Botany*, *160*, 23–35. doi:10.1016/j.sajb.2023.06.042

Eldin, I. M. T., Ahmed, E. M., & Abd, E. H. (2010). Preliminary study of the clinical hypoglycemic effects of Allium cepa [red onion] in type 1 and type 2 diabetic patients. Environmental health insights, 4, EHI-S5540.

Elgayyar, M., Draughon, F., Golden, D., & Mount, J. (2001). Antimicrobial Activity of Essential Oils from Plants against Selected Pathogenic and Saprophytic Microorganisms. *Journal of Food Protection*, *64*(7), 1019–1024. doi:10.4315/0362-028X-64.7.1019

El-Halawany, A. M., Dine, R. S. E., El Sayed, N. S., & Hattori, M. (2014). Protective Effect of Aframomum melegueta phenolics Against CCl4-Induced Rat Hepatocytes Damage; Role of Apoptosis and Pro-inflammatory Cytokines inhibition. *Scientific Reports*, *4*(1), 5880. doi:10.1038/srep05880 PMID:25077538

El-Hilaly, J., Hmammouchi, M., & Lyoussi, B. (2003). Ethnobotanical studies and economic evaluation of medicinal plants in Taounate province. *Journal of Ethnopharmacology*, *86*(2-3), 149–158. doi:10.1016/S0378-8741(03)00012-6 PMID:12738079

Elmhdwi, M. F., Muktar, M. A., & Attitalla, I. H. (2015). Hypoglycemic effects of Marrubium vulgare [Rubia] in experimentally induced autoimmune Diabetes Mellitus. *International Journal of Pharmacy & Life Sciences, 6*(4).

El-Ouady, F., Lahrach, N., Ajebli, M., Haidani, A. E., & Eddouks, M. (2020). Antihyperglycemic effect of the aqueous extract of Foeniculum vulgare in normal and streptozotocin-induced diabetic rats. *Cardiovascular & Haematological Disorders-Drug Targets, 20*(1), 54-63.

Emam & Heydari. (2008). *Effect of extract and essential oil of Coriandrum sativum seed against pentylenetetrazole-induced seizure*. Academic Press.

Eom, S.-H., Jin, C.-W., Park, H.-J., Kim, E.-H., Chung, I.-M., Kim, M.-J., ... Cho, D.-H. (2007). Far infrared ray irradiation stimulates antioxidant activity in Vitis flexuosa THUNB. Berries. *Korean Journal of Medicinal Crop Science*, *15*(5), 319–323.

Erhabor, J., Idu, M., & Udo, F. O. (2013). Ethnomedicinal Survey of Medicinal Plants Used in the Treatment of Male Infertilty among the IFA Nkari People of Ini Local Government Area of AkwaIbom State, Nigeria. *Research Journal of Recent Sciences*, *2*, 5–11.

Erinç, H., Tekin, A., & Özcan, M. (2009). Determination of fatty acid, tocopherol and phytosterol contents of the oils of various poppy (Papaver somniferum L.) seeds. *Grasas y Aceites*, *60*(4), 375–381. Advance online publication. doi:10.3989/gya.129508

Ersín, E., & Coşge Şenkal, B. (2023). Impacts of essential oil and extracts obtained from coriander cultivars (Coriandrum sativum L.) on important some pathogenic bacteria. *Current Perspectives on Medicinal and Aromatic Plants*, *6*(1), 27–37. doi:10.38093/cupmap.1281520

Evans, R. H. P. J. (2005). Product description manual: Australian macadamia kernel. Australian Macadamia Society

Evans, W. (2002). *Trease and Evans: Pharmacognocy* (Fifteenth International Edition). WB Saunders.

Facchini, P. J., & Bird, D. A. (1998). Developmental regulation of benzylisoquinoline alkaloid biosynthesis in opium poppy plants and tissue cultures. *In Vitro Cellular & Developmental Biology. Plant*, *34*(1), 69–79. doi:10.1007/BF02823126

Facchini, P. J., & De Luca, V. (1995). Phloem-Specific Expression of Tyrosine/Dopa Decarboxylase Genes and the Biosynthesis of Isoquinoline Alkaloids in Opium Poppy. *The Plant Cell*, *7*(11), 1811–1821. doi:10.2307/3870189 PMID:12242361

Fakchich, J., & Elachouri, M. (2021). An overview on ethnobotanico-pharmacological studies carried out in Morocco, from 1991 to 2015: Systematic review. *Journal of Ethnopharmacology*, *267*, 113200. doi:10.1016/j.jep.2020.113200 PMID:32750461

Fakhr, L., Chehregosha, F., Zarezadeh, M., Chaboksafar, M., & Tarighat-Esfanjani, A. (2023). Effects of fenugreek supplementation on the components of metabolic syndrome: A systematic review and dose–response meta-analysis of randomized clinical trials. *Pharmacological Research*, *187*, 106594. doi:10.1016/j.phrs.2022.106594 PMID:36470549

FAO. (2016). Food and Agriculture Organisation of United Nation. http://faostat.fao.org /site /567/default.aspx#ancor

Farh, M., Kadil, Y., Tahri, E. H., Abounasr, M., Riad, F., El Khasmi, M., & Tazi, A. (2018). Évaluation des effets anxiolytiques, antidépressifs et mnésiques de l'huile des graines de la nigelle chez le rat. *Phytothérapie*, *16*(1), 5–13. doi:10.1007/s10298-017-1095-7

Fayyadh, S. N., & Tahrim, N. A. (2020). Green nanoparticles investigation to remove water pollutants by Fenton reaction using celery leaves extract. *International Journal of Design & Nature and Ecodynamics*, *15*(3), 309–314. doi:10.18280/ijdne.150304

Fazal, S., & Singla, R. K. (2012). Review on the Pharmacognostical & Pharmacological Characterization of Apium Graveolens Linn. *Indo Global Journal of Pharmaceutical Sciences*, *2*(1), 36–42. doi:10.35652/IGJPS.2012.03

Ferguson, N. M. (1956). A Textbook of Pharmacognosy. New Delhi: Mcmillan Company.

Ferreira, P. M. P., Arcanjo, D. D. R., & Peron, A. P. (2023). Drug development, Brazilian biodiversity and political choices: Where are we heading? *Journal of Toxicology and Environmental Health. Part B, Critical Reviews*, *26*(5), 257–274. doi:10.1080/10937404.2023.2193762 PMID:36967535

Fitsiou, E., Mitropoulou, G., Spyridopoulou, K., Tiptiri-Kourpeti, A., Vamvakias, M., Bardouki, H., ... Chlichlia, K. (2016). Phytochemical profile and evaluation of the biological activities of essential oils derived from the Greek aromatic plant species Ocimum basilicum, Mentha spicata, Pimpinella anisum and Fortunella margarita. *Molecules (Basel, Switzerland)*, *21*(8), 1069. doi:10.3390/molecules21081069 PMID:27537869

Fleming, T. (2000). Article. In T. Fleming (Ed.), *PDR® for Herbal Medicines* (pp. 224–225). Medical Economics Company.

Fritschi, G., & Prescott, W. R. Jr. (1985). Morphine levels in urine subsequent to poppy seed consumption. *Forensic Science International*, *27*(2), 111–117. doi:10.1016/0379-0738(85)90173-2 PMID:3979930

G.K., P. (2017). Alzheimer's disease: A challenge in managing with certain medicinal plants - A review. *International Journal of Pharmaceutical Sciences and Research*, *4*, 4960–4972.

Galal, S. M., & Morsy, N. F. S. (2023). Cardamom in Food Applications. In M. F. Ramadan (Ed.), *Cardamom (Elettaria cardamomum): Production, Processing and Properties* (pp. 243–253). Springer International Publishing. doi:10.1007/978-3-031-35426-7_18

Gamble, J. S. (1957). Flora of the presidency of Madras. Sree Saraswaty Press Ltd.

Ganapathy, A. A., Haripriya, V. M., Acharya, N., Somappa, S. B., & Kumaran, A. (2023). Ethnobotanical significance of medicinal plants: Beta-amyloid and tau aggregation inhibitors against Alzheimer's disease. *Journal of Biochemical and Molecular Toxicology*, *37*(6), e23339. doi:10.1002/jbt.23339 PMID:37009720

Gandía-Herrero, F., Escribano, J., & García-Carmona, F. (2016). Biological activities of plant pigments betalains. *Critical Reviews in Food Science and Nutrition*, *56*(6), 937–945. doi:10.1080/10408398.2012.740103 PMID:25118005

Garg, M. L., Blake, R. J., Wills, R. B., & Clayton, E. H. (2007). Macadamia nut consumption modulates favourably risk factors for coronary artery disease in hypercholesterolemic subjects. *Lipids*, *42*(6), 583–587. doi:10.1007/s11745-007-3042-8 PMID:17437143

Garg, P., Hitawala, A. A., & Agarwal, M. (2018). Cardiotoxic Effects of Raw Opium. *Indian Journal of Critical Care Medicine : Peer-Reviewed, Official Publication of Indian Society of Critical Care Medicine*, *22*(1), 46–48. doi:10.4103/ijccm.IJCCM_427_17 PMID:29422734

Gauri, M. K., Ali, S. J., & Khan, M. S. (2015). *A Review of Apium graveolens (Karafs) special reference to Unani Medicine*. Academic Press.

Gavish, L., Korchia, G., Cohen, S., & Maor, Y. (2023). A real-world evaluation of an herbal treatment for infantile colic reported by 1218 parents in Israel. *Journal of Pediatric Nursing*, *69*, e39–e44. doi:10.1016/j.pedn.2022.12.002 PMID:36528454

Gengatharan, A., Dykes, G. A., & Choo, W. S. (2015). Betalains: Natural plant pigments with potential application in functional foods. *Lebensmittel-Wissenschaft + Technologie*, *64*(2), 645–649. doi:10.1016/j.lwt.2015.06.052

Ghazanfari, N., Tabatabaei Yazdi, F., Mortazavi, S. A., & Mohammadi, M. (2023). Using pulsed electric field pretreatment to optimize coriander seeds essential oil extraction and evaluate antimicrobial properties, antioxidant activity, and essential oil compositions. *Lebensmittel-Wissenschaft + Technologie*, *182*, 114852. doi:10.1016/j.lwt.2023.114852

Gholamnezhad, Z., Havakhah, S., & Boskabady, M. H. (2016). Preclinical and clinical effects of Nigella sativa and its constituent, thymoquinone: A review. *Journal of Ethnopharmacology*, *190*, 372–386. doi:10.1016/j.jep.2016.06.061 PMID:27364039

Ghorbani, A., & Esmaeilizadeh, M. (2017). Pharmacological properties of Salvia officinalis and its components. *Journal of Traditional and Complementary Medicine*, *7*(4), 433–440. doi:10.1016/j.jtcme.2016.12.014 PMID:29034191

Ghourri, M., Zidane, L., & Douira, A. (2013). Usage des plantes médicinales dans le traitement du Diabète Au Sahara marocain. *Journal of Animal &Plant Sciences*, *17*(1), 2388–2411.

Giannenas, I., Sidiropoulou, E., Bonos, E., Christaki, E., & Florou-Paneri, P. (2020). The history of herbs, medicinal and aromatic plants, and their extracts: Past, current situation and future perspectives. In *Feed additives*. Academic Press. doi:10.1016/B978-0-12-814700-9.00001-7

Giovannini, P., Howes, M. J. R., & Edwards, S. E. (2016). Medicinal plants used in the traditional management of diabetes and its sequelae in Central America: A review. *Journal of Ethnopharmacology*, *184*, 58–71. doi:10.1016/j.jep.2016.02.034 PMID:26924564

Girenko, M. (1982). *Initial material and basic trends in breeding of some uncommon species of vegetables*. J. Bull. VIR im. *Vavilova*, *120*, 33–37.

Gittins, D. I., Bethell, D., Nichols, R. J., & Schiffrin, D. J. (2000). Diode-like electron transfer across nanostructured films containing a redox ligand. *Journal of Materials Chemistry*, *10*(1), 79–83. doi:10.1039/a902960e

Gloor, M., Kägi, M., & Wüthrich, B. (1995). Poppyseed anaphylaxis. *Schweizerische Medizinische Wochenschrift*, *125*(30), 1434–1437. PMID:7676232

Gollucke, A., & Ribeiro, D.(2012). Use of Grape Polyphenols for Promoting Human Health: A Review of Patents. *Recent Patents on Food, Nutrition & Agriculture*, *4*(1), 26–30. doi:10.2174/2212798411204010026 PMID:22316271

Goodarzi, M. T., Khodadadi, I., Tavilani, H., & Abbasi Oshaghi, E. (2016). The role of *Anethum graveolens* L.(Dill) in the management of diabetes. *Journal of Tropical Medicine*. PMID:27829842

Goto, M., Kanda, H., Wahyudiono, & Machmudah, S. (2015). Extraction of carotenoids and lipids from algae by supercritical CO2 and subcritical dimethyl ether. *The Journal of Supercritical Fluids*, *96*, 245–251. doi:10.1016/j.supflu.2014.10.003

Grains of paradise. (2014). Encyclopædia Britannica, Inc.

Grant, J. K., Dangl, M., Ndumele, C. E., Michos, E. D., & Martin, S. S. (2023). A historical, evidence-based, and narrative review on commonly used dietary supplements in lipid-lowering. *Journal of Lipid Research*, *xx*(x), 100493. doi:10.1016/j.jlr.2023.100493 PMID:38145747

Grass, G., Rensing, C., & Solioz, M. (2011). Metallic copper as an antimicrobial surface. *Applied and Environmental Microbiology*, *77*(5), 1541–1547. doi:10.1128/AEM.02766-10 PMID:21193661

Gray, A. M., & Flatt, P. R. (1999). Insulin-releasing and insulin-like activity of the traditional anti-diabetic plant Coriandrum sativum (coriander). *British Journal of Nutrition*, *81*(3), 203–209. doi:10.1017/S0007114599000392 PMID:10434846

Gray, J. (2013). Nuts and seeds. In B. Caballero (Ed.), *Encyclopedia of Human Nutrition* (3rd ed., pp. 329–335). Academic Press. doi:10.1016/B978-0-12-375083-9.00208-7

Gröblacher, B., Maier, V., Kunert, O., & Bucar, F. (2012). Putative Mycobacterial Efflux Inhibitors from the Seeds of Aframomum melegueta. *Journal of Natural Products*, *75*(7), 1393–1399. doi:10.1021/np300375t PMID:22789014

Grosso, C., Ferraro, V., Figueiredo, A. C., Barroso, J. G., Coelho, J. A., & Palavra, A. M. (2008). Supercritical carbon dioxide extraction of volatile oil from Italian coriander seeds. *Food Chemistry*, *111*(1), 197–203. doi:10.1016/j.foodchem.2008.03.031

Grzegorz, D., Sylwester, C., & Iwona, K. (2020). Composition and quality of poppy (papaver somniferum l.) Seed oil depending on the extraction method. *Lebensmittel-Wissenschaft + Technologie*, *110167*. Advance online publication. doi:10.1016/j.lwt.2020.110167

Guenther, E. (1950). *The Essential Oils: Individual essential oils of the plant families: Gramineae, Lauraceae, Burseraceae, Myrtaceae, Umbelliferae, and Geraniaceae*. D. Van Nostrand Company.

Guerra, N. B., de Almeida Melo, E., & Mancini Filho, J. (2005). Antioxidant compounds from coriander (Coriandrum sativum L.) etheric extract. *Journal of Food Composition and Analysis*, *18*(2-3), 193–199. doi:10.1016/j.jfca.2003.12.006

Gull, J., Sultana, B., Anwar, F., Naseer, R., Ashraf, M., & Ashrafuzzaman, M. (2012). Variation in Antioxidant Attributes at Three Ripening Stages of Guava (Psidiumguajava L.) Fruit from Different Geographical Regions of Pakistan. *Molecules (Basel, Switzerland)*, *17*(3), 3165–3180. doi:10.3390/molecules17033165 PMID:22418924

Gupta, N., Bhattacharya, S., Urbanová, K., Dutta, A., Hazra, A. K., Fernández-Cusimamani, E., & Leuner, O. (2023). Systematic analysis of antimicrobial activity, phytochemistry, and in silico molecular interaction of selected essential oils and their formulations from different Indian spices against foodborne bacteria. *Heliyon*, *9*(12), e22480. doi:10.1016/j.heliyon.2023.e22480 PMID:38107328

Gupta, R., Anwer, M., & Sharma, Y. (2012). Dill. In *Handbook of Herbs and Spices* (pp. 275–285). Elsevier. doi:10.1533/9780857095671.275

Gupta, S., Singh, A. P., Singh, G., Ding, X., & Sharma, A. (2024). Plant based metabolomics: A future prospective and versatile tool for metabolite databases of *Curcuma longa. Journal of Future Foods, 4*(1), 37–47. doi:10.1016/j.jfutfo.2023.05.003

Guthrie, J., Greensill, C., Bowden, R., & Walsh, K. (2004). Assessment of quality defects in macadamia kernels using NIR spectroscopy. *Australian Journal of Agricultural Research, 55*(4), 471–476. doi:10.1071/AR03179

Guzmán-Gutiérrez, S. L., Reyes-Chilpa, R., González-Diego, L. R., Silva-Miranda, M., López-Caamal, A., García-Cruz, K. P., Jiménez-Mendoza, M. S., Arciniegas, A., & Espitia, C. (2023). Five centuries of Cirsium ehrenbergii Sch. Bip. (Asteraceae) in Mexico, from Huitzquilitl to Cardo Santo: History, ethnomedicine, pharmacology and chemistry. *Journal of Ethnopharmacology, 301*, 115778. doi:10.1016/j.jep.2022.115778 PMID:36202165

Gwaltney-Brant, S. M. (2013). Macadamia Nuts. In M. E. Peterson & P. A. Talcott (Eds.), *Small Animal Toxicology* (3rd ed., pp. 625–627). W.B. Saunders. doi:10.1016/B978-1-4557-0717-1.00056-9

Gyasi, E., Kotey, D.A., Adongo, B.A., Adams, F.K., Owusu, E.O., & Mohammed, A. (2022). Management of major seed-borne fungi of cowpea (Vigna unguiculata (L.) Walp) with four selected botanical extracts. *Advances in Agriculture*.

Haber, I., Pergolizzi, J. Jr, & LeQuang, J. A. (2019). Poppy Seed Tea: A Short Review and Case Study. *Pain and Therapy, 8*(1), 151–155. doi:10.1007/s40122-019-0113-5 PMID:30719627

Hachi, M., Ouafae, B., Hachi, T., Mohamed, E. B., Imane, B., Atmane, R., & Zidane, L. (2016). Contribution to the ethnobotanical study of antidiabetic medicinal plants of the Central Middle Atlas region. *Lazaroa, 37*, 1-11.

Hadi, S., Restuhadi, F., & Lukmanasari, P. (2023). Export competitiveness analysis of pepper (Piper Nigrum L.) commodity in the international market. *Jurnal Agrinika: Jurnal Agroteknologi dan Agribisnis, 7*(2), 134-143. doi:10.30737/agrinika.v7i2.4773

Hafeez, A., Ali, S. S., Akhtar, J., Naz, S., Alhidary, I. A., Israr, M., & Khan, R. U. (2023). Garlic (*Allium sativum*), fenugreek (*Trigonella foenum-graecum*) and coriander (*Coriandrum sativum*): Performance, nutrient digestibility and blood metabolites in broilers. *Journal of Applied Animal Research, 51*(1), 624–629. doi:10.1080/09712119.2023.2264966

Hagerman, A. E., Riedl, K. M., Jones, G. A., Sovik, K. N., Ritchard, N. T., Hartzfeld, P. W., & Riechel, T. L. (1998). High molecular weight plant polyphenolics (tannins) as biological antioxidants. *Journal of Agricultural and Food Chemistry, 46*(5), 1887–1892. doi:10.1021/jf970975b PMID:29072454

Haider, S., Batool, Z., & Haleem, D. J. (2012). Nootropic and hypophagic effects following long term intake of almonds (Prunus amygdalus) in rats. *Nutrición Hospitalaria, 27*(6), 2109–2115. doi:10.3305/nh.2012.27.6.6140 PMID:23588464

Hajji, M., Falcimaigne-Gordin, A., Ksouda, G., Merlier, F., Thomasset, B., & Nasri, M. (2021). A water-soluble polysaccharide from*Anethum graveolens* seeds: Structural characterization, antioxidant activity and potential use as meat preservative. *International Journal of Biological Macromolecules, 167*, 516–527. doi:10.1016/j.ijbiomac.2020.12.006 PMID:33279565

Halaby, M. S., Farag, M. H., & Gerges, A. H. (2014). Potential effect of date pits fortified bread on diabetic rats. *International Journal of Nutrition and Food Sciences, 3*(2), 49–59. doi:10.11648/j.ijnfs.20140302.16

Halliwell, B. (1995). Antioxidant characterization: Methodology and mechanism. *Biochemical Pharmacology, 49*(10), 1341–1348. doi:10.1016/0006-2952(95)00088-H PMID:7763275

Halvorsen, B. L., Holte, K., Myhrstad, M. C., Barikmo, I., Hvattum, E., Remberg, S. F., ... Andersen, L. F. (2002). A systematic screening of total antioxidants in dietary plants. *The Journal of Nutrition, 132*(3), 461–471. doi:10.1093/jn/132.3.461 PMID:11880572

Hamilton, R. A., & Fukunaga, E. T. (1959). *Growing macadamia nuts in Hawaii*. Academic Press.

Hamming, I., Timens, W., Bulthuis, M. L., Lely, A. T., Navis, G., & van Goor, H. (2004). Tissue distribution of ACE2 protein, the functional receptor for SARS coronavirus. A first step in understanding SARS pathogenesis. *The Journal of Pathology, 203*(2), 631–637. doi:10.1002/path.1570 PMID:15141377

Hamza, L. F. (2017). *Anethum graveolens*: Physicochemical properties, medicinal uses, antimicrobial effects, antioxidant effect, anti-inflammatory and analgesic effects: A review. *World, 5*(7).

Hamza, A. A., & Amin, A. (2007). Apium graveolens modulates sodium valproate-induced reproductive toxicity in rats. *Journal of Experimental Zoology. Part A, Ecological Genetics and Physiology, 307*(4), 199–206. doi:10.1002/jez.357

Hamza, N., Berke, B., Umar, A., Cheze, C., Gin, H., & Moore, N. (2019). A review of Algerian medicinal plants used in the treatment of diabetes. *Journal of ethnopharmacology, 238*, 111841.

Hanaa M. A., Elshafie M. A., Ismail H. A., Mahmoud M. E., & Ibrahim H.M (2015). Chemical Studies And Phytochemicalscreening Of Grape Seeds (Vitis Vinifera L.). *Minia J. of Agric. Res. & Develop., 35*(2), 313-325.

Hanausek, M., Spears, E., Walazek, Z., Kowalczyk, M. C., Wendel, C., & Slaga, T. J. (2017). Inhibition of murine skin carcinogenesis by freeze dried grape powder and other grape derived major anto-oxidants. *Nutrition and Cancer, 63*(1), 28–38. PMID:21108125

Hancock, J. F. (2022). Five Hundred Years After the Great Encounter. In J. F. Hancock (Ed.), *World Agriculture Before and After 1492: Legacy of the Columbian Exchange* (pp. 135–144). Springer International Publishing. doi:10.1007/978-3-031-15523-9_10

Hanus, M., Lafon, J., & Mathieu, M. (2004). Double-blind, randomised, placebo-controlled study to evaluate the efficacy and safety of a fixed combination containing two plant extracts (Crataegus oxyacantha and Eschscholtzia californica) and magnesium in mild-to-moderate anxiety disorders. *Current Medical Research and Opinion*, *20*(1), 63–71. doi:10.1185/030079903125002603 PMID:14741074

Hashim, M. (2005). Effect of polyphenolic compounds from Coriandrum sativum on H2O2-induced oxidative stress in human lymphocytes. *Food Chemistry*, *92*(4), 653–660. doi:10.1016/j.foodchem.2004.08.027

Hattori, H., Yamauchi, K., Onwona-Agyeman, S., & Mitsunaga, T. (2017). Effect of Grains of Paradise (GP) Extract Intake on Obesity and Sympathetic Nerve Activity. *American Journal of Plant Sciences*, *08*(2), 85–95. doi:10.4236/ajps.2017.82007

Hazarika, S., Borah, P., Deb, P. K., Venugopala, K. N., & Hemalatha, S. (2023). Icacinaceae plant family: A recapitulation of the ethnobotanical, phytochemical, pharmacological, and biotechnological aspects. *Current Pharmaceutical Design*, *29*(15), 1193–1217. doi:10.2174/1381612829666230502164605 PMID:37132105

Hazra, B., Biswas, S., & Mandal, N. (2008). Antioxidant and free radical scavenging activity of Spondias pinnata. *BMC Complementary and Alternative Medicine*, *8*(1), 122–127. doi:10.1186/1472-6882-8-63 PMID:19068130

He, G., Yang, S., & Wang, Y. (2023). The potential of Amomum tsao-ko as a traditional Chinese medicine: Traditional clinical applications, phytochemistry and pharmacological properties. *Arabian Journal of Chemistry*, *16*(8), 104936. doi:10.1016/j.arabjc.2023.104936

Helaly, A., Mady, E., Eldekashy, M. H. Z., & Craker, L. (2015). Phytochemical Analysis of Some Celery Accessions. *Journal of Medicinally Active Plants*, *1*, 1–7. doi:10.7275/R5542KJF

Hernandez, E. M. (2016). 4 - Specialty Oils: Functional and Nutraceutical Properties. In T. A. B. Sanders (Ed.), *Functional Dietary Lipids* (pp. 69–101). Woodhead Publishing. doi:10.1016/B978-1-78242-247-1.00004-1

He, X., Chen, X., Yang, Y., Xie, Y., & Liu, Y. (2024). Medicinal plants for epileptic seizures: Phytoconstituents, pharmacology and mechanisms revisited. *Journal of Ethnopharmacology*, *320*, 117386. doi:10.1016/j.jep.2023.117386 PMID:37956914

Himeshsoni. (2012). A recent update of botanicals for wound healing activity. International Research Journal of Pharmacy, 3(7).

Hogg, N. (!998). Free radicals in disease. In Seminars in reproductive endocrinology. Thieme Medical Publishers, Inc. doi:10.1055/s-2007-1016284

Hokmabadi, H., & Sedaghati, E. (2014). Safety of Food and Beverages: Nuts. In Y. Motarjemi (Ed.), *Encyclopedia of Food Safety* (pp. 340–348). Academic Press. doi:10.1016/B978-0-12-378612-8.00292-4

Hong, M. Y., Groven, S., Marx, A., Rasmussen, C., & Beidler, J. (2018). Anti-inflammatory, antioxidant, and hypolipidemic effects of mixed nuts in atherogenic diet-fed rats. *Molecules (Basel, Switzerland)*, *23*(12), 3126. doi:10.3390/molecules23123126 PMID:30501043

Hordofa, T. S., & Tolossa, T. T. (2020). Cultivation and postharvest handling practices affecting yield and quality of major spices crops in Ethiopia: A review. *Cogent Food & Agriculture*, *6*(1), 1788896. doi:10.1080/23311932.2020.1788896

Hossain, R., Khan, R. A., Mukhopadhyay, N., Jain, D., & Islam, M. T. (2024). Phytochemical Constituents 6 and Antidiabetic Features of Black Cumin (Nigella sativa L.). In Antidiabetic Medicinal Plants and Herbal Treatments (pp. 95-115). CRC Press.

Houghton, P. (1995). Herbal products. 9. Bearberry, dandelion and celery. *The Pharmaceutical Journal*, *255*(6855), 272–273.

Hsieh, L.-C., Hsieh, S.-L., Chen, C.-T., Chung, J.-G., Wang, J.-J., & Wu, C.-C. (2015). Induction of α-phellandrene on autophagy in human liver tumor cells. *The American Journal of Chinese Medicine*, *43*(01), 121–136. doi:10.1142/S0192415X15500081 PMID:25649747

Huang, C.-S., Yin, M.-C., & Chiu, L.-C. (2011). Antihyperglycemic and antioxidative potential of Psidiumguajava fruit in streptozotocin-induced diabetic rats. *Food and Chemical Toxicology*, *49*(9), 2189–2195. doi:.2011.05.032 doi:10.1016/j.fct

Huang, C., Wang, Y., Li, X., Ren, L., Zhao, J., Hu, Y., Zhang, L., Fan, G., Xu, J., Gu, X., Cheng, Z., Yu, T., Xia, J., Wei, Y., Wu, W., Xie, X., Yin, W., Li, H., Liu, M., ... Cao, B. (2020). Clinical features of patients infected with 2019 novel coronavirus in Wuhan, China. *Lancet*, *395*(10223), 497–506. doi:10.1016/S0140-6736(20)30183-5 PMID:31986264

Huh, Y. C., Solmaz, I., & Sari, N. (2018). Morphological characterization of Korean and Turkish watermelon germplasm. *Proceedings of the IXth Eucarpia meeting on genetics and breeding of Cucurbitaceae*, 327-33.

Huseini, H. F., Hasani-Rnjbar, S., Nayebi, N., Heshmat, R., Sigaroodi, F. K., Ahvazi, M., ... Kianbakht, S. (2013). Capparis spinosa L.[Caper] fruit extract in treatment of type 2 diabetic patients: A randomized double-blind placebo-controlled clinical trial. *Complementary Therapies in Medicine*, *21*(5), 447–452. doi:10.1016/j.ctim.2013.07.003 PMID:24050578

Hussain, M., Sabri, R., Zia-Ul-Haq, M., & Riaz, M. (2023). Celery. In M. Zia-Ul-Haq, A. Abdulkreem Al-Huqail, M. Riaz, & U. Farooq Gohar (Eds.), *Essentials of Medicinal and Aromatic Crops* (pp. 1165–1190). Springer International Publishing. doi:10.1007/978-3-031-35403-8_45

Hu, W., Fitzgerald, M., Topp, B., Alam, M., & O'Hare, T. J. (2019). A review of biological functions, health benefits, and possible de novo biosynthetic pathway of palmitoleic acid in macadamia nuts. *Journal of Functional Foods*, *62*, 103520. doi:10.1016/j.jff.2019.103520

Ibe, C., Jacobs, C., Imo, C., Osuocha, K. U., & Okoronkwo, M. U. (2014). Evaluation of the antioxidant activities of *PsidiumguajavaandAloe vera. Journal of Pharmaceutical Research International*, *4*(3), 271–278.

Idm'hand, E., Msanda, F., & Cherifi, K. (2020). Ethnobotanical study and biodiversity of medicinal plants used in the Tarfaya Province, Morocco. *Acta Ecologica Sinica*, *40*(2), 134–144. doi:10.1016/j.chnaes.2020.01.002

Ikechukwu, M., Maroyi, A., & Cheikhyoussef, A. (2023). Cardamom Botany, Cultivars, and Genetic Diversity. In M. F. Ramadan (Ed.), *Cardamom (Elettaria cardamomum): Production, Processing and Properties* (pp. 39–50). Springer International Publishing. doi:10.1007/978-3-031-35426-7_3

Ileke, K., Adesina, J., & Okunola, O. (2017). Larvicidal and Pupicidal Potential of Aframomum melegueta K. Schum Extracts Against Mosquito, Anopheles Species. *Journal of Entomological Research Society*, *19*, 121–127.

Ilic, N. M., Dey, M., Poulev, A. A., Logendra, S., Kuhn, P. E., & Raskin, I. (2014). Anti-inflammatory activity of grains of paradise (Aframomum melegueta Schum) extract. *Journal of Agricultural and Food Chemistry*, *62*(43), 10452–10457. doi:10.1021/jf5026086 PMID:25293633

Imo, E., Orji, J., & Nweke, C. (2018). *Corrosion and Fungal Growth Inhibiting Effects of Piper guineense Extracts*. Academic Press.

INC. (2020). *Nuts and dried fruits statistical yearbook 2019/2020*. International Nut and Dried Fruit Council Spain.

Inegbenebor, U., Amadi, K., Ebomoyi, M., & Onyia, K. (2011). Abortifacient properties of alligator pepper (Aframomum melegueta) seeds. *Journal of Pharmacy & Bioresources*, *6*(2), 6. doi:10.4314/jpb.v6i2.63327

Inegbenebor, U., & Ebomoyi, M. (2014). Effect of saline extract of alligator pepper (zingiberaceae aframomum melegueta) on serum progesterone in pregnant sprague dawley rats. *International Journal of Herbs and Pharmacological Research.*, *3*, 46–51.

Inegbenebor, U., Ebomoyi, M. I., Onyia, K. A., Amadi, K., & Aigbiremolen, A. E. (2009). Effect of alligator pepper (Zingiberaceae aframomum melegueta) on first trimester pregnancy in Sprague Dawley rats. *Nigerian Journal of Physiological Sciences; Official Publication of the Physiological Society of Nigeria*, *24*(2), 161–164. PMID:20234758

Ingallina, C., Capitani, D., Mannina, L., Carradori, S., Locatelli, M., Di Sotto, A., ... Valletta, A. (2020). Phytochemical and biological characterization of Italian "sedano bianco di Sperlonga" Protected Geographical Indication celery ecotype: A multimethodological approach. *Food Chemistry*, *309*, 125649. doi:10.1016/j.foodchem.2019.125649

Inprasit, J., Itharat, A., Ruangnoo, S., Thisayakorn, K., Sukkasem, K., Prommee, N., Khoenok, W., Sriyam, K., Pahusee, D., & Davies, N. M. (2024). Ethnopharmacological analysis based on Thai traditional medicine theory and anti-inflammatory activity of Sa-Tri-Lhung-Klod remedy as a postpartum anti-inflammatory drug. *Journal of Ethnopharmacology*, *319*, 117207. doi:10.1016/j.jep.2023.117207 PMID:37739101

Insanu, M., Hartati, R., Bajri, F., & Fidrianny, I. (2021). Macadamia genus: An updated review of phytochemical compounds and pharmacological activities. *Biointerface Research in Applied Chemistry*, *11*, 14898–14905. doi:10.33263/BRIAC116.1489814905

Iriadam, M., Musa, D., Gumushan, H., & Baba, F. (2006). Effects of two Turkish medicinal plants Artemisia herba-alba and Teucrium polium on blood glucose levels and other biochemical parameters in rabbits. *Journal of Cell and Molecular Biology*, *5*(1), 19–24.

Ishtiaque, S., Naz, S., Siddiqi, R., Jabeen, S., & Ahmed, J. (2015). Antioxidant Activity and Phenolic Contents of Ajwain, Mustard, Fenugreek and Poppy Seed. *Recent Innovations in Chemical Engineering*, *7*(2), 119–127. doi:10.2174/2405520407666150425004125

Iweala, E. J., Uche, M. E., Dike, E. D., Etumnu, L. R., Dokunmu, T. M., Oluwapelumi, A. E., Okoro, B. C., Dania, O. E., Adebayo, A. H., & Ugbogu, E. A. (2023). *Curcuma longa* (Turmeric): Ethnomedicinal uses, phytochemistry, pharmacological activities and toxicity profiles—A review. *Pharmacological Research. Modern Chinese Medicine*, *6*, 100222. doi:10.1016/j.prmcm.2023.100222

Iwu, M. M. (2014). *Handbook of African medicinal plants*. doi:10.1201/b16292

Jadid, N., Widodo, A. F., Ermavitalini, D., Sa'adah, N. N., Gunawan, S., & Nisa, C. (2023). The medicinal Umbelliferae plant Fennel (*Foeniculum vulgare* Mill.): Cultivation, traditional uses, phytopharmacological properties, and application in animal husbandry. *Arabian Journal of Chemistry*, *16*(3), 104541. doi:10.1016/j.arabjc.2023.104541

Jamila, F., & Mostafa, E. (2014). Ethnobotanical survey of medicinal plants used by people in Oriental Morocco to manage various ailments. *Journal of Ethnopharmacology*, *154*(1), 76–87. doi:10.1016/j.jep.2014.03.016 PMID:24685583

Jana, S., & Shekhawat, G. S. (2010a). *Anethum graveolens*: An Indian traditional medicinal herb and spice. *Pharmacognosy Reviews*, *4*(8), 179. doi:10.4103/0973-7847.70915 PMID:22228959

Jana, S., & Shekhawat, G. S. (2010b). Phytochemical analysis and antibacterial screening of in vivo and in vitro extracts of Indian medicinal herb:*Anethum graveolens*. *Research Journal of Medicinal Plant*, *4*(4), 206–212. doi:10.3923/rjmp.2010.206.212

Jena, A., Sachdeva, R. K., Sharma, A., & Wanchu, A. (2009). Adverse drug reactions to nonnucleoside reverse transcriptase inhibitor-based antiretroviral regimen: A 24-week prospective study. *Journal of the International Association of Physicians in AIDS Care (Chicago, Ill.)*, *8*(5), 318–322. doi:10.1177/1545109709343967 PMID:19721097

Jitngarmkusol, S., Hongsuwankul, J., & Tananuwong, K. (2008). Chemical compositions, functional properties, and microstructure of defatted macadamia flours. *Food Chemistry*, *110*(1), 23–30. doi:10.1016/j.foodchem.2008.01.050 PMID:26050161

Johnson, M. A., & Janakiraman, N. M. (2015). Studies on Antioxidant and Phytochemical Profiles of Leptochloa Uniflora Hoechst. *International Journal of Botany*, *1*, 4.

Jones, D. (1972). Phytochemical ecology. *Cyanogenic glycosides and their function*, 103-124.

Jouad, H., Haloui, M., Rhiouani, H., El Hilaly, J., & Eddouks, M. (2001). Ethnobotanical survey of medicinal plants used for the treatment of diabetes, cardiac and renal diseases in the North centre region of Morocco. *Journal of Ethnopharmacology*, *77*(2-3), 175–182. doi:10.1016/S0378-8741(01)00289-6 PMID:11535361

Jouad, H., Maghrani, M., & Eddouks, M. (2002). Hypoglycemic effect of aqueous extract of Ammi visnaga in normal and streptozotocin-induced diabetic rats. *Journal of Herbal Pharmacotherapy*, *2*(4), 19–29. doi:10.1080/J157v02n04_03 PMID:15277079

Juzeliūnas, E., Ramanauskas, R., Lugauskas, A., Leinartas, K., Samulevičienė, M., & Sudavičius, A. (2006). Influence of wild strain Bacillus mycoides on metals: From corrosion acceleration to environmentally friendly protection. Electrochimica Acta -. *Electrochimica Acta*, *51*(27), 6085–6090. doi:10.1016/j.electacta.2006.01.067

K C., M. (2021). *Material uses of nut and other fruit shells.* Retrieved from https://nbn-resolving.org/urn:nbn:de:hbz:1383-opus4-8755

Kachmar, M. R., Naceiri Mrabti, H., Bellahmar, M., Ouahbi, A., Haloui, Z., El Badaoui, K., Bouyahya, A., & Chakir, S. (2021). Traditional knowledge of medicinal plants used in the Northeastern part of Morocco. *Evidence-Based Complementary and Alternative Medicine*, *2021*, 2021. doi:10.1155/2021/6002949 PMID:34512779

Kadifkova Panovska, T., Kulevanova, S., & Stefova, M. (2005). In vitro antioxidant activity of some Teucrium species (Lamiaceae). *Acta Pharmaceutica (Zagreb, Croatia)*, *55*(2), 207–214. PMID:16179134

Kaiser, A., Kammerer, D. R., & Carle, R. (2013). Impact of blanching on polyphenol stability and antioxidant capacity of innovative coriander (Coriandrum sativum L.) pastes. *Food Chemistry*, *140*(1-2), 332–339. doi:10.1016/j.foodchem.2013.02.077 PMID:23578650

Kamte, S. L. N., Ranjbarian, F., Campagnaro, G. D., Nya, P. C. B., Mbuntcha, H., Woguem, V., Womeni, H. M., Ta, L. A., Giordani, C., Barboni, L., Benelli, G., Cappellacci, L., Hofer, A., Petrelli, R., & Maggi, F. (2017). Trypanosoma brucei Inhibition by Essential Oils from Medicinal and Aromatic Plants Traditionally Used in Cameroon (Azadirachta indica, Aframomum melegueta, Aframomum daniellii, Clausena anisata, Dichrostachys cinerea and Echinops giganteus). *International Journal of Environmental Research and Public Health*, *14*(7), 737. doi:10.3390/ijerph14070737 PMID:28684709

Karimi, K., Ahari, A. B., Weisany, W., Pertot, I., Vrhovsek, U., & Arzanlou, M. (2016). Funneliformis mosseae root colonization affects*Anethum graveolens* essential oil composition and its efficacy against Colletotrichum nymphaeae. *Industrial Crops and Products*, *90*, 126–134. doi:10.1016/j.indcrop.2016.06.024

Karovičová, O., Kohajdová, Z., Minarovičová, L., & Kuchtová, V. (2015). The chemical composition of Grape fibre. *Potravinárstvo*, *9*(1), 53–57. doi:10.5219/428

Kassem, M. A., & Jacquin, A. (2001). Somatic Embryogenesis, Rhizogenesis, and Morphinan Alkaloids Production in Two Species of Opium Poppy. *Journal of Biomedicine & Biotechnology*, *1*(2), 70–78. doi:10.1155/S1110724301000237 PMID:12488612

Katiri, A., Barkaoui, M., Msanda, F., & Boubaker, H. (2017). Ethnobotanical survey of medicinal plants used for the treatment of diabetes in the Tizi n'Test region. *Journal of Pharmacognosy & Natural Products*, *3*(1), 2472–0992. doi:10.4172/2472-0992.1000130

Katoch, P., Raina, K., Sharma, R., & Chaudhary, A. (2024). Ayush Kwath: A major contribution of ayurveda in preventing COVID-19 infection. *Current Drug Therapy*, *19*(1), 60–80. doi:10.2174/1574885518666230601150338

Kaur, C., & Kapoor, H. C. (2001). Antioxidants in fruits and vegetables, the millennium's health. *International Journal of Food Science & Technology*, *36*(7), 703–725.

Kaur, C., & Kapoor, H. C. (2002). Anti-oxidant activity and total phenolic content of some Asian vegetables. *International Journal of Food Science & Technology*, *37*(2), 153–161. doi:10.1046/j.1365-2621.2002.00552.x

Kaur, G. J., & Arora, D. S. (2009). Antibacterial and phytochemical screening of*Anethum graveolens*, Foeniculum vulgare and Trachyspermum ammi. *BMC Complementary and Alternative Medicine*, *9*(1), 30. doi:10.1186/1472-6882-9-30 PMID:19656417

Kaur, N., Chahal, K., & Singh, R. (2018). Phytochemical screening and antioxidant activity of*Anethum graveolens* L. seed extracts. *Pharma Innov. J*, *7*, 324–329.

Kaur, R., & Pujari, M. (2019). Comparative study of control of seed mycoflora of Coriandrum sativum l. by herbal and chemical methods. *Plant Archives*, *19*(2), 1082–1084.

Kavyani, Z., Musazadeh, V., Golpour-hamedani, S., Moridpour, A. H., Vajdi, M., & Askari, G. (2023). The effect of Nigella sativa (black seed) on biomarkers of inflammation and oxidative stress: An updated systematic review and meta-analysis of randomized controlled trials. *Inflammopharmacology*, *31*(3), 1149–1165. doi:10.1007/s10787-023-01213-0 PMID:37036558

Kazeem, M. I., Akanji, M. A., Hafizur, R. M., & Choudhary, M. I. (2012). Antiglycation, antioxidant and toxicological potential of polyphenol extracts of alligator pepper, ginger and nutmeg from Nigeria. *Asian Pacific Journal of Tropical Biomedicine*, *2*(9), 727–732. doi:10.1016/S2221-1691(12)60218-4 PMID:23570003

Kazemian, M., & Abad, M., Reza Haeri, M., Ebrahimi, M., & Heidari, R. (2015). Anti-diabetic effect of Capparis spinosa L. root extract in diabetic rats. *Avicenna Journal of Phytomedicine*, *5*(4), 325. PMID:26445712

Kazemi, M. (2015). Phenolic profile, antioxidant capacity and anti-inflammatory activity of*Anethum graveolens* L. essential oil. *Natural Product Research*, *29*(6), 551–553. doi:10.1080/14786419.2014.951934 PMID:25154406

Keen, R. W., Deacon, A. C., Delves, H. T., Moreton, J. A., & Frost, P. G. (1994). Indian herbal remedies for diabetes as a cause of lead poisoning. *Postgraduate Medical Journal*, *70*(820), 113–114. doi:10.1136/pgmj.70.820.113 PMID:8170881

Kehailou, F. Z., Jabari, M., Labriji, A., El Khair, M. M., Bouzoubaa, H., Ouasmyne, G., ... Mestaghanmi, H. (2021). Quality of Life Assessment with EQ-5D-3L in a Moroccan Diabetic Population. *OAlib*, *8*(5), 1–14. doi:10.4236/oalib.1107357

Kehailou, F. Z., Jabari, M., Labriji, A., & Mestaghanmi, H. (2019). Prevalence de l'anemie chez une population des diabetiques, Casablanca-maroc. *Am J Innov Res Appl Sci*, *9*(1), 109–116.

Keller, F., & Matile, P. (1989). Storage of sugars and mannitol in petioles of celery leaves. *The New Phytologist*, *113*(3), 291–299. doi:10.1111/j.1469-8137.1989.tb02406.x

Khairullah, A. R., Solikhah, T. I., Ansori, A. N. M., Hidayatullah, A. R., Hartadi, E. B., Ramandinianto, S. C., & Fadholly, A. (2021). Review on the Pharmacological and Health Aspects of Apium Graveolens or Celery: An Update. *Systematic Reviews in Pharmacy*, *12*(2).

Khalifa, A. A., Rashad, R. M., & El-Hadidy, W. F. (2021). Thymoquinone protects against cardiac mitochondrial DNA loss, oxidative stress, inflammation and apoptosis in isoproterenol-induced myocardial infarction in rats. *Heliyon*, *7*(7), e07561. doi:10.1016/j.heliyon.2021.e07561 PMID:34355083

Khalil, A., Nawaz, H., Ben Ghnia Delgay, J., Rehman, R., & Nadeem, F. (2015). Value Added Products, Chemical Constituents and Medicinal Uses of Celery (Apium graveolens L.). *RE:view*, 40–48.

Khaliq, F. A., & Mushtaq, A. (2023). The five most traded compounds worldwide: Importance, opportunities, and risks. *International Journal of Chemical and Biochemical Science*, *23*, 185–194.

Khanam, M., Dar, A. H., Beg, F., Khan, S. A., Nayik, G. A., & Karabagias, I. K. (2023). Nutmeg essential oil. In G. A. Nayik & M. J. Ansari (Eds.), *Essential Oils* (pp. 391–399). Academic Press. doi:10.1016/B978-0-323-91740-7.00012-8

Khan, M. R., & Aslam, S. (2023a). Cardamom Safety. In M. F. Ramadan (Ed.), *Cardamom (Elettaria cardamomum): Production, Processing and Properties* (pp. 227–241). Springer International Publishing. doi:10.1007/978-3-031-35426-7_17

Khan, M. R., & Aslam, S. (2023b). Plant Morphological Traits of Elettaria cardamomum. In M. F. Ramadan (Ed.), *Cardamom (Elettaria cardamomum): Production, Processing And Properties* (pp. 51–66). Springer International Publishing. doi:10.1007/978-3-031-35426-7_4

Kharchoufa, L., Merrouni, I. A., Yamani, A., & Elachouri, M. (2018). Profile on medicinal plants used by the people of North Eastern Morocco: Toxicity concerns. *Toxicon*, *154*, 90–113. doi:10.1016/j.toxicon.2018.09.003 PMID:30243516

Kilian, I. C., Swenson, S. J., Mengual, X., Gemeinholzer, B., Hamm, A., Wägele, J. W., & Peters, R. S. (2023). More complex than you think: Taxonomic and temporal patterns of plant–pollinator networks of caraway (*Carum carvi* L.). *Molecular Ecology*, *32*(13), 3702–3717. doi:10.1111/mec.16943 PMID:37004150

Kim, H., Son, K., Chang, H., & Kang, S. (2004). Anti-inflammatory Plant Flavonoids and Cellular Action Mechanism. *Journal of Pharmacological Sciences*, *96*(3), 229–245. doi:10.1254/jphs.CRJ04003X

Kim, J. H., Kim, T. I., & Ma, J.-Y. (2022). Synergistic effects of novel herbal decoctions from *Panax ginseng* and *Morus alba* on tyrosinase activity and melanogenesis in vitro. *Heliyon*, *8*(2), e08866. doi:10.1016/j.heliyon.2022.e08866 PMID:35198755

Kimoto, Y., Kessler, R., & Constantinou, C. E. (1990). Relaxation mechanisms of antispasmodics papaverine and thiphenamil on the human corpus cavernosum. *The Journal of Urology*, *144*(6), 1497–1499. doi:10.1016/S0022-5347(17)39785-9 PMID:2231955

Kim, Y., Keogh, J. B., & Clifton, P. M. (2017). Benefits of nut consumption on insulin resistance and cardiovascular risk factors: Multiple potential mechanisms of actions. *Nutrients*, *9*(11), 1271. doi:10.3390/nu9111271 PMID:29165404

Kiralan, M., Calikoglu, E., Ipek, A., Bayrak, A., & Gurbuz, B. (2009). Fatty acid and volatile oil composition of different coriander (Coriandrum sativum) registered varieties cultivated in Turkey. *Chemistry of Natural Compounds*, *45*(1), 100–102. doi:10.1007/s10600-009-9240-2

Kirimuhuzya, C., & Anywar, G. (2015). Phytochemical profile and antibacterial activity of crude extracts of the pod of Aframomum angustifolium (Sonn.) K. Schum. *European Journal of Biological Research*, *5*, 36–41.

Kmail, A., Said, O., & Saad, B. (2023). How thymoquinone from *Nigella sativa* accelerates wound healing through multiple mechanisms and targets. *Current Issues in Molecular Biology*, *45*(11), 9039–9059. doi:10.3390/cimb45110567 PMID:37998744

Knezevic, P., Aleksic, V., Simin, N., Svircev, E., Petrovic, A., & Mimica-Dukic, N. (2016). Antimicrobial activity of Eucalyptus camaldulensis essential oils and their interactions with conventional antimicrobial agents against multi-drug resistant Acinetobacter baumannii. *Journal of Ethnopharmacology*, *178*, 125–136. doi:10.1016/j.jep.2015.12.008 PMID:26671210

Knight, J. A. (1998). Free radicals: Their history and current status in aging and disease. *Annals of Clinical and Laboratory Science*, *28*(6), 331–346. PMID:9846200

Knockaert, G., Pulissery, S. K., Lemmens, L., Van Buggenhout, S., Hendrickx, M., & Van Loey, A. (2013). Isomerisation of carrot β-carotene in presence of oil during thermal and combined thermal/high pressure processing. *Food Chemistry*, *138*(2-3), 1515–1520. doi:10.1016/j.foodchem.2012.10.080 PMID:23411275

Knöss, W. (1999). Marrubium vulgare [White Horehound]: In vitro culture, and the production of diterpene marrubiin and other secondary metabolites. *Medicinal and Aromatic Plants XI*, 274-289.

Knott, E., Gürer, C. K., Ellwanger, J., Ring, J., & Darsow, U. (2008). Macadamia nut allergy. *Journal of the European Academy of Dermatology and Venereology*, *22*(11), 1394–1395. doi:10.1111/j.1468-3083.2008.02657.x PMID:18331311

Koaze, H., Karanja, P. N., Kojima, M., Baba, N., & Ishibashi, K.KOAZE. (2002). Lipid accumulation of macadamia nuts during kernel development. *Nihon Shokuhin Hozo Kagakkaishi*, *28*(2), 67–73. doi:10.5891/jafps.28.67

Kochhar, S. L. (1986). *Tropical crops: a textbook of economic botany*. Macmillan London.

Kolarovic, J., Popovic, M., Mikov, M., Mitic, R., & Gvozdenovic, L. (2009). Protective effects of celery juice in treatments with doxorubicin. *Molecules (Basel, Switzerland)*, *14*(4), 1627–1638. doi:10.3390/molecules14041627

Kooti, W., Ali Akbari, S., Asadi-Samani, M., Ghadery, H., & Ashtary-Larky, D. (2015). A review on medicinal plant of Apium graveolens. *Advanced Herbal Medicine*, *1*, 48–59.

Kooti, W., Ghasemi-boroon, M., Ghafourian, M., Asadi-Samani, M., Harizi, M., Sharafi-Ahvazi, N., & Afrisham, R. (2015). The effects of celery leave extract on male hormones in rats. *Journal of HerbMed Pharmacology*, *4*(2), 56–60.

Korshed, P., Li, L., Liu, Z., Mironov, A., & Wang, T. (2019). Size-dependent antibacterial activity for laser-generated silver nanoparticles. *Journal of Interdisciplinary Nanomedicine*, *4*(1), 24–33. doi:10.1002/jin2.54

Kostolanská, J., Jakus, V., & Barák, L. (2009). Monitoring of Early and Advanced Glycation in Relation to the Occurrence of Microvascular Complications in Children and Adolescents with Type 1 Diabetes Mellitus. *Physiological research / Academia Scientiarum Bohemoslovaca*, *58*, 553-61.

Koyyati, R. (2013). *Biological synthesis of silver nanoparticles using Raphanus sativus var. longipinnatus leaf extract and evaluation of their antioxidant and antibacterial activity*. International Journal of Medicine and Pharmaceutical Sciences, 3(4), 89-100.

Koyyati, R. (2014). Antibacterial activity of silver nanoparticles synthesized using Amaranthus viridis twig extract. *International Journal of Research in Pharmaceutical Sciences*, *5*(1), 32–39.

Kriedeman, P. (1964). Cyanide formation in Sorghum almum in relation to nitrogen and phosphorus nutrition. *Australian Journal of Experimental Agriculture*, *4*(12), 15–16. doi:10.1071/EA9640015

Krishnamurthy, K. (2008). Celery. In *Chemistry of spices* (pp. 401–412). CABI Wallingford UK. doi:10.1079/9781845934057.0401

Kudle, K. R. (2013). Fruit (epicarp and endocarp) extract mediated synthesis of silver nanoparticles from Sterculia foetida plant and evaluation of their antimicrobial activity. *Nanoscience and Nanotechnology: An International Journal*, *3*(3), 45–48.

Kudle, K. R. (2013). Green synthesis of silver nanoparticles using water soluble gum of Sterculia foetida and evaluation of its antimicrobial activity. *International Journal of Research in Pharmaceutical Sciences*, *4*(4), 563–568.

Kudle, K. R. (2013). Microwave assisted green synthesis of silver nanoparticles using Stigmaphyllon littorale leaves their characterization and anti-microbial activity. *International Journal of Nanomaterials and Biostructures*, *3*(1), 13–16.

Kulkarni, A., Nair, A., Hirode, K., Chavan, M., Jain, M., Majumdar, P., & Jamkhedkar, S. (2023). Indian Traditional Foods from an Ayurvedic Perspective. In *Ethnic Knowledge and Perspectives of Medicinal Plants* (pp. 395–424). Apple Academic Press. doi:10.1201/9781003353089-19

Kumar, A., Kennedy-Boone, D., Weisz, H., Capra, B., Uchida, T., Jennings, K., & .... (2015). Neuroprotective Effects of Aframomum melegueta Extract after Experimental Traumatic Brain Injury. *Natural Products Chemistry & Research*, *3*(1), 3. doi:10.4172/2329-6836.1000167

Kumar, C., Sarwar, M., & Dimri, B. (1977). Bulgarian coriander in India and its future prospects in export trade. *Indian Perfumer*, *21*(3), 146–150.

Kumar, M., Kaushik, D., Kaur, J., Proestos, C., Oz, F., Oz, E., Gupta, P., Kundu, P., Kaur, A., Anisha, A., & Ritika, R. (2022). A critical review on obesity: Herbal approach, bioactive compounds, and their mechanism. *Applied Sciences (Basel, Switzerland)*, *12*(16), 8342. doi:10.3390/app12168342

Kumar, V., Kumar, S., Chauhan, R. S., & Dwivedi, S. (2024). Medicinal Plants and Their Role in Mitigation of Arsenic-Induced Toxicity. In *Phytoremediation Potential of Medicinal and Aromatic Plants* (pp. 154–171). CRC Press.

Kwiecień-Obara, E., Szponar, J., Krajewska, A., Witkowska, A., Radoniewicz, A., & Szponar, M. (2016). Morphine (obtained from poppy seeds) and dextrometorfan poisoning– a case report. *Przegląd Lekarski*, *73*(8), 596–598. PMID:29677437

Laadim, M., Ouahidi, M. L., Zidane, L., El Hessni, A., Ouichou, A., & Mesfioui, A. (2017). Ethnopharmacological survey of plants used for the treatment of diabetes in the town of Sidi Slimane. *Journal of Pharmacognosy and Phytotherapy*, *9*(6), 101–110. doi:10.5897/JPP2016.0437

Lachman, J., Hamouz, K., Orsak, M., & Pivec, V. (2000). Potato tubers as a significant source of antioxidants in human nutrition. *Rostlinná Výroba*, *46*(5), 231–236.

Lahsissene, H., Kahouadji, A., & Hseini, S. (2009). Catalogue des plantes médicinales utilisées dans la région de Zaër. *Lejeunia, revue de botanique*.

Lamas, D. L., & Álvarez, S. T. (2023). Development of snacks based on Macadamia integrifolia nuts enriched with omega-3. *Food Chemistry Advances*, *3*, 100564. doi:10.1016/j.focha.2023.100564

Lamaty, G., Menut, C., Koudou, J., Régnier, P., & Bessière, J.-M. (1993). Aromatic Plants of Tropical Central Africa. XI. Essential Oils of Leaf and Seed of Aframomum melegueta (Roscoe) K. Schum. from Central African Republic. *The Journal of Essential Oil Research*, *5*(1), 81–83. doi:10.1080/10412905.1993.9698174

Lančaričová, A., Havrlentova, M., Muchová, D., & Bednárová, A. (2016). Oil content and fatty acids composition of poppy seeds cultivated in two localities of Slovakia. *Agriculture*, *62*(1), 19–27. doi:10.1515/agri-2016-0003

Lang, S., Popp, T., Kriegs, C. S., Schmidt, A., Balszuweit, F., Menacher, G., Kehe, K., Thiermann, H., Gudermann, T., & Steinritz, D. (2018). Anti-apoptotic and moderate anti-inflammatory effects of berberine in sulfur mustard exposed keratinocytes. *Toxicology Letters*, *293*, 2–8. doi:10.1016/j.toxlet.2017.09.004 PMID:28916288

Lanni, C., Govoni, S., Lucchelli, A., & Boselli, C. (2009). Depression and antidepressants: Molecular and cellular aspects. *Cellular and Molecular Life Sciences*, *66*(18), 2985–3008. doi:10.1007/s00018-009-0055-x PMID:19521663

Lara, D., Vilcacundo, E., Carrillo, C., Carpio, C., Silva, M., Alvarez, M., & Carrillo, W. (2017). Obtention of protein concentrate and polyphenols from macadamia (Macadamia integrifolia) with aqueous extraction method. *Asian Journal of Pharmaceutical and Clinical Research*, *10*(2), 138–142. doi:10.22159/ajpcr.2017.v10i2.14808

Laribi, B., Kouki, K., M'Hamdi, M., & Bettaieb, T. (2015). Coriander (Coriandrum sativum L.) and its bioactive constituents. *Fitoterapia*, *103*, 9–26. doi:10.1016/j.fitote.2015.03.012 PMID:25776008

Larsson, M., & Foley, B. (2023). The king's spice cabinet–Plant remains from Gribshunden, a 15th century royal shipwreck in the Baltic Sea. *PLoS One*, *18*(1), e0281010. doi:10.1371/journal.pone.0281010 PMID:36701280

Laura Graciela, M., Mario, S., Karen Patricia, M., Eva Eugenia Soledad, C., Edelira, V., & Laura, C. (2022). Challenges and Advances in the Production of Export-Quality Macadamia and Its Integral Use with Green Technologies. In K. Muhammad Sarwar (Ed.), Tropical Plant Species and Technological Interventions for Improvement. Rijeka: IntechOpen.

Lee, B. K., Kim, J. H., Jung, J. W., Choi, J. W., Han, E. S., Lee, S. H., Ko, K. H., & Ryu, J. H. (2005). Myristicin-induced neurotoxicity in human neuroblastoma SK-N-SH cells. *Toxicology Letters*, *157*(1), 49–56. doi:10.1016/j.toxlet.2005.01.012 PMID:15795093

Lee, Y. J., Thiruvengadam, M., Chung, I. M., & Nagella, P. (2013). Polyphenol composition and antioxidant activity from the vegetable plant'Artemisia absinthium'L. *Australian Journal of Crop Science*, *7*(12), 1921–1926.

Lehnen, T. E., da Silva, M. R., Camacho, A., Marcadenti, A., & Lehnen, A. M. (2015). A review on effects of conjugated linoleic fatty acid (CLA) upon body composition and energetic metabolism. *Journal of the International Society of Sports Nutrition*, *12*(1), 36–36. doi:10.1186/s12970-015-0097-4 PMID:26388708

Letko, M., Marzi, A., & Munster, V. (2020). Functional assessment of cell entry and receptor usage for SARS-CoV-2 and other lineage B betacoronaviruses. *Nature Microbiology*, *5*(4), 562–569. doi:10.1038/s41564-020-0688-y PMID:32094589

Leung, A. Y., & Foster, S. (1996). Encyclopedia of common natural ingredients used in food, drugs, and cosmetics. *Journal of the American Chemical Society*, *118*(37), 8988.

Li Li, W., Jiahong, X., Duoyun, Y., Jinping, L., Jun, C., & Gang, W. (2009). Effects of Quercetin on Gene and Protein Expression of NOX and NOS after Myocardial Ischemia and Reperfusion in Rabbit. *Cardiovascular Therapeutics*, *27*(1), 28–33. doi:10.1111/j.1755-5922.2009.00071.x

Liang, Y., Dai, X., Cao, Y., Wang, X., Lu, J., Xie, L., Liu, K., & Li, X. (2023). The neuroprotective and antidiabetic effects of trigonelline: A review of signaling pathways and molecular mechanisms. *Biochimie*, *206*, 93–104. doi:10.1016/j.biochi.2022.10.009 PMID:36257493

Li, J., Chen, W., Liu, H., Liu, H., Xiang, S., You, F., Jiang, Y., Lin, J., Zhang, D., & Zheng, C. (2023). Pharmacologic effects approach of essential oils and their components on respiratory diseases. *Journal of Ethnopharmacology*, *304*, 115962. doi:10.1016/j.jep.2022.115962 PMID:36529244

Li, J.-L., Wang, L., Liu, X.-Y., Zhang, Z.-P., Guo, H.-C., Liu, W.-M., & Tang, S.-H. (2009). In vitro cancer cell imaging and therapy using transferrin-conjugated gold nanoparticles. *Cancer Letters*, *274*(2), 319–326. doi:10.1016/j.canlet.2008.09.024 PMID:18977071

Li, L., & Yang, X. (2018). The Essential Element Manganese, Oxidative Stress, and Metabolic Diseases: Links and Interactions. *Oxidative Medicine and Cellular Longevity*, *7580707*, 1–11. Advance online publication. doi:10.1155/2018/7580707 PMID:29849912

Lim, T. K. (2012). *Edible medicinal and non-medicinal plants. Volume 3, Fruits*. Retrieved from https://site.ebrary.com/id/10532978 doi:10.1007/978-94-007-2534-8

Lim, T. K. (2012). *Edible medicinal and non-medicinal plants*. Springer.

Lin, T.-K., Zhong, L., & Santiago, J. L. (2017). Anti-Inflammatory and Skin Barrier Repair Effects of Topical Application of Some Plant Oils. *International Journal of Molecular Sciences*, *19*(1), 70. doi:10.3390/ijms19010070 PMID:29280987

Li, S., Huang, X., Li, Y., Ding, R., Wu, X., Li, L., ... Gu, R. (2024). Spectrum-effect relationship in Chinese herbal medicine: Current status and future perspectives. *Critical Reviews in Analytical Chemistry*, *xx*(x), 1–22. doi:10.1080/10408347.2024.2339955 PMID:38127670

Li, T., Zhu, J., Guo, L., Shi, X., Liu, Y., & Yang, X. (2013). Differential effects of polyphenols-enriched extracts from hawthorn fruit peels and fleshes on cell cycle and apoptosis in human MCF-7 breast carcinoma cells. *Food Chemistry*, *141*(2), 1008–1018. doi:10.1016/j.foodchem.2013.04.050 PMID:23790880

Little, B., & Ray, R. (2001). A review of fungal influenced corrosion. *Corrosion Reviews*, *19*(5-6), 401–417. doi:10.1515/CORRREV.2001.19.5-6.401

Liu, C., Wang, W., & Gu, J. (2023). Targeting ferroptosis: New perspectives of Chinese herbal medicine in the treatment of diabetes and its complications. *Heliyon*, *9*(12), e22250. doi:10.1016/j.heliyon.2023.e22250 PMID:38076182

Liu, H. M., & Tu, Y. K. (2002). The efficacy of papaverine administration by different routes for the treatment of experimental acute cerebral vasospasm. *Journal of Clinical Neuroscience*, *9*(5), 561–565. doi:10.1054/jocn.2001.1036 PMID:12383416

Liu, Q., Meng, X., Li, Y., Zhao, C.-N., Tang, G.-Y., & Li, H.-B. (2017). Antibacterial and Antifungal Activities of Spices. *International Journal of Molecular Sciences*, *18*(6), 1283. doi:10.3390/ijms18061283 PMID:28621716

Liu, S.-W., Katsafanas George, C., Liu, R., Wyatt Linda, S., & Moss, B. (2015). Poxvirus Decapping Enzymes Enhance Virulence by Preventing the Accumulation of dsRNA and the Induction of Innate Antiviral Responses. *Cell Host & Microbe*, *17*(3), 320–331. doi:10.1016/j.chom.2015.02.002 PMID:25766293

Liu, Y., Hu, X., Zhao, X., & Song, H. (2012). Combined effect of high pressure carbon dioxide and mild heat treatment on overall quality parameters of watermelon juice. *Innovative Food Science & Emerging Technologies*, *13*, 112–119. doi:10.1016/j.ifset.2011.11.001

Li, Y., Zheng, M. I. N., Zhai, X., Huang, Y., Khalid, A., Malik, A., ... Hou, X. (2015). Effect of Gymnema sylvestre, Citrullus colocynthis and Artemisia absinthium on blood glucose and lipid profile in diabetic human. *Acta Poloniae Pharmaceutica*, *72*(5), 981–985. PMID:26665406

Lo, D. S., & Chua, T. H. (1992). Poppy seeds: Implications of consumption. *Medicine, Science, and the Law*, *32*(4), 296–302. doi:10.1177/002580249203200403 PMID:1435125

Loo, Y. Y., Rukayadi, Y., Nor-Khaizura, M.-A.-R., Kuan, C. H., Chieng, B. W., Nishibuchi, M., & Radu, S. (2018). In Vitro Antimicrobial Activity of Green Synthesized Silver Nanoparticles Against Selected Gram-negative Foodborne Pathogens. *Frontiers in Microbiology*, *9*(1555), 1555. doi:10.3389/fmicb.2018.01555 PMID:30061871

López-Muñoz, F., & Alamo, C. (2009). Monoaminergic neurotransmission: The history of the discovery of antidepressants from 1950s until today. *Current Pharmaceutical Design*, *15*(14), 1563–1586. doi:10.2174/138161209788168001 PMID:19442174

Luan, J.-M., & Liang, T. (1983). An improved macadamia nut husker. *Journal of Agricultural Engineering Research*, *28*(4), 337–347. doi:10.1016/0021-8634(83)90065-3

Lucas-Sánchez, M., Fadhlaoui-Zid, K., & Comas, D. (2023). The genomic analysis of current-day North African populations reveals the existence of trans-Saharan migrations with different origins and dates. *Human Genetics*, *142*(2), 305–320. doi:10.1007/s00439-022-02503-3 PMID:36441222

Lugauskas, A., Leinartas, K., & Binkauskienė, E. (2008). Possibility of micromycetes detected in dust to grow on metals (Al, Fe, Cu, Zn) and on polyaniline-modified Ni. *Ekologija (Lietuvos Mokslu Akademija)*, *54*(3), 149–157. doi:10.2478/V10055-008-0023-z

Lu, Z., Li, W., & Wang, P. (2011). Chemical Composition and Ability of Scavenging DPPH Radical of Essential Oil and Residue from The Celery Seed. *Advanced Materials Research*, *183-185*, 18–21. doi:10.4028/www.scientific.net/AMR.183-185.18

Macfarlane, S., Macfarlane, G. T., & Cummings, J. H. (2006). Review article: Prebiotics in the gastrointestinal tract. *Alimentary Pharmacology & Therapeutics*, *24*(5), 701–714. doi:10.1111/j.1365-2036.2006.03042.x PMID:16918875

Machado, A. M., Lopes, V., Barata, A. M., Póvoa, O., Farinha, N., & Figueiredo, A. C. (2023). Chemical variability of the essential oils from two Portuguese Apiaceae: Coriandrum sativum L. and Foeniculum vulgare Mill. *Plants*, *12*(14), 2749. doi:10.3390/plants12142749 PMID:37514362

Madhu, S. V., Rao, P. V., Chandalia, H. B., Jothydev, K., & Gupta, A. (2023). A multicentric, randomized, controlled trial of yoga and fenugreek in prevention of type 2 diabetes mellitus: Methodological details—the Indian Prevention of Diabetes Study (IPDS). *International Journal of Diabetes in Developing Countries*, *43*(6), 875–882. doi:10.1007/s13410-023-01183-7

Mahajan, M., Vaidya, V., Farande, P., Bhagde, S., & Jadhav, R. (2023). Review on essential oils and ways to use them for the treatment of arthritis. *Biosciences Biotechnology Research Asia*, *20*(4), 1181–1194. doi:10.13005/bbra/3167

Majeed, M., & Prakash, L. (2005). Novel natural approaches to anti-aging skin care. *Cosmetics and Toiletries Manufacture Worldwide*, *1*, 11–15.

Malhotra, S. K. (2006). 15 - Caraway. In K. V. Peter (Ed.), *Handbook of Herbs and Spices* (pp. 270–298). Woodhead Publishing. doi:10.1533/9781845691717.3.270

Malick, C. P., & Singh, M. B. (1980). Estimation of Phenol. In *Plant Enzymology and histoenzymology* (p. 286). Kalyani Publishers.

Mamedov, N., & Craker, L. (2012). Man and medicinal plants: A short review. *Acta Horticulturae*, (964), 181–190. doi:10.17660/ActaHortic.2012.964.22

Mandal, S., & Mandal, M. (2015). Coriander (Coriandrum sativum L.) essential oil: Chemistry and biological activity. *Asian Pacific Journal of Tropical Biomedicine*, *5*(6), 421–428. doi:10.1016/j.apjtb.2015.04.001

Manisha, D. (2014). Microwave assisted biogenic synthesis of silver nanoparticles using dried seed extract of Coriandrum sativum, characterization and antimicrobial activity. *International Journal of Chemtech Research*, *6*, 3957–3961.

Mapeka, T., Sandasi, M., Ncube, E., Viljoen, A., & van Vuuren, S. (2024). Enhancing the antimicrobial efficacy of common herbs and spices through an optimized polyherbal approach. *South African Journal of Botany*, *164*, 91–99. doi:10.1016/j.sajb.2023.11.030

Maqbool, M., Chaudhary, K., Chalotra, R., Chauhan, S., & Dahiya, R. S. (2024). Phytopharmacology of most common Indian culinary spices and their potential in developing new pharmaceutical therapies. *Current Traditional Medicine*, *10*(2), 117–140. doi:10.2174/2215083810666230330161740

Marchant, R. (2021). Trading Language, New Crops, New Relationships: Digging Anthropocene Foundations. In R. Marchant (Ed.), *East Africa's Human Environment Interactions: Historical Perspectives for a Sustainable Future* (pp. 115–162). Springer International Publishing. doi:10.1007/978-3-030-88987-6_3

Marciano, M. A., Panicker, S. X., Liddil, G. D., Lindgren, D., & Sweder, K. S. (2018). Development of a Method to Extract Opium Poppy (Papaver somniferum L.) DNA from Heroin. *Scientific Reports*, *8*(1), 2590–2590. doi:10.1038/s41598-018-20996-9 PMID:29416103

Mariappan, B., Kaliyamurthi, V., & Binesh, A. (2023). Chapter 8 - Medicinal plants or plant derived compounds used in aquaculture. In J. Mathew, M. S. Jose, R. E.K, & A. Kumar (Eds.), Recent Advances in Aquaculture Microbial Technology (pp. 153-207). Academic Press. doi:10.1016/B978-0-323-90261-8.00003-1

Martinez, S. E., Davies, N. M., & Reynolds, J. K. (2013). *Toxicology and safety of flavonoids. Flavonoid Pharmacokinetics: Methods of Analysis*. Preclinical and Clinical Pharmacokinetics, Safety, and Toxicology.

Marzocchella, L., Fantini, M., Benvenuto, M., Masuelli, L., Tresoldi, I., Modesti, A., & Bei, R. (2011). Dietary Flavonoids: Molecular Mechanisms of Action as Anti- Inflammatory Agents. *Recent Patents on Inflammation & Allergy Drug Discovery*, *5*(3), 200–220. doi:10.2174/187221311797264937

Masihuddin, M., Jafri, M. A., Aisha, S., & Shahid, C. (n.d.). Traditional Uses, Phytochemistry and Pharmacological Activities of Papaver Somniferum with Special Reference of Unani Medicine an Updated Review. doi:10.22270/jddt.v8i5-s.2069

Matasyoh, J., Maiyo, Z. C., Ngure, R. M., & Chepkorir, R. (2009). Chemical composition and antimicrobial activity of the essential oil of Coriandrum sativum. *Food Chemistry*, *113*(2), 526–529. doi:10.1016/j.foodchem.2008.07.097

Matejka, P., Vlckova, B., Vohlidal, J., Pancoska, P., & Baumruk, V. (1992). The role of triton X-100 as an adsorbate and a molecular spacer on the surface of silver colloid: A surface-enhanced Raman scattering study. *Journal of Physical Chemistry*, *96*(3), 1361–1366. doi:10.1021/j100182a063

Matés, J. M., Pérez-Gómez, C., & De Castro, I. N. (1999). Antioxidant enzymes and human diseases. *Clinical Biochemistry*, *32*(8), 595–603. doi:10.1016/S0009-9120(99)00075-2 PMID:10638941

Máthé, Á., & Turgut, K. (2023). Introduction to Medicinal and Aromatic Plants in Türkiye. In Á. Máthé & K. Turgut (Eds.), *Medicinal and Aromatic Plants of Turkey* (pp. 1–30). Springer International Publishing. doi:10.1007/978-3-031-43312-2_1

McGee, H. (2010). *Cilantro haters, it's not your fault.* The New York Times.

Mechchate, H., Es-safi, I., Bari, A., Grafov, A., & Bousta, D. (2020). Ethnobotanical survey about the management of diabetes with medicinal plants used by diabetic patients in region of FezMeknes, Morocco. *Journal of Ethnobotany Research and Applications*.

Meena, S., Balraj, S., Dheraj, S., Ranjan, J., & Meena, R. (2013). Pre and post harvest factors effecting yield and quality of seed spices: A review. *International J. Seed Spices*, *3*(1), 1–11. doi:10.56093/IJSS.v12i1.1

Mehra, M., Pasricha, V., & Gupta, R. K. (2015). Estimation of nutritional, phytochemical and antioxidant activity of seeds of musk melon (*Cucumismelo*) and water melon (*Citrulluslanatus*) and nutritional analysis of their respective oils. *Food Chemistry*, *3*(6), 98–102.

Mejia, S. B., Kendall, C. W., Viguiliouk, E., Augustin, L. S., Ha, V., Cozma, A. I., ... Leiter, L. A. (2014). Effect of tree nuts on metabolic syndrome criteria: A systematic review and meta-analysis of randomised controlled trials. *BMJ Open*, *4*(7), e004660. doi:10.1136/bmjopen-2013-004660 PMID:25074070

Melnyk, J. P., & Marcone, M. F. (2011). Aphrodisiacs from plant and animal sources—A review of current scientific literature. *Food Research International*, *44*(4), 840–850. doi:10.1016/j.foodres.2011.02.043

Memet, I., Kirici, S., Giray, E. S., Murat, T., & Taghikhani, H. (2014). Determination of suitable coriander (*Coriandrum sativum* L.) cultivars for eastern mediterranean region. *Turkish Journal of Field Crops*, *19*(1), 1–6. doi:10.17557/tjfc.05577

Menaga, S., Kripa, K., & Sangeetha, R. (2017). Evaluation of in vitro inhibition of proton-pump by nanosilver particles synthesized using seeds of*Anethum graveolens. Evaluation*, *10*(6).

Mereles Ceuppens, L. G., Ferro, E. A., Alvarenga, N. L., Caballero, S. B., Wiszovaty, L., Piris Jara, P. A., & Michajluk, B. J. (2017). Chemical composition of Macadamia integrifolia (Maiden and Betche) nuts from Paraguay. *International Food Research Journal*, *24*(6).

Miller, R. E., & Tuck, K. L. (2013). The rare cyanogen proteacin, and dhurrin, from foliage of Polyscias australiana, a tropical Araliaceae. *Phytochemistry*, *93*, 210–215. doi:10.1016/j.phytochem.2013.03.004 PMID:23566716

Mipeshwaree Devi, A., Mutum, R. D., Lakshmipriyari Devi, M., Khomdram, K. D., Premi Devi, P., Singh, L. H., & Das, S. (2023). Traditional Practices of Ethnomedicinal Plants in the North-Eastern Region of India and Their Conservation for Sustainable Utilization. In S. Jha & M. Halder (Eds.), *Medicinal Plants: Biodiversity, Biotechnology and Conservation* (pp. 343–374). Springer Nature Singapore. doi:10.1007/978-981-19-9936-9_14

Mirunalini, S., Krishnaveni, M., & Ambily, V. (2011). Effects of raw garlic [Allium sativum] on hyperglycemia in patients with type 2 diabetes mellitus. *Pharmacologyonline*, *2*, 968–974.

Mishra, B., Rastogi, D. A., Siddiqui, A., Srivastava, M., Verma, N., Pandey, R., . . . Shukl, S. (2013). Opium Poppy: Genetic Upgradation Through Intervention of Plant Breeding Techniques. Academic Press.

Mishra, A. K., Pandey, M., Pannu, A., Dewangan, H. K., & Sahoo, P. (2024). Review on diabetes mellitus: An insight into the current scenarios, the challenges of therapy, and application of traditional drugs. *Current Traditional Medicine*, *10*(3), 107–128. doi:10.2174/2215083810666230501212125

Mk, O., Oi, A., Bo, O., Ia, A., & Bi, O. (2019). Derivatized Extracts from Aframomium melegueta K. Schum. and Vernonia amygdalina Delile Contain Organic Compounds that Showed Antiviral Effects against Atypical Fowl Pox Virus (FPV Kabete). *Journal of Antivirals & Antiretrovirals*, *11*, 1–10.

Moawad, S. A., El-Ghorab, A. H., Hassan, M., Nour-Eldin, H., & El-Gharabli, M. M. (2015). Chemical and microbiological characterization of Egyptian cultivars for some spices and herbs commonly exported abroad. *Food and Nutrition Sciences*, *6*(07), 643–659. doi:10.4236/fns.2015.67068

Mohamed, A. E. H. H., El-Sayed, M., Hegazy, M. E., Helaly, S. E., Esmail, A. M., & Mohamed, N. S. (2010). Chemical constituents and biological activities of Artemisia herba-alba. *Records of Natural Products*, 4.

Mohammad, M. K. A., Mohamed, M. I., Zakaria, A. M., Abdul Razak, H. R., & Saad, W. M. M. (2014). Watermelon (Citrulluslanatus (Thunb.) Matsum. And Nakai) Juice Modulates Oxidative Damage Induced by Low Dose X-Ray in Mice. *BioMed Research International*, *2014*, 1–6. doi:10.1155/2014/512834 PMID:24877107

Mohammed, A., Awolola, V., Koorbanally, N., & Islam, M. (2017). Inhibition of key enzymes linked to type 2 diabetes by compounds isolated from Aframomum melegueta fruit. *Pharmaceutical Biology*, *55*(1), 1010–1016. doi:10.1080/13880209.2017.1286358 PMID:28176546

Mohammed, F. A., Razvi, S. S., Abdul, W. M., Mohammed, K., Hakeem, K. R., Banaganapalli, B., & Elkady, A. I. (2019). *Protective role of medicinal herbAnethum graveolens (Dill) against various human diseases and metabolic disorders. Plant and Human Health* (Vol. 3). Pharmacology and Therapeutic Uses.

Mohanta, Y. K., Mishra, A. K., Nongbet, A., Chakrabartty, I., Mahanta, S., Sarma, B., Panda, J., & Panda, S. K. (2023). Potential use of the Asteraceae family as a cure for diabetes: A review of ethnopharmacology to modern day drug and nutraceuticals developments. *Frontiers in Pharmacology*, *14*, 14. doi:10.3389/fphar.2023.1153600 PMID:37608892

Mojekwu, T., Oshiozokhai, Y., Ojokuku, S., Oyebadejo, S., & Eboetse, Y. (2011). Hypoglyceamic effects of aqueous extract of aframomum melegueta leaf on alloxan-induced diabetic male albino rats hypoglyceamic effects of aqueous extract of aframomum melegueta leaf on alloxan-induced diabetic male albino rats. *Pacific Journal of Medical Sciences*, *8*(1).

Momin, R., & Nair, M. (2001). Mosquitocidal, Nematicidal, and Antifungal Compounds from Apium graveolens L. Seeds. *Journal of Agricultural and Food Chemistry*, *49*(1), 142–145. doi:10.1021/jf001052a

Monárrez-Espino, J., López-Alarcón, M., & Greiner, T. (2011). Randomized Placebo-Controlled Trial of Guava Juice as a Source of Ascorbic Acid to Reduce Iron Deficiency in Tarahumara Indigenous Schoolchildren of Northern Mexico. *Journal of the American College of Nutrition*, *30*(3), 191–200. doi:10.1080/07315724.2011.10719960 PMID:21896877

Monika, M., & Anna, K.-D. (2019). Nut oils and their dietetic and cosmetic significance: A review. *Journal of Oleo Science*, *68*(2), 111–120. doi:10.5650/jos.ess18216 PMID:30651417

Moreno-Pérez, A. J., Sánchez-García, A., Salas, J. J., Garcés, R., & Martínez-Force, E. (2011). Acyl-ACP thioesterases from macadamia (Macadamia tetraphylla) nuts: Cloning, characterization and their impact on oil composition. *Plant Physiology and Biochemistry*, *49*(1), 82–87. doi:10.1016/j.plaphy.2010.10.002 PMID:21071236

Motaleb, M.A., Hossain, M.K., & Office, I.B.C. (2011). *Selected medicinal plants of Chittagong Hill Tracts*. Academic Press.

Moulick, S. P., Jahan, F., Mamun, M. Z. U. A., Hossain, M. I. S., Waliullah, M., & Sathee, R. A. (2023). Analysis of indigenous spices widely consumed in Bangladesh: An assessment to explore its proximate contents, minerals, phytochemical compositions, and antioxidant activities. *Journal of Agriculture and Food Research*, *14*, 100720. doi:10.1016/j.jafr.2023.100720

Moulin, L. (2002). *Eating and drinking in Europe: a cultural history*. Mercatorfonds.

Mozaffarian, D., Micha, R., & Wallace, S. (2010). Effects on coronary heart disease of increasing polyunsaturated fat in place of saturated fat: A systematic review and meta-analysis of randomized controlled trials. *PLoS Medicine*, *7*(3), e1000252. doi:10.1371/journal.pmed.1000252 PMID:20351774

Msaada, K. (2009). *Changes in fatty acid composition of coriander (Coriandrum sativum L.) fruit during maturation*. Industrial Crops and Products, 29(2-3), 269-274.

Msaada, K., Hosni, K., Taarit, M. B., Chahed, T., Kchouk, M. E., & Marzouk, B. (2007). Changes on essential oil composition of coriander (Coriandrum sativum L.) fruits during three stages of maturity. *Food Chemistry*, *102*(4), 1131–1134. doi:10.1016/j.foodchem.2006.06.046

Msaada, K., Taarit, M. B., Hosni, K., Hammami, M., & Marzouk, B. (2009). Regional and maturational effects on essential oils yields and composition of coriander (Coriandrum sativum L.) fruits. *Scientia Horticulturae*, *122*(1), 116–124. doi:10.1016/j.scienta.2009.04.008

Msanda, F., El Aboudi, A., & Peltier, J. P. (2005). Biodiversity and biogeography of Moroccan argan tree communities. *Cahiers Agricultures*, *14*(4), 357–364.

Mukhtar, U., Mukammil, S., Khalil, U. R., Alam, R., & Mohammad, A. R. (2011). Evaluation of Local and Exotic Grapes Germ plasm at mingora, swat. – sarhad. *Nongxue Xuebao*, *27*(4), 553–556.

Mustapha, B. O., Ademoyegun, O. T., & Ahmed, R. S. (2023). Horticultural crops as natural therapeutic plants for the therapy of diabetes mellitus. *Egyptian Journal of Basic and Applied Sciences*, *10*(1), 380–395. doi:10.1080/2314808X.2023.2217387

Mutanen, M., & Pajari, A.-M. (2010). Vegetables, whole grains, and their derivatives in cancer Prevention. Dordrecht: Springer.

Mutinda, E. S., Kimutai, F., Mkala, E. M., Waswa, E. N., Odago, W. O., Nanjala, C., Ndungu, C. N., Gichua, M. K., Njire, M. M., Gituru, R. W., & Hu, G. W. (2023). Ethnobotanical uses, phytochemistry and pharmacology of pantropical genus *Zanthoxylum* L. (Rutaceae): An update. *Journal of Ethnopharmacology*, *303*, 115895. doi:10.1016/j.jep.2022.115895 PMID:36513263

Mwafy, S., & Yassin, M. (2011). Antidiabetic activity evaluation of glimepiride and Nerium oleander extract on insulin, glucose levels and some liver enzymes activities in experimental diabetic rat model. Academic Press.

NaAllah, AAyipo, Y.OKomolafe, D.ISolihu, SBamidele, BAlabi, M.A. (2021). Phytochemical screening and in silico pharmacological profiling of ethanolic extract of Aframomum melegueta for prostate carcinoma. *Journal of Applied Pharmaceutical Science*, *11*(7), 132–145.

Naceiri Mrabti, H., Bouyahya, A., Naceiri Mrabti, N., Jaradat, N., Doudach, L., & Faouzi, M. E. A. (2021). Ethnobotanical survey of medicinal plants used by traditional healers to treat diabetes in the Taza region of Morocco. *Evidence-Based Complementary and Alternative Medicine*, *2021*, 2021. doi:10.1155/2021/5515634 PMID:33986815

Nadeem, M., Muhammad Anjum, F., Issa Khan, M., Tehseen, S., El-Ghorab, A., & Iqbal Sultan, J. (2013). Nutritional and medicinal aspects of coriander (Coriandrum sativum L.). *British Food Journal*, *115*(5), 743–755. doi:10.1108/00070701311331526

Nadkarni, A. (1954). *Nadkarni's Indian Materia Medica*. Nadkarni's Indian Materia Medica.

Nagella, P., & … . (2012). Chemical constituents, larvicidal effects and antioxidant activity of petroleum ether extract from seeds of Coriandrum sativum L. *Journal of Medicinal Plants Research*, *6*(15), 2948–2954. doi:10.5897/JMPR11.992

Nair, R., & Chanda, S. (2007). Antibacterial activities of some medicinal plants of the western region of India. *Turkish Journal of Biology*, *31*(4), 231–236.

Najafipour, H., & Beik, A. (2016). The Impact of Opium Consumption on Blood Glucose, Serum Lipids and Blood Pressure, and Related Mechanisms. *Frontiers in Physiology*, *7*, 436–436. doi:10.3389/fphys.2016.00436 PMID:27790151

Nakagawa, T., Yokozawa, T., Terasawa, K., Shu, S., & Juneja, L. R. (2002). Protective Activity of Green Tea against Free Radical- and Glucose-Mediated Protein Damage. *Journal of Agricultural and Food Chemistry*, *50*(8), 2418–2422. doi:10.1021/jf011339n PMID:11929306

Nalvothula, R., & … . (2014). Biogenic synthesis of silver nanoparticles using Tectona grandis leaf extract and evaluation of their antibacterial potential. *International Journal of Chemtech Research*, *6*(1), 293–298.

Nath, M., & Debnath, P. (2023). Therapeutic role of traditionally used Indian medicinal plants and spices in combating COVID-19 pandemic situation. *Journal of Biomolecular Structure & Dynamics*, *41*(12), 5894–5913. doi:10.1080/07391102.2022.2093793 PMID:35773779

Nautiyal, O. H. (2019). Black Seed (*Nigella sativa*) Oil. In M. F. Ramadan (Ed.), *Fruit Oils: Chemistry and Functionality* (pp. 839–857). Springer International Publishing. doi:10.1007/978-3-030-12473-1_46

Naz, A., Butt, M. S., Sultan, M. T., Qayyum, M. M. N., & Niaz, R. S. (2014). Watermelon lycopene and allied health claims. *EXCLI Journal*, *13*. PMID:26417290

Nedungadi, P., Salethoor, S. N., Puthiyedath, R., Nair, V. K., Kessler, C., & Raman, R. (2023). Ayurveda research: Emerging trends and mapping to sustainable development goals. *Journal of Ayurveda and Integrative Medicine, 14*(6), 100809. doi:10.1016/j.jaim.2023.100809 PMID:37832213

Nejad Ebrahimi, S., Hadian, J., & Ranjbar, H. (2010). Essential oil compositions of different accessions of Coriandrum sativum L. from Iran. *Natural Product Research, 24*(14), 1287–1294. doi:10.1080/14786410903132316 PMID:20803372

Newberne, P., Smith, R., Doull, J., Goodman, J., Munro, I., Portoghese, P., ... Adams, T. (1999). The FEMA GRAS assessment of trans-anethole used as a flavouring substance. *Food and Chemical Toxicology, 37*(7), 789–811. doi:10.1016/S0278-6915(99)00037-X PMID:10496381

Nguyen, T., Nandasiri, R., & Eskin, N. M. (2023). Phenolics of mustard seeds: A review on composition, processing effect and their bioactvities. *Journal of the American Oil Chemists' Society, xx*(x), xxx–xxx. doi:10.1002/aocs.12783

Nijjar, G., & Singh, N. (1979). Effect of Nitrogen, Phosphorus and Potassium Fertilization on the Growth, Yield and Quality of Thompson Seedless Grapes. *Indian Journal of Horticulture, 36*(1), 114–118.

Nijveldt, R., van Nood, E., Hoorn, D., Boelens, P., Norren, K., & Leeuwen, P. (2001). Flavonoids: A review of probable mechanisms of action and potential applications. *The American Journal of Clinical Nutrition, 74*(4), 418–425. doi:10.1093/ajcn/74.4.418

Nutmakul, T., & Chewchinda, S. (2023). Synergistic effect of Trikatuk, a traditional Thai formulation, on antioxidant and alpha-glucosidase inhibitory activities. *Heliyon, 9*(1), e13063. doi:10.1016/j.heliyon.2023.e13063 PMID:36711300

Nwaehujor, C. (2014). Hepatotoxicity of Methanol Seed Extract of Aframomum melegueta [Roscoe] K. Schum. (Grains of paradise) in Sprague-Dawley Rats. *African Journal of Biomedical Research, 2*, 61–66.

Nwaichi, E. O., Chuku, L. C., & Oyibo, N. J. (2015). Original Research Article Profile of Ascorbic Acid, Beta-Carotene and Lycopene in Guava, Tomatoes, Honey and Red Wine. *International Journal of Current Microbiology and Applied Sciences, 4*(2), 39–43.

Nyaga, P. N., Kaminjolo, J. S., Mutiga, E. R., & Bebora, L. C. (1979). Occurrence of Atypical Fowlpox in Poultry Farms in Kenya. *Avian Diseases, 23*(3), 745–752. doi:10.2307/1589751 PMID:230810

Nyirenda, J., & Chipuwa, M. (2024). An ethnobotanical study of herbs and medicinal plants used in western, copperbelt, central and northern provinces of Zambia. *Phytomedicine Plus : International Journal of Phytotherapy and Phytopharmacology, 4*(1), 100514. doi:10.1016/j.phyplu.2023.100514

O'Hare, P., Bell, D., & Salmon, T. (2000). *Australian macadamia industry code of sound orchard practices*. Australian Macadamia Society.

Obuekwe, C. O. (1981). *Microbial corrosion of pipelines: University of Alberta*. Edmondton.

Odebunmi, E., Oluwaniyi, O., Awolola, V., & Adediji, O. D. (2009). Proximate and nutritional composition of kola nut (Cola nitida), bitter cola (Garcinia cola) and alligator pepper (Afromomum melegueta). *African Journal of Biotechnology*, 8.

Odetunde, A., & Avungbeto, L. (2015). Antimicrobial effect and phytochemical analysis of Aframomum melegueta on some selected bacteria and fungi. *European Journal of Biotechnology and Bioscience.*, *3*, 15–19.

Odeyemi, S., & Bradley, G. (2018). Medicinal plants used for the traditional management of diabetes in the Eastern Cape, South Africa: Pharmacology and toxicology. *Molecules (Basel, Switzerland)*, *23*(11), 2759. doi:10.3390/molecules23112759 PMID:30366359

Odo, C. (2015). Anti-bacterial Effects of Aframomum melegueta Seed Extracts on Some Enterobacteriaceae Organisms. *World Journal of Pharmaceutical Research*, *4*, 304–309.

Oelhafen, P., & Schüler, A. (2005). Nanostructured materials for solar energy conversion. *Solar Energy*, *79*(2), 110–121. doi:10.1016/j.solener.2004.11.004

Ogbunugafor, H., Ugochukwu, C., & Kyrian-Ogbonna, A. (2017). The role of spices in nutrition and health: A review of three popular spices used in Southern Nigeria. *Food Quality and Safety*, *1*(3), 1. doi:10.1093/fqsafe/fyx020

Oguzie, E. E., Iheabunike, Z. O., Oguzie, K. L., Ogukwe, C. E., Chidiebere, M. A., Enenebeaku, C. K., & Akalezi, C. O. (2013). Corrosion Inhibiting Effect of Aframomum melegueta Extracts and Adsorption Characteristics of the Active Constituents on Mild Steel in Acidic Media. *Journal of Dispersion Science and Technology*, *34*(4), 516–527. doi:10.1080/01932691.2012.682008

Oguzie, E., Ogukwe, C., Jude-anthony, O., Nwanebu, F., Adindu, B., & Udeze, I. (2012). Broad spectrum corrosion inhibition: Corrosion and microbial (SRB) growth inhibiting effects of Piper guineense extract. *Journal of Materials Science*, *47*(8), 3592–3601. doi:10.1007/s10853-011-6205-1

Ojieh, A. E., Adegor, E. C., Okolo, A. C., Lawrence, E. O., Njoku, I. P., & Onyekpe, C. U. (2015). Hypoglycemic and hypolipidaemic effect of allium cepa in streptozotocin-induced diabetes. *International Journal of Science and Engineering*, *6*(10), 23–29.

Okayasu, T., Mitani, K., & Kitahara, T. (2024). Reviewing Kampo medicine (Traditional Japanese Herbal Medicine) for otology/neurotology diseases. *Auris, Nasus, Larynx*, *51*(1), 25–30. doi:10.1016/j.anl.2023.04.009 PMID:37137794

Okhuarobo, A., Falodun, J. E., Erharuyi, O., Imieje, V., Falodun, A., & Langer, P. (2014). Harnessing the medicinal properties of Andrographis paniculata for diseases and beyond: A review of its phytochemistry and pharmacology. *Asian Pacific Journal of Tropical Disease*, *4*(3), 213–222. doi:10.1016/S2222-1808(14)60509-0

Okwu, D., & Mgbeahuruike, E. (2018). *Chemical Composition and in Vitro Antifungal Activity Screening of Seed and Leaf Extracts from Aframomum meleguata and Monodora myristica against Sclertium rolfsii of Cowpea Plant (Vigna unguiculata L. Walp)*. Academic Press.

Okwu, M. U., Olley, M., Akpoka, A. O., & Izevbuwa, O. E. (2019). Methicillin-resistant Staphylococcus aureus (MRSA) and anti-MRSA activities of extracts of some medicinal plants: A brief review. *AIMS Microbiology*, *5*(2), 117–137. doi:10.3934/microbiol.2019.2.117 PMID:31384707

Oladunmoye, M.K. (2019). Characterization of Organic Compounds in Aframomum melegueta K. Schum Using GC-MS. *Medicinal & Aromatic Plants, 8.*

Oliveira, I., Sousa, A., Ferreira, I. C. F. R., Bento, A., Estevinho, L., & Pereira, J. A. (2008). Total Phenols, Antioxidant Potential and Antimicrobial Activity of Walnut (Juglans Regia L.) Green Husks. *Food and Chemical Toxicology*, *46*(7), 2326–2331. doi:10.1016/j.fct.2008.03.017 PMID:18448225

Olumekun, V.O., Osuntokun, O. T., Ajayi, A. O., Omotuyi, I. O., & Olonisakin, A. (2020). Estimation of Vitamins in Aframomum melegueta [Roscoe] K. Schum and Its Relevance as Natural Immune Booster. *Journal of Applied Life Sciences International, 23*(7), 25-39.

Omoboyowa, D., Agha Aja, O., Eluu, F., & Ngobidi, K. (2017). Effects of Methanol Seed Extract of Aframomum melegueta (Alligator Pepper) on Wistar Rats with 2,4-Dinitrophenylhydrazine Induced Hemolytic Anemia. *Recent Advances in Biology and Medicine.*, *03*, 11. doi:10.18639/RABM.2017.03.443648

OMS. (2012).Statistiques sanitaires mondiales. OMS.

Onojah, P., & Odin, E. (2015). Cyanogenic glycoside in food plants. *Int. J. Innov. Sci. Math*, *3*, 2347–9051.

Ono, M., Kawamoto, Y., Uemori, C., Diono, W., Kanda, H., & Goto, M. (2017). Extraction of Phytochemicals from Grains of Paradise Using Supercritical Carbon Dioxide. *Engineering Journal (New York)*, *21*, 53–64.

Orch, H., Douira, A., & Zidane, L. (2015). Étude ethnobotanique des plantes médicinales utilisées dans le traitement du diabète, et des maladies cardiaques dans la région d'Izarène. *Journal of Applied Biosciences*, *86*(1), 7940–7956. doi:10.4314/jab.v86i1.3

Orouji, N., Asl, S. K., Taghipour, Z., Habtemariam, S., Nabavi, S. M., & Rahimi, R. (2023). Glucosinolates in cancer prevention and treatment: Experimental and clinical evidence. *Medical Oncology (Northwood, London, England)*, *40*(12), 344. doi:10.1007/s12032-023-02211-6 PMID:37921869

Oseni, O. A., & Okoye, V. I. (2013). Studies of Phytochemical and Antioxidant properties of the Fruit of Watermelon (*Citrulluslanatus*). *Journal of Pharmaceutical and Biomedical Sciences*, *27*(27), 508–514.

Osuntokun, O. (2020a). *Aframomum Melegueta (Grains of Paradise)*. Academic Press.

Osuntokun, O. (2020b). Synergistic efficacy of Aframomum melegueta [Roscoe] K. Schum and Spondias mombin(Linn), A predictive treatment of SARS-CoV-2 (COVID-19) Infection. *Journal of Bioscience & Biomedical Engineering*.

Osuntokun, Ot., Baraldi, C., & Gamberini, M. C. (Eds.). (2018). Evaluation of quantitative elemental compositions and antioxidant potentials of spondias mombin extracts (linn), a precursor against infectious diseases. Academic Press.

Osuntokun, O., Balogun, J., Akele, E., Adedayo, S., & Bello, O. (2021). Aframomum melegueta [Roscoe] K. Schum) ethanolic seed extract against asymptomatic bacteriuria isolates from pregnant women attending antenatal clinic in Ondo State, Nigeria. *International Journal of Pathogen Research.*, *8*(2), 12–43. doi:10.9734/ijpr/2021/v8i230200

Osuntokun, O., Olumekun, V., Ajayi, O., Omotuyi, O., & Adebisi, O. (2020). Assessment of in-vitro Antioxidant/Enzymes Inhibitory Potentials of Aframomum melegueta [Roscoe] K. Schum (Grains of Paradise) Leaf, Stem Bark, Seed Bark and Seed Extracts. *International Journal of Current Research*, *20*, 40–57.

Ouasti, I., Bussmann, R. W., & Elachouri, M. (2023). Apium Graveolens L.Apiaceae. In R. W. Bussmann, M. Elachouri, & Z. Kikvidze (Eds.), *Ethnobotany of Northern Africa and Levant* (pp. 1–8). Springer International Publishing. doi:10.1007/978-3-031-13933-8_34-1

Oudejans, L. C. J., van Velzen, M., & Dahan, A. (2016). Ketamine Analgesia. In V. R. Preedy (Ed.), *Neuropathology of Drug Addictions and Substance Misuse* (pp. 541–550). Academic Press. doi:10.1016/B978-0-12-800212-4.00050-9

Ouhaddou, H., Boubaker, H., Msanda, F., & El Mousadik, A. (2014). An ethnobotanical study of medicinal plants of the Agadir Ida Ou Tanane province. *Journal of Applied Biosciences*, *84*(1), 7707–7722. doi:10.4314/jab.v84i1.5

Owokotomo, A. (2014). Chemical Constituents of the Leaf, Stem, Root and Seed Essential Oils of Aframomum melegueta (K. Schum) from South West Nigeria. *International Research Journal of Pure and Applied Chemistry*, *4*(4), 395–401. doi:10.9734/IRJPAC/2014/7397

Oyagbemi, A. A., Saba, A. B., & Azeez, O. I. (2010). Molecular targets of [6]-gingerol: Its potential roles in cancer chemoprevention. *BioFactors (Oxford, England)*, *36*(3), 169–178. doi:10.1002/biof.78 PMID:20232343

Ozougwu, J. C. (2011). Anti-diabetic effects of Allium cepa [onions] aqueous extracts on alloxan-induced diabetic Rattus novergicus. *Journal of Medicinal Plants Research*, *5*(7), 1134–1139.

Öztürk, M., Altay, V., & Keskin, M. (2023). Folkloric Knowledge of Endemic Plants: Their Traditional Uses as Spices, Food and Herbal Teas by the Local Communities in Turkiye. In Ethnic Knowledge and Perspectives of Medicinal Plants (pp. 137-152). Apple Academic Press.

Padamanabhan, V., Manimekalai, G., Vasthi, K. E., Nirmala, A., & Jagajothi, A. (2014). Phytochemical Screening and Antioxidant Activity of Extracts of the Leaf and Bark of Albizzia Lebbeck (Benth). *Acad. J. Med. Plant*, 2(2), 26–31.

Painuli, S., Quispe, C., Herrera-Bravo, J., Semwal, P., Martorell, M., Almarhoon, Z. M., Seilkhan, A., Ydyrys, A., Rad, J. S., Alshehri, M. M., Daştan, S. D., Taheri, Y., Calina, D., & Cho, W. C. (2022). Nutraceutical profiling, bioactive composition, and biological applications of Lepidium sativum L. *Oxidative Medicine and Cellular Longevity*, *2022*, 2022. doi:10.1155/2022/2910411 PMID:35096265

Pal, D., & Lal, P. (2024). Herbal Drugs and Medicinal Plants in Controlling and Treatment of Diseases Caused by Dengue Virus (DEN-1 & 2): Ethnopharmacology, Chemistry, and Clinical and Preclinical Studies. In D. Pal (Ed.), *Anti-Viral Metabolites from Medicinal Plants* (pp. 683–746). Springer International Publishing. doi:10.1007/978-3-031-12199-9_20

Panchuk, M. (2023). Power and protest: A christian liturgical response to religious trauma. *Philosophies of Liturgy: Explorations of Embodied Religious Practice*, *xx*(x), 77. doi:10.5040/9781350349292.ch-5

Park, Y., Jin, S., Noda, I., & Jung, Y. M. (2023). Continuing progress in the field of two-dimensional correlation spectroscopy (2D-COS), part II. Recent noteworthy developments. *Spectrochimica Acta. Part A: Molecular and Biomolecular Spectroscopy*, *284*, 121750. doi:10.1016/j.saa.2022.121750 PMID:36030669

Patel, M., Vishnoi, S., & Neelima, A. (2023). An empirical review of fundamental principles of Ayurveda for women's reproductive health and diseases. *Journal of Ayurveda and Integrated Medical Sciences*, *8*(9), 105–111. doi:10.21760/jaims.8.9.17

Pathak Nimish, L. (2011). Phytopharmacological properties of Coriander sativum as a potential medicinal tree: An overview. *Journal of Applied Pharmaceutical Science*, *1*(4), 20–25.

Pathak, K., Pathak, M. P., Saikia, R., Gogoi, U., Sahariah, J. J., Das, A., Ahmad, M. Z., Paul, T., Das, J., & Alqahtani, S. A. M. (2024). Cardioprotective activities of some Indian spices: An insight into pharmacology and phytochemical investigation. *Current Traditional Medicine*, *10*(1), 52–66. doi:10.2174/2215083809666230210105442

Pathak, N., Kasture, S., & Bhatt, N. (2011). Phytochemical screening of Coriander sativum Linn. *International Journal of Pharmaceutical Sciences Review and Research*, *9*, 159–163.

Patil, R., & Lamani, B. (2015). Apigenin inhibits PMA-induced expression of pro-inflammatory cytokines and AP-1 factors in A549 cells. *Molecular and Cellular Biochemistry*, *403*(1-2), 95–106. Advance online publication. doi:10.1007/s11010-015-2340-3

Pavia, D. L. (2005). *Introduction to organic laboratory techniques: a small scale approach*. Cengage Learning.

Pavlić, B., Vidović, S., Vladić, J., Radosavljević, R., & Zeković, Z. (2015). Isolation of coriander (Coriandrum sativum L.) essential oil by green extractions versus traditional techniques. *The Journal of Supercritical Fluids*, *99*, 23–28. doi:10.1016/j.supflu.2015.01.029

Peace, C. P., Vithanage, V., Turnbull, C. G., & Carroll, B. J. (2003). A genetic map of macadamia based on randomly amplified DNA fingerprinting (RAF) markers. *Euphytica, 134*(1), 17–26. doi:10.1023/A:1026190529568

Penha, C. B., Santos, V. D. P., Speranza, P., & Kurozawa, L. E. (2021). Plant-based beverages: Ecofriendly technologies in the production process. *Innovative Food Science & Emerging Technologies*, *72*, 102760. doi:10.1016/j.ifset.2021.102760

Perkins-Veazie, P., Davis, A., & Collins, J. K. (2012). Watermelon: From dessert to functional food. *Israel Journal of Plant Sciences*, *60*(4), 395–402. doi:10.1560/IJPS.60.1.402

Pham-Huy, L. A., He, H., & Pham-Huy, C. (2008). Free radicals, antioxidants in disease and health. *International Journal of Biomedical Science; IJBS*, *4*(2), 89–96. doi:10.59566/IJBS.2008.4089 PMID:23675073

Platel, K., & Srinivasan, K. (2004). Digestive stimulant action of spices: A myth or reality? *The Indian Journal of Medical Research*, *119*(5), 167. PMID:15218978

Poddar, K. H., Sikand, G., Kalra, D., Wong, N., & Duell, P. B. (2022). Mustard oil and cardiovascular health: Why the controversy? *Journal of Clinical Lipidology*, *16*(1), 13–22. doi:10.1016/j.jacl.2021.11.002 PMID:34924350

Popovic, M., Kaurinovic, B., Trivic, S., Mimica-Dukic, N., & Bursać, M. (2006). Effect of celery (Apium graveolens) extracts on some biochemical parameters of oxidative Stress in mice treated with carbon tetrachloride. *Phytotherapy Research*, *20*(7), 531–537. doi:10.1002/ptr.1871

Postescu, I. D., Chereches, G., Tatomir, C., Daicoviciu, D., & Filip, G. A. (2012). Modulation of Doxorubicin-Induced Oxidative Stress by a Grape (Vitisvinifera L.) Seed Extract in Normal and Tumor Cells. *Journal of Medicinal Food*, *15*(7), 639–645. doi:10.1089/jmf.2011.0291 PMID:22519870

Potterat, O., & Hamburger, M. (2008). Drug discovery and development with plant-derived compounds. In *Natural Compounds as Drugs* (Vol. I, pp. 45–118). Springer. doi:10.1007/978-3-7643-8117-2_2

Poudyal, H., Kumar, S. A., Iyer, A., Waanders, J., Ward, L. C., & Brown, L. (2013). Responses to oleic, linoleic and α-linolenic acids in high-carbohydrate, high-fat diet-induced metabolic syndrome in rats. *The Journal of Nutritional Biochemistry*, *24*(7), 1381–1392. doi:10.1016/j.jnutbio.2012.11.006 PMID:23333092

Powanda, M. C., Whitehouse, M. W., & Rainsford, K. (2015). Celery seed and related extracts with antiarthritic, antiulcer, and antimicrobial activities. *Novel natural products: therapeutic effects in pain, arthritis and gastro-intestinal diseases*, 133-153.

Powell, M. A. (2009). *Predicting the geographic distributions of Macadamia integrifolia and Macadamia ternifolia*. University of the Sunshine Coast.

Powers, D., Erickson, S., & Swortwood, M. J. (2018). Quantification of Morphine, Codeine, and Thebaine in Home-Brewed Poppy Seed Tea by LC-MS/MS. *Journal of Forensic Sciences*, *63*(4), 1229–1235. doi:10.1111/1556-4029.13664 PMID:28973785

Praphasawat, R., Klundsupya, P., Muangman, T., Laovithayanggoon, S., Arunpairojana, V., & Himakoun, L. (2011). Anti-mutagenicity and anti-oxidative DNA damage properties of oligomeric pro-anthocyanidines from Thai grape seed in TK6 cells. *Asian Pacific Journal of Cancer Prevention*, *12*(5), 1317–1321. PMID:21875289

Preedy, V. R., & Watson, R. R. (2020). *Nuts and seeds in health and disease prevention*. Academic press.

Price, M. L., Van Socoyoc, S., & Butter, L. G. (1978). A critical evaluation of the vanillin reaction as an assay for tannin in sorghum grain. *Journal of Agricultural and Food Chemistry*, *26*(5), 1214–1218. doi:10.1021/jf60219a031

Priyadarshi, S., Khanum, H., Ravi, R., Borse, B. B., & Naidu, M. M. (2016). Flavour characterisation and free radical scavenging activity of coriander (Coriandrum sativum L.) foliage. *Journal of Food Science and Technology*, *53*(3), 1670–1678. doi:10.1007/s13197-015-2071-1 PMID:27570292

Probst, Y. C., Guan, V. X., & Kent, K. (2017). Dietary phytochemical intake from foods and health outcomes: A systematic review protocol and preliminary scoping. *BMJ Open*, *7*(2), e013337–e013337. doi:10.1136/bmjopen-2016-013337 PMID:28202499

Pullaiah, T. (2002). *Medicinal Plants in Andhra Pradesh*. Daya Books.

Qari, S. H., Alqethami, A., & Qumsani, A. T. (2024). Ethnomedicinal evaluation of medicinal plants used for therapies by men and women in rural and urban communities in Makkah district. *Saudi Pharmaceutical Journal*, *32*(1), 101881. doi:10.1016/j.jsps.2023.101881 PMID:38130903

Quattrocchi, U. (2012). *CRC world dictionary of medicinal and poisonous plants: common names, scientific names, eponyms, synonyms, and etymology*. CRC, Taylor & Francis Group.

Quinn, L., & Tang, H. (1996). Antioxidant properties of phenolic compounds in macadamia nuts. *Journal of the American Oil Chemists' Society*, *73*(11), 1585–1588. doi:10.1007/BF02523529

Rafieian, F., Amani, R., Rezaei, A., Karaça, A. C., & Jafari, S. M. Exploring fennel (*Foeniculum vulgare*): Composition, functional properties, potential health benefits, and safety. *Critical Reviews in Food Science and Nutrition*, 1-18. doi:10.1080/10408398.2023.2176817

Raghavan, B., Abraham, K., Shankaranarayana, M., & Koller, W. (1994). Studies on flavor changes during drying of dill (Anethum sowa Roxb.) leaves. *Journal of Food Quality*, *17*(6), 457–466. doi:10.1111/j.1745-4557.1994.tb00166.x

Rahimi, R., Amin, G., & Ardekani, M. R. S. (2012). A review on Citrullus colocynthis Schrad.: From traditional Iranian medicine to modern phytotherapy. *Journal of Alternative and Complementary Medicine (New York, N.Y.)*, *18*(6), 551–554. doi:10.1089/acm.2011.0297 PMID:22784342

Rajeshwari, U., & Andallu, B. (2011). Medicinal benefits of coriander (Coriandrum sativum L). *Spatula DD*, *1*(1), 51–58. doi:10.5455/spatula.20110106123153

Ramalingam, P. S., Mishra, R. A. K., Mekala, J. R., Gothandam, K. M., & Arumugam, S. (2024). Nigella Sativa L.–A Note on Phytochemistry, Extraction, Pharmacological Activities, and Therapeutic Utility. In Pharmacological Aspects of Essential Oils (pp. 243-264). CRC Press.

Ramezani, M., Nasri, S., & Yassa, N. (2009). Antinociceptive and anti-inflammatory effects of isolated fractions from Apium graveolens seeds in mice. *Pharmaceutical Biology*, *47*(8), 740–743. doi:10.1080/13880200902939283

Ramineedu, K., Sankaran, K. R., Mallepogu, V., Rendedula, D. P., Gunturu, R., Gandham, S., . . . Meriga, B. (2023). Thymoquinone mitigates obesity and diabetic parameters through regulation of major adipokines, key lipid metabolizing enzymes and AMPK/p-AMPK in diet-induced obese rats. *3 Biotech*, *14*(1), 16. doi:10.1007/s13205-023-03847-x

Ramírez-Alarcón, K., Martorell, M., Gürer, E. S., Laher, I., Lam, H. L., Mohieldin, E. A. M., ... Shafique, H. (2023). Myristicin: From its biological effects in traditional medicine in plants to preclinical studies and use as ecological remedy in plant protection. *eFood*, *4*(3), e90. doi:10.1002/efd2.90

Randall, K., Drew, M. D., Øverland, M., Østbye, T.-K., Bjerke, M., Vogt, G., & Ruyter, B. (2013). Effects of dietary supplementation of coriander oil, in canola oil diets, on the metabolism of [1-14C] 18: 3n-3 and [1-14C] 18: 2n-6 in rainbow trout hepatocytes. *Comparative Biochemistry and Physiology. Part B, Biochemistry & Molecular Biology*, *166*(1), 65–72. doi:10.1016/j.cbpb.2013.07.004 PMID:23867781

Rani & Vijayanchali. (2017). Phytochemical, Antioxidant Activity and Lycopene Analysis of Red Guava Fruits, Journal of Research. *Extension and Development*, *6*(2), 25–30.

Rani, D. M., Wongso, H., Purwoko, R. Y., Winarto, N. B., Shalas, A. F., Triatmoko, B., Pratama, A. N. W., Keller, P. A., & Nugraha, A. S. (2023). Anticancer bioprospecting on medicinal plants from Indonesia: A review. *Phytochemistry*, *216*, 113881. doi:10.1016/j.phytochem.2023.113881 PMID:37827225

Rashwan, H. K., Mahgoub, S., Abuelezz, N. Z., & Amin, H. K. (2023). Black cumin seed (*Nigella sativa*) in inflammatory disorders: Therapeutic potential and promising molecular mechanisms. *Drugs and Drug Candidates*, *2*(2), 516–537. doi:10.3390/ddc2020027

Rastogi, R. P., Mehrotra, B. N., Sinha, S., Pant, P., & Seth, R. (1990). Compendium of Indian medicinal plants. Lucknow: Central Drug Research Institute and Publications & Information Directorate, New Delhi Lucknow.

Rastogi, R. P., & Mehrotra, B. (1990). *Compendium of Indian medicinal plants* (Vol. 1). CSIR.

Reeves, P. M., Smith, S. K., Olson, V. A., Thorne, S. H., Bornmann, W., Damon, I. K., & Kalman, D. (2011). Variola and Monkeypox Viruses Utilize Conserved Mechanisms of Virion Motility and Release That Depend on Abl and Src Family Tyrosine Kinases. *Journal of Virology*, *85*(1), 21–31. doi:10.1128/JVI.01814-10 PMID:20962097

Regula, J., Cerba, A., Suliburska, J., & Tinkov, A. A. (2018). In Vitro Bioavailability of Calcium, Magnesium, Iron, Zinc, and Copper from Gluten-Free Breads Supplemented with Natural Additives. *Biological Trace Element Research*, *182*(1), 140–146. doi:10.1007/s12011-017-1065-4 PMID:28580502

Reiter, E., Jiang, Q., & Christen, S. (2007). Anti-inflammatory properties of alpha- and gamma-tocopherol. *Molecular Aspects of Medicine*, *28*(5-6), 668–691. doi:10.1016/j.mam.2007.01.003 PMID:17316780

Rengel, A., Pérez, E., Piombo, G., Ricci, J., Servent, A., Tapia, M. S., . . . Montet, D. (2015). *Lipid profile and antioxidant activity of macadamia nuts (Macadamia integrifolia) cultivated in Venezuela*. Academic Press.

Ren, J., Yang, L., Qiu, S., Zhang, A.-H., & Wang, X.-J. (2023). Efficacy evaluation, active ingredients, and multitarget exploration of herbal medicine. *Trends in Endocrinology and Metabolism*, *34*(3), 146–157. doi:10.1016/j.tem.2023.01.005 PMID:36710216

Rhattas, M., Douira, A., & Zidane, L. (2016). Étude ethnobotanique des plantes médicinales dans le Parc National de Talassemtane. *Journal of Applied Biosciences*, *97*(0), 9187–9211. doi:10.4314/jab.v97i1.5

Rhee, M. H., Park, H.-J., & Cho, J. Y. (2009). Salicornia herbacea: Botanical, chemical and pharmacological review of halophyte marsh plant. *Journal of Medicinal Plants Research*, *3*(8), 548–555.

Rice-Evans, C. A., Miller, N. J., Bolwell, P. G., Bramley, P. M., & Pridham, J. B. (1995). The relative antioxidant activities of plant-derived polyphenolic flavonoids. *Free Radical Research*, *22*(4), 375–383. doi:10.3109/10715769509145649 PMID:7633567

Richards, T. E., Kämper, W., Trueman, S. J., Wallace, H. M., Ogbourne, S. M., Brooks, P. R., Nichols, J., & Hosseini Bai, S. (2020). Relationships between nut size, kernel quality, nutritional composition and levels of outcrossing in three macadamia cultivars. *Plants*, *9*(2), 228. doi:10.3390/plants9020228 PMID:32053871

Rietjens, I. M. C. M., Cohen, S. M., Eisenbrand, G., Fukushima, S., Gooderham, N. J., Guengerich, F. P., Hecht, S. S., Rosol, T. J., Davidsen, J. M., Harman, C. L., & Taylor, S. V. (2023). FEMA GRAS assessment of natural flavor complexes: Allspice, anise, fennel-derived and related flavoring ingredients. *Food and Chemical Toxicology*, *174*, 113643. doi:10.1016/j.fct.2023.113643 PMID:36739890

Rizvi, S., Raza, S. T., Ahmed, F., Ahmad, A., Abbas, S., & Mahdi, F. (2014). The role of vitamin e in human health and some diseases. *Sultan Qaboos University Medical Journal*, *14*(2), e157–e165. PMID:24790736

Rizwan, D., & Masoodi, F. A. (2023). Brassica-derived isothiocyanates as anticancer therapeutic agents and their nanodelivery. *Phytotherapy Research*. doi:10.1002/ptr.8042 PMID:37882581

Roa Solís, C., Delgado Orellana, A., Fuller, D. Q., & Capparelli, A. (2023). Wild foods, woodland fuels, and cultivation through the Ceramic and Early Historical periods in Araucanía, Southern Chile (400–1850 ce). *Vegetation History and Archaeobotany*. doi:10.1007/s00334-023-00969-3

Robards, K., Kerr, A. F., & Patsalides, E. (1988). Rancidity and its measurement in edible oils and snack foods. A review. *Analyst*, *113*(2), 213–224. doi:10.1039/an9881300213 PMID:3288002

Roglic, G., & Norris, S. L. (2018). Medicines for treatment intensification in type 2 diabetes and type of insulin in type 1 and type 2 diabetes in low-resource settings: Synopsis of the World Health Organization guidelines on second-and third-line medicines and type of insulin for the control of blood glucose levels in nonpregnant adults with diabetes mellitus. *Annals of Internal Medicine*, *169*(6), 394–397. doi:10.7326/M18-1149 PMID:30178023

Romanova, D., Vachálková, A., Cipák, L., Krajcovicova, Z., & Rauko, P. (2001). Study of antioxidant effect of apigenin, luteolin and quercetin by DNA protective method. *Neoplasma*, *48*, 104–107.

Ros, E. (2010). Health benefits of nut consumption. *Nutrients*, *2*(7), 652–682. doi:10.3390/nu2070652 PMID:22254047

Ross. (1999). Medicinal plants of the world. Chemical constituents, traditional, and modern medical uses. Human Press Inc.

Rowe, P. J., Comhaire, F. H., Hargreave, T. B., & Mahmoud, A. M. A. (2000). *WHO manual for the standardized investigation, diagnosis and management of the infertile male*. World Health Organization.

Roy, K., Sarkar, C., & Ghosh, C. (2015). Apium graveolens leaf extract-mediated synthesis of silver nanoparticles and its activity on pathogenic fungi. *Digest Journal of Nanomaterials and Biostructures*, *10*(2), 393–400.

Rozek, E. (2007). Growth and yielding of leaf celery [Apium graveolens L. var. secalinum Alef.] cultivated for two-cut harvest. *Herba Polonica*, *3*(53).

Rożek, E. (2013). YIeldIng of leaf celery Apium graveolens l. var. secAlinum alef. dependIng on the nuMber of harvests and IrrIgatIon. *Modern Phytomorphology*, *3*, 83–86.

Ruch, R. T., Cheng, S. J., & Klaunig, J. E. (1984). Spin trapping of superoxide and hydroxyl radicals. J Methods in Enzymology, 105, 198-209.

Saad, B. (2024). History, Present and Prospect of Greco-Arab and Islamic Herbal Medicine. In History, Present and Prospect of World Traditional Medicine (pp. 235-300). World Scientific. doi:10.1142/9789811282171_0004

Sachan, N., & Chandra, P. (2023). A systematic review on *Sesamum indicum* L.: Bridging traditional information and pharmacological indications for future translational research. *Current Traditional Medicine*, *9*(3), 87–104. doi:10.2174/2215083808666220928142242

Sahib, A. S., Mohammed, I. H., & Sloo, S. A. (2014). Antigiardial effect of*Anethum graveolens* aqueous extract in children. *Journal of Intercultural Ethnopharmacology*, *3*(3), 109. doi:10.5455/jice.20140523104104 PMID:26401357

Saini, R. K., & Keum, Y. S. (2018). Omega-3 and omega-6 polyunsaturated fatty acids: Dietary sources, metabolism, and significance - A review. *Life Sciences*, *203*, 255–267. doi:10.1016/j.lfs.2018.04.049 PMID:29715470

Sakkaravarthy, A., Bhavadharani, P. V., Pushparaj, P., & Parameswaran, G. (2023). A review of processing novelties of extraction of essential oil from cardamom. *Lampyrid: The Journal of Bioluminescent Beetle Research*, *13*, 73–82.

Salah, A., Sleem, R., Abd-Elaziz, A., & Khalil, H. (2023). Regulation of NF-κB expression by thymoquinone; A role in regulating pro-inflammatory cytokines and programmed cell death in hepatic cancer cells. *Asian Pacific Journal of Cancer Prevention*, *24*(11), 3739–3748. doi:10.31557/APJCP.2023.24.11.3739 PMID:38019231

Salata, O. V. (2004). Applications of nanoparticles in biology and medicine. *Journal of Nanobiotechnology*, *2*(1), 3. doi:10.1186/1477-3155-2-3 PMID:15119954

Salman, T. A.-S., Ahmed, A. T., Oleiwi, G., & Al-Ibrahemi, N. (2023). Study of the Effect of Oil Extract of Dill (Anethumgraveolens L.) Plant on Oxidative Stress Parameters of Liver Enzymes in Male Rats. IOP Conference Series: Earth and Environmental Science.

Sameh, B., Bouftira, I., Mahmoud, A., Boukef, K., & Boughattas, N. (2011). Antioxidant Activity of Apium graveolens Extracts. *Journal of Biologically Active Products from Nature*, *1*(5-6), 340–343. Advance online publication. doi:10.1080/22311866.2011.10719102

Samuelsson, M. (2003). *Aquavit: And the New Scandinavian Cuisine*. Houghton Mifflin Harcourt.

Sandra, M., Barbalho, S. M., Farinazzi-Machado, F. M. V., Goulart, R. A., Brunnati, A. C. S., Ottoboni, A. M. M. B., & Nicolau, C. C. T. (2012). *PsidiumGuajava* (Guava): A Plant of Multipurpose Medicinal Applications. *Medicinal & Aromatic Plants*, *1*(4), 1–6.

Sanker, V. M., Thippeswamy, D., Chikkanna, U., & Ramakrishna, K. K. (2023). Chapter Twenty One - Ayurveda herbs and herbal preparations in the management of neurological disorders. In M. Muralidhara & P. S. Rajini (Eds.), *Ayurvedic Herbal Preparations in Neurological Disorders* (pp. 577–586). Academic Press. doi:10.1016/B978-0-443-19084-1.00028-4

Sarand, I., Traksmaa, A., Klava, D., Kunkulberga, D., Straumite, E., Galoburda, R., & Bartkiene, E. (2023). Traditional Breads from the Baltic Countries (Estonia, Latvia, Lithuania). In M. Garcia-Vaquero, K. Pastor, G. E. Orhun, A. McElhatton, & J. M. F. Rocha (Eds.), *Traditional European Breads: An Illustrative Compendium of Ancestral Knowledge and Cultural Heritage* (pp. 41–59). Springer International Publishing. doi:10.1007/978-3-031-23352-4_2

Sardella, D., Muscat, A., Brincat, J.-P., Gatt, R., Decelis, S., & Valdramidis, V. (2016). A Comprehensive Review of the Pear Fungal Diseases. *International Journal of Fruit Science*, *16*(4), 351–377. doi:10.1080/15538362.2016.1178621

Satyavati, G., Raina, M., & Sharma, M. (1987). *Medicinal plants of India* (Vol. 2). Indian Council of Medical Research.

Sawarkar, S., Deshmukh, V., & Verma, H. (2016). Nigella sativa seed, a novel beauty care ingredient: A review. *International Journal of Pharmaceutical Sciences and Research, 7*(8), 3185–3196. doi:10.13040/IJPSR.0975-8232.7(8).3185-96

Schlörmann, W., Birringer, M., Lochner, A., Lorkowski, S., Richter, I., Rohrer, C., & Glei, M. (2016). In vitro fermentation of nuts results in the formation of butyrate and c 9, t 11 conjugated linoleic acid as chemopreventive metabolites. *European Journal of Nutrition, 55*(6), 2063–2073. doi:10.1007/s00394-015-1020-0 PMID:26286349

Schmidt, J., Boettcher, C., Kuhnt, C., Kutchan, T. M., & Zenk, M. H. (2007). Poppy alkaloid profiling by electrospray tandem mass spectrometry and electrospray FT-ICR mass spectrometry after [ring-13C6]-tyramine feeding. *Phytochemistry, 68*(2), 189–202. doi:10.1016/j.phytochem.2006.10.003 PMID:17113612

Schultz, F., & Garbe, L. A. (2023). How to approach a study in ethnopharmacology? Providing an example of the different research stages for newcomers to the field today. *Pharmacology Research & Perspectives, 11*(4), e01109. doi:10.1002/prp2.1109 PMID:37497567

Schultz, S., Smith, D. R., Mock, J. J., & Schultz, D. A. (2000). Single-target molecule detection with nonbleaching multicolor optical immunolabels. *Proceedings of the National Academy of Sciences of the United States of America, 97*(3), 996–1001. doi:10.1073/pnas.97.3.996 PMID:10655473

Schwingshackl, L., & Hoffmann, G. (2014). Monounsaturated fatty acids, olive oil and health status: A systematic review and meta-analysis of cohort studies. *Lipids in Health and Disease, 13*(1), 154–154. doi:10.1186/1476-511X-13-154 PMID:25274026

Sebbagh, N., Cruciani-Guglielmacci, C., Ouali, F., Berthault, M. F., Rouch, C., Sari, D. C., & Magnan, C. (2009). Comparative effects of Citrullus colocynthis, sunflower and olive oil-enriched diet in streptozotocin-induced diabetes in rats. *Diabetes & Metabolism, 35*(3), 178–184. doi:10.1016/j.diabet.2008.10.005 PMID:19264524

Seidu, K. T., & Otutu, O. L. (2016). Phytochemical composition and radical scavenging activities of watermelon (*Citrulluslanatus*) seed constituents. *Croatian Journal of Food Science and Technology, 8*(2), 83–89. doi:10.17508/CJFST.2016.8.2.07

Sell, S. L., Boone, D. R., Weisz, H. A., Cardenas, C., Willey, H. E., Bolding, I. J., Micci, M.-A., Falduto, M. T., Torres, K. E. O., DeWitt, D. S., Prough, D. S., & Hellmich, H. L. (2019). MicroRNA profiling identifies a novel compound with antidepressant properties. *PLoS One, 14*(8), e0221163. doi:10.1371/journal.pone.0221163 PMID:31442236

Shabani, H., Karami, M. H., Kolour, J., Sayyahi, Z., Parvin, M. A., Soghala, S., Baghini, S. S., Mardasi, M., Chopani, A., Moulavi, P., Farkhondeh, T., Darroudi, M., Kabiri, M., & Samarghandian, S. (2023). Anticancer activity of thymoquinone against breast cancer cells: Mechanisms of action and delivery approaches. *Biomedicine and Pharmacotherapy, 165*, 114972. doi:10.1016/j.biopha.2023.114972 PMID:37481931

Shahabi, S., Hassan, Z. M., Mahdavi, M., Dezfouli, M., Rahvar, M. T., Naseri, M., Jazani, N. H., & Khalkhali, H. R. (2008). Hot and Cold natures and some parameters of neuroendocrine and immune systems in traditional Iranian medicine: A preliminary study. *Journal of Alternative and Complementary Medicine (New York, N.Y.)*, *14*(2), 147–156. doi:10.1089/acm.2007.0693 PMID:18307392

Shah, M., Abbas, M., Ullah, W., & Imtiaz, M., Islam, B., Said, F., Jalal, F., & Hussain, S. (2019). Pheno-Physiological Assessment of Grapes (Vitisvinifera) Germplasm. *Acta Scientific Agriculture*, *3*(3), 37–42.

Shahzad, A. N. (2019). S. Shakil-ur-Rehman, and N. Rafique, *Effectiveness of eccentric and concentric muscle energy techniques on hamstring length in healthy population. Rawal Medical Journal*, *44*(2), 350–352.

Shaik, M. I., Hamdi, I. H., & Sarbon, N. M. (2023). A comprehensive review on traditional herbal drinks: Physicochemical, phytochemicals and pharmacology properties. *Food Chemistry Advances*, *3*, 100460. doi:10.1016/j.focha.2023.100460

Shankar, K., & Liao, L. P. (2004). Traditional systems of medicine. *Phys Med Rehabil Clin N Am, 15*(4), 725-747. doi:10.1016/j.pmr.2004.03.006

Sharma, M., & Bais, R. T. (2015). Preliminary Phytochemical Screening of Methanolic Extracts of Green Grapes and Black Grapes. *International Journal of Scientific Research*, *4*(10), 2015.

Sharma, V., Yngard, R., & Lin, Y. (2008). Silver nanoparticles: Green synthesis and their antimicrobial activities. Adv. Colloid Interface Sci. 145, 83-96. *Advances in Colloid and Interface Science*, *145*, 83–96. doi:10.1016/j.cis.2008.09.002 PMID:18945421

Shayani Rad, M., Moohebati, M., & Mohajeri, S. A. (2023). Beneficial effects of celery seed extract (Apium graveolens), as a supplement, on anxiety and depression in hypertensive patients: A randomized clinical trial. *Inflammopharmacology*, *31*(1), 395–410. doi:10.1007/s10787-022-01083-y PMID:36334223

Shiraishi, Y., & Toshima, N. (2000). Oxidation of ethylene catalyzed by colloidal dispersions of poly (sodium acrylate)-protected silver nanoclusters. *Colloids and Surfaces. A, Physicochemical and Engineering Aspects*, *169*(1-3), 59–66. doi:10.1016/S0927-7757(00)00417-9

Shivatare, R. S., Musale, R., Bhutale, N. K., Kewatkar, S. M., Tare, H. L., Chatur, V., Khatawakar, A. N., & Suryavanshi, D. S. (2022). Phytochemicals: A novel approach for the Management of Coronavirus Disease 2019. *Indian Journal of Pharmaceutical Sciences*, *84*(3), 519–531. doi:10.36468/pharmaceutical-sciences.946

Shulgin, A. T., & Sargent, T. (1967). Psychotropic phenylisopropylamines derived from apiole and dillapiole. *Nature*, *215*(5109), 1494–1495. doi:10.1038/2151494b0 PMID:4861200

Siddiqui, R. A., Xu, Z., Harvey, K. A., Pavlina, T. M., Becker, M. J., & Zaloga, G. P. (2015). Comparative study of the modulation of fructose/sucrose-induced hepatic steatosis by mixed lipid formulations varying in unsaturated fatty acid content. *Nutrition & Metabolism*, *12*(1), 1–14. doi:10.1186/s12986-015-0038-x PMID:26583036

Sidhu, H. S. (2023). Transforming India's Agrarian Society in a Post-COVID World: Prospects and Challenges. In A. Fazli & A. Kundu (Eds.), *Reimagining Prosperity: Social and Economic Development in Post-COVID India* (pp. 133–152). Springer Nature Singapore. doi:10.1007/978-981-19-7177-8_8

Sikarwar, M. S., Patil, M. B., Kokate, C. K., Sharma, S., & Bhat, V. (2009). Antidiabetic activity of Nerium indicum leaf extract in alloxan-induced diabetic rats. *Journal of Young Pharmacists : JYP*, *1*(4), 330. doi:10.4103/0975-1483.59323

Silva, F., Ferreira, S., Duarte, A., Mendonça, D. I., & Domingues, F. C. (2011). Antifungal activity of Coriandrum sativum essential oil, its mode of action against Candida species and potential synergism with amphotericin B. *Phytomedicine*, *19*(1), 42–47. doi:10.1016/j.phymed.2011.06.033 PMID:21788125

Silva, F., Ferreira, S., Queiroz, J. A., & Domingues, F. C. (2011). Coriander (Coriandrum sativum L.) essential oil: Its antibacterial activity and mode of action evaluated by flow cytometry. *Journal of Medical Microbiology*, *60*(10), 1479–1486. doi:10.1099/jmm.0.034157-0 PMID:21862758

Singhal, M., Paul, A., Singh, H. P., Dubey, S. K., & Gaur, K. (2011). Article. *Journal of Chemical and Pharmaceutical Research*, *3*(3), 639–645.

Singh, G., Maurya, S., De Lampasona, M., & Catalan, C. (2005). Chemical constituents, antimicrobial investigations, and antioxidative potentials of*Anethum graveolens* L. essential oil and acetone extract: Part 52. *Journal of Food Science*, *70*(4), M208–M215. doi:10.1111/j.1365-2621.2005.tb07190.x

Singh, N., Yadav, S. S., Kumar, S., & Narashiman, B. (2021). A review on traditional uses, phytochemistry, pharmacology, and clinical research of dietary spice Cuminum cyminum L. *Phytotherapy Research*, *35*(9), 5007–5030. doi:10.1002/ptr.7133 PMID:33893678

Singh, P., & Tripathy, D. N. (2003). Fowlpox Virus Infection Causes a Lymphoproliferative Response in Chickens. *Viral Immunology*, *16*(2), 223–227. doi:10.1089/088282403322017956 PMID:12828873

Singh, S., Chaurasia, P. K., & Bharati, S. L. (2023). Hypoglycemic and hypocholesterolemic properties of Fenugreek: A comprehensive assessment. *Applied Food Research*, *3*(2), 100311. doi:10.1016/j.afres.2023.100311

Sipailiene, A., Venskutonis, R., Sarkinas, A., & Cypiene, V. (2005). Composition and antimicrobial activity of celery (Apium graveolens) leaf and root extracts obtained with liquid carbon dioxide. *Acta Horticulturae*, (677), 71–77. doi:10.17660/ActaHortic.2005.677.9

Skalli, S., Hassikou, R., & Arahou, M. (2019). An ethnobotanical survey of medicinal plants used for diabetes treatment in Rabat, Morocco. *Heliyon*, *5*(3), e01421. doi:10.1016/j.heliyon.2019.e01421 PMID:30976694

Skeels, S. (2017). *Northern Africa*. John Wiley & Sons Ltd. doi:10.1002/9781118961933

Skoog, D. A., Holler, F. J., & Crouch, S. R. (2007). Instrumental analysis. Brooks/Cole, Cengage Learning.

Słupski, J., Lisiewska, Z., & Kmiecik, W. (2005). Contents of macro and microelements in fresh and frozen dill (Anethum graveolens L.). *Food Chemistry*, *91*(4), 737–743. doi:10.1016/j.foodchem.2004.06.046

Soares, B. V., Morais, S. M., dos Santos Fontenelle, R. O., Queiroz, V. A., Vila-Nova, N. S., Pereira, C. M. C., Brito, E. S., Neto, M. A. S., Brito, E. H. S., Cavalcante, C. S. P., Castelo-Branco, D. S. C. M., & Rocha, M. F. G. (2012). Antifungal activity, toxicity and chemical composition of the essential oil of Coriandrum sativum L. fruits. *Molecules (Basel, Switzerland)*, *17*(7), 8439–8448. doi:10.3390/molecules17078439 PMID:22785271

Sobhy, H. A., Elissawy, A. E., Eldahshan, O. A., & Singab, A. N. B. (2024). Role of Egyptian Essential Oils in Health Care and Industry. In *Pharmacological Aspects of Essential Oils* (pp. 107–119). CRC Press.

Sogi, D. S., Oberoi, D. P. S., & Malik, S. (2010). Effect of Particle Size, Temperature, and Total Soluble Solids on the Rheological Properties of Watermelon Juice: A Response Surface Approach. *International Journal of Food Properties*, *13*(6), 1207–1214. doi:10.1080/10942910903013373

Soleymani, A. R., Rafigh, S. M., & Hekmati, M. (2020). Green synthesis of RGO/Ag: As evidence for the production of uniform mono-dispersed nanospheres using microfluidization. *Applied Surface Science*, *518*, 146264. doi:10.1016/j.apsusc.2020.146264

Solomon, U. (2005). Anti-stress potential of aqueous seed extract of Aframomum melegueta. *African Journal of Biomedical Research*, 8.

Solomon, U., & Ashorobi, B. (2008). Further pharmacological studies on aqueous seed extract of Aframomum melegueta in rats. *Journal of Ethnopharmacology*, *115*(3), 489–493. doi:10.1016/j.jep.2007.10.019 PMID:18060710

Solomon, U., & Ashorobi, R. (2007). Further studies on the antinociceptive action of aqueous seed extract of Aframomum melegueta. *Journal of Ethnopharmacology*, *109*(3), 501–504. doi:10.1016/j.jep.2006.08.025 PMID:17023130

Somova, L. I., Shode, F. O., Ramnanan, P., & Nadar, A. (2003). Antihypertensive, antiatherosclerotic and antioxidant activity of triterpenoids isolated from Olea europaea, subspecies africana leaves. *Journal of Ethnopharmacology*, *84*(2-3), 299–305. doi:10.1016/S0378-8741(02)00332-X PMID:12648829

Sorce, C., Lorenzi, R., & Ranalli, P. (1997). The effects of (S)-(+)-carvone treatments on seed potato tuber dormancy and sprouting. *Potato Research*, *40*(2), 155–161. doi:10.1007/BF02358241

Sorkin, B., & Gahche, J. (2020). Application of p-curve analysis to dietary supplement clinical trials: Case study of trials of curcumin products for arthritis. *Current Developments in Nutrition, 4*. doi:10.1093/cdn/nzaa056_026

Soumya, V., & Ramana-Rao, T.V. (2014). Nutritional quality evaluation of four icebox cultivars of watermelon fruit during their development and ripening. *International Food Research Journal, 21*(2).

Sourmaghi, M. H. S., Kiaee, G., Golfakhrabadi, F., Jamalifar, H., & Khanavi, M. (2015). Comparison of essential oil composition and antimicrobial activity of Coriandrum sativum L. extracted by hydrodistillation and microwave-assisted hydrodistillation. *Journal of Food Science and Technology, 52*(4), 2452–2457. doi:10.1007/s13197-014-1286-x PMID:25829632

Sowbhagya, H. (2014). Chemistry, technology, and nutraceutical functions of celery (Apium graveolens L.): An overview. *Critical Reviews in Food Science and Nutrition, 54*(3), 389–398. doi:10.1080/10408398.2011.586740

Sowbhagya, H., Srinivas, P., & Krishnamurthy, N. (2010). Effect of enzymes on extraction of volatiles from celery seeds. *Food Chemistry, 120*(1), 230–234. doi:10.1016/j.foodchem.2009.10.013

Spence, C. (2024). Cinnamon: The historic spice, medicinal uses, and flavor chemistry. *International Journal of Gastronomy and Food Science, 35*, 100858. doi:10.1016/j.ijgfs.2023.100858

Sreejayan, N., & Rao, M. N. A. (1996). Free radical scavenging activity of curcuminoids. *Journal of Drug Research, 46*, 169. PMID:8720307

Sreelatha, S., & Inbavalli, R. (2012). Antioxidant, antihyperglycemic, and antihyperlipidemic effects of Coriandrum sativum leaf and stem in alloxan-induced diabetic rats. *Journal of Food Science, 77*(7), T119–T123. doi:10.1111/j.1750-3841.2012.02755.x PMID:22671941

Sreelatha, S., Padma, P., & Umadevi, M. (2009). Protective effects of Coriandrum sativum extracts on carbon tetrachloride-induced hepatotoxicity in rats. *Food and Chemical Toxicology, 47*(4), 702–708. doi:10.1016/j.fct.2008.12.022 PMID:19146910

Srikanth, N., & Goel, S. (2023). Systematic Documentation and Drug Development from Local Health Traditions (LHTs) and Ethnomedical Practices (EMPs): Challenges and Way Forward. In S. Reddy, N. Guite, & B. Subedi (Eds.), *Ethnomedicine and Tribal Healing Practices in India: Challenges and Possibilities of Recognition and Integration* (pp. 237–249). Springer Nature Singapore. doi:10.1007/978-981-19-4286-0_15

Sriti, J., Msaada, K., Talou, T., Faye, M., Vilarem, G., & Marzouk, B. (2012). Coupled extruder-headspace, a new method for analysis of the essential oil components of Coriandrum sativum fruits. *Food Chemistry, 134*(4), 2419–2423. doi:10.1016/j.foodchem.2012.04.029 PMID:23442704

Sriti, J., Talou, T., Faye, M., Vilarem, G., & Marzouk, B. (2011). Oil extraction from coriander fruits by extrusion and comparison with solvent extraction processes. *Industrial Crops and Products, 33*(3), 659–664. doi:10.1016/j.indcrop.2011.01.005

Steinmetz, K. A., & Potter, J. D. (1996). Vegetables, fruit, and cancer prevention: A review. *Journal of the American Dietetic Association*, *96*(10), 1027–1039. doi:10.1016/S0002-8223(96)00273-8 PMID:8841165

Stephenson, R. A. (2003). NUTS. In B. Caballero (Ed.), *Encyclopedia of Food Sciences and Nutrition* (2nd ed., pp. 4205–4212). Academic Press. doi:10.1016/B0-12-227055-X/00848-8

Stephenson, R., & Gallagher, E. (1986a). Effects of night temperature on floral initiation and raceme development in macadamia. *Scientia Horticulturae*, *30*(3), 213–218. doi:10.1016/0304-4238(86)90099-3

Stephenson, R., & Gallagher, E. (1986b). Effects of temperature during latter stages of nut development on growth and quality of macadamia nuts. *Scientia Horticulturae*, *30*(3), 219–225. doi:10.1016/0304-4238(86)90100-7

Stephenson, R., Gallagher, E., & Doogan, V. (2003). Macadamia responses to mild water stress at different phenological stages. *Australian Journal of Agricultural Research*, *54*(1), 67–75. doi:10.1071/AR02108

Stolp, L. J., & Kodali, D. R. (2022). Naturally occurring high-oleic oils: Avocado, macadamia, and olive oils. In F. J. Flider (Ed.), *High Oleic Oils* (pp. 7–52). AOCS Press. doi:10.1016/B978-0-12-822912-5.00003-4

Struempler, R. E. (1987). Excretion of Codeine and Morphine Following Ingestion of Poppy Seeds. *Journal of Analytical Toxicology*, *11*(3), 97–99. doi:10.1093/jat/11.3.97 PMID:3599922

Studzińska-Sroka, E., Galanty, A., Gościniak, A., Wieczorek, M., Kłaput, M., Dudek-Makuch, M., & Cielecka-Piontek, J. (2021). Herbal infusions as a valuable functional food. *Nutrients*, *13*(11), 4051. doi:10.3390/nu13114051 PMID:34836310

Sugihara, N., Arakawa, T., Ohnishi, M., & Furuno, K. (2000). Anti and pro-oxidative effects of flavonoids on metal induced lipid hydroperoxide dependent lipid peroxidation in cultured hepatocyte loaded with a-Linolenic acid. *Free Radical Biology & Medicine*, *27*(11-12), 1313–1323. doi:10.1016/S0891-5849(99)00167-7

Sulieman, A. M. E., Abdallah, E. M., Alanazi, N. A., Ed-Dra, A., Jamal, A., Idriss, H., Alshammari, A. S., & Shommo, S. A. M. (2023). Spices as sustainable food preservatives: A comprehensive review of their antimicrobial potential. *Pharmaceuticals (Basel, Switzerland)*, *16*(10), 1451. doi:10.3390/ph16101451 PMID:37895922

Sultan, M. T., Saeed, F., Raza, H., Ilyas, A., Sadiq, F., Musarrat, A., Afzaal, M., Hussain, M., Raza, M. A., & Al Jbawi, E. (2023). Nutritional and therapeutic potential of nutmeg (Myristica fragrans): A concurrent review. *Cogent Food & Agriculture*, *9*(2), 2279701. doi:10.1080/23311932.2023.2279701

Šunić, L., Ilić, Z. S., Stanojević, L., Milenković, L., Stanojević, J., Kovač, R., Milenković, A., & Cvetković, D. (2023). Comparison of the essential oil content, constituents and antioxidant activity from different plant parts during development stages of wild fennel (*Foeniculum vulgare* Mill.). *Horticulturae*, *9*(3), 364. doi:10.3390/horticulturae9030364

Sunil, C., Agastian, P., Kumarappan, C., & Ignacimuthu, S. (2012). In vitro antioxidant, antidiabetic and antilipidemic activities of Symplocos cochinchinensis (Lour.) S. Moore bark. *Food and Chemical Toxicology*, *50*(5), 1547–1553. doi:10.1016/j.fct.2012.01.029 PMID:22306921

Sun, Y., Yang, T., & Wang, C. (2022). Capparis spinosa L. as a potential source of nutrition and its health benefits in foods: A comprehensive review of its phytochemistry, bioactivities, safety, and application. *Food Chemistry*, 135258. PMID:36587515

Suttisansanee, U., Thiyajai, P., Inthachat, W., Pruesapan, K., Wongwathanarat, K., Charoenkiatkul, S., Sahasakul, Y., & Temviriyanukul, P. (2023). Exploration of the nutritional and carotenoids profiles of vegetables in Thai cuisine as potential nutritious ingredients. *Heliyon*, *9*(5), e15951. doi:10.1016/j.heliyon.2023.e15951 PMID:37205996

Svoboda, K. P., & Hampson, J. B. (n.d.). Bioactivity of essential oils of selected temperate aromatic plants: antibacterial, antioxidant, antiinflammatory and other related pharmacological activities. Academic Press.

Swanston-Flatt, S., Day, C., Bailey, C. J., & Flatt, P. R. (1990). Traditional plant treatments for diabetes. Studies in normal and streptozotocin diabetic mice. *Diabetologia*, *33*(8), 462–464. doi:10.1007/BF00405106 PMID:2210118

Szychowski, K. A., Rybczynska-Tkaczyk, K., Gawel-Beben, K., Swieca, M., Karas, M., Jakuczyk, A., ... Gminski, J. (2018). Characterization of active compounds of different garlic [Allium sativum L.] cultivars. *Polish Journal of Food and Nutrition Sciences*, *68*(1), 68. doi:10.1515/pjfns-2017-0005

Tabaza, Y., & Aburjai, T. (2024). Ethnopharmacological survey of medicinal plants used by local herbalists and traditional healers for the treatment of cancer in Jordan. *Current Traditional Medicine*, *10*(3), 6–19. doi:10.2174/2215083810666230515161319

Tabiri, B., Agbenorhevi, J. K., Wireko-Manu, F. D., & Ompouma, E. I. (2016). Watermelon Seeds as Food: Nutrient Composition, Phytochemicals and Antioxidant Activity. *International Journal of Nutrition and Food Sciences*, *5*(2), 139–144. doi:10.11648/j.ijnfs.20160502.18

Tahiliani, A. G., & Beinlich, C. J. (1991). Pantothenic acid in health and disease. *Vitamins and Hormones*, *46*, 165–228. doi:10.1016/S0083-6729(08)60684-6 PMID:1746161

Tahraoui, A., El-Hilaly, J., Israili, Z. H., & Lyoussi, B. (2007). Ethnopharmacological survey of plants used in the traditional treatment of hypertension and diabetes in south-eastern Morocco. *Journal of Ethnopharmacology*, *110*(1), 105–117. doi:10.1016/j.jep.2006.09.011 PMID:17052873

Tahri, N., El Basti, A., Zidane, L., Rochdi, A., & Douira, A. (2012). Etude ethnobotanique des plantes medicinales dans la province de Settat. Kastamonu University Journal of Forestry Faculty, 12(2), 192-208.v

Takooree, H., Aumeeruddy, M. Z., Rengasamy, K. R. R., Venugopala, K. N., Jeewon, R., Zengin, G., & Mahomoodally, M. F. (2019). A systematic review on black pepper (*Piper nigrum* L.): from folk uses to pharmacological applications. *Critical Reviews in Food Science and Nutrition, 59*(sup1), S210-S243. doi:10.1080/10408398.2019.1565489

Talhaoui, N., Trabelsi, N., Taamalli, A., Verardo, V., Gomez-Caravaca, A. M., Fernández-Gutiérrez, A., & Arraez-Roman, D. (2018). Olea europaea as potential source of bioactive compounds for diseases prevention. *Studies in Natural Products Chemistry, 57*, 389-411.

Tananuwong, K., & Jitngarmkusol, S. (2019). *Macadamia flours: Nutritious ingredients for baked goods. In Flour and Breads and their Fortification in Health and Disease Prevention.* Elsevier.

Taniguchi, M., Yanai, M., Xiao, Y. Q., Kido, T., & Baba, K. (1996). Three isocoumarins from Coriandrum sativum. *Phytochemistry, 42*(3), 843–846. doi:10.1016/0031-9422(95)00930-2

Tapiero, H., Tew, K. D., Ba, G. N., & Mathé, G. (2002). Polyphenols: Do they play a role in the prevention of human pathologies? *Biomedicine and Pharmacotherapy, 56*(4), 200–207. doi:10.1016/S0753-3322(02)00178-6 PMID:12109813

Tarazona-D'ıaz, J., Viegas, J., Moldao-Martins, M., & Aguayo, E. (2017). Bioactive Compounds from Flesh and by-product of fresh-cut Watermelon cultivars. *Journal of the Science of Food and Agriculture, 91*(5), 805–812. doi:10.1002/jsfa.4250 PMID:21384347

Tastekin, D., Atasever, M., Adiguzel, G., Keles, M., & Tastekin, A. (2006). Hypoglycaemic effect of Artemisia herba-alba in experimental hyperglycaemic rats. *Bulletin of the Veterinary Institute in Pulawy*, 50.

Tejeda, L., Vasquez Iriarte, P. E., Ortiz, J. V., Aliaga-Rossel, E., Mollinedo, P., & Peñarrieta, J. M. (2022). Characterization of Bolivian chili peppers; antioxidant capacity, total phenolic compounds, capsaicin and β-carotene concentration. *Nutrition & Food Science, 52*(8), 1314–1327. doi:10.1108/NFS-11-2021-0341

Thahira Banu, A., & Aswini, M. (2023). Chapter 4 - Role of Piper nigrum in functional food and nutraceuticals. In A. Amalraj, S. Kuttappan, K. Varma A.C, & A. Matharu (Eds.), Herbs, Spices and Their Roles in Nutraceuticals and Functional Foods (pp. 41-60). Academic Press. doi:10.1016/B978-0-323-90794-1.00019-3

Thakran, S., Siddiqui, M. R., & Baquer, N. Z. (2004). Trigonella foenum graecum seed powder protects against histopathological abnormalities in tissues of diabetic rats. *Molecular and Cellular Biochemistry, 266*(1/2), 151–159. doi:10.1023/B:MCBI.0000049153.14295.0d PMID:15646037

Thema, F. T., Beukes, P., Gurib-Fakim, A., & Maaza, M. (2015). Green synthesis of Monteponite CdO nanoparticles by Agathosma betulina natural extract. *Journal of Alloys and Compounds, 646*, 1043–1048. doi:10.1016/j.jallcom.2015.05.279

Thirumurugan, G., Shaheedha, S. M., & Dhanaraju, M. D. (2009). In vitro evaluation of antibacterial activity of silver nanoparticles synthesised by using Phytophthora infestans. *International Journal of Chemtech Research*, *1*, 714–716.

Thoppil, R. J., & Bishayee, A. (2011). Terpenoids as potential chemopreventive and therapeutic agents in liver cancer. *World Journal of Hepatology*, *3*(9), 228. doi:10.4254/wjh.v3.i9.228 PMID:21969877

Thovhogi, N., Diallo, A., Gurib-Fakim, A., & Maaza, M. (2015). Nanoparticles green synthesis by Hibiscus Sabdariffa flower extract: Main physical properties. *Journal of Alloys and Compounds*, *647*, 392–396. doi:10.1016/j.jallcom.2015.06.076

Tijjani, H., & Luka, C. (2013). Effects of Afromomum melegueta, Zingiber officinale and Piper nigrum on Some Biochemical and Haematological Parameters in Rats Fed with High Lipid Diet. *International Journal of Pure & Applied Bioscience.*, *1*, 61–67.

Tlili, I., Hdider, C., Lenucci, M. S., Illahy, R., Jebari, H., & Dalessandro, G. (2011). Bioactive compounds and antioxidant activities during fruit ripening of watermelon cultivars. *Journal of Food Composition and Analysis*, *24*(7), 923–928. doi:10.1016/j.jfca.2011.03.016

Tohamy, A. A., Mohammed, R. S., Abdalla, M. S., Ibrahim, A. K., & Mahran, K. F. (2013). The effect of Lupinus Albus [Termis] and Hyphaene thebaica [doum] on some biochemical parameters in streptozotocin-induced diabetic rats. *The Egyptian Journal of Hospital Medicine*, *53*(1), 789–794. doi:10.12816/0001640

Toh, E. Y. S., Chooi Ling, L., Ling, A., Soi Moi, C., & Koh, R. Y. (2019). Overview of the pharmacological activities of aframomum melegueta. *Pertanika. Journal of Tropical Agricultural Science*, *42*, 1–13.

Trifan, A., Zengin, G., Korona-Glowniak, I., Skalicka-Woźniak, K., & Luca, S. V. (2023). Essential oils and sustainability: In vitro bioactivity screening of *Myristica fragrans* Houtt. postdistillation byproducts. *Plants*, *12*(9), 1741. doi:10.3390/plants12091741 PMID:37176799

Tsumura, T. (1988). *Factors Affecting Macadamia Nut Stability*. University of Hawai'i at Manoa.

Tu, X. H., Wu, B., Xie, Y., Xu, S. L., Wu, Z. Y., Lv, X., Wei, F., Du, L.-Q., & Chen, H. (2021). A comprehensive study of raw and roasted macadamia nuts: Lipid profile, physicochemical, nutritional, and sensory properties. *Food Science & Nutrition*, *9*(3), 1688–1697. doi:10.1002/fsn3.2143 PMID:33747479

Uemori, C. (2018). *Extraction and nanoparticle preparation of natural materials by using supercritical carbon dioxide*. Graduate School of Engineering, Nagoya University.

Uitterhaegen, E., Sampaio, K., Delbeke, E., De Greyt, W., Cerny, M., Evon, P., Merah, O., Talou, T., & Stevens, C. (2016). Characterization of French coriander oil as source of petroselinic acid. *Molecules (Basel, Switzerland)*, *21*(9), 1202. doi:10.3390/molecules21091202 PMID:27617992

Ukeh, D. (2008). Bioactivities of Essential Oils of Aframomum melegueta and Zingiber officinale both (Zingiberaceae) Against Rhyzopertha dominica (Fabricius). *Journal of Entomology*, *5*(3), 193–199. doi:10.3923/je.2008.193.199

Ukeh, D., Birkett, M., Pickett, J., Bowman, A., & Luntz, A. (2009). Repellent activity of alligator pepper, Aframomum melegueta, and ginger, Zingiber officinale, against the maize weevil, Sitophilus zeamais. *Phytochemistry*, *70*(6), 751–758. doi:10.1016/j.phytochem.2009.03.012 PMID:19394981

Umer, A., Naveed, S., Ramzan, N., Rafique, M. S., & Imran, M. (2014). A green method for the synthesis of copper nanoparticles using L-ascorbic acid. *Matéria (Rio de Janeiro)*, *19*(3), 197–203. doi:10.1590/S1517-70762014000300002

Umukoro, S., & Ashorobi, R. B. (2005). Effect of Aframomum melegueta. Seed Extract on Castor Oil-Induced Diarrhea. *Pharmaceutical Biology*, *43*(4), 330–333. doi:10.1080/13880200590951748 PMID:28925843

Urabe, A. A., & Aziz, W. J. (2019). Biosynthesis of cobalt oxide (Co3O4) nanoparticles using plant extract of Camellia sinensis (L.) Kuntze and Apium graveolens L. as the antibacterial application. *World News of Natural Sciences*, (24), 356–364.

Usman, I., Saif, H., Imran, A., Afzaal, M., Saeed, F., Azam, I., Afzal, A., Ateeq, H., Islam, F., Shah, Y. A., & Shah, Y. A. (2023). Innovative applications and therapeutic potential of oilseeds and their by-products: An eco-friendly and sustainable approach. *Food Science & Nutrition*, *11*(6), 2599–2609. doi:10.1002/fsn3.3322 PMID:37324916

Usmani, K., Jain, S. K., & Yadav, S. (2023). Mechanism of action of certain medicinal plants for the treatment of asthma [Review]. *Journal of Ethnopharmacology*, *317*, 116828. doi:10.1016/j.jep.2023.116828 PMID:37369335

Valko, M., & Jomova, K. (2011). Advances in metal-induced oxidative stress and human disease. *Toxicology*, *283*(2), 65–87. PMID:21414382

Vallet, A., Locatelli, B., Valdivia-Díaz, M., Quispe Conde, Y., Matencio García, G., Ramos Criales, A., Valverde Huamanñahui, F., Ramos Criales, S., Makowski, D., & Lavorel, S. (2023). Knowledge coproduction to improve assessments of nature's contributions to people. *Conservation Biology*, *14182*(6), e14182. Advance online publication. doi:10.1111/cobi.14182 PMID:37889094

Vanderveen, J. E. (1994). Regulatory history for stearic acid. *The American Journal of Clinical Nutrition*, *60*(6, Suppl), 983s–985s. doi:10.1093/ajcn/60.6.983S PMID:7977156

Vannabhum, M., Ziangchin, N., Thepnorarat, P., & Akarasereenont, P. (2023). Metabolomic analysis of Thai herbal analgesic formula based on ultrahigh-performance liquid chromatography-quadrupole time-of-flight mass spectrometry. *Heliyon*, *9*(7), e18296. doi:10.1016/j.heliyon.2023.e18296 PMID:37539319

Varghese, R., & Ray, J. G. (2023). Sustainable agriculture of tropical spices: Arbuscular mycorrhizal fungi as an ecotechnological tool – A critical review. *Journal of Applied Research on Medicinal and Aromatic Plants, 36*, 100507. doi:10.1016/j.jarmap.2023.100507

Vats, A., & Sharma, P. (2012). Formulation and evaluation of topical anti acne formulation of coriander oil. *Int J Pharm Pharm Sci Res, 2*, 61–66.

Vats, A., & Sharma, P. (2013). Comparative study, and analysis of release kinetics of coriander formulations. *Indo American Journal of Pharmaceutical Research, 3*(1), 1334–1348.

Vaughan, J., & Geissler, C. (2009). *The New Oxford Book of Food Plants*. OUP Oxford.

Vayalil, P. K. (2012). Date fruits [Phoenix dactylifera Linn]: An emerging medicinal food. *Critical Reviews in Food Science and Nutrition, 52*(3), 249–271. doi:10.1080/10408398.2010.499824 PMID:22214443

Vickers, A., Zollman, C., & Lee, R. (2001). Herbal medicine. *The Western Journal of Medicine, 175*(2), 125–128. doi:10.1136/ewjm.175.2.125 PMID:11483560

Vieira, L. V., & Juvenato, M. E., Krause, M., Heringer, O. A., Ribeiro, J. S., Brandão, G. P., . . . Carneiro, M. T. W. D. (2022). The effects of drying methods and harvest season on piperine, essential oil composition, and multielemental composition of black pepper. *Food Chemistry, 390*, 133148. doi:10.1016/j.foodchem.2022.133148 PMID:35551027

Vigushin, D. M., Poon, G. K., Boddy, A., English, J., Halbert, G. W., Pagonis, C., ... Committee, C. R. C. P. I. I. C. T. (1998). Phase I and pharmacokinetic study of D-limonene in patients with advanced cancer. *Cancer Chemotherapy and Pharmacology, 42*(2), 111–117. doi:10.1007/s002800050793 PMID:9654110

Vogel, A. I. (1989). *Vogel's textbook of practical organic chemistry* (Vol. 5). Longman Scientific & Technical New York.

von Maydell, D., Beleites, C., Stache, A.-M., Riewe, D., Krähmer, A., & Marthe, F. (2024). Genetic variation of annual and biennial caraway (Carum carvi) germplasm offers diverse opportunities for breeding. *Industrial Crops and Products, 208*, 117798. doi:10.1016/j.indcrop.2023.117798

Waheed, A., Miana, G. A., Ahmad, S. I., & Khan, M. A. (2006). Clinical investigation of hypoglycemic effect of Coriandrum sativum in type-2 [NIDDM] diabetic patients. *Pakistan Journal of Pharmacology, 23*(1), 7–11.

Wallace, H. M., & Walton, D. (2011). *Macadamia (Macadamia integrifolia, Macadamia tetraphylla and hybrids). In Postharvest biology and technology of tropical and subtropical fruits*. Elsevier.

Wallace, H. M., & Walton, D. A. (2011). 19 - Macadamia (Macadamia integrifolia, Macadamia tetraphylla and hybrids). In E. M. Yahia (Ed.), *Postharvest Biology and Technology of Tropical and Subtropical Fruits* (pp. 450–474e). Woodhead Publishing. doi:10.1533/9780857092885.450

Wall, M. (2013). *Improving the quality and safety of macadamia nuts. In Improving the safety and quality of nuts*. Elsevier.

Wall, M. M. (2010). Functional lipid characteristics, oxidative stability, and antioxidant activity of macadamia nut (Macadamia integrifolia) cultivars. *Food Chemistry*, *121*(4), 1103–1108. doi:10.1016/j.foodchem.2010.01.057

Wall, M. M. (2013). 12 - Improving the quality and safety of macadamia nuts. In L. J. Harris (Ed.), *Improving the Safety and Quality of Nuts* (pp. 274–296). Woodhead Publishing. doi:10.1533/9780857097484.2.274

Walton, D. (2005). *Anatomy and handling: Implications for macadamia nut quality*. University of the Sunshine Coast.

Walton, D. A., & Wallace, H. M. (2010). Dropping macadamia nuts-in-shell reduces kernel roasting quality. *Journal of the Science of Food and Agriculture*, *90*(13), 2163–2167. doi:10.1002/jsfa.4063 PMID:20632387

Wanasundara, P., Shahidi, F., & Shukla, V. (1997). Endogenous antioxidants from oilseeds and edible oils. *Food Reviews International*, *13*(2), 225–292. doi:10.1080/87559129709541106

Wangensteen, H., Samuelsen, A. B., & Malterud, K. E. (2004). Antioxidant activity in extracts from coriander. *Food Chemistry*, *88*(2), 293–297. doi:10.1016/j.foodchem.2004.01.047

Wang, M., Wang, L., Pan, X.-J., & Zhang, H. (2012). Monocytic differentiation of K562 cells induced by proanthocyanidins from grape seeds. *Archives of Pharmacal Research*, *35*(1), 129–135. doi:10.1007/s12272-012-0114-y PMID:22297751

Wang, R., van Welie, N., van Rijswijk, J., Johnson, N. P., Norman, R. J., Dreyer, K., Mijatovic, V., & Mol, B. W. (2019). Effectiveness on fertility outcome of tubal flushing with different contrast media: Systematic review and network meta-analysis. *Ultrasound in Obstetrics & Gynecology*, *54*(2), 172–181. doi:10.1002/uog.20238 PMID:30740799

Wang, W., Nguyen, K. T. K., Zhao, C., & Hung, H. (2023). Earliest curry in Southeast Asia and the global spice trade 2000 years ago. *Science Advances*, *9*(29), eadh5517. Advance online publication. doi:10.1126/sciadv.adh5517 PMID:37478176

Wardani, R. S., Schellack, N., Govender, T., Dhulap, A. N., Utami, P., Malve, V., & Wong, Y. C. (2023). Treatment of the common cold with herbs used in Ayurveda and Jamu: Monograph review and the science of ginger, liquorice, turmeric and peppermint. *Drugs in Context*, *12*, 1–10. Advance online publication. doi:10.7573/dic.2023-2-12 PMID:37342461

Warinhomhoun, S., Sanpinit, S., Chonsut, P., Ngamdokmai, N., Ahmad, A. R., & Wetchakul, P. (2023). Effects of black seed oil combined with olive oil or honey on antioxidant activities, phenolic content, and identification and quantification of thymoquinone, a key bioactive compound. *Journal of Agriculture and Food Research*, *14*, 100891. doi:10.1016/j.jafr.2023.100891

Warrier, P., Nambiar, V., Ramankutty, C., & Vaidya, S. A. (1995). *Indian Medicinal Plants*. Orient Longman.

Waswa, E. N., Ding, S.-X., Wambua, F. M., Mkala, E. M., Mutinda, E. S., Odago, W. O., Amenu, S. G., Muthui, S. W., Linda, E. L., Katumo, D. M., Waema, C. M., Yang, J.-X., & Hu, G.-W. (2024). The genus Actinidia Lindl. (Actinidiaceae): A comprehensive review on its ethnobotany, phytochemistry, and pharmacological properties. *Journal of Ethnopharmacology, 319*, 117222. doi:10.1016/j.jep.2023.117222 PMID:37793579

Wells, M. L. (2013). 1 - Agricultural practices to reduce microbial contamination of nuts. In L. J. Harris (Ed.), *Improving the Safety and Quality of Nuts* (pp. 3–21). Woodhead Publishing. doi:10.1533/9780857097484.1.3

Weluwanarak, T., Changbunjong, T., Leesombun, A., Boonmasawai, S., & Sungpradit, S. (2023). Effects of *Piper nigrum* L. fruit essential oil toxicity against stable fly (Diptera: Muscidae). *Plants, 12*(5), 1043. doi:10.3390/plants12051043 PMID:36903904

Wheeler, A. G., J. (2019). Bugs and Poppies: Heteroptera Associated with Developing and Mature Capsules (Fruits) of <i>Papaver somniferum</i> L. (Papaveraceae). *Proceedings of the Entomological Society of Washington, 121*(3), 511-524. 10.4289/0013-8797.121.3.511

White, N., & Hanan, J. (2014). *A model of macadamia with application to pruning in orchards.* Paper presented at the XXIX International Horticultural Congress on Horticulture: Sustaining Lives, Livelihoods and Landscapes (IHC2014).

Whitfield, K. C., Bourassa, M. W., Adamolekun, B., Bergeron, G., Bettendorff, L., Brown, K. H., Cox, L., Fattal-Valevski, A., Fischer, P. R., Frank, E. L., Hiffler, L., Hlaing, L. M., Jefferds, M. E., Kapner, H., Kounnavong, S., Mousavi, M. P. S., Roth, D. E., Tsaloglou, M.-N., Wieringa, F., & Combs, G. F. Jr. (2018). Thiamine deficiency disorders: Diagnosis, prevalence, and a roadmap for global control programs. *Annals of the New York Academy of Sciences, 1430*(1), 3–43. doi:10.1111/nyas.13919 PMID:30151974

Wichtl, M. (1994). *Coriandri fructus. Herbal Drugs and Phytopharmaceuticals.* CRC Press.

Wijayagunawardanea, M. P. B., Wijerathnea, C. U. B., & Herathb, C. B. (2015). Indigenous herbal recipes for treatment of liver cirrhosis. *Procedia Chemistry, 14*, 270–276. doi:10.1016/j.proche.2015.03.038

Wiseman, S. A., Balentine, D. A., & Frei, B. (1997). Antioxidants in tea. *Critical Reviews in Food Science and Nutrition, 37*(8), 705–718. doi:10.1080/10408399709527798 PMID:9447271

Wood, L. G., & Garg, M. L. (2011). *Macadamia Nuts (Macadamia integrifolia and tetraphylla) and their Use in Hypercholesterolemic Subjects. In Nuts and Seeds in Health and Disease Prevention.* Elsevier.

Wu, J., Wan, Z., Yi, J., Wu, Y., Peng, W., & Wu, J. (2012). Investigation of the Extracts from Bidenspilosa Linn. var. radiata Sch. Bip. for Antioxidant activities and Cytotoxicity against human Tumor cells. *Journal of Natural Medicines, 67*(1), 17–26. doi:10.1007/s11418-012-0639-x PMID:22382861

Wu, K., Dong, X., Zhu, J., Wu, P., Liu, C., Wang, Y., Wu, J., Hou, J., Liu, Z., & Guo, X. (2018). Designing biomimetic porous celery: TiO2/ZnO nanocomposite for enhanced CO2 photoreduction. *Journal of Materials Science*, *53*(16), 11595–11606. doi:10.1007/s10853-018-2397-y

Xia, Y., Rogers, J. A., Paul, K. E., & Whitesides, G. M. (1999). Unconventional Methods for Fabricating and Patterning Nanostructures. *Chemical Reviews*, *99*(7), 1823–1848. doi:10.1021/cr980002q PMID:11849012

Xu, H., Zhong, L., Deng, J., Peng, J., Dan, H., Zeng, X., Li, T., & Chen, Q. (2020). High expression of ACE2 receptor of 2019-nCoV on the epithelial cells of oral mucosa. *International Journal of Oral Science*, *12*(1), 8. doi:10.1038/s41368-020-0074-x PMID:32094336

Yadav, M., Chatterji, S., Gupta, S. K., & Watal, G. (2014). Preliminary phytochemical screening of six medicinal plants used in traditional medicine. *International Journal of Pharmacy and Pharmaceutical Sciences*, *6*(5), 539–542.

Yadav, S., Tyagi, A., Kumari, R., Srivastava, H., Rangan, P., Wankhede, D., & Gaikwad, A. B. (2023). Comparative transcriptome profiling of fruit tissue provides novel insights into piperine biosynthesis in black pepper (*Piper nigrum* L.). *Scientia Horticulturae*, *322*, 112451. doi:10.1016/j.scienta.2023.112451

Yagoub, S. O. (2023). Chapter four - Black cumin: morphology, physiology, growth, and agricultural yield. In A. A. Mariod (Ed.), *Biochemistry, Nutrition, and Therapeutics of Black Cumin Seed* (pp. 19–25). Academic Press. doi:10.1016/B978-0-323-90788-0.00016-0

Yamagata, K. (2022). Chapter 1 - Effects of nut consumption on the inhibition of obesity-related endothelial cell dysfunction and prevention of cardiovascular diseases. In Studies in Natural Products Chemistry (Vol. 75, pp. 1-39). Elsevier.

Yang, J., Liu, R. H., & Halim, L. (2009). Antioxidant and antiproliferative activities of common edible nut seeds. *Lebensmittel-Wissenschaft + Technologie*, *42*(1), 1–8. doi:10.1016/j.lwt.2008.07.007

Yang, J., Wang, H. P., Zhou, L., & Xu, C. F. (2012). Effect of dietary fiber on constipation: A meta analysis. *World Journal of Gastroenterology*, *18*(48), 7378–7383. doi:10.3748/wjg.v18.i48.7378 PMID:23326148

Yang, J., & Xiao, Y.-Y. (2013). Grape Phytochemicals and Associated Health Benefits. *Critical Reviews in Food Science and Nutrition*, *53*(11), 1202–1225. doi:10.1080/10408398.2012.692408 PMID:24007424

Yan, S., Liu, Y., Zhao, W., Zhao, H., & Xue, X. (2023). Chemical markers of a rare honey from the traditional spice plant Amomum tsao–ko Crevost et Lemarié, via integrated GC–MS and LC–MS approaches. *Food Research International*, *172*, 113234. doi:10.1016/j.foodres.2023.113234 PMID:37689964

Yap, P. S. X., Yiap, B. C., Ping, H. C., & Lim, S. H. E. (2014). Essential oils, a new horizon in combating bacterial antibiotic resistance. *The Open Microbiology Journal*, *8*(1), 6–14. doi:10.2174/1874285801408010006 PMID:24627729

Yaser, A., Muneer, A., Abdelhafid, B., & Fawzia, A. (2014). Chemical composition, phytochemical constituents and antioxidant activities of the seeds extract of Apiumgraveoleus L from Yemen. *Research Journal of Pharmaceutical, Biological and Chemical Sciences*, *5*(1), 1074–1087.

Yazdanparast, R., & Alavi, M. (2001). Antihyperlipidaemic and antihypercholesterolaemic effects of*Anethum graveolens* leaves after the removal of furocoumarins. *Cytobios*, *105*(410), 185. PMID:11409638

Yili, A., Yimamu, H., Maksimov, V., Aisa, H., Veshkurova, O., & Salikhov, S. I. (2006). Chemical composition of essential oil from seeds of*Anethum graveolens* cultivated in China. *Chemistry of Natural Compounds*, *42*(4), 491–492. doi:10.1007/s10600-006-0190-7

Yılmaz, C., Taş, N. G., Kocadağlı, T., & Gökmen, V. (2019). Determination of serotonin in nuts and nut containing products by liquid chromatography tandem mass spectrometry. *Food Chemistry*, *272*, 347–353. doi:10.1016/j.foodchem.2018.08.064 PMID:30309554

Young, E., & Tarawou, T. (2016). Determination of PAHs in Fresh and Smoke-dried Alligator Pepper. *International Journal of Scientific Research in Science and Technology*, *2*, 204–207.

Zafar, S., Khan, M. K., Perveen, S., Iqbal, M., & Al-Huqail, A. A. (2023). Fennel. In M. Zia-Ul-Haq, A. Abdulkreem Al-Huqail, M. Riaz, & U. Farooq Gohar (Eds.), *Essentials of Medicinal and Aromatic Crops* (pp. 483–514). Springer International Publishing. doi:10.1007/978-3-031-35403-8_19

Zaidi, S. F., Yamada, K., Kadowaki, M., Usmanghani, K., & Sugiyama, T. (2009). Bactericidal activity of medicinal plants, employed for the treatment of gastrointestinal ailments, against Helicobacter pylori. *Journal of Ethnopharmacology*, *121*(2), 286–291. doi:10.1016/j.jep.2008.11.001 PMID:19041711

Zarghami, M. (2015). Iranian Common Attitude Toward Opium Consumption. *Iranian Journal of Psychiatry and Behavioral Sciences*, *9*(2), e2074–e2074. doi:10.5812/ijpbs.2074v3 PMID:26288642

Zavala, J. S., Hernández-Martínez, D., & Osorio-Revilla, G. (2023). Elaboración de una bebida de nuez de macadamia. *Investigación y Desarrollo en Ciencia y Tecnología de Alimentos*, *8*(1), 471–476. doi:10.29105/idcyta.v8i1.63

Zeng, M., Zhang, M., Chen, J., He, Z., Qin, F., Hu, C., Xu, H., Tao, G., Zhang, S., & Chen, J. (2017). UPLC–MS/MS and multivariate analysis of inhibition of heterocyclic amine profiles by black pepper and piperine in roast beef patties. *Chemometrics and Intelligent Laboratory Systems*, *168*, 96–106. doi:10.1016/j.chemolab.2017.06.012

Zentai, A., Sali, J., Szeitzné-Szabó, M., Szabó, I. J., & Ambrus, Á. (2012). Exposure of consumers to morphine from poppy seeds in Hungary. *Food Additives & Contaminants. Part A, Chemistry, Analysis, Control, Exposure & Risk Assessment*, *29*(3), 403–414. doi:10.1080/19440049.2011.636762 PMID:22165856

Zhang, H., Liu, Y., & Sun, S. (2010). Synthesis and assembly of magnetic nanoparticles for information and energy storage applications. *Frontiers of Physics in China*, *5*(4), 347–356. doi:10.1007/s11467-010-0104-9

Zhang, L., Hogan, S., Li, J., Sun, S., Canning, C., Zheng, S. J., & Zhou, K. (2011). Grape skin Extract inhibits Mammalian Intestinal α-glucosidase Activity and Suppresses postprandial Glycemic response in Streptozocin-treated mice. *Food Chemistry*, *126*(2), 466–471. doi:10.1016/j.foodchem.2010.11.016

Zhang, Y. (2021). Diet According to Traditional Chinese Medicine for Health and Longevity. In S. I. S. Rattan & G. Kaur (Eds.), *Nutrition, Food and Diet in Aging and Longevity* (pp. 331–356). Springer International Publishing. doi:10.1007/978-3-030-83017-5_17

Zhang, Y.-H., Park, Y.-S., Kim, T.-J., Fang, L.-H., Ahn, H.-Y., Hong, J., Kim, Y., Lee, C.-K., & Yun, Y. (2001). Endothelium-dependent vasorelaxant and antiproliferative effects of apigenin. *General Pharmacology*, *35*(6), 341–347. doi:10.1016/S0306-3623(02)00113-1

Zhang, Y., Jiao, X., Liu, J., Feng, G., Luo, X., Zhang, M., Zhang, B., Huang, L., & Long, Q. (2024). A new direction in Chinese herbal medicine ameliorates for type 2 diabetes mellitus: Focus on the potential of mitochondrial respiratory chain complexes. *Journal of Ethnopharmacology*, *321*, 117484. doi:10.1016/j.jep.2023.117484 PMID:38012971

Zhao, W., Lv, P., & Gu, H. (2013). Studies on carotenoids in watermelon flesh. *Agricultural Sciences*, *4*(7), 13–20. doi:10.4236/as.2013.47A003

Zheng, G. Q., Kenney, P. M., Zhang, J., & Lam, L. K. (1993). Chemoprevention of benzo[a]pyrene-induced forestomach cancer in mice by natural phthalides from celery seed oil. *Nutrition and Cancer*, *19*(1), 77–86. doi:10.1080/01635589309514238

Zhong, Y., Xiang, X., Wang, X., Zhang, Y., Hu, M., Chen, T., & Liu, C. (2020). Fabrication and characterization of oil-in-water emulsions stabilized by macadamia protein isolate/chitosan hydrochloride composite polymers. *Food Hydrocolloids*, *103*, 105655. doi:10.1016/j.foodhyd.2020.105655

Zhou, K., & Raffoul, J. J. (2012). Potential Anticancer Properties of Grape Antioxidants. *Journal of Oncology*, *2012*, 1–8. doi:10.1155/2012/803294 PMID:22919383

Zidorn, C., Jöhrer, K., Ganzera, M., Schubert, B., Sigmund, E. M., Mader, J., Greil, R., Ellmerer, E. P., & Stuppner, H. (2005). Polyacetylenes from the Apiaceae vegetables carrot, celery, fennel, parsley, and parsnip and their cytotoxic activities. *Journal of Agricultural and Food Chemistry*, *53*(7), 2518–2523. doi:10.1021/jf048041s

Ziyyat, A., Legssyer, A., Mekhfi, H., Dassouli, A., Serhrouchni, M., & Benjelloun, W. (1997). Phytotherapy of hypertension and diabetes in oriental Morocco. *Journal of Ethnopharmacology*, *58*(1), 45–54. doi:10.1016/S0378-8741(97)00077-9 PMID:9324004

Zoubiri, S., & Baaliouamer, A. (2010). Essential oil composition of Coriandrum sativum seed cultivated in Algeria as food grains protectant. *Food Chemistry*, *122*(4), 1226–1228. doi:10.1016/j.foodchem.2010.03.119

[illegible]

[illegible]

About the Contributors

Sara Musaddiq is serving as Associate Professor and Chairperson of Department of Chemistry, The Women University Multan. She has also served as Director Quality Enhancement Cell. She earned her PhD degree in Natural Product Chemistry in 2014 from the Islamia University Bahawalpur, Pakistan. Dr Sara is founding Editor-in-Chief, of the Journal Frontiers in Chemical Sciences. Being an energetic young scientist, she is playing her role in community services by conducting research to explore the potential of precious natural resources. Her research remains focused on unraveling the mysteries of nature's chemical repertoire, with a particular emphasis on exploring the medicinal potential of local herbs as well as use of natural resources in synthesis of nanomaterials. In addition to her research endeavors, Sara Musaddiq is deeply committed to mentorship and education, inspiring the next generation of scientists to explore the wonders of natural product chemistry. In order to sustain her research group, Dr. Sara has received several competitive research grants including research grant under National Research Program for Universities from HEC Pakistan. She has supervised various MPhil and PhD theses.

Imama Fayyaz is currently passionately immersed in her Doctoral studies in Chemistry at The Women University Multan, Pakistan. With gold medals secured in both her MSc and M. Phil., her academic excellence is evident. Her research focuses on nanocatalysis and organic synthesis, particularly in exploring coupling reactions. With a strong dedication to advancing scientific knowledge, Imama's work demonstrates her expertise and commitment to pushing the boundaries of chemical science. Additionally, she specializes in utilizing natural products for the formulation of various health and skincare products. Imama has also made contributions to the field through her previous publication on phytopharmaceuticals as crucial anticancer therapeutic agents. Her expertise extends to practical applications on an industrial scale, where she seeks to implement her research findings for real-world impact across diverse sectors.

Kiran Mustafa is a distinguished chemical scientist specializing in advanced polymeric membranes for water treatment. Her expertise significantly advances the fields of membrane science, separation technology, and nanotechnology. Dr. Kiran Mustafa is acclaimed for her extensive research and scholarly publications and has made notable strides in both academic and industry collaborations. As a dedicated lecturer, she has effectively integrated innovative teaching methodologies and technology to elevate student learning. Her leadership in securing research grants underscores her prowess in proposal writing and project management. Dr. Kiran Mustafa is also the author of "Nanotechnology and the Generation of Sustainable Hydrogen," highlighting her pivotal contributions to environmental remediation and sustainable technology development.

* * *

Zahra Asghar is a Ph.D Scholar under the supervision of Dr. Sidra Mubeen. Her field of specialization is Organic chemistry. She has done her Master of Philosophy from Women University Multan, Punjab, Pakistan under the supervision of Dr. Samina Aslam with research interests that includes " Amino Acids Catalysed Synthesis Of 5-Arylfuran-2-Carbaldehyde Oximes". She has got 3rd position in M.phil. She has done her BS Chemistry from " Ghazi University D.G.Khan. She has done Intermediate and Matric from BISE D.G.Khan. She has also the degree of B.ed. She has an experience in writeup. She has attended a one day symposium on " Role of Researcher to link academia and industry" in 2019. Due to excellent academic records she has been awarded with laptop from Govt of Punjab, Pakistan. She has a 1 year teaching experience as NSB teacher in Govt School and 2 years experience in PEF School. Other interests are Graphic designing, Corel draw, Adobe Photoshop. Her hobby is gardening.

Tahira Batool is the student of 4th semester of PhD in Organic Chemistry at The Women University Multan. She is working under the supervision of Dr. Sidra Mubeen with the scientific interest that focuses on the novel applications of different types of Hydrotalcite (anionic clay or Mg-Al alloy). In current research Tahira Batool is working on the purification of oil by using hydrotalcite. She has conferred the degree of Master of Philosophy under the supervision under the supervision of Dr. Sara Musaddiq (Head of Department of Chemistry, The Women University Multan) and Dr. Nadeem Iqbal (Director of Microtech Chemicals & Minerals Pvt. Ltd. Lahore) with research interests that motif on the modified synthesis of Hydrotalcite on commercial applications like in PVC pipes formation as thermal stabilizers. She has passed her matriculation examination from Faiz-e-Aam High School Vehari. Then she went to Muzaffargarh from where she passed her FSc.

Examination. She came back and done her BSc. from Punjab college Vehari. She completed her degrees of Msc. and M.phil from The Women University Multan. Due to brilliant research experience in M.phil she got scholarship from industry Microtech & Minerals Pvt. Ltd. Lahore, Punjab, Pakistan. She has 2.5 years teaching experience in private sector. Other interests include creative designing, coral draw. Her hobby is book reading.

Mhmmed El Khasmi is a professor in the Department of Bi-ology, Ben M'Sik Faculty of Sciences, Hassan II University of Ca-sablanca, MOROCCO. Holder of a PhD on phosphocalcic and bone metabolism of the dromedary, the author has more than 30 years of experience in scientific research, teaching and community services. He has published a large number of articles in indexed journals, par-ticipated in several national and international meetings, and super-vised the research work of a large number of master's and doctoral students. His main interests are metabolism, endocrinology and the impact of stress on the homeostasis and products quality of camel.

Hrharan G. is working as an Assistant Professor in the Department of Biochemistry at "Srimad Andavan Arts & Science College (Autonomous)", Affiliated to Bharathidasan University, Thiruvanaikovil, Trichy. He is pursuing his Ph.D., in the field of Diabetes and Pharmacology from Bharathidasan University, Tiruchirappalli. He has 8 years of teaching experience and 9 years of research experience. His area of research interests are Diabetes, Cancer biology, Bioinformatics, Phytochemistry and Arthritis. He has published 12 research articles in national and international journals. He has also published 2 book chapters from this one from Taylor and Francis group. He has supervised 09 M.Sc., students. He has 2 conference proceedings at International level conference. ORCID: 0000-0003-0805-9245.

Shobana G. formerly worked as an assistant professor in the Department of Biochemistry, Srimad Andavan Arts and Science College (Autonomous), Trichy, Tamilnadu. She received her Ph.D -Biochemistry from the Bharathidasan university in the field of Wound Healing using herbal medicine. She has 8 years of teaching experience. Her area of research interests are pharmacology, phytochemistry, bioinformatics etc. To her credit she has published 25 research articles and also written eight book chapters and has five conference proceedings.She completed one Minor Research Project Funded by UGC in the year of 2014-2016. She has guided 20 Ug and Pg students. ORCID: 0000-0002-4793-6178.

Sdia Iqbal is a Lecturer of Department of Chemistry at the women university Multan. Broadly, her research uses polymer and surface chemistry principles to fabricate interfaces, thermodynamics of hybrid systems. Organic-inorganic relationships in hybrid materials are an underlying research theme. By understanding the kinetic behavior of hybrid materials, advanced materials are created by controlling surface properties and dispersion/aggregation of nanoparticles. Applications range from tuning optical properties of metal nanoparticles and their use as a catalyst. Advanced characterization tools handled include X-ray diffraction, Fourier transform spectrometer and UV-visible spectrophotometer. At LUMS Dr. Sadia served as Research Assistant (2013-2014) and worked on a project. She is currently working as lecturer in this university.

Sna Mahboob received her Bachelor's degree in chemistry (Honours) in 2019 from Govt. Emerson College Multan, Master's degree in 2021, from The Women University Multan. She secured first position in Master's degree and received Gold Medal. She started her teaching profession in 2022 and at present is senior teacher in department of chemistry, Aligharh Public School and College, Manga Mandi, Lahore, Pakistan.

Houriya Mestaghanmi is a distinguished teacher-researcher at the Faculty of Sciences Ben M'Sik, affiliated with Hassan II University in Morocco. After completing her Master's degree (DEA) in Endocrinology, Dr. Mestaghanmi pursued doctoral studies in Nutrition and Food at the esteemed University of Bordeaux 1, graduating in 1986. Following her academic achievements, she was appointed as an Assistant Professor at the Faculty of Sciences Ben M'Sik, where she continues to excel in her role. In 2003, Dr. Mestaghanmi undertook a State Doctorate in Toxicology in collaboration with Professor Dauça from Nancy, France, further enhancing her expertise in the field. Her illustrious career spans over 37 years, marked by dedicated service and exemplary contributions to academia and research. Dr. Mestaghanmi's research interests are diverse and multidisciplinary, encompassing areas such as nutrition, metabolism, and toxicology. She has conducted extensive investigations into the toxicity of electromagnetic radiation and heavy metals, as well as exploring the intricate relationship between nutrition, environment, reproduction, and cancer. Her pioneering work extends to genetic polymorphisms and their implications for chronic pathologies, notably diabetes. Dr. Mestaghanmi's research methodology integrates epidemiological, clinical, and molecular approaches, resulting in a comprehensive understanding of complex health issues. Dr. Mestaghanmi's scholarly endeavors have yielded numerous publications in esteemed scientific journals. Her research output encompasses a wide range of topics, including the impact of vitamin D derivatives on insulin secretion, the developmental effects of aluminum chloride, and the health implications of electromagnetic radiation from telecommunications infrastructure. She has also investigated the association between dietary habits, student behavior, and obesity, contributing valuable insights to public health discourse. Additionally, Dr. Mestaghanmi's epidemiological studies on diabetes within the Moroccan population have provided critical insights into the molecular mechanisms underlying this prevalent condition.

Aleena Munir is a dedicated scholar pursuing her Ph.D. in the Department of Chemistry at The Women University Multan, Pakistan. Her focus lies in the intricate realm of Super Absorbent Polymers (SAP) and their composite synthesis. With a passion for innovation and a commitment to advancing scientific knowledge, Aleena's research endeavors contribute significantly to the understanding and application of SAP technology. Her academic journey showcases a relentless pursuit of excellence and a deep-seated curiosity to unravel the complexities of polymer chemistry. Through her work, Aleena continues to make valuable contributions to both academia and industry, shaping the future of materials science.

Anel Arul John N. currently works as an Associate Professor and is also the Research Director in Srimad Andavan Arts & Science college(Autonomous), Trichy, Tamilnadu. He has 25 years of teaching experience and has won the Best Faculty award by Novel Research Academy in 2019. To his credit he has published 50 research articles. He has guided 30 M. Phil students and guided 5 Ph.D students. He also written many edited book chapters. He is an reviewer of various national and international journals. ORCID: 0000-0003-0626-531X.

Zukhruf Nadeem, currently pursuing an MPhil degree at "The Women University Multan". My research focuses on the modification of antifouling PVDF membranes by incorporating metal oxide nanoparticles.

Fkhr Un Nisa is a PhD student at Women University Multan, focusing on wastewater treatment using nanotechnology, membrannes and green methods. She has a M.Phi and BS degree, and has experience in scientific research and writing. Fakhar has presented her research on dye removal from agricultural waste at international conferences. She has received a PEEF scholarship and laptop from Govt of Punjab, Pakistan, and has over 50 extracurricular certificates. She also enjoys calligraphy, graphic designing, and volunteer work.

Wajiha Noor, an Mphil scholar at The Women University Multan, Pakistan. The research area concerns about the magnetized nano catalysts for C-C cross coupling reactions.

Bnha Yingngam is working as Associate Professor (Pharmaceutical Sciences and Technology) in the Faculty of Pharmaceutical Sciences at Ubon Ratchathani University, Ubon Ratchathani, Thailand. He received his B.Pharm. and Ph.D. (Pharmaceutical Sciences) at the Ubon Ratchathani, and postdoctoral fellowship at Institute of Pharmaceutical Sciences, University of Graz, Graz, Austria. His areas of interest include artificial intelligence, optimization methods, green chemistry, natural product chemistry, cosmeceuticals, and pharmaceutical nanotechnology. He has authored and co-authored many scientific papers published in international journals, and proceedings of international and national scientific conferences. He has also authored more than 30 book chapters and edited three textbooks.

Index

N

O

P

S

T

Premier Reference Source
The Use of Artificial Intelligence in Digital Marketing
Competitive Strategies and Tactics
IGI Global

Premier Reference Source
Innovations in Materials Chemistry, Physics, and Engineering Research
IGI Global

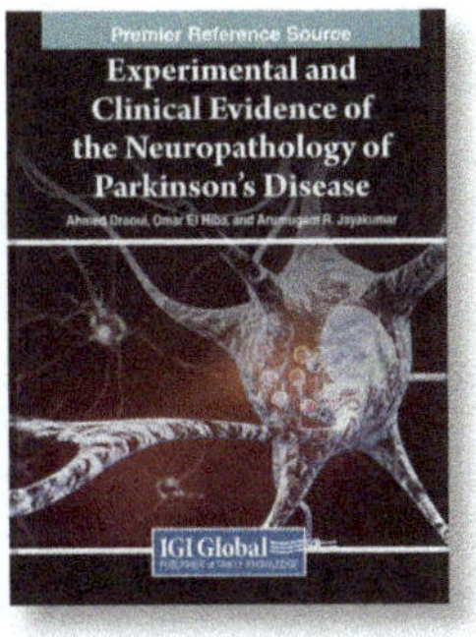

Premier Reference Source
Experimental and Clinical Evidence of the Neuropathology of Parkinson's Disease
IGI Global

Premier Reference Source
Balance and Boundaries in Creating Meaningful Relationships in Online Higher Education
IGI Global

www.ingramcontent.com/pod-product-compliance
Lightning Source LLC
LaVergne TN
LVHW081545300926
849161LV00006B/1519